ENGINEERING MATHEM

[VOLUME I]

ENGINEERING MATHEMATICS

[VOLUME I]

Dr B. K. Kar

MSc, Pure Mathematics, University Medallist, CU

PhD, Appd. Maths, ISM

NEW CENTRAL BOOK AGENCY (P) LTD

8/1 Chintamoni Das Lane, Kolkata 700 009

INDIA

ENGINEERING MATHEMATICS
[Volume I]

First Published: July 2007

Publisher: New Central Book Agency (P) Ltd
8/1 Chintamoni Das Lane, Kolkata 700 009

Typesetter: Anin
BC 97, Sector I, Salt Lake, Kolkata 700 064

Printer: New Central Book Agency (P) Ltd
Web-Offset Division, Dhulagarh, Sankrail, Howrah

Cover Designer: Soumen Paul

Cover Printer: Biswanath Studio
3 Garanhatta Street, Kolkata 700 006

Proof-checking: Pradip Kr Chowdhury and the Author

Project Team: Prabhat Jas and Subir Pal

ISBN: 81-7381-543-7

Price: Rs 185.00 [Rupees One Hundred and Eighty-five only]

In memory of

my

Parents

Contents

Preface

This book has been written initially in two volumes based on the revised syllabus of B.Tech. courses prescribed by the West Bengal University of Technology. It is intended for the two-semester students of all branches of Engineering and Technology willing to grasp the ideas of mathematical methods and apply the techniques to solve problems efficiently and smartly. To achieve that goal extensive use of mathematical symbols and abbreviations has been made. Not all these symbols are uncommon. However, meanings of the symbols and abbreviations used in this book have been given in the Ready Reference in the beginning of the book. Some useful properties of summation ($\sum$) and multiplication ($\prod$) which have been used are also explained in the Ready Reference. A special attention to these properties is drawn before going through the topics where these symbols have been used.

The first of the two volumes contains seven chapters dealing with Differential and Integral Calculus, Infinite Sequence and Series, Vector Analysis and Three-Dimensional Coordinate Geometry. Both geometrical and physical interpretations of important theorems, which are essential for the would-be engineers, are given adequately. The 3-D coordinate geometry has not been treated in a stereotyped way. To stimulate interest in and dispel the boredom of the subject the Set Theory of Modern Algebra and principles of Vector Algebra have been used.

The second volume contains four chapters which deal with Linear Algebra, Differential Equations, Laplace Transformation and Numerical Methods including the Estimation of Errors.

Although proofs of many theorems and results are not required for the purpose of examination they are not totally omitted. They are given separately in the appendices at the ends of the chapters with a view to avoid interruption in the study during preparation for the examination. Moreover, a student having a keen interest and inclination in the subject may not be satisfied in accepting a result without a valid proof.

An adequate number of problems have been solved to help the students know how and where a particular mathematical technique is to be applied promptly in solving a problem. A considerable number of problems has been given in exercises following each chapter for self-assessment on the topics discussed in the chapter.

There are doubtless some residual errors for which I am solely responsible and I shall be grateful to the readers who will help me to correct them and I shall consider it a privilege to have comments and suggestions from the punctilious readers—they will help me in further improving its contents.

This book is addressed to those who intend to apply the subject at the proper place and time while keeping him/her aware to the needs of the society where he/she can lend his/her expert service; and also to those who can be useful to the community without even going through the formal process of drilling through rigorous treatment of mathematics. If the book can help in the realisation of these aspects even to a small extent, I shall consider my labour fully rewarded.

B. K. KAR

June 2007
Salt Lake, Kolkata

Acknowledgements

I am grateful to the Indian School of Mines University, a well-known institute of national importance, where I had the opportunity to serve as a teacher, teaching almost all the topics of the book in the undergraduate and postgraduate levels. I am also indebted to the Scottish Church College, Kolkata, where I started my career as a lecturer, teaching in the undergraduate classes, both Pass and Honours, for about a decade.

I owe a debt of gratitude too large to measure to Dr R. K. Kar, my capable younger brother and author of a number of books, who persuaded me in writing the book which, he had thought, might be helpful for the students.

In course of preparation of this book, I have drawn liberally from contemporary literature, the ideas and materials relevant to those included in it. I take the opportunity to acknowledge my indebtedness to all the authors of these publications, a list of whose names has been given in the Bibliography at the end of the book.

I owe a special debt of gratitude to Dr Amaresh Chattopadhyay, Professor, ISM, my young friend; guide and colleague, Dr S. Dey, Professor (Retd.), ISM, Dr Parijat De, Director, Technical Education and Training, W.B., who directly helped me in the preparation of this book.

I am also indebted to Shri Sanat Kumar Kar, Principal (Retd.), Beldanga College, my second brother, who was always beside me during my critical time. I also wish to thank my wife Nibha, whose incessant inspiration was a guiding force to me and who spared no pains in relieving me from all those matters which might hinder the progress in preparing the manuscript.

I am particularly grateful to Mr Pradip Kr. Chowdhury for accurate and meticulous proofreading and Mr Amitabha Sen, Director, New Central Book Agency (P) Ltd, Mr Prabhat Jas, and Mr Soumen Paul for publishing the book in such a short time.

June 2007 **B. K. KAR**
Salt Lake, Kolkata

Acknowledgements

I am grateful to the Indian School of Mines University, a well-known institute of national importance, where I had the opportunity to serve as a teacher, teaching almost all the topics of the book at the undergraduate and postgraduate levels. I am also indebted to the Scottish Church College, Kolkata, where I started my career as a lecturer teaching in the undergraduate classes, both Pass and Honours, for about a decade.

I owe a debt of gratitude too large to measure to Dr R. N. Nag, my [illegible] and author of a number of books, who persuaded me to write the book which he had thought might be helpful for the students.

In course of preparation of this book, I have drawn liberally from contemporary literature, treatises and materials relevant to those included in it. I take the opportunity to acknowledge my indebtedness to all the authors of these publications, a list of which has been given in the Bibliography at the end of the book.

I owe a special debt of gratitude to Dr [illegible] Chattopadhyay, Professor, ISM, my young friend, guide and colleague, Dr S. Das, Professor [illegible], ISM, Dr [illegible] De, Director, Technical Education and Training, W.B., who directly helped me in the preparation of this book.

I am also indebted to [illegible] Kumar Kar, Principal (Retd.), [illegible] College, my second brother, who was always beside me during my critical time. I also wish to thank my wife [illegible], whose incessant inspiration was a guiding force to me and who spared me [illegible], relieving me from all those mundane works under the process of preparing the manuscript.

I am particularly grateful to Mr Pradip Kr. Choudhury for advice and meticulous proofreading, and Mr Amitabha Sen, Director, New Central Book Agency (P) Ltd., Mr [illegible], and Mr Sameer Paul for publishing the book in such a short time.

June 2007
Salt Lake, Kolkata

B. K. KAR

Ready Reference

Symbols and Their Meanings

I. Logical Symbols

In the following p and q are logical statements which are either true or false and $P(x)$ is a property involving x.

$p \Rightarrow q$: if p then q or p implies q.

$p \not\Rightarrow q$: p does not imply q or if p then q is not necessary.

$p \Leftrightarrow q$ or p iff q: p if and only if q or if p then q and if q then p.

: , /, s.t.: such that.

$\forall x$: for every x or for all x.

$\forall x, y$: for every pair x and y.

$\forall x, y, z$: for all x, y and z.

$\exists y$: there exists a y.

E!y : there exists a unique y.

Thus the sentence: 'For every positive ϵ there is a positive δ such that for every x belonging to X, if the numerical value of the difference of x from a is less than δ then the numerical value of the difference of $f(x)$ from l is less than ϵ' is symbolised:

$\forall \epsilon > 0 \exists \delta > 0 : \forall x \in X, \ |x - a| < \delta \Rightarrow |f(x) - l| < \epsilon.$

II. Set Theoretical Symbols and Abbreviations

Capital letters like A, B, C, X, Y, etc., will, in general, be used to denote sets, whereas small letters like a, b, c, x, y, etc., will be used to represent elements.

$a \in A$: a belongs to A, or sometimes a belonging to A.

$a \notin A$: a does not belong to A, or a not belonging to A.

$A \subseteq B$: A is a subset of B so that $x \in A \Rightarrow x \in B$.

$A \subset B$: A is a proper subset of B so that $\exists x : x \in B$ but $x \notin A$.

$S = \{x \in U : P(x)\} \Rightarrow S$ is the set of all those elements of U, which satisfy the property P.

U = the universal set containing all the subsets in a discourse.

$A \cup B = \{x \in U$: either $x \in A$ or $x \in B\}$, the union of A and B.

$A \cap B = \{x \in U$: both $x \in A$ and $x \in B\}$, the intersection of A and B.

$\emptyset$ stands for the null set, or empty set, containing no elements of U.

$A - B$ or $A \backslash B$: the difference of B from A and is the set : $\{x \in A$ and $x \notin B\}$.

A' or A^c : the complement of A in $U = U - A$.

$A \times B$ = the Cartesian product of A and $B = \{(x, y) : x \in A \text{ and } y \in B\}$.

$A \times B \times C$ = the Cartesian product of A, B and $C = \{(x, y, z) : x \in A,\ y \in B,\ z \in C\}$.

Special Sets

$\mathbb{N} = \{1, 2, 3, \ldots\}$, the set of all the natural numbers.

$\mathbb{W} = \{0, 1, 2, 3, \ldots\}$, the set of all the whole numbers.

$\mathbb{Z} = \{0, \pm 1, \pm 2, \ldots\}$, the set of all integers.

$\mathbb{J}_n = \{1, 2, 3, \ldots, n\} = \{r\}_1^n$, the set of first n natural numbers.

$\mathbb{Q}$ = the set of all the rational numbers in the form of vulgar fraction, finite decimal or recurring decimal.

$\mathbb{Q}'$ = the set of all the irrational numbers whether algebraic or transcendental.

$\overline{\mathbb{Q}}_A$ = the set of all the algebraic irrationals.

$\overline{\mathbb{Q}}_T$ = the set of all the transcendental numbers.

$\mathbb{Q}' = \mathbb{Q}'_A \cup \mathbb{Q}'_T$.

$\mathbb{R}$ = the set of all the real numbers $= \mathbb{Q} \cup \mathbb{Q}'$.

$\mathbb{R}^+$ = the set of all the positive real numbers.

$\mathbb{R}_0$ = the set of all the non-zero real numbers $= \mathbb{R} - \{0\}$.

$\mathbb{R}^2 = \mathbb{R} \times \mathbb{R} = \{(x, y) : \text{both } x \text{ and } y \in \mathbb{R}\}$, $\mathbb{R}^3 = \{(x, y, z) : x, y \text{ and } z \text{ all} \in \mathbb{R}\}$.

$\mathbb{C}$ = the set of all the complex numbers.

$[a, b] = I_c$; the closed interval $= \{x \in \mathbb{R} : a \leq x \leq b\}$.

(a, b) or $]a, b[= I_o$, the open interval $= \{x \in \mathbb{R} : a < x < b\}$.

$[a, b)$ or $[a, b[$ = the closed open interval $= \{x \in \mathbb{R} : a \leq x < b\}$.

$(a, b]$ or $]a, b]$ = the open closed interval $= \{x \in \mathbb{R} : a < x \leq b\}$.

Abbreviations

A.S. : Arithmetic Sequence (or Series).

G.S. : Geometric Sequence (or Series).

H.S. : Harmonic Sequence (or Series).

sqce : sequence.

nbd : neighbourhood.

lt.pt. : limit point.

LDE : Linear Differential Equation.

HLE : Homogeneous LDE.

NLE : Nonhomogeneous LDE.

ODE : Ordinary Differential Equation.

LEC : LDE with Constant Coefficients.

HEC : HLE with Constant Coefficients.

NEC : NLE with Constant Coefficients.

LEV : LDE with Variable Coefficients.

HEV: HLE with Variable Coefficients.

NEV: NLE with Variable Coefficients.

Dom f: Domain of the function f.

Ran f: Range of the function f.

III. Useful Properties of Summation and Multiplication Using the Summation and Multiplication Symbols $\sum$ and $\prod$

(a) Summation involving three symbols

(i) $\sum a = a+b+c$, $\sum b = b+c+a$ and $\sum c = c+a+b \Rightarrow \sum a = \sum b = \sum c$.

(ii) $\sum (b-c) = \sum b - \sum c = \sum a - \sum a = 0$.

(iii) $\sum (b+c-a) = \sum b + \sum c - \sum a = \sum a + \sum a - \sum a = \sum a$.

(iv) $\sum bc = \sum ca = \sum ab = bc + ca + ab$.

(v) $\sum a(b-c) = \sum ab - \sum ac = \sum bc - \sum bc = 0$.

(vi) $\sum b^2c = \sum c^2a = \sum a^2b = b^2c + c^2a + a^2b$.

(vii) $\sum a^2(b-c) = \sum a^2b - \sum a^2c = \sum b^2c - \sum c^2b = \sum (b^2c - c^2b) = \sum bc(b-c)$.

For a triangle ABC:

(viii) $\sum A = A + B + C = \sum B = \sum C = 180°$ or π^c, according as the measures of the angles are in degrees or in radians,

(ix) $\sum \sin A = \sin A + \sin B + \sin C$, $\sum \cos A = \cos A + \cos B + \cos C$, etc.,

(x) $\sum \sin A \cos B \cos C = \sin A \cos B \cos C + \sin B \cos C \cos A + \sin C \cos A \cos B$.

(b) Summation involving n symbols

The sum of n symbols $a_1, a_2, \ldots, a_n$ will be denoted by

$$\sum_{r=1}^{n} a_r \quad \text{or} \quad \sum_{r=1}^{n} a_r \quad \text{or} \quad \sum_{1}^{n} a_r \quad \text{or} \quad \sum_{1 \le r \le n} a_r \quad \text{or} \quad \sum a_r,\ 1 \le r \le n.$$

In the above the symbol r, taking the natural number values from 1 to n, appearing as the lower index in a_r, is called the **dummy index**. This may be replaced by any other symbol without affecting the sum. For example,

$$\sum_{i=1}^{n} a_i = \sum_{p=1}^{n} a_p = \sum_{\alpha=1}^{n} a_\alpha.$$

The numbers 1 and n are respectively called the **lower** and the **upper limits** of the index or of the summation.

It might be verified that

(i) $\sum_{r=1}^{n} a_r$, $\sum_{i=1}^{n} a_i$, etc., all equal to $a_1 + a_2 + a_3 + \cdots + a_n$,

(ii) If k is a constant, independent of the dummy index $\sum ka_r = k\sum a_r$,

***(iii)** $\sum_{1}^{n} a_r = \sum_{1}^{n} a_{n-r+1}$,

so that **if the terms of a sum of n numbers $a_1, a_2, \ldots, a_n$ are to be written in the reversed order, the dummy index r is to be replaced by the difference of the sum of the limits of summation and the index r itself.**

So,

$$\sum_{r=m}^{n} a_r = \sum_{r=m}^{n} a_{m+n-r}.$$

***(iv)** $\sum_{1}^{n} a_r = \sum_{0}^{n-1} a_{r+1} = \sum_{-1}^{n-2} a_{r+2} = \sum_{p}^{n+p-1} a_{r-p+1}$.

The sum remains unchanged if both the upper and lower limits of summation are increased (or decreased) by the same amount and at the same time the dummy index is decreased (or increased) by the same amount.

(v) $\sum_{1}^{n}(a_r \pm b_r) = \sum_{1}^{n} a_r \pm \sum_{1}^{n} b_r$.

***(vi)** $\sum_{1}^{n}(a_{r+1} - a_r) = a_{n+1} - a_1$.

The sum of the first n differences of two consecutive terms of a sequence is the difference of the two terms of the sequence obtained by replacing the dummy index r by the upper and lower limits in the greater and smaller indicial values respectively.

So, $\sum_{p}^{p+q}(a_{r+1} - a_r) = a_{p+q+1} - a_p$, $\sum_{p}^{p+q}(a_i - a_{i-1}) = a_{p+q} - a_{p-1}$.

Similarly,

(vii) $\sum_{1}^{n}(a_{2r+2} - a_{2r}) = \sum_{1}^{n}[a_{2(r+1)} - a_{2(r)}] = a_{2(n+1)} - a_{2(1)} = a_{2n+2} - a_2$.

(viii) $\sum_{1}^{n}(a_{2r+1} - a_{2r-1}) = \sum_{1}^{n}[a_{2r+1} - a_{2(r-1)+1}] = a_{2n+1} - a_1$.

(ix) $\sum_{p}^{q}(a_{r+1} - a_r) = a_{q+1} - a_p$, where $p < q$.

(c) Multiplication involving three symbols

(i) $\prod a = abc$, $\prod b = bca$, $\prod c = cab \Rightarrow \prod a = \prod b = \prod c$.

(ii) $\prod(b - c) = (b - c)(c - a)(a - b) = \prod(c - a) = \prod(a - b)$.

(iii) $\prod \sin A = \sin A \sin B \sin C$, $\prod \cos A = \cos A \cos B \cos C$, etc.

(iv) $\prod \tan A = \prod\{(\sin A)/\cos A\} = (\prod \sin A)\,/\,(\prod \cos A)$.

(d) Multiplication involving n symbols

The product of n quantities $a_1, a_2, \ldots, a_n$ will be denoted by

(i) $\prod_{r=1}^{n} a_r$ or $\prod_{r=1}^{n} a_r$ or $\prod_{1}^{r} a_r$ or $\prod_{1\le r\le n} a_r$ or $\prod a_r$, $1 \le r \le n$, so that each is equal to $a_1 \cdot a_2 \cdot a_3 \cdots a_n$.

Here r is the dummy index. As in summation this dummy index may be replaced by any other symbol.

(ii) $\prod_{1}^{n}(ka_r) = \left(\prod_{1}^{n} k\right)\left(\prod_{1}^{n} a_r\right) = k^n \prod_{1}^{n} a_r$, where k is independent of r.

(iii) $\prod_{1}^{n} a_r = \prod_{1}^{n} a_{n-r+1}$,

so that **if the factors of a product of n quantities $a_1, a_2, \ldots, a_n$ are to be written in the reversed order, the dummy index r is to be replaced by the difference of the sum of the limits of multiplication and the index r itself.**

So, just like summation

$$\prod_{r=m}^{n} a_r = \prod_{r=m}^{n} a_{m+n-r}.$$

(iv) $\prod_{1}^{n} a_r = \prod_{0}^{n-1} a_{r+1} = \prod_{-1}^{n-2} a_{r+2} = \prod_{p}^{n+p-1} a_{r-p+1}$.

(v) $\prod_{1}^{n}(a_r b_r) = \left(\prod_{1}^{n} a_r\right)\left(\prod_{1}^{n} b_r\right)$

and $\prod(1/a_r) = \prod a_r^{-1} = 1/\prod a_r$, $\prod(a_r/b_r) = \left(\prod a_r\right)/\left(\prod b_r\right)$.

***(vi)** $\prod_{1}^{n}(a_{r+1}/a_r) = a_{n+1}/a_1$, $\prod_{p}^{p+q}(a_{r+1}/a_r) = a_{p+q+1}/a_p$, a property similar to (b) **(vi)** in summation.

(vii) $\prod_{1}^{n}(a_{2r+2}/a_{2r}) = a_{2n+2}/a_2$.

(viii) $\prod_{1}^{n}(a_{2r+1}/a_{2r-1}) = a_{2n+1}/a_1$.

(ix) $\prod_{p}^{q}(a_{r+1}/a_r) = a_{q+1}/a_p$, where $p < q$.

(e) Properties involving both $\sum$ and $\prod$

(i) For three symbols

$$\sum a^2(b-c) = -\prod(b-c) = \sum bc(b-c).$$

$$\sum a(b^2 - c^2) = \prod (b - c).$$

$$\sum a^3(b - c) = -\left(\sum a\right)\prod (b - c).$$

$$\sum a^3 - 3\prod a = \left(\sum a\right)\left[\sum a^2 - \sum bc\right].$$

(ii) For n quantities

$\prod_1^n (x - \alpha_r) = \sum_0^n (-1)^r s_r x^{n-r}$, where s_r = the sum of all the products of the n quantities $\alpha_1, \alpha_2, \ldots, \alpha_n$ taken r at a time so that there will be ${}^nC_r$ such product terms in s_r.

$$\left(\sum_1^n a_i\right)^2 = \sum_1^n a_i \sum_1^n a_j = \sum_{i=1}^n \sum_{j=1}^n a_i a_j = \sum\sum_{i=j} a_i a_j + \sum\sum_{i \neq j} a_i a_j$$

$$= \sum_1^n a_i^2 + \sum\sum_{i<j} a_i a_j + \sum\sum_{i>j} a_i a_j = \sum_1^n a_i^2 + 2\sum\sum_{i<j} a_i a_j.$$

IV. Greek Letters of Alphabet

Cap.	Small	Name	Cap.	Small	Name	Cap.	Small	Name
A	α	Alpha	I	ι	Iota	P	ρ	Rho
B	β	Beta	K	κ	Kappa	Σ	σ	Sigma
Γ	γ	Gamma	Λ	λ	Lambda	T	τ	Tau
Δ	δ	Delta	M	μ	Mu	Υ	υ	Upsilon
E	ϵ	Epsilon	N	ν	Nu	Φ	ϕ	Phi
Z	ζ	Zeta	Ξ	ξ	Xi	X	χ	Chi
H	η	Eta	O	o	Omicron	Ψ	ψ	Psi
Θ	θ	Theta	Π	π	Pi	Ω	ω	Omega

Some Important Points to Remember

V. Algebra

1. Quadratic Equation: $Q = 0$, where $Q = ax^2 + bx + c$.

The roots are $[-b \pm \sqrt{D}]/2a$, where $D = b^2 - 4ac$.

s = the sum of the roots $= -b/a$,

p = the product of the roots $= c/a$.

If d = the difference of the roots, then $|d| = \sqrt{D}/|a|$.

2. Progressions: $\{a_r\}$ is an arithmetic sequence (A.S.) if $a_r = a + \overline{r-1}d$, where a is the 1st term and d is the common difference (C.D.). The terms $a, a+d, a+2d, \ldots$ are said to be in A.P.

$\{g_r\}$ is a geometric sequence (G.S.) if $g_r = g.x^{r-1}$, where $g(\neq 0)$ is the 1st term and $x(\neq 0)$ is the common ratio (C.R.).

$\{h_r\}$ is a harmonic sequence (H.S.) if $h_r = 1/(a + \overline{r-1}d)$, i.e, the rth term is the reciprocal of the rth term of an arithmetic sequence.

$\{r\}$ is the **standard A.S.**, $\{x^{r-1}\}$ is the **standard G.S.** and $\{1/r\}$ is the **standard H.S.**

$$\sum_1^n a_r = \frac{n}{2}(2a + \overline{n-1}d), \quad \sum_1^n g_r = g(1-x^n)/(1-x)$$

and $\sum_1^\infty g_r = g/(1-x)$ if $-1 < x < 1$.

3. Logarithm: The logarithm of a number x with respect to a base y will be denoted by $\log_y x$, so that a logarithm is a function of two real variables x and y. For a unique real value of the logarithm $x > 0$, and $y > 0$ but $\neq 1$. The base y satisfying this condition will said to be an **admissible base.** By the symbol $\log x$ we mean Logarithm of x with respect to some admissible base.

The system of logarithm whose base is e defined by the limit : $\underset{n\to\infty}{\text{Lt}}(1 + 1/n)^n$, $n \in \mathbb{N}$, is called the **natural** or **Napierean** system. Such a logarithm will be denoted by **ln** in place of $\log_e$.

Properties

(i) $\log_y x = z \Leftrightarrow x = y^z$.

(ii) $\log 1 = 0$.

(iii) $\log 0$ is undefined, although $\underset{x\to 0}{\text{Lt}} \log x = -\infty$ or $+\infty$ according as the base is $>$ or $<$ 1.

(iv) $\log_x x = 1$.

(v) $\log m + \log n = \log(mn)$, $\sum \log m_r = \log \prod m_r$.

(vi) $\log(1/m) = -\log m$, $\log m - \log n = \log(m/n)$.

(vii) $\log m^n = n \log m$.

(viii) $\log_y x = (\log x)/(\log y)$, bases in the numerator and denominator being the same.

***(ix)** $a^b = c^{\log_c a^b} = c^{b \log_c a} = e^{b \ln a}$.

4. Remainder Theorem: $\forall$ polynomial $P(x), P(a)$ is the remainder when it is divided by $x - a$, so that $P(a) = 0 \Rightarrow (x - a)$ is a factor of $P(x)$.

$(x-a)^2$ is a factor of $P(x)$ if both $P(a)$ and $P'(a)$, the derivative of $P(x)$ at $x = a$, are zero.

5. Permutation and Combination: The symbol ${}^nP_r$ stands for the number of permutations of n dissimilar things taken r at a time. It is also **the number of ordered r-tuples that can be formed from a set of n dissimilar objects.** So, n and r must be natural numbers and $n \geq r$ and

$${}^nP_r = n(n-1)\cdots(n-r+1) = \prod_{q=1}^{r}(n-q+1) = n!/(n-r)!$$

Assumed as a **function** of two integral variables,

$${}^nP_r = 1 \quad \text{when} \quad r = 0$$
$$= 0 \quad \text{when} \quad r > n, \text{ whence } 1/(-n)! = 0 \text{ if } n \in \mathbb{N}.$$

Similarly, ${}^nC_r$ stands for the number of combinations of n dissimilar objects taken r at a time. It is also **the number of different subsets each of r elements that can be formed**

from a set of n dissimilar elements. So here again n and r are natural numbers and $n \geq r$.
Also

$$^nC_r = \frac{1}{r!}\ ^nP_r = \frac{n(n-1)\cdots(n-1+1)}{r(r-1)\cdots 3\cdot 2\cdot 1} = \frac{n!}{r!(n-r)!}.$$

Assumed as a **function** of two integral variables,

$$^nC_r = {}^nC_{n-r} = 1 \quad \text{when} \quad r = 0 \ \text{ or } \ n,$$
$$= 0 \quad \text{when} \quad r > n.$$

6. Binomial Theorem: (i) $\forall n \in \mathbb{N}, (x+y)^n = \sum_{r=0}^{n} {}^nC_r x^{n-r} y^r$ and $(1+x)^n = \sum_{r=0}^{n} {}^nC_r x^r$.

(ii) If n is a negative integer or a non-integral rational number and $|x| < 1$, then

$$(1+x)^n = \sum_{0}^{\infty} \binom{n}{r} x^r, \text{ if } x \neq 0$$
$$= 1 \text{ if } x = 0,$$

where $\binom{n}{r} = 1$ if $r = 0$,

$$= n(n-1)\cdots(n-r+1) \div r! \text{ for } r \in \mathbb{N}.$$

In particular,

$$\binom{-1}{r} = (-1)^r, \quad \binom{-2}{r} = (-1)^r(r+1), \ \forall n \in \mathbb{N} \text{ and}$$

$$\binom{-n}{r} = (-1)^r\ {}^{n+r-1}C_r, \quad \binom{-1/2}{r} = (-1)^r \frac{(2r)!}{2^{2r}(r!)^2}.$$

VI. Trigonometry

1. $\pi = c/d$, where c and d are the lengths of the circumference and the diameter of a circle.

***2.** $\pi^c = \pi$ radians $= 180°$, $\pi^c/2 = 90°$, $\pi^c/3 = 60°, \ldots$.

* $\sin\theta = \sin\theta^c$ and not $\sin\theta°$.

* $\sin 90 \neq 1$ and $\sin\pi°/2 \neq 1$.

3. If $T(\theta)$ is the trigonometric ratio of the angle θ^c, then

$$T(\pi/2 - \theta) = CT(\theta),$$

i.e., $\cos(\pi/2-\theta) = \sin\theta$, $\sin(\pi/2-\theta) = \cos\theta$.

So, if $T = \cos$, $CT = \sin$ and if $T = \sin$, $CT = \cos$, etc.

$$T(n\pi/2 \pm \theta) = \pm T(\theta) \quad \text{or} \quad \pm CT(\theta),$$

according as n is even or odd, the sign being determined by the sign-rule: All, Sine, Tan, Cos.

4. $\cos(A+B) = \cos A\cos B - \sin A \sin B$,

$\sin(A+B) = \sin A\cos B + \cos A\sin B$.

Knowing the values of $\frac{\sin}{\cos}(-B)$, $\cos(A-B)$ and $\sin(A-B)$ can be found out.

* $\cos\sum A = \cos(A+B+C) = \prod \cos A - \sum \cos A \sin B \sin C$,

* $\sin\sum A = \sin(A+B+C) = \sum \sin A\cos B\cos C - \prod \sin A$,

$* \tan \sum A = \left(\sum \tan A - \prod \tan A\right) / \left(1 - \sum \tan B \tan C\right)$

and $* \cot \sum A = \left(\sum \cot A - \prod \cot A\right) / \left(1 - \sum \cot B \cot C\right)$.

5. $\forall a, b$ and $x \in \mathbb{R}\ \exists r$ and $\theta \in \mathbb{R} : a\cos x + b\sin x = r\cos(x+\theta)$, where $r = \sqrt{(a^2+b^2)}$ and θ is the common solution of

$$a = r\cos\theta \quad \text{and} \quad b = r\sin\theta.$$

$\theta = \tan^{-1}(b/a)$ if $a > 0$,
$= \pi + \tan^{-1}(b/a)$ if $a < 0$,
$= \pm\pi/2$ according as $b >$ or < 0 and $a = 0$.

6. Inverse Circular Functions (ICF): $\theta = T^{-1}(x)$

ICF	Domain	Range
$\sin^{-1}$	$[-1, 1]$	$[-\pi/2, \pi/2]$
$\cos^{-1}$	"	$[0, \pi]$
$\tan^{-1}$	$\mathbb{R}$	$]-\pi/2, \pi/2[$
$\cot^{-1}$	"	$]0, \pi[$
$\sec^{-1}$	$\mathbb{R}-]-1, 1[$	$[0, \pi] - \{\pi/2\}$
cosec^{-1}	"	$[-\pi/2, \pi/2] - \{0\}$

7. Properties of Triangle

(i) $a/\sin A = b/\sin B = c/\sin C$,

(ii) $\cos A = (b^2 + c^2 - a^2)/(2bc), \ldots$,

(iii) $a = b\cos C + c\cos B, \ldots$.

8. Sums of Some Series[1]

(i) Geometric Series: $\sum_1^\infty x^{n-1} = 1/(1-x),\ |x| < 1$.

(ii) Arithmetico-geometric Series: $\sum_1^\infty n x^{n-1} = 1/(1-x)^2,\ |x| < 1$.

(iii) Binomial Series: $\sum_0^\infty \binom{n}{r} x^r = (1+x)^n \forall x \in \mathbb{R}$ if $n \in \mathbb{N}$ and $\forall x \in]-1, 1[$ if $n \notin \mathbb{N}$ and $= 1$ if $n = 0$.

(iv) Exponential Series: $\sum_0^\infty x^n/n! = e^x \forall x \in \mathbb{R}$.

(v) Logarithmic Series: **(a)** $\sum_1^\infty (-1)^{n-1} x^n/n = \ln(1+x),\ -1 < x \le 1$.

---

[1] For $x = 0$, the first term of some of the series takes the form 0^0. In such cases the corresponding sum has been assumed to be 1.

(b) $\sum_{1}^{\infty} x^{2n-1}/(2n-1) = (1/2)\ln[(1+x)/(1-x)],\ -1 < x < 1.$

(vi) Sine Series: $\sum_{0}^{\infty}(-1)^n x^{2n+1}/(2n+1)! = \sin x \forall x \in \mathbb{R}.$

(vii) Cosine Series: $\sum_{0}^{\infty}(-1)^n x^{2n}/(2n)! = \cos x \forall x \in \mathbb{R}.$

9. Complex Numbers

(a) $z = x + iy = r(\cos\theta + i\sin\theta) = re^{i\theta},\ x, y, r, \theta \in \mathbb{R},$

$r = |z| =$ mod z and $\theta = amz,\ -\pi < \theta \le \pi.$

(b) Euler's Formulae

$e^{i\theta} = \cos\theta + \sin\theta,\ e^{-i\theta} = \cos\theta - i\sin\theta,\ e^{2n\pi i} = 1,\ e^{\pi i} = -1,$

$\cos\theta = (e^{i\theta} + e^{-i\theta})/2,\ \sin\theta = (e^{i\theta} - e^{-i\theta})/(2i),\ \ln(-1) = \pi i.$

(c) De Moivre's Theorem

(i) $\forall n \in \mathbb{N},\ \cos n\theta + i\sin n\theta = (\cos\theta + i\sin\theta)^n.$

If $n = p/q$, where $p \in \mathbb{Z},\ q \in \mathbb{N}$,

$$(\cos\theta + i\sin\theta)^n = (\cos\theta + i\sin\theta)^{p/q} = \{(\cos\theta + i\sin\theta)^p\}^{1/q} = (\cos p\theta + i\sin p\theta)^{1/q},$$

so that $\cos(p\theta/q) + \sin(p\theta/q)$ is one of the values of $(\cos\theta + i\sin\theta)^n$. All the q values are obtained from $\cos\{(2n\pi + p\theta)/q\} + i\sin\{(2n\pi + p\theta)/q\}$ by assigning any q consecutive integral values to n.

(d) If $z = e^{i\theta},\ z^{-1} = e^{-i\theta}$. Hence $2\cos\theta = z + z^{-1}$ and $2i\sin\theta = z - z^{-1}$.

$\cos^n\theta = 2^{-n+1}\sum {}^nC_r\cos(n-2r)\theta,\ 0 \le r \le n/2.$

$\sin^n\theta = (-1)^{(n-1)/2}2^{-n+1}\sum(-1)^r\, {}^nC_r\sin(n-2r)\theta,\ 0 \le r \le n/2$ or,

$= (-1)^{n/2}2^{-n+1}\sum(-1)^r\, {}^nC_r\cos(n-2r)\theta,\ 0 \le r \le n/2,$

according as n is odd or even.

(e) Hyperbolic Functions

(i) $\cosh x = (e^x + e^{-x})/2,$

(ii) $\sinh x = (e^x - e^{-x})/2,\ \tanh x = \sinh x/\cosh x,\ \coth x = 1/\tan x, \text{sech}\, x = 1/\cosh x,$ $\text{cosech}\, x = 1/\sinh x, \cosh^2 x - \sinh^2 x = 1, \text{sech}^2 x + \tanh^2 x = 1, \coth^2 x - \text{cosech}^2 x = 1.$

Inverse Hyperbolic Functions

(i) $\cosh^{-1} x = \ln(x + \sqrt{(x^2-1)})$, **(ii)** $\sinh^{-1} x = \ln(x + \sqrt{(x^2+1)})$, $\tanh^{-1} x = \dfrac{1}{2}\ln\dfrac{1+x}{1-x}.$

VII. Coordinate Geometry

(i) Collinearity of three points $P_i(x_i, y_i),\ i = 1, 2, 3$: $\sum x_1(y_2 - y_3) = 0$ or $\sum(x_2y_3 - x_3y_2) = 0.$

(ii) The area of the triangle with vertices $P_i(x_i, y_i)$:

$$\Delta = \frac{1}{2}\left|\sum x_1(y_2 - y_3)\right| = \frac{1}{2} \text{ mod } \begin{vmatrix} x_1 & y_1 & 1 \\ x_2 & y_2 & 1 \\ x_3 & y_3 & 1 \end{vmatrix}.$$

(iii) A locus is a set of points. If $\Gamma = \{(x, y) : F(x, y) = \text{ or } > \text{ or } < \text{ or } \geq \text{ or } \leq 0\}$, then Γ is a plane locus whose equation or inequation is $F(x, y) = \text{ or } > \text{ or } < \text{ or } \geq \text{ or } \leq 0$.

Straight line is a locus for which $F(x, y) = ax + by + c$, where a and b are not both zero.

Conic Section is a locus for which

$$F(x, y) = ax^2 + 2hxy + by^2 + 2gx + 2fy + c,$$

where a, h, b are not all zero. This conic is degenerate, i.e., a pair of straight lines or a real point if

$$D = 0, \text{ where } D = \begin{vmatrix} a & h & g \\ h & b & f \\ g & f & c \end{vmatrix} = abc + 2fgh - af^2 - ag^2 - ch^2$$

and it is non-degenerate if $D \neq 0$. A non-degenerate conic is a parabola, an ellipse or a hyperbola according as C, the cofactor of c in the expansion of D is $=, >$ or < 0.

(iv) For the straight line Λ, given by $F(x, y) = 0$, where $F(x, y) = ax + by + c$, a point $P \in \Lambda$ or $\notin \Lambda$ according as $F(P)$ is or is not zero.

The distance $\delta(P, \Lambda) = |F(P)|\sqrt{a^2 + b^2}$.

If $F_i = 0$ for $i = 1, 2$ are the equations of two straight lines Λ_i, $F = F_1 + \lambda F_2 = 0$, where λ is a parameter, represents a one-parameter family of (i) concurrent straight lines if Λ_1 and Λ_2 intersect at a point and (ii) parallel straight lines if $\Lambda_1 || \Lambda_2$ but $\Lambda_1 \neq \Lambda_2$.

(v) Family of Tangents

If m is a parameter and a and b are positive constants, then

(a) $y = mx + a\sqrt{1 + m^2}$ represents a **family of tangents** to the **upper semi-circle** and, $y = mx - a\sqrt{1 + m^2}$ represents a **family of tangents to the lower semi-circle**, where the complete circle is

$$x^2 + y^2 = a^2.$$

(b) $y = mx \pm \sqrt{a^2m^2 + b^2}$ represent **two families of tangents** to the ellipse $x^2/a^2 + y^2/b^2 = 1$, touching respectively the **upper** and **lower halves**.

(c) $y = mx \pm \sqrt{a^2m^2 - b^2}$ represent two families of tangents to the hyperbola $x^2/a^2 - y^2/b^2 = 1$ touch the both branches **below** and **above** the **x-axis respectively**.

VIII. Differential Calculus

(a) Standard Limits

(i) $\underset{x \to a}{\text{Lt}} [(x^n - a^n)/(x - a)] = na^{n-1}$, where $a \neq 0$ if $n < 1$.

(ii) $\underset{x \to 0}{\text{Lt}} [\sin x/x] = \underset{x \to 0}{\text{Lt}} [(\sin x^c)/x] = 1, \ \underset{x \to 0}{\text{Lt}} [(\sin mx)/x] = m$.

(iii) $\underset{x \to 0}{\text{Lt}} (1 + x)^{1/x} = \underset{x \to \pm\infty}{\text{Lt}} (1 + 1/x)^x = e, \ \underset{x \to 0}{\text{Lt}} (1 + mx)^{1/x} = \underset{x \to \pm\infty}{\text{Lt}} (1 + m/x)^x = e^m$.

If $\underset{x \to a}{\text{Lt}}\ u = 0, \ \underset{x \to a}{\text{Lt}}\ v = \pm\infty$ and $\underset{x \to a}{\text{Lt}} (uv) = l$, then $\underset{x \to a}{\text{Lt}} (1 + u)^v = e^l$.

(iv) $\underset{x\to 0}{\text{Lt}}\,(1/x)\ln(1+x) = \underset{x\to\pm\infty}{\text{Lt}}\, x\ln(1+1/x) = 1.$

$\underset{x\to 0}{\text{Lt}}\,(1/x)\ln(1+mx) = \underset{x\to\pm\infty}{\text{Lt}}\, x\ln(1+m/x) = m.$

(b) Formulae for Derivatives

$d/dx = D = {}'$, $d/dt = \cdot$

(i) $(u \pm v)' = u' \pm v'$.

(ii) $(cy)' = cy'$, $c=$ constant.

(iii) $(uv)' = u'v + uv'$.

(iv) $(1/v)' = -1/v^2$.

$(u/v)' = (u'v - uv')/v^2$.

$\left(\prod u_i\right)' = \left(\prod u_i\right)\sum (u_i'/u_i)$.

(v) $(u^v)' = (e^{v\ln u})' = u^v[v'\ln u + (v/u)u']$.

(vi) If $x = x(t)$, $y = y(t)$ represent y as a function of x, then $y' = \dot{y}/\dot{x}$ and $y'' = (\dot{x}\ddot{y} - \ddot{x}\dot{y})/\dot{x}^3$.

If $F(x, y) = 0$ represents y as some derivable function of x, then $y' = -F_x/F_y$.

(c) Derivatives of Elementary Functions

$f(x)$	$f'(x)$	$f(x)$	$f'(x)$
x^n	nx^{n-1}	$\ln x$	$1/x$
$\sin x$	$\cos x$	$\sin^{-1} x$	$1/\sqrt{(1-x^2)}$
$\cos x$	$-\sin x$	$\cos^{-1} x$	$-1/\sqrt{(1-x^2)}$
$\tan x$	$\sec^2 x$	$\tan^{-1} x$	$1/(1+x^2)$
$\cot x$	$-\operatorname{cosec}^2 x$	$\cot^{-1} x$	$-1/(1+x^2)$
$\sec x$	$\sec x \tan x$	$\sec^{-1} x$	$1/\lvert x\rvert\sqrt{(x^2-1)}$
$\operatorname{cosec} x$	$-\operatorname{cosec} x \cot x$	$\operatorname{cosec}^{-1} x$	$-1/\lvert x\rvert\sqrt{(x^2-1)}$
e^x	e^x	$\sinh^{-1} x$ or	
$\sinh x$ or		$\ln(x+\sqrt{(x^2+1)})$	$1/\sqrt{(x^2+1)}$
$(e^x - e^{-x})/2$	$\cosh x$	$\cosh^{-1} x$ or	
$\cosh x$ or		$\ln(x+\sqrt{(x^2-1)})$	$1/\sqrt{(x^2-1)}$
$(e^x + e^{-x})/2$	$\sinh x$		

(d) Formula for nth Derivatives

(i) $D^n(ax+b)^m = \binom{m}{n} n!a^n(ax+b)^{m-n}$

$= n!a^n$ if $m = n$.

(ii) $D^n e^{ax} = a^n e^{ax}$, $D^n(a^x) = D^n(e^{x\ln a}) = a^x(\ln a)^n$.

(iii) $D^n \ln(ax+b) = aD^{n-1}(ax+b)^{-1} = (-1)^{n-1}(n-1)!a^n/(ax+b)^n$.

(iv) $D^n \begin{pmatrix} \sin \\ \cos \end{pmatrix} (ax+b) = a^n \begin{pmatrix} \sin \\ \cos \end{pmatrix} (ax+b+n\pi/2).$

(v) $D^n e^{ax} \begin{pmatrix} \cos \\ \sin \end{pmatrix} (bx+c) = (a^2+b^2)^{n/2} e^{ax} \begin{pmatrix} \sin \\ \cos \end{pmatrix} (bx+c+n\tan^{-1}(b/a)).$

(vi) Leibnitz's Theorem: $D^n(u/v) = (uv)_n = \sum_0^n {}^nC_r u_{n-r} v_r = \sum_0^n {}^nC_r D^{n-r} u D^r v.$

IX. Integral Calculus

(a) Some Formulae for Integration

(i) $DF(x)$ or $F'(x) = f(x) \Rightarrow D^{-1}f(x) = F(x) + c$ or

$\int f(x)dx = F(x) + c$, where c is a parameter.

(ii) $\int (u \pm v)dx = \int udx \pm \int vdx, \ \int \sum u_i dx = \sum \int u_i dx.$

(iii) $\int cf(x)dx = c\int f(x)dx.$

(iv) $\int f(x)dx = F(x) \Rightarrow \int f(ax+b)dx = (1/a)F(ax+b).$

(v) If $f(x) = u/v$, where $u = v'$, then $\int f(x)dx = \ln v.$

(vi) Integration by parts: $\int uvdx = u\int vdx - \int u'(\int vdx)dx.$

Repeated integration by parts:

$$\int uvdx = uv_1 - u^{(1)}v_2 + u^{(2)}v_3 - \cdots = \sum_0^\infty (-1)^r u^{(r)} v_{r+1},$$

$$= \sum_0^{n-1} (-1)^r u^{(r)} v_{r+1} + (-1)^n \int u^{(n)} v_n dx,$$

where $u^{(r)} = d^r u/dx^r, \ v_r = \int v_{r-1}dx, \ u^{(0)} = u, \ v_0 = v.$

(b) A Table of Integrals of Some Functions, where $L = px+q, \ p \neq 0, \ Q = ax^2+bx+c, \ a \neq 0, \ D = b^2 - 4ac, \ Q' = dQ/dx$. The parameter of Integration is omitted.

Sl. No.	$f(x)$	$I = \int f(x)dx$
1.	x^n	$x^{n+1}/(n+1), n \neq -1$, $\ln x$, when $n = -1$
2.	$\cos x$	$\sin x$
3.	$\sin x$	$-\cos x$
4.	$\tan x$	$\ln \sec x$
5.	$\cot x$	$\ln \sin x$

Contd.

Sl. No.	$f(x)$	$I = \int f(x)dx$
6.	$\sec x$	$\ln(\sec x + \tan x) = \ln \tan(\pi/4 + x/2)$
7.	$\operatorname{cosec} x$	$\ln(\operatorname{cosec} x - \cot x) = \ln \tan(x/2)$
8.	e^x	e^x
9.	a^x	$a^x / \ln a$
10.	$\ln x$	$x \ln x - x$
11.	$\log_a x$	$(x \ln x - x)/ \ln a$
12.	$\sinh x$	$\cosh x$
13.	$\cosh x$	$\sinh x$
14.	$\sec^2 x$	$\tan x$
15.	$\operatorname{cosec}^2 x$	$-\cot x$
16.	$\sec x \tan x$	$\sec x$
17.	$\operatorname{cosec} x \cot x$	$-\operatorname{cosec} x$
18.	$1/(1+x^2)$	$\tan^{-1} x$ or $-\cot^{-1} x$
19.	$1/(1-x^2)$	$(1/2) \ln[(1+x)/(1-x)]$ or $\tanh^{-1} x$
20.	$1/\sqrt{(1-x^2)}$	$\sin^{-1} x$ or $-\cos^{-1} x$
21.	$1/\sqrt{(1+x^2)}$	$\ln(x + \sqrt{(1+x^2)})$ or $\sinh^{-1} x$
22.	$1/\sqrt{(x^2-1)}$	$\ln(x + \sqrt{(x^2-1)})$ or $\cosh^{-1} x$
23.	$1/\|x\|\sqrt{(x^2-1)}$	$\sec^{-1} x$ or $-\operatorname{cosec}^{-1} x$
24.	$\cos^{-1} x$	$x \cos^{-1} x - \sqrt{(1-x^2)}$
25.	$\sin^{-1} x$	$x \sin^{-1} x + \sqrt{(1-x^2)}$
26.	$\tan^{-1} x$	$x \tan^{-1} x - (1/2) \ln(1+x^2)$
27.	$\cot^{-1} x$	$x \cot^{-1} x + (1/2) \ln(1+x^2)$
28.	$\sec^{-1} x$	$x \sec^{-1} x \mp \ln(x + \sqrt{(x^2-1)})$ according as $x > 1$ or < -1
29.	$\operatorname{cosec}^{-1} x$	$x \operatorname{cosec}^{-1} x \pm \ln(x + \sqrt{(x^2-1)})$ according as $x > 1$ or < -1
30.	$1/(x^2+a^2)$	$(1/a) \tan^{-1}(x/a),\ a \neq 0$
31.	$1(x^2-a^2)$	$(1/2a) \ln[(x-a)/(x+a)],\ a \neq 0$
32.	$1/\sqrt{(x^2 \pm a^2)}$	$\ln(x + \sqrt{(x^2 \pm a^2)}), a \neq 0$ $\pm \ln x$ according as $x >$ or < 0 and $a = 0$
*33.	$1/\sqrt{(a^2-x^2)}$	$\sin^{-1}(x/\|a\|),\ a \neq 0$
34.	$\sqrt{(x^2 \pm a^2)}$	$(1/2)[x\sqrt{(x^2 \pm a^2)} \pm a^2 \ln(x + \sqrt{(x^2 \pm a^2)})],\ a \neq 0$ $\pm x^2/2$ according as $x >$ or < 0 and $a = 0$
35.	$\sqrt{(a^2-x^2)}$	$(1/2)[x\sqrt{(a^2-x^2)} + a^2 \sin^{-1}(x/\|a\|)]$

Contd.

Sl. No.	$f(x)$	$I = \int f(x)dx$
36.	L^n	$L^{n+1}/[p(n+1)], \quad n \neq -1$ $(\ln L)/p, \quad n = -1$
*37.	$1/Q$	$(1/\sqrt{D}) \ln[(Q' - \sqrt{D})/(Q' + \sqrt{D})], \quad$ if $D > 0$ $-2/Q', \quad$ if $D = 0$ $[-2/\sqrt{(-D)}] \tan^{-1}(Q'/\sqrt{(-D)}], \quad$ if $D < 0$
*38.	$1/\sqrt{Q}$	$(1/\sqrt{a}) \ln(Q' + 2\sqrt{aQ}), a > 0, D \neq 0$ $\pm(1/\sqrt{a}) \ln Q'$ according as $Q' >$ or $< 0, a > 0, D = 0$ $-(1/\sqrt{-a}) \sin^{-1}(Q'/\sqrt{D})$ if $a < 0, \ D > 0$
*39.	$\sqrt{Q}$	$[1/(2\sqrt{a})^3] \times [Q' 2\sqrt{aQ} - D \ln(Q' + 2\sqrt{aQ}]$ if $a > 0, \ D \neq 0$; $\pm(Q')^2/(2\sqrt{a})^3$ according as $Q' >$ or $< 0, \ a > 0, D = 0$; $-[1/\left(2\sqrt{-a}\right)^3] \times \left[Q' \cdot 2\sqrt{-aQ} + D \sin^{-1}(Q'/\sqrt{D})\right]$ if $a < 0, D > 0$
*40.	$1/Q\sqrt{Q}$	$-2Q'/(D\sqrt{Q}), \ D \neq 0$ $-2\sqrt{a}/(Q')^2, a > 0, D = 0$
*41.	$Q\sqrt{Q}$	$(1/16a)\left[2Q^{3/2}Q' - 3D \int \sqrt{Q} \times dx\right]$ or $\sinh^{-1} x$
42.	$\dfrac{1}{a \cos x + b}$	$\dfrac{1}{\sqrt{a^2 - b^2}} \ln \dfrac{(a-b)t + \sqrt{a^2 - b^2}}{(a-b)t - \sqrt{a^2 - b^2}}, \quad$ if $a^2 > b^2$; $\dfrac{2}{\sqrt{b^2 - a^2}} \tan^{-1} \dfrac{(b-a)t}{\sqrt{b^2 - a^2}}, \quad$ if $a^2 < b^2$; $t/a, \quad$ if $b = a$ $1/(at), \quad$ if $b = -a$, where $t = \tan(x/2)$
43.	$\dfrac{1}{a \sin x + b}$	$\dfrac{1}{\sqrt{a^2 - b^2}} \ln \dfrac{a + bt - \sqrt{a^2 - b^2}}{a + bt + \sqrt{a^2 - b^2}}, \quad a^2 > b^2$; $\dfrac{2}{\sqrt{b^2 - a^2}} \tan^{-1} \dfrac{a + bt}{\sqrt{b^2 - a^2}}, \quad a^2 < b^2$ $-2/[a(t+1)], \quad b = a$ $2/[a(t-1)], \quad b = -a$, where $t = \tan(x/2)$

Contd.

Sl. No.	$f(x)$	$I = \int f(x)dx$
44.	$a^{ax}\cos(bx+c)$	$\frac{e^{ax}}{a^2+b^2}[a\cos(bx+c)+b\sin(bx+c)]$ or $\frac{e^{ax}}{\sqrt{a^2+b^2}}\cos\left(bx+c-\tan^{-1}\frac{b}{a}\right)$
45.	$e^{ax}\sin(bx+c)$	$\frac{e^{ax}}{a^2+b^2}[a\sin(bx+c)-b\cos(bx+c)]$ or $\frac{e^{ax}}{\sqrt{a^2+b^2}}\sin\left(bx+c-\tan^{-1}\frac{b}{a}\right)$

***The correction in the formula for the integral of $1/\sqrt{(a^2-x^2)}$ and the derivation of the formulae 37 to 41 have been made by the author.**

1

Differential Calculus

Functions of a Single Variable

Function, Limit, Continuity and Differentiability

1.1 Function—Definitions

Definition 1.1. Given two non-empty sets A and B, a function is defined as a rule of correspondence that assigns a unique element y of B to every element x of A.

But the concept of rule is not always very clear. For example, if $A = \{1, 2, 3\}$ and $B = \{a, b, c, d\}$ and a corresponds to both 1 and 3 and d corresponds to 2, then f is a function from A to B. But what is the rule that assigns a to 1 and 3, and d to 2?

A better definition is—

Definition 1.2. Given two non-empty sets A and B, a function f mapping A into B is defined to be a subset of $A \times B$ such that $\forall x \in AE!(x, y) \in f$.

Such a function is symbolised as

$$f : A \to B.$$

Since $f \subseteq A \times B$, every element of f is an ordered pair (x, y) where $x \in A$ and $y \in B$. So, if $P \in f$, then $\exists x \in A$ and $y \in B$ such that $P = (x, y)$ and $\forall x \in A$, $(x, y) \in f$ and $(x, y') \in f \Rightarrow y = y'$.

If $(x, y) \in f$, y is called **the f-image** or simply **the image** of x denoted by $f(x)$ and x is **a pre-image of** y. There may be other pre-images. The set A, every element of which has its unique image in B, is called the **domain of f** (Dom f). The set of all the images is called the **range of f** (Ran f) or $f(A)$.

Following the second definition it is not at all difficult to have an idea of a function. It is just a collection of ordered pairs of elements chosen in a specific way from two sets. In the above example, we may write

$$f = \{(1, a), (2, d), (3, a)\}.$$

Then since for every element of A we get a unique ordered pair in f and hence a unique element of B, f is a function mapping A into B.

If x varies, taking different values from A, its image y may remain fixed or may vary, taking different values from B. Thus to every x in A corresponds a unique $y (= f(x))$ in B and that is why sometimes we say that **y, called the dependent variable, is a function of another variable x, called the independent variable or argument.** It should be remembered that, $y = f(x)$ is **not** a function of x but, regarding x and y as real or complex variables, it is an equation which defines y as a function of x and such a function is sometimes expressed as $f : x \to y$.

Definition 1.3. (Sequence). A sequence (sqce.) is defined to be a function whose domain is either $\mathbb{J}_n$ or $\mathbb{N}$.

If

$$f : \mathbb{J}_n \text{ or } \mathbb{N} \to A$$

then the f-image of a natural number r is sometimes denoted by $a_r, b_r, c_r, u_r, v_r, w_r$ or S_r in place of $f(r)$.

The corresponding sequence is often denoted by $\{a_r\}, \{b_r\}, \{c_r\}$ etc., in place of f. The elements of f are (r, a_r) for all r belonging to the domain of f. Thus

$$f : r \to a_r \quad \text{or,} \quad f : n \to a_n.$$

Classification of Functions—Types of Function

Primarily, a function may be—

I(a) One-one or one-to-one or injective, or **(b)** Many-one

or,

II(a) Onto or surjective, or **(b)** Into.

Let f be a function mapping A into B. Then—

I(a) f is one-one or injective, if distinct elements of A have distinct f-images in B.

I(b) f is many-one if there exist at least two distinct elements of A having the same image in B.

II(a) f is onto or surjective if every element of B is the f-image of some element of A.

II(b) f is into if there exists at least one element in B which is not the image of any element of A.

A function may be a combination of types I and II. The most important combination is that of I(a) and II(a). Such a function is called a **one-one and onto** or a **bijective function** or a **bijection**. A function is therefore bijective if it is both injective and surjective.

Identity Function. A function $f : A \to A$ is an identity function in A if $f(x) = x \ \forall x \in A$. Such a function is denoted by I_A.

Equality of Functions. Two functions f and g are said to be equal if—

1. $\text{Dom} f = \text{Dom} g = X$

and **2.** $f(x) = g(x) \ \forall x \in X$.

Composite Function. Given two functions

$$f : A \to B \quad \text{and} \quad g : C \to D$$

we can combine them to form a function

$$h : A_1 \to D,$$

where $A_1 (\subseteq A)$ is such that $f(A_1) = C \cap f(A)$ and $\forall x \in A_1$

$$h(x) = g(f(x)).$$

This function h is denoted by $g \circ f$ so that

$$g \circ f : x \to y,$$

where $x \in A_1$ and $y \in D$ and

$$g \circ f(x) = g(f(x)) = y.$$

Here h or $g \circ f$ which has been composed of f and g is called the **composite function.**

It might be noted that the composite function $g \circ f$ exists with a non-null domain $A_1 \subseteq A$ if $C \cap f(A) \neq \phi$. Then $f(A_1) = C \cap f(A)$ and A_1 becomes the domain of $g \circ f$. For example, if

$$f(x) = \sin x, \ \text{Dom}\, f = \mathbb{R} = A, \ f(A) = [-1, 1],$$

$$g(x) = \ln x \text{ or } \log_e x, \ \text{Dom}\, g = \mathbb{R}^+ = C \text{ then}$$

$g \circ f(x) = \ln(\sin x)$. If $\text{Dom}\, g \circ f = A_1$, then

$$g \circ f(A_1) = f(A) \cap C =]0, 1]$$

$$\therefore \quad A_1 = \bigcup_{n \in \mathbb{Z}}]2n\pi, (2n+1)\pi[.$$

Inverse Function. Let f be an injective function whose domain is X and range is Y so that $Y = f(X)$. Then $\forall y \in Y E!x \in X$ such that $y = f(x)$. So we get a function $g : Y \to X$ such that $\forall y \in Y E!x$ where $x = g(y)$.

$$\therefore \quad x = g(y) = g(f(x)) = g \circ f(x)$$

$$\therefore \quad \forall x \in X \quad g \circ f(x) = x.$$

i.e., $g \circ f = I_X$.

This function g is called the **inverse of** f and it is denoted by f^{-1}. Thus

$$f^{-1} \circ f = I_X.$$

Similarly, $f \circ f^{-1} = I_Y$,

since $\quad y = f(x) = f(g(y)) = f \circ g(y)$.

1.2 Real Function

A function $f : X \to Y$ is said to be real if both X and $Y \subseteq \mathbb{R}$.

The codomain of a real function is usually taken as $\mathbb{R}$. We shall be mainly concerned with real functions.

1.3 Graphical Representation

If $f : X \to Y$ is a real function, taking the rectangular Cartesian xy-frame, the elements of X can be taken on the x-axis and those of Y on the y-axis. In the Fig. 1.1, the set of all the points (x, y), where $x \in X$ and $y \in Y$ lying within and on the sides of the rectangle $ABCD$ is $X \times Y$. The set of encircled points may be such a function $f : X \to Y$.

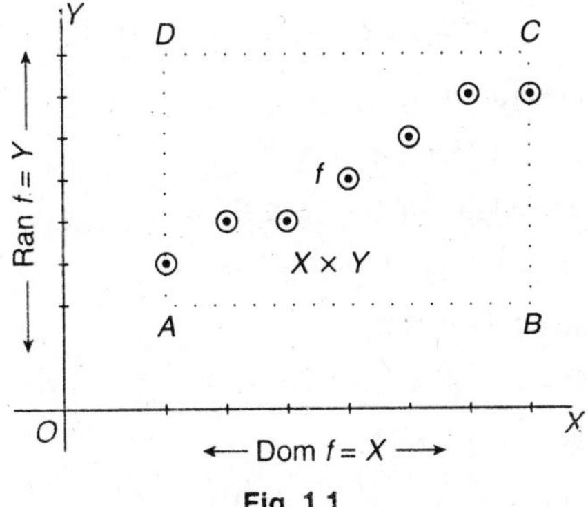

Fig. 1.1

1.4 Algebraic Operations

Let X_i be the domain of a real function f_i for $i = 1, 2, 3, \ldots$

1. Sum Function: The sum of two functions f_1 and f_2 is defined to be a function s whose domain is $X = X_1 \cap X_2$ and $\forall x \in X, s(x) = f_1(x) + f_2(x)$. Symbolically,

$$s = f_1 + f_2.$$

Thus, $f_1 + f_2 : x \to f_1(x) + f_2(x), \forall x \in X_1 \cap X_2.$

The sum of n functions f_i, for $i \in \mathbb{J}_n$ is $\sum_{i=1}^{n} f_i$ which maps x onto $\sum_{i=1}^{n} f_i(x)\ \forall x \in \bigcap_{i=1}^{n} X_i \neq \phi$.

With this definition of addition the functions will form an additive Abelian group. The **additive identity** or **the null function** is

$$0 : x \to 0 \quad \forall x \in \mathbb{R}.$$

The additive inverse of a real function f is the function $-f : x \to -f(x)$.

2. Product Function: The product of two functions f_1 and f_2 is a real function p whose domain is $X = X_1 \cap X_2$ and $p(x) = f_1(x) f_2(x)\ \forall x \in X_1 \cap X_2$. The function p is denoted by $f_1 f_2$. Thus,

$$f_1 f_2 : x \to f_1(x) f_2(x)\ \forall x \in X.$$

The product of n functions f_i for $i \in \mathbb{J}_n$ is the function $\prod_{i=1}^{n} f_i$, where

$$\prod_{i=1}^{n} f_i : x \to \prod_{i=1}^{n} f_i(x).$$

$\forall f \neq 0$, its multiplicative inverse $\frac{1}{f}$ is a function which maps x into $\frac{1}{f(x)}$, where $f(x) \neq 0$. The domain of $\frac{1}{f} = \text{Dom}\, f - \{x \in \text{Dom}\, f : f(x) = 0\}$.

3. Difference Function and Quotient Function: The difference of f_1 and f_2 denoted by $f_1 - f_2$ is a function d where $d = f_1 + (-f_2)$, so that $d : x \to f_1(x) - f_2(x)\ \forall x \in X$.

The quotient of two functions f_1 and f_2 is a function

$$q : (X - N) \to \mathbb{R}$$

such that $\forall x \in X - N,\ q(x) = f_1(x)/f_2(x)$, where

$$N = \{x \in X : f_2(x) = 0\}.$$

The quotient function q is denoted by f_1/f_2.

Examples 1.1. **1.** *Let* $f_1(x) = x$ *for* $1 \le x \le 3$

and $f_2(x) = -x$ *for* $0 \le x \le 2$.

Then $s(x) = (f_1 + f_2)(x) = f_1(x) + f_2(x)$ *for* $1 \le x \le 2$

Here $X_1 = \text{Dom}\, f_1 = \{x \in \mathbb{R} : 1 \le x \le 3\}$

$X_2 = \text{Dom}\, f_2 = \{x \in \mathbb{R} : 0 \le x \le 2\}$

and $X = X_1 \cap X_2 = \{x \in \mathbb{R} : 1 \le x \le 2\}$.

Also $p(x) = (f_1 f_2)(x) = f_1(x) f_2(x) = -x^2\ \forall x \in X$.

2. *If* $f_1(x) = \sin x$ *and* $f_2(x) = \cos x\ \forall x \in \mathbb{R}$, *then* $X_1 = Dom\, f_1 = X_2 = Dom f_2 = \mathbb{R}$. *The quotient function is given by*

$$q(x) = (f_1/f_2)(x) = (\sin / \cos)x = (\sin x)/\cos x = \tan x\ \forall x \in X - N,$$

where $X = X_1 \cap X_2 = \mathbb{R}$, *so that*

$X - N = \mathbb{R} - N$, N *being* $\{x \in \mathbb{R} : \cos x = 0\}$.

$\therefore \quad N = \{(2n+1)\pi/2\ \forall n \in \mathbb{Z}\}$.

1.5 Elements of a Point Set

Interval of a variable x. Given two real numbers a and b where $a < b$, by an **interval** we mean any one of the following four sets: (i) $\{x \in \mathbb{R} : a \leq x \leq b\}$, (ii) $\{x \in \mathbb{R} : a < x < b\}$, (iii) $\{x \in \mathbb{R} : a \leq x < b\}$, (iv) $\{x \in \mathbb{R} : a < x \leq b\}$. They are denoted by $[a,b], (a,b)$ or $]a,b[, [a,b)$ or $[a,b[$ and $(a,b]$ or $]a,b]$ respectively and are called the **closed, open, closed-open and open-closed** intervals. The numbers a and b are respectively called the **left end point** and the **right end point**, and the positive number $b-a$ is called the **length** of each of the intervals.

Neighbourhood (nbd). Given a real number a, any set S of real numbers containing an open interval N containing a is called a neighbourhood of a. The open interval $(a-\delta, a+\delta)$ is called the **delta neighbourhood (δ-nbd)** of a and is denoted by $N(\delta, a)$. Then

$$N(\delta, a) = \{x \in \mathbb{R} : a - \delta < x < a + \delta \text{ or } |x-a| < \delta\}.$$

The set $\{x \in \mathbb{R} : 0 < |x-a| < \delta\}$ is called the **deleted δ-nbd of a**. It is denoted by $N(\delta, \not a)$ which is therefore $N(\delta, a) - \{a\}$.

Delta Neighbourhood of ∞ and $-\infty$

$\mathbb{R}$, the set of all real numbers, is a **field** so that the elements of $\mathbb{R}$ satisfy certain specific properties with regard to addition and multiplication. The number 0 is called the **additive identity** and 1, the **multiplicative identity**.

Two symbols ∞ and $-\infty$, called respectively infinity (or plus infinity) and minus infinity, are introduced into $\mathbb{R}$ and a new set $\mathbb{R}^*$, called the **extended field** of real numbers is formed. The numbers of $\mathbb{R}$ may be called **ordinary real numbers**, whereas the symbols $\pm\infty$ may be called **extraordinary real numbers**. The properties, which $\pm\infty$ are supposed to satisfy, are given below:

(a) If $x \in \mathbb{R}$, then $-\infty < x < \infty$, and $x + \infty = \infty$, $x - \infty = -\infty$, $x/\infty = x/(-\infty) = 0$.

(b) If $x > 0$, then $x \cdot \infty = \infty$, $x \cdot (-\infty) = -\infty$ and $x/0 = \infty$.

(c) If $x < 0$, then $x \cdot \infty = -\infty$, $x \cdot (-\infty) = \infty$ and $x/0 = -\infty$.

(d) $\infty + \infty = \infty$, $-\infty - \infty = -\infty$, $\infty \cdot \infty = (-\infty)(-\infty) = \infty$, $\infty \cdot (-\infty) = -\infty \cdot \infty = -\infty$.

The combinations: $0.\infty, 0.(-\infty), \infty - \infty, \infty/\infty$ and $1^{\pm\infty}$ are meaningless and therefore they have no numerical value.

The **δ-nbd of ∞** is defined to be

$N(\delta, \infty) = \{x \in \mathbb{R} : x > \delta\}$.

The **δ-nbd of $-\infty$** is defined to be

$N(\delta, -\infty) = \{x \in \mathbb{R} : x < -\delta\}$.

Limit Point or Limiting Point of a Set. Let $S \subseteq \mathbb{R}$ and $a \in \mathbb{R}$. **Then the point (number) a is said to be a limit point (lt pt) of S if**

$$\forall \delta > 0 \, \exists x \in S : x \in N(\delta, \not a),$$

i.e., if every deleted nbd of a contains at least one point of S.

If x is a real variable taking values from S only, we say that **x approaches a** or **x tends to a** or $x \to a$ provided a is a lt pt of S. If $\forall \delta > 0$, $\exists x \in S$ such that $x > a$ and $x \in N(\delta, \not a)$, we say that **x can approach a from the right-hand side (RHS).** Symbolically, $x \to a+$ or $x \to a + 0$. Similarly, if $\forall \delta > 0$, $\exists x \in S : x < a$ and $x \in N(\delta, \not a)$, then we say that **x can approach a from the left-hand side (LHS).** Symbolically, $x \to a-$ or $x \to a - 0$. Thus, if a is a limit point of $S, x(\in S)$ may approach a either from the LHS or from the RHS or from both. And it should be noted that a limit point may or may not belong to the set S.

If ∞ is a limit point of S, then

$$\forall \delta > 0 \, \exists x \in S : x \in N(\delta, \infty)$$

so that $x > \delta$ and $x \to \infty$. In this case every nbd of ∞ contains a point of S.

Similarly, if $-\infty$ is a limit point of S, then

$$\forall \delta > 0 \, \exists x \in S : x \in N(\delta, -\infty)$$

so that $x < -\delta$ and $x \to -\infty$.

Isolated Point of a Set. Given $S \subseteq \mathbb{R}$ and $a \in S$, the point a is said to be an **isolated point** of S if it is not a limit point of S. So, if a is an isolated point of S, it is possible to find a deleted δ-nbd $N(\delta, \not a)$ which does not contain any point of S.

Examples 1.2. 1. *If $S = \{1, 1/2, 1/3, \ldots\}$, then $S \subseteq \mathbb{R}$. If $x \in S$, then $x = 1/r$ for some $r \in \mathbb{N}$.*

S has only one limit point, viz., 0. Hence x can approach 0 ($x \to 0$). All other points are isolated points. Hence $x \not\to a$ for any $a \in S$. Here the only limit point 0 does not belong to the set S.

2. *If $S = \mathbb{N}$, then $S \subseteq \mathbb{R}$. It has only one limit point ∞, an extraordinary real number. So, if the variable $n \in \mathbb{N}$, n can approach ∞ only and no other number. All the points of $\mathbb{N}$ are isolated points.*

Note. In finding the **limit of a sequence**, which is nothing but a function of the variable $n \in \mathbb{N}$, we **always take $n \to \infty$.**

3. *If $S = (0, 1]$, then although $0 \notin S$, it is a limit point of S. Not only that, all the points of S are limit points and hence it has no isolated point.*

Also $x \to 0+$ and $x \to 1-$. Consequently, x remaining in S can approach both 0 and 1.

1.6 Limit of a Function

Finite Limit of a Function at an Ordinary Point

Let $f : X \to Y$ be a real function so that

$$X = \text{Dom}\, f \subseteq \mathbb{R} \text{ and } Y = f(X) \subseteq \mathbb{R}.$$

Let a and l both $\in \mathbb{R}$ and let x be a real variable taking values from X only. Then $\forall x \in X \, E!y \in Y$, where $y = f(x)$. Also let a be a limit point of X so that x can approach a.

Then **the number l is called the limit of f at a if $\forall \epsilon > 0 \exists \delta > 0 : y \in N(\epsilon, l)$ whenever $x \in X \cap N(\delta, \not{a})$.** Or, which is the same as

$$x \in X \cap N(\delta, \not{a}) \Rightarrow y, \text{ i.e., } f(x) \in N(\epsilon, l).$$

Or, $x \in X$ and $0 < |x - a| < \delta \Rightarrow |y - l|$ or $|f(x) - l| < \epsilon$.

Loosely speaking, a finite number l is defined to be the limit of f at a if y or $f(x)$ is very very close to or exactly equal to l whenever $x(\in X)$ is very very close to but not exactly equal to a. If l is the limit of f at a, it is expressed symbolically as

$$f(x) \to l \text{ as } x \to a \text{ or } \underset{x \to a}{\text{Lt}} f(x) = l.$$

It might be noted that $\underset{x \to a}{\text{Lt}} f(x) = l$ if $\forall \epsilon > 0 \exists \delta > 0 : |f(x) - l| < \epsilon \, \forall x \in X$, satisfying $0 < |x - a| < \delta$.

Note 1. The limit of a function, when it exists, is unique.

Note 2. For the existence of limit at a, the point a may or may not belong to X, the domain of f but it is necessary for a to be a limit point of X so that x may approach a. The point l may or may not belong to the range Y of f and it **may or may not be a limit point of Y.**

Example 1.3. *If $X = \{1, 1/2, 1/3, \ldots\}$ and f is a real function with X as its domain such that $\forall x \in X$*

$$y = f(x) = 2x + 3$$

then although

$$\underset{x \to 0}{\text{Lt}} f(x) = \underset{x \to 0}{\text{Lt}} (2x + 3) = 3$$

but $\underset{x \to .001}{\text{Lt}} f(x) = \underset{x \to .001}{\text{Lt}} (2x + 3) \neq 3.002.$

For, 0 is limit point of X but .001, although belongs to X, is an isolated point and not a limit point of X and $x \not\to .001$.

Infinite Limits and Limits at Infinity

(i) Infinite Limit at an Ordinary Point. If a is a limit point of the domain X of a function f, then f is said to approach infinity as $x(\in X)$ tends to a if

$$\forall \epsilon > 0 \exists \delta > 0 : f(x) \in N(\epsilon, \infty) \forall x \in X \cap N(\delta, \not{a}).$$

Symbolically, $\underset{x \to a}{\text{Lt}} f(x) = \infty$ or $f(x) \to \infty$ as $x \to a$.

Note. Since equating something to ∞ is not desirable, it is better to use the symbol.

$$f(x) \to \infty \text{ as } x \to a$$

to indicate that ∞ is the limit (?) of f at a.

Similarly, f is said to approach minus infinity as $x(\in X)$ tends to a $(\in \mathbb{R})$ if

$$\forall \epsilon > 0 \exists \delta > 0 : f(x) \in N(\epsilon, -\infty) \forall x \in X \cap N(\delta, \not{a}).$$

Symbolically, $\underset{x \to a}{\text{Lt}} f(x) = -\infty$ or $f(x) \to -\infty$ as $x \to a$.

(ii) Finite Limit at Infinity. A number $l(\in \mathbb{R})$ is said to be the limit of f at infinity if ∞ is a limit point of the domain X of f and if

$$\forall \epsilon > 0 \exists \delta > 0 : f(x) \in N(\epsilon, l) \, \forall x \in X \cap N(\delta, \infty).$$

Symbolically, $\underset{x \to \infty}{\text{Lt}} f(x) = l.$

Similarly, l is the limit of f at minus infinity if $-\infty$ is a limit point of X and if

$$\forall \epsilon > 0 \, \exists \, \delta > 0 : f(x) \in N(\epsilon, l) \, \forall \, x \in X \cap N(\delta, -\infty).$$

Symbolically, $\underset{x \to -\infty}{\text{Lt}} f(x) = l.$

(iii) Infinite Limits at Infinity. (a) A function f is said to approach infinity as x tends to infinity if ∞ is a limit point of the domain X of f and if

$$\forall \epsilon > 0 \, \exists \, \delta > 0 : f(x) \in N(\epsilon, \infty) \, \forall \, x \in X \cap N(\delta, \infty).$$

Symbolically, $f(x) \to \infty$ as $x \to \infty$ or $\underset{x \to \infty}{\text{Lt}} f(x) = \infty.$

(b) $f(x) \to -\infty$ as $x \to \infty$ or $\underset{x \to \infty}{\text{Lt}} f(x) = -\infty,$

(c) $f(x) \to \infty$ as $x \to -\infty$ or $\underset{x \to -\infty}{\text{Lt}} f(x) = \infty,$

and **(d)** $f(x) \to -\infty$ as $x \to -\infty$ or $\underset{x \to -\infty}{\text{Lt}} f(x) = -\infty$

can be similarly defined.

1.7 Basic Theorems on Limits

Let $f_r(x) \to l_r$ for $r = 1, 2, 3, \ldots$ as $x \to a$ and $X_r = \text{Dom } f_r$. Then as $x \to a$, where a is a limit point of the corresponding intersection of the domains,

1. $(f_1 \pm f_2)(x) \to l_1 \pm l_2$.

2. $\left(\sum f_r\right)(x) \to \sum l_r, \; \sum = \sum_{r=1}^{n}$

3. $(1/f_1)(x) \to 1/l_1$, provided $l_1 \neq 0$

4. $(f_1 f_2)(x) \to l_1 l_2$

5. $\left(\prod f_r\right)(x) \to \prod l_r, \; \prod = \prod_{r=1}^{n}$

6. $(f_1/f_2)(x) \to l_1/l_2$, provided $l_2 \neq 0$,

7. $f_1(x) < f_2(x)$ or $f_1(x) \leq f_2(x) \; \forall \, x \in N(\delta, \not{a}) \Rightarrow l_1 \leq l_2$

8. If $\forall \, x \in N(\delta, \not{a})$

either $f_1(x) < f_2(x) < f_3(x)$ or $f_1(x) \leq f_2(x) < f_3(x)$

or $f_1(x) < f_2(x) \leq f_3(x)$ or $f_1(x) \leq f_2(x) \leq f_3(x)$

and $l_1 = l_3 = l$ say,

then $l_2 = l$ also.

This is what is known as the **Sandwich Theorem.**

9. Let $X = \text{Dom } f_2 \circ f_1, u = f_1(x) \; \forall x \in X$ and l_1 be a limit point of $f_1(X)$ and $f_2(u) \to m$ as $u \to l_1$, then

$$f_2 \circ f_1(x) \text{ or } f_2(f_1(x)) \to m \text{ as } x \to a,$$

provided a is also a limit point of X.

Converses of the Theorems are not necessarily true.

Examples 1.4. 1. *If $f(x) = 1/x$ and $X = \text{Dom} f = \mathbb{R}_0$, then 0 is a limit point of X. But still $\underset{x\to 0}{\text{Lt}} f(x)$ does not exist. For, $\underset{x\to 0+}{\text{Lt}} f(x) = \infty$, $\underset{x\to 0-}{\text{Lt}} f(x) = -\infty$.*

2. *If $f(x) = 1/x^2$, $\underset{x\to 0}{\text{Lt}} f(x) = \infty$.*

3. *If $f_1(x) = 1/x$, $f_2(x) = -1/x$, then the limit of neither $f_1(x)$ nor $f_2(x)$ exists at $x = 0$. But*

$$\underset{x\to 0}{\text{Lt}}(f_1 + f_2)(x) = \underset{x\to 0}{\text{Lt}}[f_1(x) + f_2(x)] = \underset{x\to 0}{\text{Lt}}(1/x - 1/x)$$

exists and it is 0.

4. *If $l_1 = 0 = l_2$, $\underset{x\to a}{\text{Lt}}(f_1/f_2)(x)$ may or may not exists. For example,*

(i) *if $f_1(x) = x = f_2(x)$, then $\underset{x\to 0}{\text{Lt}}(f_1/f_2)(x) = \underset{x\to 0}{\text{Lt}}[f_1(x)/f_2(x)] = \underset{x\to 0}{\text{Lt}}(x/x) = 1$,*

(ii) *if $f_1(x) = x, f_2(x) = x^2$, $\underset{x\to 0}{\text{Lt}}(f_1/f_2)(x)$ does not exist,*

(iii) *if $f_1(x) = x, f_2(x) = x^3$, $\underset{x\to 0}{\text{Lt}}(f_1/f_2)(x) = \infty$,*

(iv) *if $f_1(x) = -x$ and $f_2(x) = x^3$, $\underset{x\to 0}{\text{Lt}}(f_1/f_2)(x) = -\infty$.*

1.8 Fundamental Limits

The following limits are regarded as fundamental, because the limits of functions are calculated using these limits:

1. If $n \in \mathbb{N}$, $\underset{n\to\infty}{\text{Lt}} x^n = 0$ if $|x| < 1$

$= 1$ if $x = 1$.

For any other value of x the limit does not exist as a finite number.

2. If $n \in \mathbb{N}$, $\underset{x\to\infty}{\text{Lt}} x^n/n! = 0 \ \forall x \in \mathbb{R}$.

3. $\underset{n\to\infty}{\text{Lt}}(1 + 1/n)^n = \underset{n\to\infty}{\text{Lt}} \sum_{r=1}^{n} 1(r-1)! = e$, where $n \in \mathbb{N}$.

$$\therefore \quad \sum_{r=1}^{\infty} 1/(r-1)! = e.$$

4. $\underset{x\to a}{\text{Lt}} \dfrac{x^n - a^n}{x - a} = \underset{h\to 0}{\text{Lt}} \dfrac{(a+h)^n - a^n}{h} = na^{n-1}$, $\forall n, a \in \mathbb{R}$ except $a = 0$ when $n \le 1$ and the limit is 1 when $a = 0$ and $n = 1$.

5. $\underset{\theta\to 0}{\text{Lt}} \dfrac{\sin\theta^c}{\theta} = \underset{\theta\to 0}{\text{Lt}} \dfrac{\sin\theta}{\theta} = 1$, where $\theta^c = \theta$ radians and $\sin\theta = \sin\theta^c$.

6. $\underset{x\to\pm\infty}{\text{Lt}}(1 + 1/x)^x = \underset{x\to 0}{\text{Lt}}(1 + x)^{1/x} = e$.

7. $\underset{x\to 0}{\text{Lt}} \dfrac{1}{x}\ln(1 + x) = \underset{x\to\pm\infty}{\text{Lt}} x \ln\left(1 + \dfrac{1}{x}\right) = 1$, where $\ln x = \log_e x$, the natural logarithm of x.

8. $\underset{x\to 0}{\text{Lt}} \dfrac{e^x - 1}{x} = \underset{x\to\pm\infty}{\text{Lt}} x(e^{1/x} - 1) = 1$.

Exercises 1.1

1. If $\sum p_r a^{\alpha_r} = 0 = \sum q_r a^{\beta_r}$ but $\sum q_r \beta_r a^{\beta_r - 1} \neq 0$, then show that
$$\operatorname*{Lt}_{x \to a} \left[\sum p_r x^{\alpha_r} / \sum q_r x^{\beta_r}\right] = \sum p_r \alpha_r a^{\alpha_r - 1} / \sum q_r \beta_r a^{\beta_r - 1}, \text{ where } 1 \leq r \leq n.$$
[**Hint.** If l is the required limit and $\sum p_r a^{\alpha_r} = 0 = \sum q_r a^{\beta_r}, l = \sum p_r\{(x^{\alpha_r} - a^{\alpha_r})/(x-a)\}/\sum q_r\{(x^{\beta_r} - a^{\beta_r})/(x-a)\}$.]

2. Evaluate the following limits, assuming them to exist without using *L'* Hospital's rule or expansion of a function:

   (a) $\operatorname*{Lt}_{x \to 0} \{(\sin x)/x - 1\}/x^3$.

   (b) $\operatorname*{Lt}_{x \to 1} [p/(1 - x^p) - q/(1 - x^q)]$.

   [**Hint.** (a) $l = \operatorname*{Lt}_{\theta \to 0} \{(\sin 3\theta)/3\theta - 1\}/(3\theta)^3$,

   (b) $l = \operatorname*{Lt}_{x \to 1} [\{p(1 - x^p) + px^p\}/(1 - x^p) - \{q(1 - x^p) + qx^q\}/(1 - x^q)]$
   $= (p - q) - \operatorname*{Lt}_{x \to 1} [p/(x^{-p} - 1) + q/(x^{-q} - 1)]$
   $= (p - q) - \operatorname*{Lt}_{y \to 1} \{p/(1 - y^p) + q/(1 - y^q)\}$, where $y = x^{-1}$.]

3. Evaluate $\operatorname*{Lt}_{n \to \infty} \Pi\{(r^3 - 8)/(r^3 + 8)\}, 3 \leq r \leq n$.
   [**Hint.** If $f(n) = \Pi u_r$, where $u_r = v_r/v_{r+1}$, then $f(n) = v_1/v_{n+1}$. Here
   $u_r = (r - 2)(r^2 + 2r + 4)/(r + 2)(r^2 - 2r + 4)$
   $= \{(r - 2)/(r - 1)\}\{(r - 1)/r\}\{r/(r + 1)\}\{(r + 1)/(r + 2)\}$
   $\{r^2 + 2r + 4)/(r^2 + 3)\}\{(r^2 + 3)/(r^2 - 2r + 4)\}$.]

4. Evaluate $\operatorname*{Lt}_{x \to \infty} [x^3/(3x^2 - 4) - x^2/(3x + 2)]$.
   [**Hint.** $l = \operatorname*{Lt}_{x \to \infty} [x^2\{x(3x + 2) - (3x^2 - 4)\}/\{(3x^2 - 4)(3x + 2)\}]$
   $= \operatorname*{Lt}_{x \to \infty} [x^2(2x + 4)/\{(3x^2 - 4)(3x + 2)\}]$.]

5. Evaluate $\operatorname*{Lt}_{x \to \pi/6} [(2\sin^2 x + \sin x - 1)/(2\sin^2 x - 3\sin x + 1)]$.
   [**Hint.** Put $z = \sin x$.]

6. $\operatorname*{Lt}_{x \to 0} (1 + x)^{1/x} = \operatorname*{Lt}_{x \to \pm\infty} (1 + 1/x)^x = e$.
   Derive from it $\operatorname*{Lt}_{x \to a} (1 + u)^v = e^l$, where
   $\operatorname*{Lt}_{x \to a} u = 0$ and $\operatorname*{Lt}_{x \to a} v = \infty($ or $-\infty)$ and $\operatorname*{Lt}_{x \to a} (uv) = l$.
   Also derive $\operatorname*{Lt}_{x \to a} u^v = e^l$, if $\operatorname*{Lt}_{x \to a} u = 1$, $\operatorname*{Lt}_{x \to a} v = \infty($ or $-\infty)$ and $\operatorname*{Lt}_{x \to a} (u - 1)v = l$.

7. Evaluate the following limits:
   (a) $\operatorname*{Lt}_{x \to \infty} [x/(1 + x)]^x$, **(b)** $\operatorname*{Lt}_{x \to 0} [(1 + \tan x)/(1 + \sin x)]^{\operatorname{cosec} x}$.

8. Evaluate $\operatorname*{Lt}_{x \to \pi/4} [(\ln \tan x)/(1 - \cot x)]$.

9. Show that $\sin x$ does not tend to any limit as $x \to \infty$.

10. Show that

$$\underset{x\to 2}{\text{Lt}}[\{(\cos\alpha)^x - (\sin\alpha)^x - \cos 2\alpha\}]/(x-2)] = \cos^2\alpha \ln\cos\alpha - \sin^2\alpha \ln\sin\alpha.$$

Answers

2. **(a)** $-1/6$; **(b)** $(p-q)/2$. **3.** 2/7. **4.** 2/9. **5.** -3.

7. **(a)** e^{-1}; **(b)** 1. **8.** 1.

1.9 Continuity at a Point

A function $f: X \to Y$, where $X = \text{Dom}\, f$ and $Y = \text{Ran}\, f$, is said to be continuous at a point $a \in X$ if

$$\forall \in > 0 \,\exists\, \delta > 0 : f(x) \in N(\epsilon, f(a)) \;\forall x \in X \cap N(\delta, a).$$

According to this definition, the point a need not be a limit point of X. Hence the condition that

$$\underset{x\to a}{\text{Lt}}\, f(x) = f(a)$$

is not necessary for f to be continuous at a. But, if a happens to be a limit point of X, then the above condition follows automatically from the definition.

It might be noted that every function is continuous at an isolated point of X.

Continuity in the Domain. A function is said to be continuous in its domain if it is continuous pointwise at every point of its domain.

All the elementary functions are continuous in their respective domains.

Continuity in an Interval. Let I be an interval such that $I \subseteq X = \text{Dom}\, f$. Then f is said to be continuous in I if it is continuous at every point of I.

Sequential Criterion for Continuity. A function $f: X \to \mathbb{R}$ is continuous at a point $a \in X$ if $\forall \{x_n\}$ in X having its limit a the sequence $\{y_n\}$ has the limit $f(a)$, where $y_n = f(x_n)$.

This criterion is known as **Heine's criterion** and it is normally used to prove the discontinuity of a function.

Algebraic Operations on Continuous Functions. Let $f_r : X_r \to \mathbb{R}$ for $r = 1, 2, 3, \ldots$ be continuous in X_r. Then $\sum f_r$ and Πf_r are continuous in $\cap X_r$, $-f_r$ is continuous in X_r and $1/f_r$ is continuous in X_r except at the points where $f_r(x)$ vanishes.

The composite function $f_2 \circ f_1$ is continuous at $a \in \text{Dom}\, f_2 \circ f_1$ if f_1 is continuous at a and f_2 is continuous at $f_1(a)$.

1.10 Derivability and Differentiability

Derivability and Derivative. The function $f: X \to \mathbb{R}$ is said to be derivable at $a \in X$ if the limit

$$\underset{x\to a}{\text{Lt}} \frac{f(x) - f(a)}{x - a}$$

exists. This limit is called the **derivative** of f at a and it is denoted by symbols like

$$f'(a), y'(a), y1(a), dy|dx|_{x=a}, Df(a) \text{ etc.}$$

The derivative at $x \in X$ is the limit

$$\operatorname*{Lt}_{t \to x} \frac{f(t) - f(x)}{t - x} \quad \text{or,} \quad \operatorname*{Lt}_{h \to 0} \frac{f(x+h) - f(x)}{h}$$

and it is denoted by $Df(x), f'(x), y', y_1, dy/dx$, if $y = f(x)$.

Differentiability and Differential. $f : X \to \mathbb{R}$ is said to be **differentiable** at $x \in X$ if the difference

$$\Delta y = \Delta f(x) = f(x + \Delta x) - f(x)$$

can be expressed as $A\Delta x + \eta \Delta x$, where A is independent of the increment Δx in x and η, depending on Δx, tends to zero as $\Delta x \to 0$.

The part $A\Delta x$, called the **principal part** of the difference, is defined to be the **differential** of y (or $f(x)$) and it is denoted by dy or $df(x)$. Thus

$$dy \text{ or } df(x) = A\Delta x. \tag{1}$$

Since $\Delta y = A\Delta x + \eta \Delta x, \; \dfrac{\Delta y}{\Delta x} = A + \eta.$

And $\dfrac{dy}{dx} = \operatorname*{Lt}_{\Delta x \to 0} \dfrac{\Delta y}{\Delta x} = \operatorname*{Lt}_{\Delta x \to 0} (A + \eta) = \operatorname*{Lt}_{\Delta x \to 0} A + \operatorname*{Lt}_{\Delta x \to 0} \eta = A.$

Thus A, in the differential dy, is the coefficient of the arbitrary increment Δx in x. That is why A, which is nothing but the derivative dy/dx of f at x, is called the **differential coefficient** of y with respect to x. Thus

$$A = dy/dx = f'(x) \quad \text{and} \quad dy = f'(x)\Delta x.$$

To find the differential of x we take $y = x$, so that $f(x) = x$. Then knowing that $f'(x) = 1$

$$dx = f'(x)\Delta x = 1 \cdot \Delta x = \Delta x,$$

so that the differential of the dependent variable x is nothing but an arbitrary increment Δx in x. So from (1)

$$A = dy/\Delta x = dy \div dx.$$

Thus on the basis of the definition of differential of a function the derivative $f'(x) (= A)$ is the **quotient** of the differential of the dependent and the independent variables.

If $f'(x) = \frac{d}{dx} f(x) = Df(x)$, then d/dx or D is regarded as an operator called **differentiator**.

1.10.1 Geometrical Interpretation of the Derivative

If $f : X \to \mathbb{R}$ is a real function, then we know that $f \subseteq X \times \mathbb{R}$ and the set f consists of the elements each of which is an ordered pair (x, y) of real numbers. Each such ordered pair represents, in the xy-plane a point P. So geometrically f is set of points in the xy-plane and this set of points is called the **graph of the function** or **of the equation** $y = f(x)$. The derivative $f'(x)$, which is the limit

$$\operatorname*{Lt}_{h \to 0} \frac{f(x+h) - f(x)}{h}$$

is the **gradient of the tangent to the graph of** f **at** $P(x)$. If ψ^c is the angle of inclination of the tangent, then

$$f'(x) = \tan \psi.$$

This gradient is also called the **gradient of the curve at** $P(x)$.

An important property of a **differentiable function** is that **it is continuous at every point where it is differentiable.**

1.11 Derivatives of Functions Obtained Algebraically from Other Functions and Composite Functions

Let $f_r : x_r \to \mathbb{R}$ for $r = 1, 2, 3, \ldots$ be differentiable at a common point x belonging to the intersection of the domains, and $f_r'(x)$ represents the derivative of f_r at x. Then

1. $\left(\sum f_r\right)'(x) = \sum f_r'(x)$,
2. $\left(\prod f_r\right)'(x) = \prod f_r(x) \sum [f_r'(x)/f_r(x)], f_r(x) \neq 0$,
3. $(-f_r)'(x) = -f_r'(x)$,
4. $(1/f_r)'(x) = -f_r'(x)/f_r^2(x),\ f_r(x) \neq 0$, whence

$$\begin{aligned}(f_1/f_2)'(x) = (f_1 \cdot (1/f_2))'(x) &= f_1(x) \cdot 1/f_2(x) \left[\frac{f_1'(x)}{f_1(x)} + \frac{(1/f_2)'(x)}{(1/f_2)(x)}\right] \text{ by (2)} \\ &= \frac{f_1'(x) f_2(x) - f_1(x) f_2'(x)}{f_2^2(x)},\end{aligned}$$

5. $(f_2 \circ f_1)'(x) = \dfrac{df_2(x)}{df_1(x)} \cdot \dfrac{df_1(x)}{dx}$. (Chain rule)

1.12 Derivative of a Function Defined Implicitly

Implicit Function. Let the equation

$$F(x, y) = 0 \tag{1}$$

be such that $\forall x \in X \subseteq \mathbb{R} \, \exists y \in \mathbb{R}$ such that $F(x, y) = 0$. Then (1) is said to define a **real relation** between the real variables x and y. Then X is said to be the **domain** and the set of all the corresponding values of y, the **range** Y of the relation. The value of y corresponding to $x \in X$ may or may not be unique. If the value of y is unique, then the equation (1) is said to define a function called **implicit function** of x. If the value of y is not unique, then the range Y may be restricted to a set Y_1, such that $\forall x \in X E! y \in Y_1$, such that (x, y) satisfies (1). Then also we are getting a function $f : X \to Y$. This function is also an implicit function defined by the equation (1).

For example, the equation

$$xy - y + 1 = 0, \tag{2}$$

which is the same as

$$y = \frac{1}{1 - x}, x \neq 1, \tag{2'}$$

defines a real relation with $\mathbb{R} - \{1\} = X$ as its domain and range $Y = \mathbb{R} - \{0\}$. So there is a function

$$f : X \to Y$$

such that (x, y) or $(x, f(x))$ will satisfy (2). This function f is the unique implicit function defined by (2), which can be made explicit by the equation (2′).

Similarly

$$y^2 - 4x = 0 \tag{3}$$

defines a real relation with domain $X = \mathbb{R}^+ \cup \{0\}$ and range $Y = \mathbb{R}$. Here for every $x(\neq 0) \in X$ we do not get a unique value of y in Y, but two values of y. But if we take $Y_1 = \mathbb{R}^+ \cup \{0\} = X$, then $Y_1 \subseteq Y$ and $\forall x \in X$ we get a unique $y \in Y_1$ given by

$$y = 2\sqrt{x} \geq 0.$$

So if we take $f(x) = 2\sqrt{x}\ \forall x \in X$ we get a real function

$$f : X \to Y_1$$

such that (x, y) or $(x, 2\sqrt{x}) \in f$ satisfies (3).

Such a function is an implicit function defined by (3). Similarly restricting Y to $Y_2 = \{x \in \mathbb{R} : x \leq 0\}$ we get another real function

$$g : X \to Y_2$$

defined by (3).

To get the derivative of a function f, defined implicitly by (1) at a point $P(x, y)$, where

$$F(P) = F(x, y) = 0$$

and $x \in X$ and $y = f(x) \in Y$ we take the derivative of the function $F(x, y)$ with respect to x regarding y constant. Let this derivative be denoted by $F_x(x, y)$ or simply by F_x or $\partial F/\partial x$. Similarly, we take the derivative of $F(x, y)$ with respect to y, regarding x constant. Let this be denoted by $F_y(x, y)$ or F_y or $\partial F/\partial y$. Then it will be seen later in § 4.9.1 that

$$f'(x) \quad \text{or} \quad dy/dx = -F_x/F_y, \quad \text{where } F_y \neq 0.$$

1.13 Derivative of a Function Defined Parametrically

Let $\phi : T \to \mathbb{R}$ and $\psi : T \to \mathbb{R}$ be two functions having their common domain $T \subseteq \mathbb{R}$. Then $\forall t \in T$ there must exist a unique ϕ-image $x(= \phi(t))$ and a unique ψ-image $y(= \psi(t))$. The set of all these ordered pairs (x, y) may not give a real function, since corresponding to every x the value of y may not be unique. But if ϕ is a bijective function, it is invertible and if $X = \text{Ran}\,\phi$,

$$\phi^{-1} : X \to T$$

is a function and $\forall x \in X$, $t = \phi^{-1}(x)$. Then

$$y = \psi(t) = \psi(\phi^{-1}(x)) = (\psi \circ \phi^{-1})(x) = f(x),$$

so that y is a function of x, called the **parametric function.**

$$f'(x) = dy/dx = (dy/dt)/(dx/dt) = \dot{\psi}(t)/\dot{\phi}(t),$$

where a dot represents differentiation with respect to t.

1.14 Differentiation of an Exponential Function

If $y = u^v$, where u and v are both differentiable, then $y' = u^v[v' \ln u + (v/u)u']$, $u > 0$.

For $\quad y = u^v = e^{v \ln u} = e^w$ say.

$\therefore \quad y' = dy/dx = (dy/dw)(dw/dx) = e^w w'.$

But $\because$ $w = v \ln u$, $w' = v' \ln u + (v/u)u'$.

Hence the formula.

Example. If $y = (\sin x)^{\cos x}$, $u = \sin x$ and $v = \cos x$, $u' = \cos x$ and $v' = -\sin x$. So

$$y' = (\sin x)^{\cos x}[-\sin x \ln \cos x + \cot x \cdot \cos x].$$

Examples 1.5.1. 1. *If $y = \sin^{-1}(\sin x)$ $\forall x \in \mathbb{R}$, show that although $dy/dx = 1$ at $x = 0$, but it does not exist at $x = \pi/2$. Find $y'(\pi)$ if it exists.*

$$y = \sin^{-1}(\sin x) = x \quad \text{if } -\pi/2 < x \leq \pi/2 \text{ and}$$
$$= \pi - x \quad \text{if } \pi/2 < x \leq 3\pi/2$$

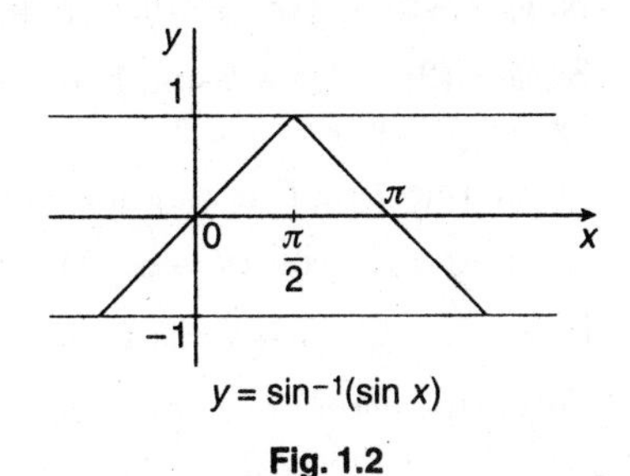

Fig. 1.2

$$\therefore \quad y'(0) = \left.\frac{d}{dx}x\right|_{x=0} = 1.$$

But since $y(x) = \pi - x$ for $\pi/2 < x \leq 3\pi/2$,

$$y'(\pi) = \left.\frac{d}{dx}(\pi - x)\right|_{x=\pi} = -1.$$

Note that $y'(\pi/2)$ does not exist. For,

$$y'(\pi/2 - 0) = 1 \text{ whereas } y'(\pi/2 + 0) = -1.$$

2. *Does the equation $xy = (x+y)^2$ define y as a real function of x? If so, find the derivative of that function at a point within its domain.*

Here $xy = (x+y)^2 \Rightarrow x^2 + xy + y^2 = 0$.

$\therefore \quad y = (-x \pm \sqrt{-3x^2})/2 = (1 \pm i\sqrt{3})x/2.$

y is real only when $x = 0$. Since 0 is not the limit point of any real domain of y, y cannot be derivable at this point and at nowhere else.

Note. As a complex function of the real variable x, y is derivable at any real point including $x = 0$.

Before finding the derivative of a function defined implicitly it is necessary to check whether that equation defines one of the variables as a real function of the other in a certain open interval on the real axis.

3. *Find dy/dx from $(\sin x)^y = x - y$ at $x = \pi/2$, assuming that the equation defines y as a real function of x in a certain neighbourhood of $\pi/2$.*

Since $\sin x > 0$ in a certain neighbourhood of $\pi/2$ taking $u = \sin x$ and $v = y$, and using the formula:

$$(u^v)' = u^v[v' \ln u + (v/u)u'],$$

on differentiating both sides of $(\sin x)^y = x - y$ with respect to x we have

$$(\sin x)^y[y' \ln \sin x + (y/\sin x)\cos x] = 1 - y'$$

or, $\quad y' = -\{y(x-y)\cot x - 1\}/\{(x-y)\ln \sin x + 1\}. \qquad \text{(i)}$

Alternatively, taking $F(x, y) = 0$, where

$$F(x, y) = (\sin x)^y - x + y,$$

$F_x = y(\sin x)^{y-1}\cos x - 1 = y(x-y)\cot x - 1$

and $F_y = (\sin x)^y \ln\sin x + 1 = (x-y)\ln\sin x + 1.$

$\therefore$ $y' = -F_x/F_y =$ R.H.S. of (i)

and $y' = 1$ at $x = \pi/2$.

4. *If* $\sqrt{(1-x^2)} + \sqrt{(1-y^2)} = a(x-y)$, *show that* $y' = \sqrt{(1-y^2)/(1-x^2)}$.

Let $x = \sin\theta$ and $y = \sin\phi$, where $-\pi/2 < \theta,\ \phi < \pi/2$ and $\theta \geq$ or $\leq \phi$ according as $a >$ or < 0. Then $\cos\theta = \sqrt{(1-x^2)}$ and $\cos\phi = \sqrt{(1-y^2)}$. From the given equation

$$\cos\theta + \cos\phi = a(\sin\theta - \sin\phi) \quad \text{(i)}$$

or, $2\cos\{(\theta+\phi)/2\}\cos\{(\theta-\phi)/2\} = 2a\sin\{(\theta-\phi)/2\}\cos\{(\theta+\phi)/2\}$.

Now $\cos(\theta+\phi)/2 = 0 \Rightarrow \theta + \phi = (2n+1)\pi, n \in \mathbb{Z}$.

So $\phi = (2n+1)\pi - \theta$ and hence $\sin\phi = \sin\theta$ or $y = x$, which does not satisfy the given equation except when $x = y = 1$.

So rejecting this relation between θ and ϕ we get

$$\cot\{(\theta-\phi)/2\} = a \text{ or, } \theta - \phi = 2\cot^{-1} a. \quad \text{(ii)}$$

Now $$dy/dx = (dy/d\phi)(d\phi/d\theta)/(dx/d\theta). \quad \text{(iii)}$$

From (ii), $d\phi/d\theta = 1$. So from (iii)

$$dy/dx = \cos\phi/\cos\theta = \sqrt{(1-y^2)}/\sqrt{(1-x^2)}.$$

Alternatively, $F(x,y) = \sqrt{(1-x^2)} + \sqrt{(1-y^2)} - a(x-y)$, $F(x,y) = 0 \Rightarrow y' = -F_x/F_y$.

Here $F_x = -x/\sqrt{(1-x^2)} - a$ and $F_y = -y/\sqrt{(1-y^2)} + a$.

$$\therefore \quad dy/dx = -F_x/F_y = -\frac{x + a\sqrt{(1-x^)}}{y - a\sqrt{(1-y^2)}} \cdot \frac{\sqrt{(1-y^2)}}{\sqrt{(1-x^2)}}.$$

From the given equation

$$\frac{(1-x^2)-(1-y^2)}{\sqrt{(1-x^2)}-\sqrt{(1-y^2)}} = a(x-y)$$

or, $$\frac{x^2-y^2}{x-y} = a\{\sqrt{(1-y^2)} - \sqrt{(1-x^2)}\}$$

or, $x + a\sqrt{(1-x^2)} = -\{y - a\sqrt{(1-y^2)}\}$.

$\therefore$ $-(x + a\sqrt{1-x^2})/(y - a\sqrt{1-y^2}) = 1$.

Hence $dy/dx = \sqrt{(1-y^2)}/\sqrt{(1-x^2)}$.

5. *Find* dy/dx, *where* $y = x^{\sin x} + (\sin x)^x$.

Applying the formula: $(u^v)' = u^v[v'\ln u + (v/u)u']$,

$y' = x^{\sin x}[\cos x \ln x + x^{-1}\sin x\cos x] + (\sin x)^x[1\ln\sin x + x\operatorname{cosec} x\cos x]$.

6. *Differentiate* $\sin^{-1}[x\sqrt{(1-x)} - \sqrt{x}\sqrt{(1-x^2)}]$ *with respect to* x.

If $y = \sqrt{x}$, then both x and $y \geq 0$ and hence $xy \geq 0$.

$$\therefore \quad \sin^{-1}[x\sqrt{(1-x)} - \sqrt{x}\sqrt{(1-x^2)}] = \sin^{-1}[x\sqrt{(1-y^2)} - y\sqrt{(1-x^2)}]$$
$$= \sin^{-1} x - \sin^{-1} y = \sin^{-1} x - \sin^{-1}\sqrt{x}.$$

So, the required derivative is $1/\sqrt{(1-x^2)} - (1/2)/\sqrt{(x-x^2)}$.

7. *Show that the function defined by $f(x) = x^2 \sin(1/x), x \neq 0$ and $f(0) = 0$ is derivable at every real point x, but the derivative is not continuous at $x = 0$. Does $f''(0)$ exist?*

[AMIE 1996]

$$\text{If } x \neq 0,\ f'(x) = 2x\sin(1/x) + x^2\cos(1/x)(-1/x^2)$$
$$= 2x\sin(1/x) - \cos(1/x),$$

which exists everywhere else. So, f is continuous at every point except $x = 0$. Now

$$f'(0) = \underset{x\to 0}{\text{Lt}}\{f(x) - f(0)\}/x = \underset{x\to 0}{\text{Lt}}\ x\sin(1/x) = 0,$$

since $\forall\, \epsilon > 0, |x\sin(1/x)| \leq |x| < \delta$ or $|x - 0| < \delta$, where $\delta = \epsilon$.

Since $f'(0)$ exists, $f(x)$ is continuous at $x = 0$. But $f'(x) = 2x\sin(1/x) - \cos(1/x)$ is not continuous at $x = 0$.

For, although $\underset{x\to 0}{\text{Lt}}\ x\sin(1/x)$ exists and equals 0, but $\underset{x\to 0}{\text{Lt}}\ \cos(1/x)$ does not exist. Since $\underset{x\to 0}{\text{Lt}}\ f'(x)$ does not exist, $f'(x)$ is not continuous at $x = 0$.

Since $f'(x)$ is discontinuous at $x = 0, f''(0)$ does not exist.

Exercises 1.2

1. Let $X = \{1, 2, 3, 4\}$ and $f : X \to \mathbb{R} : \forall x \in X : f(x) = 1/x$. Show that f is continuous on X.

[**Hint.** Every point of X is an isolated point.]

2. If $f(x) = \sin(\pi/x)\ \forall x (\neq 0) \in \mathbb{R}$ and $f(0) = 0$, show that f is not continuous at 0.

[**Hint.** If $x_n = 1/n$ and $x'_n = 2/(4n+1)\ \forall n \in \mathbb{N}$, then $y_n = \sin(\pi/x_n)$ and $y'_n = \sin(\pi/x'_n)$ and they approach different limits.]

3. If $f(x) = x\sin(\pi/x)\ \forall x(\neq 0) \in \mathbb{R}$ and $f(0) = 0$, show that f is continuous at $x = 0$.

4. Find the points of discontinuity of—

(i) the signum function: $\text{sgn}\, x = |x|/x$ when $x \neq 0$ and $\text{sgn} 0 = 0$; and
(ii) the greatest integer function : $[x] = n$ if $n \leq x < n + 1$, where $n \in \mathbb{Z}$.

5. If $f(x) = x^{-1}\sin x$ for $x \neq 0$ and $f(0) = 1$, show that f is continuous at $x = 0$.

6. Examine the continuity and differentiability of the function given by

$$f(x) = 1 \quad \text{when } x \leq 0$$
$$= 1 + \sin x \quad \text{when } x > 0,$$

at $x = 0$. **[W.B. Sem. Exam. 2003]**

7. Discuss the continuity of the function f defined by $f(x) = (e^{1/x} - 1)/(e^{1/x} + 1),\ x \neq 0$ and $f(0) = 0$.

8. Show that the function f defined by

$$f(x) = x \quad \text{if } x \text{ is irrational,}$$
$$= -x \quad \text{if } x \text{ is rational,}$$

is continuous only at $x = 0$.

9. Show that the function f is not continuous at $x = 0$, where $f(x) = 1/(1 + e^{1/x})$ when $x \neq 0$ and $f(0) = 0$. **[BIT 1999]**

10. Show that f is continuous everywhere, where

$$f(x) = \lim_{n\to\infty} \sum_{r=1}^{n} [2rx]/n^2,$$

and $[x]$ is the greatest integer function of $x \in \mathbb{R}$.

11. Differentiate the following with respect to x:

(i) $\tan^{-1}(\operatorname{cosec} x + \cot x)$ if $0 < x < 2\pi$;

(ii) $\cos^{-1}(\cos x)$ if $\pi < x < 2\pi$;

(iii) $\sin^{-1}\{(1 - \tan^2 x)/(1 + \tan^2 x)\}$ at $x = 4$;

[**Hint.** $\sin^{-1} \cos u = \sin^{-1} \sin(\pi/2 - u) = 5\pi/2 - u$ if $2\pi < u \leq 3\pi$.]

(iv) $\cos^{-1} \sqrt{\{(1 - \cos x)/2\}}$ at $x = -\pi/2$.

12. If $f(x) = x \sin(\pi/x)$ for $x \neq 0$ and $f(0) = 0$, show that $f'(0)$ does not exist although f is continuous at $x = 0$.

13. Examine the continuity and differentiability of the function f given by

$$\begin{aligned} f(x) &= 1 && \text{when } x \leq 0 \\ &= 1 + \sin x && \text{when } x > 0. \end{aligned}$$

14. Test the continuity and differentiability of the following function at $x = 0$ and $x = \pi/2$:

$$\begin{aligned} f(x) &= 1 && \text{for } x < 0 \\ &= 1 + \sin x && \text{for } 0 \leq x < \pi/2 \\ &= 2 + (x - \pi/2)^2 && \text{for } \pi/2 \leq x. \end{aligned}$$

15. Differentiate $\tan^{-1}[\{\sqrt{(1 + x^2)} - 1\}/x]$ with respect to $\tan^{-1} x$.

16. Find the derivative of each of the following with respect to x:

(i) $\ln\{x + \sqrt{(x^2 + a^2)}\}$;

(ii) $\ln[\{\sqrt{(x^2 + a^2)} + x\}/\{\sqrt{(x^2 + a^2)} - x\}]$;

(iii) $e^{\sin \sqrt{x}}$;

(iv) $e^{\sqrt{\cot x}}$;

(v) $\cos^{-1}\{(a \cos x + b)/(a + b \cos x)\}$.

17. Regarding y as a real function of x defined implicity by the following equations, find dy/dx:

(i) $\sqrt{x} + \sqrt{y} = \sqrt{a}$;

(ii) $ax^2 + 2hxy + by^2 + 2gx + 2fy + c = 0$;

(iii) $x^m y^n = (x - y)^{m+n}$;

(iv) $x \cos y + y \cos x = \tan(x + y)$.

18. If $\cos y = x \cos(a + y)$, prove that $dy/dx = \{\cos^2(a + y)\}/\sin x$.

19. If $\sqrt{(1 + x^2)} + \sqrt{(1 + y^2)} = a(x - y)$, show that $dy/dx = \sqrt{\{(1 + y^2)/(1 + x^2)\}}$.

20. Find dy/dx, when—

(i) $x = a(\theta + \sin\theta)$, $y = a(1 - \cos\theta)$;

(ii) $x = a(\theta - \sin\theta)$, $y = a(1 - \cos\theta)$;

(iii) $x = 2t/(1 + t^2)$, $y = (1 - t^2)/(1 + t^2)$;

(iv) $x = a\cos^3\theta$, $y = a\sin^3\theta$ at $\theta = \pi/4$;

(v) $x = 3at/(1 + t^2)$, $y = 3at^2/(1 + t^2)$ at $t = 1/2$.

21. Find the derivatives, with respect to x, of the following:

(i) $\sin^3 x \cos^5 x$;

(ii) $\cos x \cos 2x \cos 3x \cos 4x$;

(iii) $\{(x + 1)^2\sqrt{(x - 1)}\}/\{(x + 4)^3 e^x\}$.

[**Hint.** Use logarithmic differentiation.]

22. Find the derivatives, with respect to x, of each of the following functions y, defined implicity or explicitly by:

(i) $y = x^x + x^{1/x}$;

(ii) $y = x^{\tan x} + (\tan x)^x$;

(iii) $x^y = y^x$;

(iv) $(\cos x)^y = (\sin y)^x$.

[**Hint.** Use the formula: $(u^v)' = (e^{v \ln u})' = u^v\{v' \ln u + (v/u)u'\}$.]

Answers

4. (i) 0; **(ii)** $n \in \mathbb{Z}$. **11. (i)** $-1/2$; **(ii)** -1;
(iii) -2; **(iv)** $1/2$. **15.** $1/2$. **16. (i)** $1/\sqrt{(x^2 + a^2)}$;
(ii) $2/\sqrt{(x^2 + a^2)}$; **(iii)** $e^{\sin\sqrt{x}}(\cos\sqrt{x})/(2\sqrt{x})$;
(iv) $-e^{\sqrt{\cot x}}(\operatorname{cosec}^2 x)/(2\sqrt{\cot x})$; **(v)** $\sqrt{(a^2 - b^2)}/(a + b\cos x)$.

17. (i) $-\sqrt{y/x}$. **(ii)** $-(ax + hy + g)/(hx + by + f)$; **(iii)** y/x;
(iv) $\{\sec^2(x + y) - \cos y + y\sin x\}/\{\cos x - x\sin y - \sec^2(x + y)\}$.

20. (i) $\tan\theta/2$; **(ii)** $\cot\theta/2$; **(iii)** $-2t/(1 - t^2)$;0 **(iv)** -1;
(v) $4/3$. **21. (i)** $\sin^3 x \cos^5 x(3\cot x - 5\tan x)$;
(ii) $-\cos x \cos 2x \cos 3x \cos 4x(\tan x + 2\tan 2x + 3\tan 3x + 4\tan 4x)$;
(iii) $[(x + 1)^2\sqrt{x - 1}/\{(x + 4)^3 e^x\}] \div [2/(x + 1) + 1/\{2(x - 1)\} - 3/(x + 4) - 1]$.

22. (i) $x^x(1 + \ln x) + x^{1/x}(1 - \ln x)/x^2$;
(ii) $x^{\tan x}\{(\tan x)/x + \sec^2 x \ln x\} + (\tan x)^x(2x \operatorname{cosec} 2x + \ln\tan x)$;
(iii) $y(x\ln y - y)/[x(y\ln x - x)]$; **(iv)** $(\ln\sin y + y\tan x)/(\ln\cos x - x\cot y)$.

1.15 Successive Differentiation

Let $f : X \to \mathbb{R}$ be a differentiable function and $X_1 = \{x \in X : f'(x) \text{ exists}\}$. Then $X_1 (\subseteq X)$ is the domain of differentiation of f.

If $y = f(x)\ \forall x \in X$, y' or $dy/dx = f'(x)\ \forall x \in X_1$.

So, f' can be regarded as a function whose domain is X_1:

$$f' : X_1 \to \mathbb{R}.$$

Let f' itself be differentiable at $x \in X_1$. Then if the derivative of f' at x exists for some values of x in X_1, this derivative of f' at x is denoted by $(f')'(x)$ or $f''(x)$, which is called the **second derivative** of f at $x \in X_2 \subseteq X_1$.

Since $\quad f'(x) = dy/dx = y' \quad$ or $\quad y_1$,

$$f''(x) = \frac{d}{dx}(f'(x)) = \frac{d}{dx}\left(\frac{dy}{dx}\right),$$

which is denoted by $d^2y/dx^2 = y''$ or y_2.

Similarly, $f'''(x) = \dfrac{d}{dx}\left(\dfrac{d^2y}{dx^2}\right) = d^3y/dx^3 = y''' = y_3$, called the **third derivative** of f at x.

This is also the second derivative of f' at x and the first derivative of f'' at x.

In place of $f', f'', f''', \ldots$ the symbols $f^{(1)}, f^{(2)}, f^{(3)}, \ldots$ are convenient to use.

The nth order derivative of f at x is therefore denoted by $f^{(n)}(x)$ or d^ny/dx^n or $y^{(n)}$ or y_n.

Here d^ny is the ***n*th order differential of *y***, whereas dx^n is actually $(dx)^n$, the nth power of the differential dx.

Using the operator D, the nth order derivative is $D^nf(x)$ or D^ny. Thus $D^ny = d^ny/dx^n$, so that $D^n \equiv (d/dx)^n = d^n/dx^n$.

For uniformity in notation we shall take

$$D^0y = y_0 = y = f(x),$$

the underived function.

Examples 1.6. **1.** *If* $y = c\ \forall x \in \mathbb{R}$, $D^0y = c$, $D^ny = 0\ \forall n \in \mathbb{N}$.

2. *If* $y = x^m$, $y_1 = mx^{m-1}$, $y_2 = m(m-1)x^{m-2}$.

$$\therefore \quad y_n = m(m-1)\ldots(m-n+1)x^{m-n} = \binom{m}{n} n! x^{m-n}$$

$$= m! \text{ if } n = m \in \mathbb{N}$$

$$= 0 \quad \text{if } n > m \text{ and } m \in \mathbb{N}.$$

In particular when $m = -1$,

$$y = x^{-1} = 1/x \quad \text{and} \quad y_n = (-1)(-2)\ldots(-n)x^{-1-n} = (-1)^n n! x^{-(n+1)}$$

and if $y = \ln x$, $x > 0$

$$y_1 = 1/x = x^{-1}$$

$$\therefore \quad y_{n+1} = (-1)^n n! x^{-(n+1)}\ \forall n \in \mathbb{W}$$

and $\quad y_n = (-1)^{n-1}(n-1)! x^{-n}\ \forall n \in \mathbb{N}$.

1.16 *n*th Order Derivatives of Some Functions

(1) If $y = (ax+b)^m$, then $\forall\, n \in \mathbb{N}$

$$y_n = m(m-1)\cdots(m-n+1)a^n(ax+b)^{m-n} = \binom{m}{n} n!a^n(ax+b)^{m-n}$$
$$= m!a^m \quad \text{if } n = m \in \mathbb{N}$$
$$= 0 \quad \text{if } n > m \in \mathbb{N}.$$

For, $y_1 = ma(ax+b)^{m-1}$, by direct differentiation.

Let $y_r = m(m-1)\ldots(m-r+1)a^r(ax+b)^{m-r}$.

Since $y_{r+1} = m(m-1)\cdots(m-r+1)a^r(m-r)a(ax+b)^{m-r-1}$

$$= m(m-1)\cdots(m-\overline{r+1}+1)a^{r+1}(ax+b)^{m-\overline{r+1}},$$

by induction the nth derivative is as given above.

When $m \in \mathbb{N}$ and $n > m$, $m(m-1)\cdots(m-n+1)$ or $\binom{m}{n}$ is 0.

So $y_n = 0$.

Corollaries.

1. $D^n x^m = \binom{m}{n} n!x^{m-n}$
2. $D^n \ln x = (-1)^{n-1}(n-1)!x^{-n}$
3. $D^n \ln(ax+b) = (-1)^{n-1}(n-1)!a^n/(ax+b)^n$.

(2) If $y = e^{ax}$, $y_n = a^n e^{ax}$.

For $y_1 = ae^{ax}$, $y_2 = a^2e^{ax}, \ldots, y_n = a^n e^{ax}$.

(3) $D^n \sin(ax+b) = a^n \sin(ax+b+n\pi/2)$, $\forall n \in \mathbb{N}$,

$$= (-1)^{(n-1)/2}a^n \cos(ax+b) \quad \text{if } n \text{ is odd},$$
$$= (-1)^{n/2}a^n \sin(ax+b) \quad \text{if } n \text{ is even}.$$

For, $D\sin(ax+b) = a\cos(ax+b) = (-1)^{(1-1)/2}a^1\cos(ax+b)$

$$= a\sin(ax+b+1\cdot\pi/2).$$
$$D^2\sin(ax+b) = -a^2\sin(ax+b) = (-1)^{2/2}a^2\sin(ax+b)$$
$$= a^2\sin(ax+b+2\cdot\pi/2).$$

So by induction, the above formula is true for all $n \in \mathbb{N}$.

Cor. $D^n \sin x = \sin(x + n\pi/2)\ \forall n \in \mathbb{N}$,

$$= (-1)^{(n-1)/2}\cos x \quad \text{if } n \text{ is odd},$$
$$= (-1)^{n/2}\sin x \quad \text{if } n \text{ is even}.$$

(4) $D^n \cos(ax+b) = a^n\cos(ax+b+n\pi/2)$, $\forall n \in \mathbb{N}$,

$$= (-1)^{(n+1)/2}a^n\sin(ax+b), \quad \text{if } n \text{ is odd},$$
$$= (-1)^{n/2}a^n\cos(ax+b), \quad \text{if } n \text{ is even}.$$

(5) $\forall n \in \mathbb{N}$ $D^n[e^{ax}\cos(bx+c)] = (a^2+b^2)^{n/2}e^{ax}\cos(bx+c+n\theta)$ where $\tan\theta = b/a$.

For, $D[e^{ax}\cos(bx+c)] = ae^{ax}\cos(bx+c) - be^{ax}\sin(bx+c)$

$$= e^{ax}[a\cos(bx+c) - b\sin(bx+c)].$$

If $a = r\cos\theta$ and $b = r\sin\theta$, then $r = (a^2+b^2)^{1/2}$ and $\tan\theta = b/a$.

$$\therefore \quad D[e^{ax}\cos(bx+c)] = re^{ax}\cos(bx+c+\theta)$$

$$\begin{aligned} D^2[e^{ax}\cos(bx+c)] &= rD[e^{ax}\cos(bx+c+\theta)] \\ &= r[ae^{ax}\cos(bx+c+\theta) - e^{ax}b\sin(bx+c+\theta)] \\ &= re^{ax}[a\cos(bx+c+\theta) - b\sin(bx+c+\theta)] \\ &= r^2e^{ax}[\cos\theta\cos(bx+c+\theta) - \sin\theta\sin(bx+c+\theta)] \\ &= r^2e^{ax}\cos(bx+c+2\theta). \end{aligned}$$

By induction,

$$\begin{aligned} D^n &= r^n e^{ax}\cos(bx+c+n\theta) \\ &= (a^2+b^2)^{n/2}e^{ax}\cos(bx+c+n\theta), \text{ where } \tan\theta = b/a. \end{aligned}$$

(6) $\forall n \in \mathbb{N}, D^n[e^{ax}\sin(bx+c)] = (a^2+b^2)^{n/2}e^{ax}\sin(bx+c+n\theta)$ where $\tan\theta = b/a$.

Combining the formulae (5) and (6) we find that $\forall n \in \mathbb{N}$

$$D^n\left[e^{ax}\begin{matrix}\sin\\ \cos\end{matrix}(bx+c)\right] = (a^2+b^2)^{n/2}e^{ax}\begin{matrix}\sin\\ \cos\end{matrix}(bx+c+n\theta).$$

(7) $D^n\left(\dfrac{x}{x^2+a^2}\right) = (-1)^n n!(x^2+a^2)^{-(n+1)/2}\cos(n+1)\theta,\ \forall\, n \in \mathbb{N}$, where $\tan\theta = a/x$.

$$\because \quad D\left(\frac{x}{x^2+a^2}\right) = D\left[\frac{1}{2}\left(\frac{1}{x-ai}+\frac{1}{x+ai}\right)\right] = \frac{1}{2}[D(x-ai)^{-1} + D(x+ai)^{-1}],$$

$$\begin{aligned} \therefore \quad D^n\left(\frac{x}{x^2+a^2}\right) &= \frac{1}{2}[D^n(x-ai)^{-1} + D^n(x+ai)^{-1}] \\ &= \frac{1}{2}(-1)^n n!\left[\frac{1}{(x-ai)^{n+1}} + \frac{1}{(x+ai)^{n+1}}\right], \text{ by (1).} \end{aligned}$$

Taking $x = r\cos\theta$, $a = r\sin\theta$, $r = (x^2+a^2)^{1/2}$ and $\tan\theta = \dfrac{a}{x}$,

$$(x-ai)^{-(n+1)} = \{r(\cos\theta - i\sin\theta)\}^{-(n+1)} = r^{-(n+1)}\{\cos(n+1)\theta + i\sin(n+1)\theta\}$$

and $(x+ai)^{-(n+1)} = r^{-(n+1)}\{\cos(n+1)\theta - i\sin(n+1)\theta\}$.

$$\therefore \quad \frac{1}{(x-ai)^{n+1}} + \frac{1}{(x+ai)^{n+1}} = 2r^{-(n+1)}\cos(n+1)\theta = 2(x^2+a^2)^{-n/2}\cos(n+1)\theta.$$

Hence the required formula.

(8) $D^n\left(\dfrac{1}{x^2+a^2}\right) = \dfrac{(-1)^n n!}{a^{n+2}}\sin(n+1)\theta\sin^{n+1}\theta,\ \forall n \in \mathbb{N}$, where $\tan\theta = \dfrac{a}{x}$.

$$\begin{aligned} D^n\left(\frac{1}{x^2+a^2}\right) &= D^n\left[\frac{1}{2ai}\left(\frac{1}{x-ai}-\frac{1}{x+ai}\right)\right] = \frac{1}{2ai}[D^n(x-ai)^{-1} - D^n(x+ai)^{-1}] \\ &= \frac{(-1)^n n!}{2ai}[(x-ai)^{-(n+1)} - (x+ai)^{-(n+1)}], \text{ by (1).} \end{aligned}$$

As in (7), if $x = r\cos\theta$ and $a = r\sin\theta$,

$$D^n\left(\frac{1}{x^2+a^2}\right) = \frac{(-1)^n n!}{2ai}[r^{-(n+1)}\{(\cos\theta - i\sin\theta)^{-(n+1)} - (\cos\theta + i\sin\theta)^{-(n+1)}\}]$$

$$= \frac{(-1)^n n!}{2ai}\ \frac{1}{r^{n+1}}[\cos(n+1)\theta + i\sin(n+1)\theta - \cos(n+1)\theta + i\sin(n+1)\theta]$$

$$= \frac{(-1)^n n!}{a}\left(\frac{\sin\theta}{a}\right)^{n+1}\sin(n+1)\theta = \frac{(-1)^n n!}{a^{n+2}}\sin(n+1)\theta\sin^{n+1}\theta.$$

Cor. $D^n\left(\tan^{-1}\frac{x}{a}\right) = \frac{(-1)^{n-1}(n-1)!}{a^n}\sin n\theta\sin^n\theta.$

Formulae for the nth derivatives, in general, can be obtained only for a limited number of functions. The following example will make it clear:

If $y = \tan x$, $y_1 = \sec^2 x$, $y_2 = 2\sec^2 x\tan x = 2[\tan x + \tan^3 x]$,

$y_3 = 2[\sec^2 x + 3\tan^2 x\sec^2 x] = 2[3\tan^4 x + 4\tan^2 x + 1]$.

Higher order derivatives will be much more involved and no simple formula can be found out for the nth order derivative of $\tan x$.

Examples 1.7. 1. *If* $y = x^2/[(x-1)(x-2)(x-3)]$, *find* y_n. **[Sem. Exam. B.Tech. 2001]**

Resolving into partial fractions, let

$$y = \frac{A}{x-1} + \frac{B}{x-2} + \frac{C}{x-3}$$

where, by the method of suppression,

$$A = \left.\frac{x^2}{(x-2)(x-3)}\right|_{x=1} = \frac{1}{2},$$

obtained by suppressing the factor $x-1$ from the given fraction and putting $x = 1$, for which that particular suppressed factor vanishes.

Similarly, $B = \left.\frac{x^2}{(x-1)(x-3)}\right|_{x=2} = -4$

and $C = \left.\frac{x^2}{(x-1)(x-2)}\right|_{x=3} = \frac{9}{2}.$

So, $$y_n = (-1)^n n!\left[\frac{1}{2}(x-1)^{-n-1} - 4(x-2)^{-n-1} + \frac{9}{2}(x-3)^{-n-1}\right].$$

2. *Find* y_n, *where* $y = \sin^2 x\cos^3 x$.

Here, $y = \frac{1}{2}(1-\cos 2x)\frac{1}{4}(3\cos x + \cos 3x)$

$$= \frac{1}{8}[3\cos x + \cos 3x - 3\cos x\cos 2x - \cos 2x\cos 3x]$$

$$= \frac{1}{8}[3\cos x + \cos 3x - \frac{3}{2}(\cos x + \cos 3x) - \frac{1}{2}(\cos x + \cos 5x)]$$

$$= \frac{1}{16}[2\cos x - \cos 3x - \cos 5x].$$

$$\therefore \quad y_n = \frac{1}{16}[2\cos(x + n\pi/2) - 3^n \cos(3x + n\pi/2) - 5^n \cos(5x + n\pi/2)]$$
$$= \frac{1}{16}[2\cos(x + n\pi/2) - 3^n \cos(3x + n\pi/2) - 5^n \cos(5x + n\pi/2)].$$

3. *Find y_n if $y = e^{2x}\cos^2 x$.*

Here $y = \frac{1}{2}e^{2x}(1 + \cos 2x) = \frac{1}{2}[e^{2x} + e^{2x}\cos 2x]$.

If $u = e^{2x}$, $u_n = 2^n e^{2x}$ and if $v = e^{2x}\cos 2x$, $v_n = e^{2x}(2^2 + 2^2)^{n/2}\cos(2x + n\theta) = 8^{n/2}e^{2x}\cos(2x + n\pi/4)$, where $\tan\theta = 2/2 = 1$ or, $\theta = \pi/4$.

$$\therefore \; y_n = \frac{1}{2}[u_n + v_n] = \frac{1}{2}[2^n e^{2x} + 2^{3n/2}e^{2x}\cos(2x + n\pi/4)]$$
$$= e^{2x} \cdot 2^{n-1}\{1 + 2^{n/2}\cos(2x + n\pi/4)\}.$$

4. *Find the nth derivative of y with respect to x, where $y = \sin^{-1}\dfrac{2x}{1+x^2}$ if $1 \le x < \infty$.*

If $x = \tan\theta$, where $-\pi/2 < \theta < \pi/2$, $\theta = \tan^{-1} x$,

$$2x/(1 + x^2) = 2\tan\theta/(1 + \tan^2\theta) = \sin 2\theta.$$

$$\therefore \quad y = \sin^{-1}(\sin 2\theta) = 2\theta = 2\tan^{-1} x,$$

provided $-\pi/2 < 2\theta < \pi/2$ or $-\pi/4 < \theta < \pi/4$ or $-1 < x < 1$.

$y = \pi - 2\theta$ if $\pi/2 < \pi - 2\theta < \pi/2$ or $\pi/4 \le \theta \le 3\pi/4$ or $\pi/4 \le \theta < \pi/2$ or $1 \le x < \infty$.

And $y = -\pi - 2\theta$ if $-\pi/2 < -\pi - 2\theta < \pi/2$ or $-3\pi/4 \le \theta \le -\pi/4$ or $-\pi/2 < \theta \le -\pi/4$ or $-\infty < x < -1$.

Thus
$$y = 2\tan^{-1} x \quad \text{if } -1 < x < 1,$$
$$= \pi - 2\tan^{-1} x \quad \text{if } 1 \le x < \infty,$$
$$= -\pi - 2\tan^{-1} x \quad \text{if } -\infty < x \le -1.$$

But $D^n \tan^{-1} x = (-1)^{n-1}(n-1)!\sin n\phi \sin^n \phi$, where $\cot\phi = x$ by the corollary, formula (8).

So, when $1 \le x < \infty$,

$$y_n = D^n[\pi - 2\tan^{-1} x] = (-1)^n 2(n-1)!\sin n\phi \sin^n \phi,$$

where $\phi = \cot^{-1} x$.

Note. It is a normal tendency to take $2\tan^{-1} x$ for $\sin^{-1}[2x/(1+x)^2]$ for all $x \in \mathbb{R}$. But it is not always true.

5. *If $x(t)$ and $y(t)$ are twice differentiable with respect to t and they represent y as a function $f(x)$ of x, then show that*

$$f''(x) = (\dot{x}\ddot{y} - \ddot{x}\dot{y})/\dot{x}^3,$$

where $\dot{u} = du/dt$ and $\ddot{u} = d^2u/dt^2$.

We know that $f'(x) = dy/dx = \dot{y}/\dot{x}$.

But $\dot{y}/\dot{x}$ is a function of t. Let

$$\phi(t) = \dot{y}/\dot{x}.$$

Then $\quad \dot{\phi} = (\ddot{y}\dot{x} - \ddot{x}\dot{y}/\dot{x}^2.$

But $f''(x) = \dfrac{d}{dx}f'(x) = \dfrac{d}{dx}\phi(t) = \dfrac{d}{dt}\phi(t)/(dx/dt)$

$$= \frac{\dot{x}\ddot{y} - \ddot{x}\dot{y}}{\dot{x}^2 \cdot \dot{x}} = (\dot{x}\ddot{y} - \ddot{x}\dot{y})/\dot{x}^3.$$

6. *If the equation $ax^2 + 2hxy + by^2 + 2gx + 2fy + c = 0$ defines y as a twice differentiable function $\phi(x)$ of x, show that*

$$\phi''(x) = D(hx + by + f)^{-3},$$

where $D = \begin{vmatrix} a & h & g \\ h & b & f \\ g & f & c \end{vmatrix}$.

Suppose $b \neq 0$. Then the given equation is quadratic in y. So, solving for y we get two functions:

$$y = \phi(x) = -\frac{1}{b}[hx + f - \sqrt{\{(hx + f)^2 - b(ax^2 + 2gx + c)\}}]$$

and $\quad y = \psi(x) = -\dfrac{1}{b}[hx + f + \sqrt{\{(hx + f)^2 - b(ax^2 + 2gx + c)\}}].$

If A, B, C, F, G and H are the cofactors of a, b, c, f, g and h respectively in the expansion of D, then we know that

$A = bc - f^2,\ B = ca - g^2,\ C = ab - h^2,\ F = gh - af,\ G = hf - bg$ and $H = fg - ch$.

Now the expression under the radical sign for $\phi(x)$ or $\psi(x)$ is $-Cx^2 + 2Gx - A$.

$\because \quad by = -(hx + f) \pm \sqrt{(-Cx^2 + 2Gx - A)},$

$\therefore \quad (hx + by + f)^2 = -Cx^2 + 2Gx - A. \qquad (1)$

Now differentiating both sides of (1) with respect to x and treating y as either $\phi(x)$ or $\psi(x)$ we get

$$(hx + by + f)(h + by') = -Cx + G. \qquad (2)$$

$\therefore \quad h + by' = \dfrac{-Cx + G}{hx + by + f}. \qquad (3)$

Differentiating both sides of (2) with respect to x, after multiplying both sides of (3) by $hx + by + f$,

$$(h + by')^2 + (hx + by + f)by'' = -C.$$

Using (3),

$$\begin{aligned}(hx + by + f)by'' &= -C - \frac{(-Cx + G)^2}{(hx + by + f)^2} \\ &= -C - \frac{(-Cx + G)^2}{-Cx^2 + 2Gx - A}, \text{ by (1)} \\ &= \frac{CA - G^2}{(hx + by + f)^2}. \qquad (4)\end{aligned}$$

$\because \quad (CA - G^2)D = \begin{vmatrix} A & H & G \\ 0 & 1 & 0 \\ G & F & C \end{vmatrix} \begin{vmatrix} a & h & g \\ h & b & f \\ g & f & c \end{vmatrix} = \begin{vmatrix} D & 0 & 0 \\ h & b & f \\ 0 & 0 & D \end{vmatrix} = bD^2,$

$\therefore \quad CA - G^2 = bD$ and

$$y'' = \frac{D}{(hx + by + f)^3}.$$

7. *If $x = 2\cos t - \cos 2t$ and $y = 2\sin t - \sin 2t$, find the value of d^2y/dx^2 when $t = \pi/2$.*

We know that $d^2y/dx^2 = (\dot{x}\ddot{y} - \ddot{x}\dot{y})/\dot{x}^3$.

Now $\quad x = 2\cos t - \cos 2t \Rightarrow \dot{x} = -2\sin t + 2\sin 2t = -4$ at $t = \pi/2$

and $\quad \ddot{x} = -2\cos t + 4\cos 2t = -4$ at $t = \pi/2$.

Similarly, $\quad y = 2\sin t - \sin 2t \Rightarrow \dot{y} = 2\cos t - 2\cos 2t = 2$ at $t = \pi/2$

and $\quad \ddot{y} = -2\sin t + 4\sin 2t = -2$ at $t = \pi/2$.

$\therefore$ at $t = \pi/2$,

$$d^2y/dx^2 = [(-2)(-2) - (-4)^2]/(-2)^3 = 3/2.$$

8. *If $y = \sin(\sin x)$, show that $y'' + y' \tan x + y\cos^2 x = 0$.*

Since $\quad y = \sin(\sin x), \qquad (1)$

$$y' = \cos(\sin x) \cdot \cos x \qquad (2)$$

and $\quad y'' = -\sin(\sin x)\cos^2 x - \cos(\sin x)\sin x$

$$= -y\cos^2 x - y' \tan x, \text{ using (1) and (2).}$$

$\therefore \quad y'' + y' \tan x + y\cos^2 x = 0.$

1.17 Leibnitz's Theorem

If u and v are both functions of x possessing derivatives of order up to n, then the nth derivative of their product is given by

$$(uv)_n = \sum_{r=0}^{n} {}^nC_r u_{n-r} v_r \quad \text{or} \quad D^n(uv) = \sum_{r=0}^{n} {}^nC_r (D^{n-r}u)(D^r v).$$

The theorem can be proved easily by induction. In the above it is assumed that $u_0 = D^0 u = u$ and $v_0 = D^0 v = v$.

Examples 1.8. 1. *Show that the nth order derivative of $e^{ax}\cos(bx + c)$ is*

$$e^{ax} \sum_{r=0}^{n} {}^nC_r a^{n-r} b^r \cos(bx + c_r), \text{ where } c_r = c + r\pi/2.$$

Taking $u = e^{ax}$ and $v = \cos(bx + c)$ and knowing that

$$u_r = a^r e^{ax} \quad \text{and} \quad v_r = b^r \cos(bx + c + r\pi/2) = b^r \cos(bx + c_r),$$

by Leibnitz's theorem:

$$(uv)_n = \sum_0^n {}^nC_r u_{n-r} v_r = \sum_0^n {}^nC_r a^{n-r} e^{ax} b^r \cos(bx + c_r)$$

$$= e^{ax} \sum_0^n {}^nC_r a^{n-r} b^r \cos(bx + c_r).$$

2. *If* $y = x^2 e^{ax} \cos bx$, *show that*

$$y_n = (a^2 + b^2)^{(n-2)/2} e^{ax}[(a^2 + b^2)\cos(bx + n\theta) + 2n(a^2 + b^2)^{1/2}\cos(bx + \overline{n-1}\theta) + n(n-1)\cos(bx + \overline{n-2}\theta)],$$

where $\tan\theta = b/a$.

Let $u = x^2$ and $v = e^{ax}\cos bx$.

Then $y = uv$ and $y_n = (uv)_n = \sum_0^n {}^nC_r u_{n-r} v_r = \sum_0^n {}^nC_{n-r} u_r v_{n-r} = \sum_0^n {}^nC_r u_r v_{n-r}$,

where we have used the property of summation: $\sum_{r=0}^n a_r = \sum_{r=0}^n a_{n-r}$.

$$\therefore \quad y_n = \sum_0^n {}^nC_r u_r v_{n-r} = {}^nC_0 u_0 v_n + {}^nC_1 u_1 v_{n-1} + {}^nC_r u_2 v_{n-2} + \cdots.$$

But since $u = x^2$, $u_1 = 2x$, $u_2 = 2$ and $u_r = 0$ for $r \geq 3$.

By (5), §1.16, $v_r = (a^2 + b^2)^{r/2} e^{ax}\cos(bx + c_r)$, where $c_r = c + r\theta = r\theta$, since $c = 0$, $\tan\theta = b/a$.

$$\begin{aligned}\therefore \quad y_n &= x^2(a^2 + b^2)^{n/2} e^{ax}\cos(bx + c_n) + n2x \cdot (a^2 + b^2)^{(n-1)/2} e^{ax}\cos(bx + c_{n-1}) \\ &\quad + n(n-1)(a^2 + b^2)^{(n-2)/2} e^{ax}\cos(bx + c_{n-2}) \\ &= (a^2 + b^2)^{(n-2)/2} e^{ax}[(a^2 + b^2)x^2\cos(bx + c_n) + 2n(a^2 + b^2)^{1/2} x\cos(bx + c_{n-1}) \\ &\quad + n(n-1)\cos(bx + c_{n-2})].\end{aligned}$$

3. *If* $y = a\cos(\ln x) + b\sin(\ln x)$, *show that*

$$x^2 y_{n+2} + (2n+1)x y_{n+1} + (n^2 + 1)y_n = 0.$$

Let $a = r\cos\theta$ and $b = r\sin\theta$. Then $r^2 = (a^2 + b^2)$ and $\tan\theta = b/a$. So

$$y = r\cos(\ln x - \theta) \text{ and } y_1 = -\frac{r}{x}\sin(\ln x - \theta).$$

$$\therefore \quad x y_1 = -r\sin(\ln x - \theta).$$

Differentiating both sides w.r.t. x

$$x y_2 + y_1 = -\frac{r}{x}\cos(\ln x - \theta) = -\frac{y}{x}$$

or, $\quad x^2 y_2 + x y_1 + y = 0.$

Taking the nth derivatives of both sides

$$D^n(x^2 y_2) + D^n(x y_1) + D^n y = 0.$$

Using Leibnitz's theorem

$$({}^nC_0 x^2 y_{n+2} + {}^nC_1 2x y_{n+1} + {}^nC_2 2y_n) + {}^nC_0 x y_{n+1} + {}^nC_1 y_n + y_n = 0$$

or, $\quad x^2 y_{n+2} + (2n+1)x y_{n+1} + (n^2 + 1)y_n = 0.$

4. *If* $y = e^{m\sin^{-1}x}$, *find* $y_n(0)$.

$$y = e^{m\sin^{-1}x} \Rightarrow y_1 = \frac{m}{\sqrt{1-x^2}} e^{m\sin^{-1}x} = \frac{my}{\sqrt{1-x^2}}.$$

$$\therefore \quad (1 - x^2)y_1^2 = m^2 y^2.$$

Differentiating w.r.t. x

$$2y_1y_2(1-x^2)+y_1^2(-2x)=m^2 2yy_1.$$

Cancelling $2y_1$

$$y_2(1-x^2)-y_1x-m^2y=0. \tag{1}$$

Taking the nth derivatives of both sides and using Leibnitz's theorem

$$\{y_{n+2}(1-x^2)+ny_{n+1}(-2x)+\frac{n(n-1)}{2}(-2)y_n\}-\{y_{n+1}x+ny_n.1\}-m^2y_n=0.$$

Putting $x=0$ and taking a_r for $y_r(0)$, we have

$$a_{n+2}-n(n-1)a_n-na_n-m^2a_n=0$$

or, $$a_{n+2}-(n^2+m^2)a_n=0\ \forall n\in\mathbb{N}. \tag{2}$$

Replacing n by $2r$

$$a_{2(r+1)}=(4r^2+m^2)a_{2r} \tag{3}$$

for $r=0,1,2,3,\ldots$.

When $r=0$, L.H.S. $=a_2=y_2(0)$. By (1), when $x=0$

$$y_2(0)-m^2y(0)=0\Rightarrow a_2=m^2a_0$$

and R.H.S. $=m^2a_0$ when $r=0$.

From (3)

$$\prod_{r=0}^{n-1}\frac{a_{2(r+1)}}{a_{2r}}=\prod_{r=0}^{n-1}(4r^2+m^2)$$

or, $$\frac{a_{2n}}{a_0}=\prod_{r=0}^{n-1}(4r^2+m^2)$$

or, $$y_{2n}(0)=\prod_{r=0}^{n-1}(4r^2+m^2),\ \because\ y(0)=1.$$

Replacing n by $n/2$

$$y_n(0)=\prod_{r=0}^{(n-2)/2}(4r^2+m^2), \tag{4}$$

when n is even.

Again replacing n by $2r-1$ in (2)

$$a_{2r+1}=\{(2r-1)^2+m^2\}a_{2r-1} \tag{5}$$

for $r=1,2,3,\ldots$.

So, $$\prod_{r=1}^{n}\frac{a_{2r+1}}{a_{2r-1}}=\prod_{r=1}^{n}\{(2r-1)^2+m^2\}$$

or, $$\frac{a_{2n+1}}{a_1}=\prod_{r=1}^{n}\{(2r-1)^2+m^2\}.$$

But $a_1 = y_1(0)$ and

$$y_1(x) = \frac{m}{\sqrt{1-x^2}}\, e^{m \sin^{-1} x} = m \text{ when } x = 0.$$

$$\therefore \qquad y_{2n+1}(0) = m \prod_{r=1}^{n} \{(2r-1)^2 + m^2\}.$$

Replacing n by $(n-1)/2$

$$y_n(0) = m \prod_{r=1}^{(n-1)/2} \{(2r-1)^2 + m^2\}, \tag{6}$$

when n is odd.

From (4) and (6)

$$y_n(0) = m \prod_{r=1}^{(n-1)/2} \{(2r-1)^2 + m^2\} \text{ or, } \prod_{r=0}^{(n-2)/2} (4r^2 + m^2)$$

according as n is odd or even.

5. *If* $y = \sin(m \sin^{-1} x)$, *show that*

$$(1-x^2)y_{n+2} = (2n+1)xy_{n+1} + (n^2 - m^2)y_n$$

and hence find $y_n(0)$.

$$y = \sin(m \sin^{-1} x) \Rightarrow a_0 = y(0) = 0. \tag{1}$$

Also $\quad \sin^{-1} y = m \sin^{-1} x.$

Differentiating w.r.t. x, we get

$$\frac{1}{\sqrt{1-y^2}} y_1 = \frac{m}{\sqrt{1-x^2}}, \tag{2}$$

whence $\quad a_1 = y_1(0) = m$, since $a_0 = y(0) = 0$.

Squaring both sides of (2) and simplifying

$$y_1^2(1-x^2) = m^2(1-y^2).$$

Differentiating

$$2y_1y_2(1-x^2) - 2y_1^2 x = -2m^2 y y_1.$$

Cancelling $2y_1$

$$y_2(1-x^2) = y_1 x + m^2 y = 0, \tag{3}$$

whence $\quad a_2 = y_2(0) = -m^2 y(0) = 0. \qquad (4)$

Differentiating both sides of (3) n times and using Leibnitz's theorem we get

$$\{y_{n+2}(1-x^2) + ny_{n+1}(-2x) + \frac{n(n-1)}{2} y_n(-2)\} - \{y_{n+1}\, x + ny_n\} + m^2 y_n = 0$$

or, $\quad y_{n+2}(1-x^2) - (2n+1)xy_{n+1} - (n^2 - m^2)y_n = 0. \qquad (5)$

Replacing x by 0 and $y_r(0)$ by a_r for $r = 0, 1, 2, 3, \ldots$

$$a_{n+2} - (n^2 - m^2)a_n = 0. \tag{6}$$

Replacing n by $2r$

$$a_{2r+2} - (4r^2 - m^2)a_{2r} = 0.$$

$$\therefore \quad \prod_{r=1}^{n-1} \frac{a_{2r+2}}{a_{2r}} = \prod_{r=1}^{n-1} (4r^2 - m^2)$$

$$\text{or,} \quad a_{2n} = a_2 \prod_{r=1}^{n-1} (4r^2 - m^2) = 0, \because \; a_2 = 0.$$

$$\therefore \quad y_{2n} = 0 \,\forall\, n \in \mathbb{N}$$

or, $y_n = 0$ when $n(\in \mathbb{N})$ is even.

Again replacing n by $2r - 1$ in (6)

$$a_{2r+1} - \{(2r-1)^2 - m^2\}a_{2r-1} = 0 \text{ for } r = 1, 2, 3, \ldots.$$

$$\therefore \quad \prod_{r=1}^{n} \frac{a_{2r+1}}{a_{2r-1}} = \prod_{r=1}^{n} \{(2r-1)^2 - m^2\}$$

$$\text{or,} \quad \frac{a_{2n+1}}{a_1} = \prod_{1}^{n} \{(2r-1)^2 - m^2\}.$$

By (2) $a_1 = m$.

$$\text{So,} \quad y_{2n+1}(0) = m \prod_{1}^{n} \{(2r-1)^2 - m^2\}.$$

Replacing n by $(n-1)/2$

$$y_n(0) = m \prod_{1}^{(n-1)/2} \{(2r-1)^2 - m^2\}, \text{ when } n \text{ is odd.}$$

6. *If* $y = \ln[x + \sqrt{1+x^2}]$, *find* $y_n(0)$.

$$y = \ln[x + \sqrt{1+x^2}] \Rightarrow a_0 = y(0) = 0, \tag{1}$$

$$y_1 = \frac{1}{x + \sqrt{1+x^2}} \cdot \left(1 + \frac{x}{\sqrt{1+x^2}}\right) = \frac{1}{\sqrt{1+x^2}} \Rightarrow a_1 = y_1(0) = 1. \tag{2}$$

Squaring and simplifying

$$y_1^2(1+x^2) = 1, \text{ whence } 2y_1y_2(1+x^2) + y_1^2 \cdot 2x = 0$$

$$\text{or,} \quad y_2(1+x^2) + xy_1 = 0, \tag{3}$$

$$\text{whence} \quad a_2 = y_2(0) = 0. \tag{4}$$

Differentiating both sides of (3) w.r.t. x n times and using Leibnitz's theorem

$$y_{n+2}(1+x^2) + ny_{n+1}2x + 2n(n-1)y_n 2 + y_{n+1}x + xny_n = 0.$$

Putting $x = 0$ and a_r for $y_r(0)$

$$a_{n+2} = -n^2 a_n \quad \text{or } a_{n+2}/a_n = -n^2. \tag{5}$$

Replacing n by $2r$ and multiplying

$$\prod_{r=0}^{n-1} \frac{a_{2r+2}}{a_{2r}} = \prod_{r=0}^{n-1} (-4r^2)$$

or, $$a_{2n} = a_0 \prod_{r=0}^{n-1} (-4r^2) = 0, \quad \because \quad a_0 = 0.$$

$\therefore$ $$y_n = 0 \ \forall \text{ even } n \in \mathbb{N}.$$

Again replacing n by $2r - 1$ in (5) and multiplying

$$\prod_{r=1}^{n} \frac{a_{2r+1}}{a_{2r-1}} = \prod_{r=1}^{n} \{-(2r-1)^2\}$$

or, $$a_{2n+1} = a_1(-1)^n \prod_{r=1}^{n} (2r-1)^2.$$

$\therefore$ $$y_{2n+1}(0) = (-1)^n \prod_{r=1}^{n} (2r-1)^2, \quad \because \quad a_1 = 1 \text{ by (2)}.$$

Replacing n by $(n-1)/2$

$$y_n(0) = (-1)^{(n-1)/2} \prod_{r=1}^{(n-1)/2} (2r-1)^2, \text{ when } n(\in \mathbb{N}) \text{ is odd.}$$

7. *Show that* $\dfrac{d^n}{dx^n}\left(\dfrac{\ln x}{x}\right) = (-1)^n \dfrac{n!}{x^{n+1}}\left[\ln x - \sum_{r=1}^{n} \dfrac{1}{r}\right].$

Let $u_0 = u = \ln x$ and $v_0 = v = x^{-1}$.

Then $u_r = (-1)^{r-1}(r-1)!x^{-r}$ and $v_r = (-1)^r r! x^{-(r+1)} \ \forall r \in \mathbb{N}$.

Now since $\dfrac{\ln x}{x} = uv$, by Leibnitz's theorem

$$\begin{aligned}
\frac{d^n}{dx^n}\left(\frac{\ln x}{x}\right) = (uv)_n &= \sum_{0}^{n} {}^nC_r u_{n-r} v_r \\
&= \sum_{0}^{n} {}^nC_r (-1)^{n-r-1}(n-r-1)!x^{-(n-r)}(-1)^r r! x^{-(r+1)} \\
&= \sum_{0}^{n-1} (-1)^{n-1} \frac{n!}{r!(n-r)!}(n-r-1)!r!x^{-(n+1)} + (-1)^n n! x^{-(n+1)} \ln x \\
&= \frac{(-1)^n n!}{x^{n+1}}\left[\ln x - \sum_{r=1}^{n} \frac{1}{r}\right].
\end{aligned}$$

8. *If* $u = (x^2 - 1)^n \ \forall n \in \mathbb{N}$, *show that*

$$(1 - x^2)u_{n+2} - 2xu_{n+1} + n(n+1)u_n = 0.$$

Hence show that y *satisfies the Legendre equation*

$$(1 - x^2)y_2 - 2xy_1 + n(n+1)y = 0,$$

where $y = D^n(x^2 - 1)^n$.

Given $u = (x^2 - 1)^n$, $\ln u = n \ln(x^2 - 1)$.

Differentiating once w.r.t. x

$$\frac{u_1}{u} = \frac{2nx}{x^2 - 1}, \text{ whence } u_1(1 - x^2) + 2nux = 0.$$

By Leibnitz's theorem

$$\{u_{n+1}(1 - x^2) + nu_n(-2x) + \frac{n(n-1)}{2}u_{n-1}(2)\} + 2n(u_n x + nu_{n-1}.1) = 0$$

or, $\quad u_{n+1}(1 - x^2) + n(n+1)u_{n-1} = 0.$

Differentiating once more

$$(1 - x^2)u_{n+2} - 2xu_{n+1} + n(n+1)u_n = 0. \qquad (1)$$

Since $\quad u_n = D^n(x^2 - 1)^n$, $y = u_n$, $y_1 = u_{n+1}$ and $y_2 = u_{n+2}$.

So, from (1)

$$(1 - x^2)y_2 - 2xy_1 + n(n+1)y = 0,$$

showing that y satisfies the Legendre equation.

9. *If $I_n = D^n(x^n \ln x)$ $\forall n \in \mathbb{N}$, prove that $I_n = nI_{n-1} + (n-1)!$*

and hence show that

$$I_n = n!\left(\ln x + \sum_1^n \frac{1}{r}\right).$$

$$I_n = D^n(x^n(\ln x) \Rightarrow I_{n-1} = D^{n-1}(x^{n-1} \ln x).$$

But $\quad I_n = D^{n-1}[D(x^n \ln x)] = D^{n-1}[nx^{n-1} \ln x + x^{n-1}]$

$$= nI_{n-1} + D^{n-1}x^{n-1} = nI_{n-1} + (n-1)!.$$

Multiplying both sides with $1/n!$

$$\frac{I_n}{n!} = \frac{I_{n-1}}{(n-1)!} + \frac{1}{n} \text{ for } n = 2, 3, 4, \ldots.$$

Replacing n by r and $I_r/r!$ by a_r

$$a_r = a_{r-1} + 1/r \quad \text{or,} \quad \sum_{r=2}^n (a_r - a_{r-1}) = \sum_2^n (1/r) \quad \text{or,} \quad a_n - a_1 = \sum_2^n \frac{1}{r}.$$

But $\quad a_1 = \dfrac{I_1}{1!} = I_1 = D(x \ln x) = \ln x + 1.$

$$\therefore \quad a_n = \ln x + 1 + \sum_2^n \frac{1}{r} = \ln x + \sum_1^n \frac{1}{r}.$$

$$\therefore \quad I_n = n!\left(\ln x + \sum_1^n \frac{1}{r}\right).$$

10. *If $x + y = 1$, prove that $D^n(x^n y^n) = n! \sum_0^n (-1)^r ({}^nC_r)^2 x^r y^{n-r}$.*

Let $u = x^n$ and $v = y^n = (1 - x)^n$.

Then $u_r = \dfrac{n!}{(n-r)!}x^{n-r}$ and $v_r = (-1)^r \dfrac{n!}{(n-r)!}(1-x)^{n-r} = \dfrac{(-1)^r n!}{(n-r)!}y^{n-r}.$

So, by Leibnitz's theorem

$$D^n(x^n y^n) = (uv)_n = \sum_0^n {}^nC_r u_{n-r} v_r = \sum_0^n {}^nC_r \frac{n!}{r!} x^r (-1)^r \frac{n!}{(n-r)!} y^{n-r}$$

$$= n! \sum_0^n (-1)^r \left({}^nC_r\right)^2 x^r y^{n-r}.$$

Exercises 1.3

1. If $x^3 + y^3 = 3axy$, prove that $(y^2 - ax)^3 y_2 + 2a^2 xy = 0$.

2. Change the independent variable x to a new variable z in the equation $y'' + y' \cot x + 4y \operatorname{cosec}^2 x = 0$, by means of the transformation $z = \ln \tan(x/2)$ and show that the transformed equation is $d^2y/dz^2 + 4y = 0$.

3. If $ax^2 + 2hxy + by^2 = 1$, prove that $y_2 = (h^2 - ab)/(hx + by)^3$.

4. If $x = a(\cos t + t \sin t)$ and $y = a(\sin t - t \cos t)$, find y''.

5. Find y'' when $x = a \cos^3 \theta$, $y = b \sin^3 \theta$.

6. Find y_n where y is

(a) $(x^2 - 6x + 8)^{-1}$; **(b)** $\tan^{-1}\{2x/(1 - x^2)\}$;

(c) $\tan^{-1}\{(1 + x)/(1 - x)\}$, **[DU 1951]**; **(d)** $e^{ax} \cos^2 x \sin x$.

7. If $y = (\sin^{-1} x)^2$, show that $(1 - x)^2 y_{n+2} - (2n + 1)xy_{n+1} - n^2 y_n = 0$. Hence find $y_n(0)$.

8. If $y^{1/m} + y^{-1/m} = 2x$, prove that

$$(x^2 - 1)y_{n+2} + (2n + 1)xy_{n+1} - (n^2 - m^2)y_n = 0.$$ **[DCE 2002]**

9. If $y = (x^2 - 1)^n$, prove that $(x^2 - 1)y_{n+2} + 2xy_{n+1} - n(n + 1)y_n = 0$. **[AMIE 1996]**

10. If $\cos^{-1}(ay) = \ln(x/n)^n$, prove that $x^2 y_{n+2} + (2n + 1)xy_{n+1} + 2n^2 y_n = 0$.

11. If $y = \{x + \sqrt{(x^2 + 1)}\}^m$, prove that

(a) $(x^2 + 1)y_2 + xy_1 - m^2 y = 0$;

(b) $y_{n+2} + (n^2 - m^2)y_n = 0$ at $x = 0$.

Hence find $y_n(0)$.

12. If $y = \tan^{-1} x$, show that $y_n = (n - 1)! \cos\{ny + (n - 1)\pi/2\} \cos^n y$.

13. Differentiate the terms of the differential equation $(1 - x^2)y_2 - xy_1 + a^2 y = 0$, n times with respect to x and simplify.

14. If $y = e^{m \cos^{-1} x}$, show that $(1 - x^2)y_{n+2} - (2n + 1)xy_{n+1} - (n^2 + m^2)y_m = 0$. Find $y_n(0)$.

15. If $y = D^n(x^2 - 1)^n$, show that $(x^2 - 1)y_{n+2} + 2xy_{n+1} - n(n + 1)y_n = 0$.

16. Find $D^n\{\sin(ax + b)\}$. Show that if $u = \sin ax + \cos ax$, $D^n u = a^n\{1 + (-1)^n \sin 2ax\}^{1/2}$, where $D = d/dx$. **[W.B. Sem. Exam. 2003]**

17. If $y = x^{n-1} \ln x$, show that $y_n = (n - 1)!/x$.

Answers

4. $1/(at\cos^3 t)$. **5. (a)** $b(\sec^4\theta\,\mathrm{cosec}\,\theta)/(3a^2)$.

6. (a) $[(-1)^n n!/2][(x-4)^{-n-1}-(x+2)^{-n-1}]$;

(b) $2(-1)^{n-1}(n-1)!\sin nx\sin^n\alpha$, where $\cot\alpha = x$;

(c) $(-1)^n(1/2)(n-1)!\sin n\theta\sin^n\theta$, where $\cot\theta = x$;

(d) $(1/4)(a^2+1)^{n/2}e^{ax}\sin(x+n\cot^{-1}a)$.

7. $y_n(0)=0$, where n is odd, $=2$ when $n=2$ and $=2^{n+1}\Pi_2^{n/2}(r-1)^2$, when $n(>2)$ is even.

Basic Theorems on Differentiable Functions

1.18 Rolle's Theorem

If a function $f: I \to \mathbb{R}$, where $I=[a,b]$, is

(i) continuous in I,

(ii) differentiable in $I_0=(a,b)$

and (iii) $f(a)=f(b)$,

then there exists at least one point ξ in I_0 such that

$$f'(\xi)=0.$$

The conditions imposed on the function are only sufficient for the vanishing of the derivative at a point in I_0. So, if any of the three conditions are violated even then a point may exist in I_0, whereat the derivative vanishes.

1.18.1 Geometrical Interpretation

Geometrically, the function $f: I \to \mathbb{R}$ where I is the closed interval $[a,b]$, is a subset of $I\times\mathbb{R}$, where $I\times\mathbb{R}$ is the collection of all points in the xy-plane between the lines $x=a$ and $x=b$.

The set of points $\in f$ is the graph of f with the end points A and B which correspond, respectively, to $x=a$ and $x=b$.

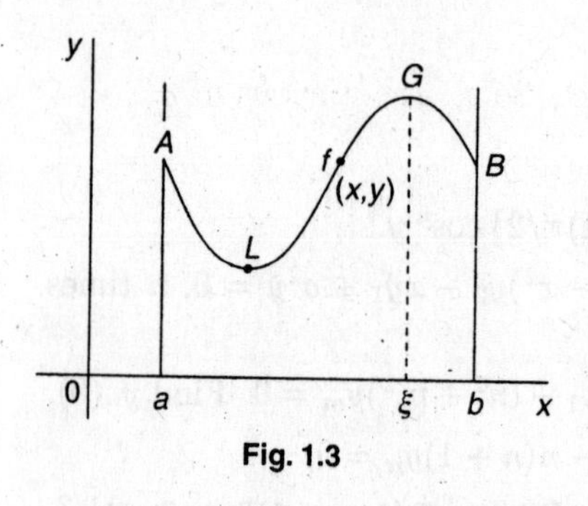

Fig. 1.3

Now the first condition in Rolle's theorem viz., the continuity of f in I ensures gaplessness of the graph between A and B. The second condition ensures the existence of a unique tangent at every point on the graph between A and B. Physically, this corresponds to smoothness of the graph or curve. The third condition ensures us that the end points A and B are on the same side of and at the same distance from the x-axis. Or, loosely, the end points A and B are the same height (depth). Lastly, vanishing of the derivative implies that the tangent is parallel to the x-axis. So the geometrical interpretation of the Rolle's theorem is that,

if (i) the graph of f is gapless (i.e., continuous),

(ii) the graph possesses a unique tangent at every point within the end points

and (iii) the end points of the graph lie on the same side of the x-axis and are at the same distance from the x-axis, then there exists at least one point where the tangent to the curve is parallel to the x-axis.

An alternative form of the statement of Rolle's theorem:

If (i) $f(x)$ is continuous in $I = [a, a+h]$,

(ii) $f'(x)$ exists in $I_0 = (a, a+h)$

and (iii) $f(a) = f(a+h)$,

then there always exists one number θ in $(0,1)$ such that $f'(a+\theta h) = 0$.

Since in the first definition $a < b$, if $b = a+h$, $b > a \Rightarrow a+h > a \Rightarrow h > 0$.

Now, $\xi \in I_0 \Rightarrow a < \xi < b \Rightarrow a < \xi < a+h \Rightarrow 0 < \xi - a < h \Rightarrow 0 < (\xi - a)/h < 1$.

So, taking

$$\theta = (\xi - a)/h$$

we find that $\theta \in (0,1)$. Now $\theta = (\xi - a)/h \Rightarrow \xi = a + \theta h$.

Hence the result.

Examples 1.9. 1. *Verify whether all the three conditions of Rolle's theorem are satisfied by the following functions in the corresponding intervals:*

(i) $1 - x^2$, $I = [-1, 1]$,

(ii) $|x-1|$, $I = [0, 2]$,

(iii) $e^x \sin x$, $I = [0, \pi]$,

(iv) $1 - \sqrt[3]{x^2}$, $I = [-1, 1]$,

(v) $\ln[(x^2 + ab)/(a+b)x]$, $I = [a, b]$,

(vi) $(x-a)^m (x-b)^n$, $I = [a, b]$, $m, n \in \mathbb{N}$.

(i) Let $f(x) = 1 - x^2 \ \forall x \in I = [-1, 1]$.

Since $f(x)$ is a polynomial function, it is continuous in $\mathbb{R}$. So it is continuous in I.

Since $f'(x) = -2x \ \forall x \in I_0 = (-1, 1)$, so f is derivable in I_0.

Since $f(-1) = 1 - (-1)^2 = 0$ and $f(1) = 1 - 1^2 = 0$,

$\therefore \quad f(-1) = f(1)$.

Thus all the conditions of Rolle's theorem are satisfied.

(ii) If $f(x) = |x-1|, \forall x \in [0, 2]$,

$$\begin{aligned} f(x) &= x - 1 && \text{if} \quad x - 1 > 0 \text{ or } x > 1 \text{ or } 1 < x \le 2, \\ &= 0 && \text{when } x = 1, \\ &= -(x-1) && \text{when } x - 1 < 0 \text{ or } x < 1 \text{ or } 0 \le x < 1. \end{aligned}$$

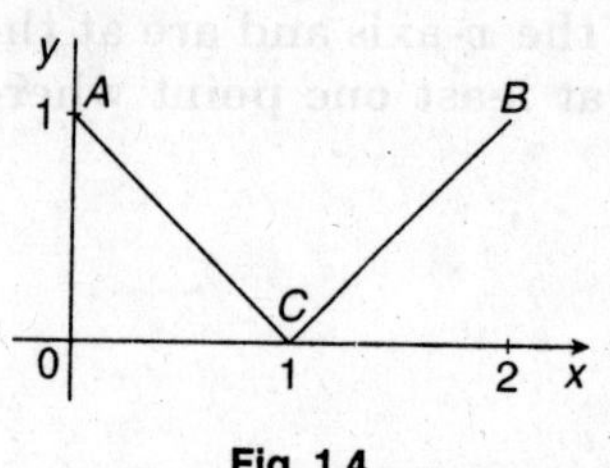

Fig. 1.4

Graph of f consists of two line segments AC and BC where A and B are respectively $(0, 1)$ and $(2, 1)$, and C is the vertex at $(1, 0)$ of the V-shaped curve. Since the graph is gapless, it is continuous in I. The graph is smooth everywhere except at C. So at C the graph has no unique tangent and therefore the second condition of Rolle's theorem is violated.

$$f(0) = |0 - 1| = 1 \text{ and } f(2) = |2 - 1| = |1| = 1.$$

So $f(0) = f(2)$ and the third condition, viz., that A and B are at the same height, is satisfied.

Since one of the conditions of Rolle's theorem has been violated, there is no guarantee that a point such as ξ exists in I_0, where $f'(\xi) = 0$. In fact, there is no such point in I_0 where the tangent is parallel to the x-axis.

(iii) Since e^x and $\sin x$ are both continuous in $\mathbb{R}$, their product $e^x \sin x$, i.e., $f(x)$ is continuous in $\mathbb{R}$ and hence in $I = [0, \pi]$.

Next $f'(x) = e^x \sin x + e^x \cos x = e^x(\sin x + \cos x)\ \forall\, x \in \mathbb{R}$. So $f'(x)$ exists in $I_0 = (0, \pi)$ also.

Lastly, $f(0) = f(\pi)$, both being 0.

Thus all the three conditions are satisfied. Hence there must exist at least one point in I_0, where the derivative vanishes.

$f'(x) = 0$ when $\sin x + \cos x = 0$ or $\tan x = -1$

or, $x = n\pi - \pi/4\ \forall\, n \in \mathbb{Z}$. When $n = 1$, $x = 3\pi/4$ which $\in I_0$.

(iv) Given $f(x) = 1 - \sqrt[3]{x^2}\ \forall\, x \in I = [-1, 1]$.

$\sqrt[3]{x^2}$ or $x^{2/3}$ is continuous in $\mathbb{R}$ and 1 is also continuous in $\mathbb{R}$. So their difference $1 - \sqrt[3]{x^2}$ or $f(x)$ is continuous in $\mathbb{R}$ and hence in I. Since

$$f'(x) = -\frac{2}{3}x^{-1/3} = -\frac{2}{3x^{1/3}},$$

so $f'(x)$ exists everywhere except $x = 0$ and $0 \in I_0$. Thus the second condition is violated.

The third condition is satisfied since $f(-1) = 0 = f(1)$.

Since one of the three conditions has not been satisfied, there is no guarantee that $f'(x) = 0$ for some $x \in I_0$. In fact,

$f'(x) = -\dfrac{2}{3x^{1/3}} \neq 0$ anywhere in I_0.

(v) When $f(x) = \ln[(x^2 + ab)/(a + b)x]\ \forall\, x \in I = [a, b]$, $(x^2 + ab)/(a + b)x > 0\ \forall\, x \in I$. Assuming $0 < a < b$ this inequality implies that

$$(x^2 + ab)/x > 0.$$

But $\quad x^2 + ab > 0 \forall\, x \in \mathbb{R} \Rightarrow x > 0.$

Since $0 < a < b$ and $a \leq x \leq b$, the condition that $x > 0$ is satisfied.

Since $\quad f(x) = \ln(x^2 + ab) - \ln(a + b) - \ln x,$

$$\therefore \quad f'(x) = \frac{2x}{x^2 + ab} - \frac{1}{x} = \frac{x^2 - ab}{x(x^2 + ab)},$$

which exists $\forall\, x > 0$ and hence $\forall\, x \in I_0 = (a, b)$.

Lastly, $f(a) = \ln \dfrac{a^2 + ab}{(a+b)a} = \ln 1 = 0$

and $\quad f(b) = \ln \dfrac{b^2 + ab}{(a+b)b} = \ln 1 = 0.$

$\therefore \quad f(a) = f(b).$

Thus all the conditions of Rolle's theorem are satisfied.

So there must exist at least one point ξ such that $f'(\xi) = 0$.

$$f'(x) = 0 \Rightarrow \frac{x^2 - ab}{x(x^2 + ab)} = 0 \Rightarrow x^2 - ab = 0$$

$\Rightarrow \quad x = \pm\sqrt{ab}$. So $\xi = \sqrt{ab}$, which lies in I_0 if

$a < \sqrt{ab} < b$ or $a^2 < ab < b^2$

$\because \quad a^2 < ab,\ a < b$ and $\because\ ab < b^2, a < b.$

which are true. Alternatively, the G.M. of any two distinct positive numbers always lies between the numbers.

(vi) If $f(x) = (x-a)^m(x-b)^n\ \forall\, x \in I = [a,b]$, f is obviously continuous in I, since $f(x)$ is a polynomial.

$$f'(x) = m(x-a)^{m-1}(x-b)^n + n(x-a)^m(x-b)^{n-1}$$

which exists $\forall\, x \in \mathbb{R}$, since $m, n \in \mathbb{N}$.

$\therefore \quad f$ is derivable in $I_0 = (a,b)$.

Lastly, $f(a) = 0 = f(b)$.

Thus all the three conditions of Rolle's theorem are satisfied and hence there must exist a point ξ in I_0 such that $f'(\xi) = 0$.

Now $f'(x) = (x-a)^{m-1}(x-b)^{n-1}[m(x-a) + n(x-b)]$.

$\therefore \quad f'(x) = 0$ for $x \in I_0$, when

$m(x-a) + n(x-b) = 0$

or, $\quad x = \dfrac{ma + nb}{m+n}.$

Since m and n are positive, x divides the line segment joining the points a and b on the x-axis internally in the ratio $m : n$. So the point $\xi = (ma+nb)/(m+n) \in I_0$.

1.19 Lagrange's First Mean Value Theorem

If $f : I \to \mathbb{R}$, where $I = [a,b]$,

(i) is continuous in I,

(ii) is derivable in $I_0 = (a,b)$,

then there always exists at least one point ξ in I_0 such that

$$\frac{f(b) - f(a)}{b - a} = f'(\xi).$$

Proof. We construct the function $\phi : I \to \mathbb{R}$ such that $\phi(x) = f(x) + Ax\ \forall\, x \in I$, A being a constant. Since $f(x)$ and x are both continuous in I, $\phi(x)$ is also continuous in I for every value of A. Since $f(x)$ and x are both derivable in $I_0, \phi(x)$ is also so for every value of A.

Now, if we impose the condition

$$\phi(a) = \phi(b),$$

then we shall get a definite value of A.

$$\phi(a) = \phi(b) \Rightarrow f(a) + Aa = f(b) + Ab$$

$$\Rightarrow \quad A = -[f(b) - f(a)]/(b - a).$$

With this definite value of A, ϕ satisfies all the conditions of Rolle's theorem. Hence there must exist a point ξ in I_0 such that $\phi'(\xi) = 0$.

But $\phi'(x) = f'(x) + A$.

So, $\phi'(\xi) = 0 \Rightarrow f'(\xi) + A = 0$

or, $-A = f'(\xi)$

or, $\dfrac{f(b) - f(a)}{b - a} = f'(\xi).$

If $b = a + h$, then a value of θ in $(0, 1)$ can be found out such that $\xi = a + \theta h$ and the Mean Value Theorem (M.V.Th.) gives

$$\boldsymbol{f(a + h) = f(a) + hf'(a + \theta h),}$$

which is the first step to the expansion of $f(x)$ about the point a. For $x = a + h \Rightarrow h = x - a$ and we get

$$f(x) = f(a) + (x - a)f'(a + \theta\overline{x - a}).$$

1.19.1 Geometrical Interpretation of the First Mean Value Theorem

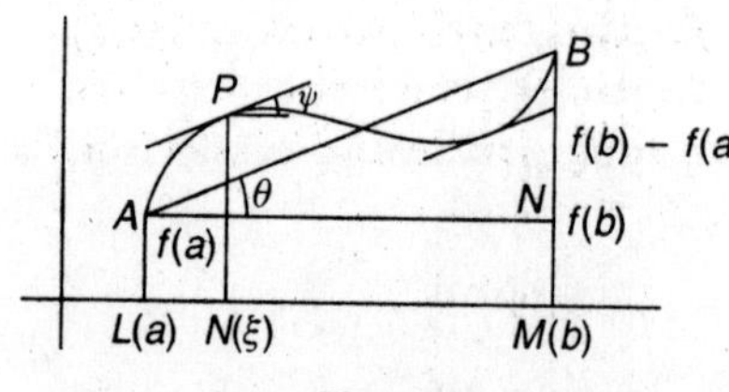

Fig. 1.5

As explained in Rolle's theorem, the first two conditions give, respectively, the gaplessness or so-called continuity of the curve and the existence of the unique tangent at every point on the curve between A and B or the smoothness of the curve.

In the Fig. 1.5, since $f(b) = BM$, $f(a) = AL$ and $b - a = LM$,

$$\therefore \quad \frac{f(b) - f(a)}{b - a} = \frac{BM - AL}{LM} = \frac{BN}{AN} = \tan\theta,$$

the slope of the chord AB, of the curve, joining the end points A and B.

The derivative $f'(\xi)$ represents the slope $\tan\psi$ of the tangent at P which corresponds to $x = \xi$. So, from the M.V.Th.

$$\tan\theta = \tan\psi \quad \text{or,} \quad \theta = \psi,$$

which implies that the tangent at P is parallel to the chord joining the end points.

Therefore, the geometrical interpretation of the M.V.Th. is that—

if a curve with distinct end points is gapless and smooth, then there always exists a point, on the curve, between the end points, where the tangent is parallel to the chord joining the end points.

1.19.2 Physical Interpretation of Lagrange's M.V.Th.

Let the curve of $f : I \to \mathbb{R}$, joining the end points A and B represent the path of a moving particle. Let the position vector (p.v.) of the particle at any instant t be $\overline{r}(t)$. So if P is the position of the particle at t, then

$$\overline{r} = \overline{r}(t) = \overline{OP}.$$

Let A and B correspond to $t = \alpha$ and $t = \beta$ respectively. Then

$$\overline{r}_A = \overline{r}(\alpha) = \overrightarrow{OA} \text{ and } \overline{r}_B = \overline{r}(\beta) = \overrightarrow{OB}.$$

The displacement of the particle in the time interval $[\alpha, t]$ is

$$\overrightarrow{AP} = \overline{r}_{PA} = \overline{r}_P - \overline{r}_A = \overline{r}(t) - \overline{r}(\alpha).$$

Now $\overline{v} = \dot{\overline{r}} = \dfrac{d\overline{r}}{dt}$ represents the velocity at $P(t)$.

The direction of the velocity of a particle moving along a smooth curve is always tangential to the path.

$$\text{Now} \quad \frac{\overline{r}(\beta) - \overline{r}(\alpha)}{\beta - \alpha} = \frac{\overline{OB} - \overline{OA}}{\beta - \alpha} = \frac{1}{\beta - \alpha}\overline{AB}, \tag{1}$$

where $\overline{AB}$ is the displacement of the particle during the time interval $\beta - \alpha$. R.H.S. of (1) is a vector parallel to the vector $\overrightarrow{AB}$ representing the average velocity during the time interval $[\alpha, \beta]$.

$$\text{If} \quad \frac{\overline{r}(\beta) - \overline{r}(\alpha)}{\beta - \alpha} = \dot{\overline{r}}(\gamma), \text{ where } \alpha < \gamma < \beta,$$

$$\text{then} \quad \frac{1}{\beta - \alpha}\overline{AB} = \overline{v}(\gamma).$$

$\therefore$ The M.V.Th. can be physically interpreted in the following way:

If a particle moves from the initial position $A(\alpha)$ to the final position $B(\beta)$ along a path without jumping (ensuring continuity of motion), then there exists an instant (γ) when the instantaneous velocity is equal, in magnitude and direction, to the average velocity.

Similarly, since $\dfrac{\dot{\overline{r}}(\beta) - \dot{\overline{r}}(\alpha)}{\beta - \alpha} =$ the average acceleration

and $\ddot{\overline{r}}(\gamma) =$ the instantaneous acceleration,

the consequence of the M.V.Th., viz.

$$\frac{\dot{\overline{r}}(\beta) - \dot{\overline{r}}(\alpha)}{\beta - \alpha} = \ddot{\overline{r}}(\gamma)$$

gives the following physical interpretation:

If the velocity of a particle during the time interval $[\alpha, \beta]$ is continuous and the acceleration exists during the motion, then there is some instant when the instantaneous acceleration is equal, in magnitude and direction, to the average acceleration.

Note. As in the case of Rolle's theorem, the two conditions given in Lagrange's M.V.Th. are sufficient only. If any one of the conditions is not satisfied even then a point such as ξ may exist in I_0 such that

$$[f(b) - f(a)]/(b - a) = f'(\xi).$$

For example, if

$$f(x) = 4x^2 \;\forall\, x \in [0, 2] \text{ except } x = 1/2$$

and $f(1/2) = 0$

then obviously both the conditions of the M.V.Th. are violated. The function is neither continuous nor differentiable at 1/2. But still there exists a point ξ in $(0, 2)$ where

$$\frac{f(2) - f(0)}{2 - 0} = f'(\xi), \tag{1}$$

$$[f(2) - f(0)]/2 = (16 - 0)/2 = 8$$

and $f'(x) = 8x,\ x \neq 1/2.$

$\therefore\ f'(\xi) = 8\xi$ if $\xi \neq 1/2$.

So (1) $\Rightarrow\ 8 = 8\xi$ or, $\xi = 1 \in (0, 2)$.

Examples 1.10. 1. *Verify Lagrange's M.V.Th. for*

(i) $(x - 1)(x - 2)$ *in* $[0, 1]$,

(ii) $|x|$ *in* $[-1, 1]$,

(iii) $\ln x$ *in* $[1, e]$, **[Burdwan 2003]**

(iv) $lx^2 + mx + n$ *in* $[a, b]$

and **(v)** *the function defined implicitly by* $x^{2/3} + y^{2/3} = 1$ *in* $[-1/2, 1]$, *where* $y > 0$.

(i) Let $f(x) = (x - 1)(x - 2)\ \forall\, x \in I = [0, 1]$.

Then $f(x)$ being a polynomial is continuous in I and differentiable in I_0. So $f(x)$ satisfies both the conditions of the M.V.Th. So there exists a point ξ such that

$$f'(\xi) = \frac{f(1) - f(0)}{1 - 0} = \frac{0 - 2}{1} = -2.$$

Now $f'(x) = 1(x - 2) + (x - 1).1 = 2x - 3.$

$\therefore\quad f'(\xi) = -2 \Rightarrow 2\xi - 3 = -2$ or, $\xi = 1/2 \in (0, 1)$.

(ii) If $f(x) = |x|$, it is continuous everywhere and hence in $[-1, 1]$. But $f'(x)$ does not exist at $x = 0$ although $f'(x) = 1$ for $x > 0$ and $f'(x) = -1$ for $x < 0$. Thus the second condition is violated. Hence there is no guarantee that there is a point ξ such that

$$f'(\xi) = \frac{f(1) - f(-1)}{1 - (-1)} = 0$$

And, in fact, if $\xi \neq 0$,

$$f'(\xi) = -1 \quad \text{if } \xi > 0,$$
$$\qquad = -1 \quad \text{if } \xi < 0$$

and $f'(\xi) = 0$ nowhere.

(iii) If $f(x) = \ln x\ \forall\, x \in [1, e]$, then since $f(x)$ is continuous and differentiable for all $x > 0$, it is continuous in $[1, e]$ and differentiable in $(1, e)$.

Since both the conditions of the M.V.Th. are satisfied, there must exist a point ξ in $(1, e)$ such that

$$f'(\xi) = \frac{f(e) - f(1)}{e - 1} \quad \text{or,} \quad f'(\xi) = \frac{1 - 0}{e - 1} = \frac{1}{e - 1}.$$

Now $\quad f(x) = \ln x \Rightarrow f'(x) = \dfrac{1}{x} \Rightarrow f'(\xi) = \dfrac{1}{\xi}$.

$$\therefore \quad \frac{1}{\xi} = \frac{1}{e - 1} \Rightarrow \xi = e - 1.$$

Since $\; 2 < e < 3,\; 1 < e - 1 < 2 < e$ and $1 < \xi < e$.

Thus $\; \xi \in (1, e)$.

(iv) Since $f(x) = lx^2 + mx + c$, a polynomial in x, f is continuous and differentiable in $\mathbb{R}$ and hence in $I = [a, b]\ \forall\, a, b \in \mathbb{R}$ with $a < b$. So, by the M.V.Th., there exists at least one point ξ such that

$$\frac{f(b) - f(a)}{b - a} = f'(\xi). \tag{1}$$

$$\text{L.H.S.} = \frac{(lb^2 + mb + c) - (la^2 + ma + c)}{b - a} = l(b + a) + m$$

and $\;$ R.H.S. $= 2l\xi + m$.

So, (1) gives $l(b + a) + m = 2l\xi + m$ or, $\xi = \dfrac{a + b}{2}$, which being the A.M. between two numbers a and b always lies between a and b.

(v) Since $y > 0$, we get from $x^{2/3} + y^{2/3} = 1$

$$f(x) = y = +\sqrt{(1 - x^{2/3})^3}\ \forall x \in I = [-1/2, 1].$$

Here $f(x)$, although irrational, is continuous in I.

$\because \; f'(x) = -\dfrac{\sqrt{1 - x^{2/3}}}{x^{1/3}}$, $f'(0)$ does not exist and $0 \in I_0 = (-1/2, 1)$.

Thus the second condition of the M.V.Th. is violated. So, it is uncertain whether there exists a point ξ in I_0 such that

$$[f(1) - f(-1/2)]/(1 - (-1/2)) = -\frac{\sqrt{1 - \xi^{2/3}}}{\xi^{1/3}} = -\sqrt{\xi^{2/3} - 1}.$$

Assuming that such a point exists we have

$$-\sqrt{\xi^{2/3} - 1} = \frac{0 - [1 - (-1/2)^{2/3}]^{3/2}}{1 - (-1/2)} = -\frac{2}{3}\left[1 - \frac{1}{4^{1/3}}\right]^{3/2} = -\frac{1}{3}(4^{1/3} - 1)^{3/2}$$

or, $\quad \xi^{2/3} - 1 = \dfrac{1}{9}(4^{1/3} - 1)^3$

or, $\quad \xi^{2/3} = \dfrac{9}{9 + (4^{1/3} - 1)^3}$

or, $\quad \xi = \pm\left\{\dfrac{9}{9 + (4^{1/3} - 1)^3}\right\}^{3/2}$

Taking $\xi = +\left\{\dfrac{9}{9 + (4^{1/3} - 1)^3}\right\}^{3/2}$

obviously $\xi > 0$ and since

$$\frac{9}{9+(4^{1/3}-1)^3} < 1, \xi < 1.$$

Thus $\quad 0 < \xi < 1$ and hence $-1 < \xi < 1$, a fortiori.

Here although a condition of the M.V.Th. is not satisfied but still there exists a point ξ such that

$$[f(b) - f(a)]/(b-a) = f'(\xi).$$

1.20 Cauchy's Mean Value Theorem

If the functions f and g, having $I = [a, b]$ as their common domain, are both

(i) continuous in I,

(ii) differentiable in $I_0 = (a, b)$

and (iii) $g'(x) \neq 0$ in I_0,

then there always exists at least one point ξ such that

$$\frac{f(b)-f(a)}{g(b)-g(a)} = \frac{f'(\xi)}{g'(\xi)}.$$

Proof. Let us set $\phi(x) = f(x) = +Ag(x)\ \forall x \in I$, where A is a parameter.

Then the function $\phi : I \to \mathbb{R}$ satisfies the first two conditions of Rolle's theorem.

Imposing the third condition of Rolle's theorem, viz.: $\phi(a) = \phi(b)$ we evaluate the parameter A.

$\because \quad \phi(a) = \phi(b) \Rightarrow f(a) + Ag(a) = f(b) + Ag(b),$

$\therefore \quad A = -[f(b) - f(a)]/[g(b) - g(a)]. \qquad (1)$

With this choice of A, ϕ satisfies all the conditions of the Rolle's theorem. So there exists a point ξ in I_0 such that

$$\phi'(\xi) = 0 \quad \text{or,} \quad f'(\xi) + Ag'(\xi) = 0,$$

whence $\quad A = -\dfrac{f'(\xi)}{g'(\xi)}. \qquad (2)$

From (1) and (2) we get the required result.

Note. Since $g'(x)$ does not vanish anywhere in I_0, $g(x)$ is monotonic and hence $g(a) \neq g(b)$. So A, determined by both (1) and (2), exists.

Taking $\xi = a + \theta h$, where $\theta \in (0, 1)$, Cauchy's formula for the M.V.Th. becomes

$$[f(b) - f(a)]/[g(b) - g(a)] = f'(a+\theta h)/g'(a+\theta h).$$

1.20.1 Geometrical Interpretation of Cauchy's M.V.Th.

The equations $y = f(t)$ and $x = g(t)\ \forall t \in [\alpha, \beta]$ may represent an arc of a curve Γ joining the end points $A(t = \alpha)$ and $B(t = \beta)$. Then the Cartesian coordinates of A and B are $(g(\alpha), f(\alpha))$

and $(g(\beta), f(\beta))$ respectively. The slope m of the chord AB is given by

$$m = [f(\beta) - f(\alpha)]/[g(\beta) - g(\alpha)].$$

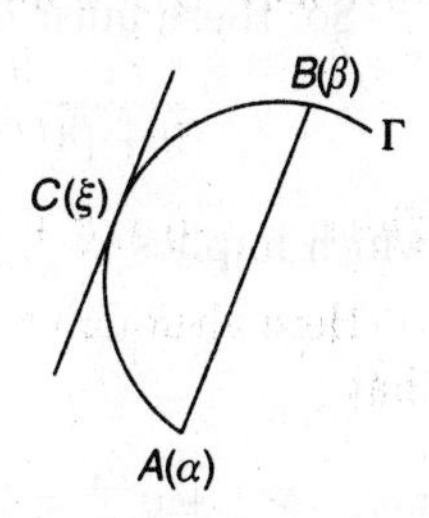

Fig. 1.6

The derivative $dy/dx = (dy/dt)/(dx/dt) = f'(t)/g'(t)$.

So, the slope m' of Γ at a point $C(t = \xi)$ is given by $m' = f'(\xi)/g'(\xi)$, where $g'(\xi) \neq 0$.

Now the continuity of f and g in $[\alpha, \beta]$ ensures the continuity of the arc AB and $g'(t) \neq 0$ for any t in (α, β) implies the smoothness of the curve between A and B.

Thus we find that, **if f and g satisfy all the conditions of Cauchy's M.V.Th., then geometrically $m = m'$, i.e., the slope of the chord AB is equal to that of the curve at an intermediate point C on the curve.**

Also, we find that by taking $x = g(t)$ and $y = f(t)$ as the parametric equations of curve, the geometrical interpretations of Cauchy's M.V.Th. and Lagrange's M.V.Th. are the same.

Examples 1.11. 1. *Verify Cauchy's M.V.Th. for the functions e^x and e^{-x} in the interval $I = [a, b]$ and verify that ξ is the midpoint of I.*

Let $f(x) = e^x$ and $g(x) = e^{-x}$ $\forall x \in I$.

Then obviously f and g satisfy the conditions of Cauchy's M.V.Th. Hence there exists a point ξ in I_0 such that

$$\frac{f(b) - f(a)}{g(b) - g(a)} = \frac{f'(\xi)}{g'(\xi)}. \tag{1}$$

Now L.H.S. $= \dfrac{e^b - e^a}{e^{-b} - e^{-a}} = -e^{a+b}$ and R.H.S. $= \dfrac{e^\xi}{-e^{-\xi}} = -e^{2\xi}$.

So from (1), $-e^{a+b} = -e^{2\xi}$ or $a + b = 2\xi$ or $\xi = (a+b)/2$.

Thus ξ is the midpoint of I.

2. *If $f(x) = \sqrt{x}$ and $g(x) = 1/\sqrt{x}$ $\forall x \in I = [a, b]$, where $0 < a < b$, then show that f and g both satisfy all the conditions of Cauchy's M.V.Th. Show that ξ is the G.M. between the end points of I.*

Since $0 < a < b$, $0 \notin I$ and $f'(x) = 1/2\sqrt{x}$ and $g'(x) = -1/2\sqrt{x^3}$, f and g both satisfy all the conditions of Cauchy's M.V.Th. theorem. $f'(x)$ and $g'(x)$ both exist $\forall x > 0$. So there must exist a $\xi \in I_0$ such that

$$[f(b) - f(a)]/[g(b) - g(a)] = f'(\xi)/g'(\xi).$$

The L.H.S. $= \dfrac{\sqrt{b} - \sqrt{a}}{\frac{1}{\sqrt{b}} - \frac{1}{\sqrt{a}}} = -\sqrt{ab}$ and the R.H.S. $= \dfrac{1/(2\sqrt{\xi})}{-1/(2\xi\sqrt{\xi})} = -\xi$.

$\therefore$ $\xi = \sqrt{ab}$, the G.M. of a and b, and so it lies between a and b.

3. *If $f(x) = x^{-2}$ and $g(x) = x^{-1}$ and $0 < a < b$, then show that f and g both satisfy all the conditions of Cauchy's M.V.Th. in $I = [a, b]$ and the point ξ is the H.M. between a and b.*

Since $0 \notin I$, f and g are both continuous in I.

$f'(x) = -\frac{2}{x^3}$ and $g'(x) = \frac{-1}{x^2}$, they do not exist at $x = 0$. But $0 \notin I_0$. So, f and g are both differentiable in I_0 and moreover $g'(x) \neq 0$ in I_0.

So, there must exist a ξ in I_0 such that

$$[f(b) - f(a)]/[g(b) - g(a)] = f'(\xi)/g'(\xi),$$

which implies

$$\frac{b^{-2} - a^{-2}}{b^{-1} - a^{-1}} = \frac{-2/\xi^3}{-1/\xi^2}$$

i.e., $b^{-1} + a^{-1} = 2/\xi$ or, $\xi = \frac{1}{2}\left(\frac{1}{a} + \frac{1}{b}\right)$.

Thus ξ is the H.M. between a and b, and as a result ξ lies between a and b.

Exercises 1.4

1. Verify Rolle's Theorem for the following functions in the corresponding intervals:

(i) $x^3(x-1)^2$ in $[0, 1]$;
(ii) $(x-1)(x-2)(x-3)$ in $[1, 3]$;
(iii) $\sin x + \cos x$ in $[0, 2\pi]$;
(iv) $\sin^2 x$ in $[0, \pi]$;
(v) $x(x+3)e^{-x/2}$ in $[-3, 0]$.

2. Are all conditions of Rolle's Theorem satisfied by the function $f(x) = |x|$ in $[-1, 1]$? Give reasons.

3. Verify Lagrange's M.V.Th. for the following functions in the corresponding intervals and find the point ξ when the theorem is applicable:

(i) $\sqrt{(x-1)}$ in $[1, 3]$;
(ii) $\ln|x|$ in $[-1, -1/2]$;
(iii) $x^3 - 2x^2 - x + 3$ in $[0, 1]$;
(iv) $|x-1|$ in $[0, 2]$.

4. Using Lagrange's M.V.Th. show that

(i) $|\sin\alpha - \sin\beta| \leq |\alpha - \beta|, \forall \alpha, \beta \in \mathbb{R}$;
(ii) $(\beta - \alpha)/(1 + \beta^2) < \tan^{-1}\beta - \tan^{-1}\alpha < (\beta - \alpha)/(1 + \alpha^2)$, where $|\alpha| < |\beta|$.

5. State Cauchy's M.V.Th. Apply the theorem to the functions $f(x) = e^x$ and $g(x) = e^{-x}$ in $[x, x+h]$ and obtain the value of θ. Interpret your result. **[W.B. Exam. 2003]**

6. If $f(x)$ and $g(x)$ are both continuous in $[a, b]$ and derivable in (a, b), then show that $\exists\, \xi$ in (a, b) such that

$$\begin{vmatrix} f(a) & f(b) \\ g(a) & g(b) \end{vmatrix} = (b-a)\begin{vmatrix} f(a) & f'(\xi) \\ g(a) & g'(\xi) \end{vmatrix}.$$ **[CH 1968]**

7. Use Lagrange's M.V.Th. to prove that

(i) $\sqrt{101}$ lies between 10 and 10.05;
(ii) $\sin 46°$ is approximately equal to $(1 + \pi/180)/\sqrt{2}$.

Is the estimate high or low? **[WBUT Sem. 2003]**

8. If $f'(x) = 1/(1+x^2)\ \forall x \in \mathbb{R}$ and $f(0) = 0$, using M.V.Th. in $[0,2]$, show that $0.4 < f(2) < 2$.

9. Use M.V.Th. to prove the following inequalities:

(i) $0 < x^{-1}\ln[(e^x - 1)/x] < 1$;
[**Hint.** $f(x) = e^x$ in $[0, x]$.]

(ii) $0 < 1/\ln(1+x) - 1/x < 1$.
[**Hint.** $f(x) = \ln(1+x)$ in $[0,x]$. $\{f(x) - f(0)\}/(x-0) = f'(\xi)$ $\Rightarrow x^{-1}\ln(1+x) = (1+\xi)^{-1}$, where $0 < \xi < x$. So $1 < 1+\xi < 1+x$ and $(1+x)^{-1} < (1+\xi)^{-1} < 1$.]

Answers

2. No. $f'(0)$ does not exist. **3.(i)** $3/2$; **(ii)** $-1/\ln 16$; **(iii)** $1/3$; **(iv)** not applicable.

5. $1/2$. Mean value is obtained at the midpoint of the interval $[x, x+h]$.

7. (ii) high.

1.21 Expansion of a Function—Taylor's and Maclaurin's Series

Lagrange's M.V.Th. in the form

$$f(a+h) = f(a) + hf'(a+\theta h), \text{ where } \theta \in (0,1)$$

is the First M.V.Th. The Second M.V.Th. is

$$f(a+h) = f(a) + hf'(a) + \frac{h^2}{2!}f'(a+\theta h), \text{ where } \theta \in (0,1).$$

So a natural generalisation is

$$f(a+h) = f(a) + hf'(a) + \frac{h^2}{2!}f''(a) + \ldots + \frac{h^{n-1}}{(n-1)!}f^{n-1}(a) + R_n,$$

where $R_n = \frac{h^n}{n!}f^{(n)}(a+\theta h), \theta \in (0,1)$, which, in fact, is **Taylor's Theorem** or the **Generalised M.V.Th.** with **Lagrange's or Taylor's form of remainder.**

Taylor's Theorem. If $f : I \to \mathbb{R}$, where $I = [a, a+h]$ is such that

(i) $f^{(n-1)}$ is continuous in I

and (ii) $f^{(n)}$ exists in $I_0 = (a, a+h)$,

then there always exists a number θ in $(0,1)$ such that

$$f(a+h) = \sum_{r=0}^{n-1} \frac{h^r}{r!} f^{(r)}(a) + R_n$$

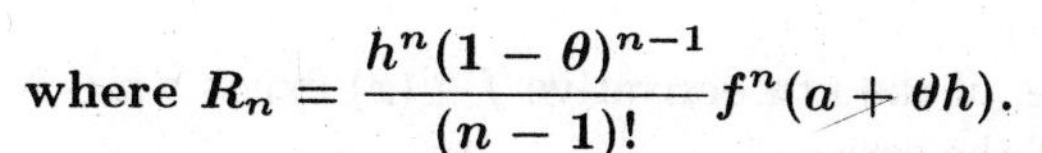

where $R_n = \dfrac{h^n(1-\theta)^{n-1}}{(n-1)!} f^n(a+\theta h).$

In the above summation h^0 has been taken as 1 even for $h = 0$ and $f^{(0)}(a) = f(a)$.

The theorem in this form is known as **Taylor's theorem in finite form** with **Cauchy's form of remainder**. This theorem also gives the expansion of $f(x)$.

Taking $x = a + h,\ h = x - a$ we get

$$f(x) = \sum_{r=0}^{n-1} \frac{(x-a)^r}{r!} f^{(r)}(a) + R_n,$$

where $R_n = \dfrac{(x-a)^n}{n!} f^{(n)}(a + \theta\overline{x-a})$ or, $\dfrac{(x-a)^n(1-\theta)^{n-1}}{(n-1)!} f^{(n)}(a + \theta\overline{x-a})$.

The first term $(x-a)^0 f^{(0)}(a)/0!$ has been assumed to be $f(a)$ even if $x = a$.

This gives us an expansion of $f(x)$ about $x = a$ in the form of a finite series of $(n+1)$ terms with the $(n+1)$th term R_n as the remainder after the n terms.

Putting $a = 0$ we get, for $x > 0$

$$f(x) = \sum_{r=0}^{n-1} \frac{x^r}{r!} f(0) + R_n,$$

where $R_n = \dfrac{x^n}{n!} f^n(\theta x)$ or, $\dfrac{x^n(1-\theta)^{n-1}}{(n-1)!} f^n(\theta x)$,

which is the expansion of $f(x)$ about $x = 0$ in the form of a finite series. This is **Maclaurin's expansion** of $f(x)$ about 0 in finite form.

Note. We find that there are two forms of remainder after n terms. Both of them involve θ, which lies in $(0, 1)$. But the value of θ in both the forms of the remainder for the same function need not be the same. Moreover, there is a more general form of remainder. If

$$R_n = \frac{h^n(1-\theta)^{n-p}}{p(n-1)!} f^{(n)}(a + \theta h),$$

where $p(\leq n) \in \mathbb{N}$, then it is called **Schlőmilch's** or **Roche's form of remainder**. If $p = 1$, we get Cauchy's form and if $p = n$, we get Lagrange's form.

Taking $\phi(t) = \sum_{r=0}^{n-1} \frac{(x-t)^r}{r!} f^{(r)}(t) + A(x-t)^p$, where $t \in [a, x]$ and $p(\leq n) \in \mathbb{N}$, $\phi(t)$ can be shown to satisfy the first two conditions of Rolle's theorem. Imposing the third condition, viz., $\phi(x) = \phi(a)$, there exists a $\theta \in (0, 1)$ such that

$$\phi'(a + \theta\overline{x-a}) = 0,$$

which will give Taylor's theorem with the remainder in Schlőmilch's form.

Maclaurin's expansion of $f(x)$ about $x = 0$ in finite form can be expressed as

$$f(x) = S_n(x) + R_n, \tag{1}$$

where R_n may be in any form. Then $S_n(x)$ is the sum of the series $\sum_0^{n-1} \frac{x^r}{r!} f^{(r)}(0)$ having n terms.

Now if f possesses derivatives of all orders, i.e., if the nth derivative $f^{(n)}(x)$ exists $\forall\, n \in \mathbb{N}$ in some neighbourhood of the origin, then we get the series

$$\sum_0^{\infty} (x^r/r!) f^{(r)}(0) \tag{2}$$

associated with $f(x)$. Since (2) is an infinite series, the question of convergence arises. Suppose the series converges $\forall\, x$ is some nbd $N(\delta, 0)$ of the origin. Then

$$\underset{n\to\infty}{\text{Lt}}\ S_n(x)$$

exists $\forall\, x \in N(\delta, 0)$. But the limit $S(x)$ may not be equal to $f(x)$. Suppose $S(x) = f(x)$ i.e.,

$$\underset{n\to\infty}{\text{Lt}}\ S_n(x) = f(x).$$

Then from (1) we get

$$\underset{n\to\infty}{\text{Lt}}\ R_n = \underset{n\to\infty}{\text{Lt}}\ [f(x) - S_n(x)] = 0,\ \forall\, x \in N(\delta, 0).$$

Thus, if $\underset{n\to\infty}{\text{Lt}}\ R_n(x) = 0,\ \forall\, x \in N(\delta, 0)$, then we get

$$f(x) = \underset{n\to\infty}{\text{Lt}}\ S_n(x) = \underset{n\to\infty}{\text{Lt}} \sum_{0}^{n-1} \frac{x^r}{r!} f^{(r)}(0) = \sum_{0}^{\infty} \frac{x^r}{r!} f^{(r)}(0),$$

which is **Maclaurin's theorem in infinite form** and the infinite series is called **the infinite power series expansion of $f(x)$.**

1.21.1 Expansions of Some Elementary Functions

1. Expand e^x in an infinite power series in x.

If $f(x) = e^x$ we know that $f^{(r)}(x) = e^x\ \forall\, r \in \mathbb{W}$ and $\forall\, x \in \mathbb{R}$. Taking Lagrange's form of remainder

$$R_n = \frac{x^n}{n!} f^{(n)}(\theta x) = \frac{x^n}{n!} e^{\theta x},$$

we know that for any fixed $x, x^n/n! \to 0$ as $n \to \infty$.

So, $R_n \to 0$ as $n \to \infty$. So, we get

$$f(x) = \sum_{r=0}^{\infty} \frac{x^r}{r!} f^r(0) = \sum_{r=0}^{\infty} \frac{x^r}{r!}$$

or, $$\boldsymbol{e^x = \sum_{r=0}^{\infty} \frac{x^r}{r!} = \sum_{n=0}^{\infty} \frac{x^n}{n!}}.$$

Note. For $x = 0$, the first term of the series is $\frac{0^0}{0!} = 0^0$ which is meaningless. In that case we write

$$f(x) = f(0) + \sum_{r=1}^{\infty} \frac{x^r}{r!} f^{(r)}(0).$$

$$\therefore \quad e^x = 1 + \sum_{r=1}^{\infty} \frac{x^r}{r!}$$

or, $$e^x = \sum_{r=0}^{\infty} \frac{x^r}{r!} \quad \text{if } x \neq 0$$
$$= 1 \quad \text{if } x = 0.$$

[If we agree to take $0^0 = 1$ here, then $e^x = \sum_{0}^{\infty} x^r/r!\ \forall\, x \in \mathbb{R}$.]

2. Use Maclaurin's expansion to show that

$$\textbf{(a)}\ \cos x = 1 + \sum_{r=1}^{\infty}(-1)^r x^{2r}/(2r)!\ \forall x \in \mathbb{R}$$

and $$\textbf{(b)}\ \sin x = \sum_{r=0}^{\infty}(-1)^r x^{2r+1}/(2r+1)!\,\forall x \in \mathbb{R}.$$

(a) Taking $f(x) = \cos x$ we know that

$$f^{(r)}(x) = \cos(x + r\pi/2)\ \forall r \in \mathbb{W}$$

and $f^{(r)}(0) = \cos r\pi/2,$

which is 0 or $(-1)^{r/2}$ according as r is odd or even.

$\therefore\quad f^{(2r+1)}(0) = 0\ \forall r \in \mathbb{W}$

and $f^{(2r)}(0) = (-1)^r.$

Now it can be shown that $R_n \to 0$ as $n \to \infty$.

$$\text{Now } f(x) = f(0) + \sum_{1}^{\infty}\frac{x^r}{r!}f^{(r)}(0) = f(0) + \sum_{1}^{\infty}\frac{x^{2r}}{(2r)!}f^{(2r)}(0) + \sum_{0}^{\infty}\frac{x^{2r+1}}{(2r+1)!}f^{(2r+1)}(0)$$

$$= f(0) + \sum_{1}^{\infty}\frac{x^{2r}}{(2r!)}f^{(2r)}(0), \quad \because f^{(2r+1)}(0) = 0.$$

$$\therefore \qquad \cos x = 1 + \sum_{1}^{\infty}(-1)^r\ x^{2r}/(2r)!$$

$$\text{or,} \qquad \cos x = \sum_{0}^{\infty}(-1)^r x^{2r}/(2r)! \text{ if } x \neq 0$$

$$= 1 \qquad \text{if } x = 0.$$

(b) Taking $f(x) = \sin x,\ f^{(r)}(x) = \sin(x + r\pi/2).$

$\therefore\quad f^{(r)}(0) = \sin(r\pi/2).$

$\therefore\quad f^{(2r)}(0) = \sin r\pi = 0$

and $f^{(2r+1)}(0) = \sin(r\pi + \pi/2) = (-1)^r.$

$$\text{Since}\quad f(x) = \sum_{0}^{\infty}\frac{x^r}{r!}f^{(r)}(0) \quad \text{if} \quad x \neq 0, \text{ and}$$

$$= f(0) \qquad \text{for } x = 0.$$

$$\sin x = \sum_{0}^{\infty}\frac{x^r}{r!}f^{(r)}(0) = \sum_{0}^{\infty}\frac{x^{2r}}{(2r)!}f^{(2r)}(0) + \sum_{0}^{\infty}\frac{x^{2r+1}}{(2r+1)!}f^{(2r+1)}(0)$$

$$= 0 + \sum_{0}^{\infty}(-1)^r\frac{x^{2r+1}}{(2r+1)!}.$$

3. Use Maclaurin's expansion to show that

$$(1+x)^m = 1+\sum_{1}^{m} {}^mC_r x^r \ \forall x \text{ if } m \in \mathbb{N}, \text{ and}$$

$$= 1+\sum_{1}^{m} \binom{m}{r} x^r \text{ if } m \text{ is a negative integer or a non-integral rational number.}$$

Taking $f(x) = (1+x)^m$, we have already seen that

$$f^{(r)}(x) = \binom{m}{r} r!(1+x)^{m-r},$$

where $\binom{m}{r} = 1$ if $r = 0$

$$= \frac{m(m-1)\dots(m-r+1)}{r!} \text{ for } r \in \mathbb{N}$$

$$= {}^mC_r \text{ if } m(\geq r) \in \mathbb{N}$$

$$= 0 \text{ if } m < r.$$

Case I. $m \in \mathbb{N}$

$$f^{(r)}(0) = \binom{m}{r} r! = {}^mC_r \ \forall r \leq m.$$

If $r = 0$, taking $f(0) = f^{(0)}(0)$, and knowing that $f(0) = (1+0)^m = 1$,

$$f^{(r)}(0) = {}^mC_r r! \ \forall r \leq m,$$

$$= 0 \text{ if } r > m.$$

So, the series $f(0) + \sum_{1}^{\infty} (x^r/r!) f^{(r)}(0)$ terminates and we get

$$f(0) + \sum_{1}^{m} {}^mC_r x^r.$$

Thus when $m \in \mathbb{N}$,

$$(1+x)^m = 1 + \sum_{1}^{m} {}^mC_r x^r,$$

which is the well-known binomial expansion of $(1+x)^m$.

Case II. m is a negative integer or a non-integral rational number.

We shall see that R_n, in Cauchy's form, in Maclaurin's finite form of expansion, will tend to zero as $n \to \infty$ if $-1 < x < 1$ or $|x| < 1$, and we shall get

$$(1+x)^m = f(0) + \sum_{1}^{\infty} (x^r/r!) f^{(r)}(0) \ \forall x \in (-1,1)$$

$$= 1 + \sum_{1}^{\infty} \binom{m}{r} x^r.$$

Note. It can be shown that the above expansion is valid for $x = 1$ if $m > -1$ and it is valid for $x = -1$ if $m > 0$. So when $m = -\frac{1}{2} > -1$, the infinite series

$1+\sum_{1}^{\infty}\binom{-1/2}{r}$ is convergent and its sum is $(1+1)^{-1/2}$, i.e., $\frac{1}{\sqrt{2}}$,

i.e., $1+\binom{-1/2}{1}+\binom{-1/2}{2}+\cdots=\frac{1}{\sqrt{2}}$

or, $1-\frac{1}{2}+\frac{1}{2!}\left(-\frac{1}{2}\right)\left(-\frac{1}{2}-1\right)+\cdots=\frac{1}{\sqrt{2}}$.

Similarly taking $m=\frac{1}{2}$ and $x=-1$,

$$1+\sum_{1}^{\infty}\binom{1/2}{r}=(1-1)^{\frac{1}{2}}=0,$$

i.e., $1+\frac{1}{2}+\frac{1}{2!}\cdot\frac{1}{2}\left(\frac{1}{2}-1\right)+\cdots=0$

or, $1+\frac{1}{2}-\frac{1}{2\cdot 4}+\frac{1\cdot 3}{2\cdot 4\cdot 6}-\frac{1\cdot 3\cdot 5}{2\cdot 4\cdot 6\cdot 8}+\cdots=0.$

Cor. Since $(a+x)^n=a^n(1+x/a)^n$, $a\neq 0$

$$(a+x)^n=a^n\cdot\left[1+\sum_{1}^{\infty}{}^nC_r(x/a)^r\right] \text{ if } n\in\mathbb{N}$$

$$=a^n\left[1+\sum_{1}^{\infty}\binom{n}{r}(x/a)^r\right]$$

$$=\sum_{0}^{\infty}\binom{n}{r}a^{n-r}x^r \text{ if } |x/a|<1, \text{ i.e., if } |x|<|a|$$

and n is a negative integer or a non-integral rational number. If $|x|>|a|$

$$(a+x)^n=\sum_{0}^{\infty}\binom{n}{r}a^r x^{n-r}.$$

4. Show, by using Maclaurin's expansion, that

$$\ln(1+x)=\sum_{1}^{\infty}(-1)^{r-1}x^r/r \text{ when } -1<x\leq 1.$$

If $f(x)=\ln(1+x)$, $f'(x)=\frac{1}{1+x}$, $f'(0)=1$

and $f^{(r)}(x)=\frac{(-1)^{r-1}(r-1)!}{(1+x)^r}\ \forall r\in\mathbb{N},\ x>-1.$

So, $f^{(r)}(0)=(-1)^{r-1}(r-1)!$ for $r\in\mathbb{N}$.

When $0\leq x\leq 1$, by taking R_n in Lagrange's form it can be shown that $R_n\to 0$ as $n\to\infty$ and we shall get

$$\ln(1+x)=f(0)+\sum_{1}^{\infty}(x^r/r!)f^{(r)}(0)=\ln(1+0)+\sum_{1}^{\infty}(-1)^{r-1}x^r/r$$

$$=\sum_{1}^{\infty}(-1)^{r-1}x^r/r.$$

When $-1 < x < 0$, by taking Lagrange's form it cannot be shown that $R_n \to 0$ as $n \to \infty$. But if we take Cauchy's form, it can be shown that $R_n \to 0$ as $n \to \infty$ and again we get the same expansion. Hence the result.

Since the expansion is valid for $x = 1$, we get

$$\ln 2 = \sum_1^\infty (-1)^{r-1}/r = 1 - 1/2 + 1/3 - 1/4 + \cdots.$$

Examples 1.12. **1.** *Using Maclaurin's expansion show that*

$$\tan^{-1} x = \sum_1^\infty (-1)^{n-1} x^{2n-1}/(2n-1).$$

If $f(x) = \tan^{-1} x$, $f(0) = 0$.
$f'(x) = 1/(1+x^2) = 1$ when $x = 0$

$$f^{(n)}(x) = (-1)^{n-1}(n-1)! \sin n\theta \sin^n \theta,$$

where $\cos\theta = x/\sqrt{1+x^2}$ and $\sin\theta = 1/\sqrt{1+x^2}$.

$\therefore$ $f^{(n)}(0) = (-1)^{n-1}(n-1)! \sin n\pi/2$ so that
$f^{(n)}(0) = 0$ when n is even.
$f^{(2n-1)}(0) = (2n-2)! \sin(2n-1)\pi/2 = (-1)^{n-1}(2n-2)!$.

$$\therefore \quad \tan^{-1} x = \sum_1^\infty x^{2n-1} f^{(2n-1)}(0)/(2n-1)! = \sum_1^\infty (-1)^{n-1} x^{2n-1}/(2n-1).$$

2. *Expand $e^{m \sin^{-1} x}$ as a power series in x.*

Let $y = e^{m\sin^{-1} x}$. Then

$$y_1 = y \cdot m/\sqrt{(1-x)^2} \quad \text{or,} \quad (1-x^2)y_1^2 - m^2y^2 = 0. \tag{1}$$

Differentiating with respect to x

$$(1-x^2) \cdot 2y_1y_2 - 2xy_1^2 - m^2 2yy_1 = 0 \tag{2}$$

or, $\quad (1-x^2)y_2 - xy_1 - m^2y = 0.$

Taking the nth order derivatives

$$(1-x^2)y_{n+2} - 2nxy_{n+1} - n(n-1)y_n - xy_{n+1} - ny_n - m^2y_n = 0.$$

When $x = 0$, $y_{n+2}(0) = (n^2 + m^2)y_n(0)$.

Taking $y_0 = y = e^{m\sin^{-1} x}, y_0(0) = 1$

$$y_1(0) = my_0(0) = m,$$
$$y_2(0) = m^2y_0(0) = m^2,$$
$$y_3(0) = (1^2 + m^2)y_1(0) = m(m^2 + 1^2),$$
$$y_4(0) = (m^2 + 2^2)y_2(0) = m^2(m^2 + 2^2).$$

$$\therefore \quad y = f(x) = \sum x^n f^{(n)}(0)/n! = \sum x^n y_n(0)/n!$$
$$= 1 + mx + m^2x^2/2! + m(m^2 + 1^2)x^3/3! + m^2(m^2 + 2^2)x^4/4! + \cdots.$$

3. *Show, by using Maclaurin's expansion, that*

$$\sin^{-1} x = x + 1 \cdot x^3/2 \cdot 3 + 1 \cdot 3 \cdot x^5/2 \cdot 4 \cdot 5 + 1 \cdot 3 \cdot 5 \cdot x^7/2 \cdot 4 \cdot 6 \cdot 7 + \cdots.$$

Let $y = f(x) = \sin^{-1} x$. Then $y(0) = f(0) = \sin^{-1} 0 = 0$ (1)

and $y_1(x) = f'(x) = 1/\sqrt{(1-x^2)} \Rightarrow y_1(0) = 1$ (2)

and $(1-x^2)y_1^2 = 1$.

Differentiating with respect to x and simplifying

$$(1-x^2)y_2 - xy_1 = 0.$$

Differentiating n times successively with respect to x

$$(1-x^2)y_{n+2} - 2nxy_{n+1} - n(n-1)y_n - xy_{n+1} - ny_n = 0.$$

$\therefore$ $y_{n+2}(0) - n^2 y_n(0) = 0.$ (3)

Replacing n by $2r-1$

$$y_{2r+1}(0)/y_{2r-1}(0) = (2r-1)^2.$$

$$\therefore \quad \prod_1^n [y_{2r+1}(0)/y_{2r-1}(0)] = \prod_1^n (2r-1)^2$$

or, $y_{2n+1}(0)/y_1(0) = \prod_1^n (2r-1)^2$

or, $y_{2n+1}(0) = \prod_1^n (2r-1)^2, \because y_1(0) = 1$ by (2). (4)

Replacing n by $2r-2$ in (3)

$y_{2r}(0) = 4(r-1)^2 y_{2r-2}(0).$ (5)

$\therefore$ $y_2(0) = 4(1-1)^2 y_0(0) = 0.$

$\because$ $y_0(0) = 0,\ y_2(0) = 0$, let $y_{2r}(0) = 0$ for some $r \in \mathbb{N}$.

From (5), $y_{2r+2}(0) = 4r^2 y_{2r}(0) = 0$.

Hence, by induction $y_{2n}(0) = 0\ \forall n \in \mathbb{N}$.

So,
$$\sin^{-1} x = \sum_0^\infty (x^n/n!) f^{(n)}(0) = \sum_0^\infty \{x^{2n+1}/(2n+1)!\} f^{(2n+1)}(0)$$

$$= x + \sum_1^\infty \frac{\prod_1^n (2r-1)^2}{\prod_1^n (2r) \prod_1^n (2r-1)} \frac{x^{2n+1}}{2n+1} = x + \sum_1^\infty \frac{\prod_1^n (2r-1)}{\prod_1^n (2r)} \frac{x^{2n+1}}{2n+1}$$

$$= x + \frac{1}{2} \cdot \frac{x^3}{3} + \frac{1 \cdot 3}{2 \cdot 4} \cdot \frac{x^5}{5} + \frac{1 \cdot 3 \cdot 5}{2 \cdot 4 \cdot 6} \cdot \frac{x^7}{7} + \cdots.$$

4. *Expand* $\sin x$ *about* $\pi/2$. *Hence find the value of* $\sin 91°$, *correct to* 4 *decimal places.* **[Kanpur 1996]**

The Taylor expansion of $f(x)$ about a is given by $f(a+h) = \sum_0^\infty (h^n/n!) f^{(n)}(a)$.

Let $f(x) = \sin x$. Then $f^{(n)}(x) = \sin(x + n\pi/2)$.

$$\therefore \quad \sin(a+h) = \sum_0^{\infty}(h^n/n!)\sin(a+n\pi/2). \qquad \text{(i)}$$

Now taking $a = \pi/2$ and $h = \pi/180$, $(a+h)^c = (\pi/2+\pi/180)^c = 91°$.

$$\therefore \quad \sin 91° = \sin(\pi/2+\pi/180) = \sum_0^{\infty}(\pi^n/n!180^n)\sin(\pi/2+n\pi/2), \text{ by (i)}.$$

But $\sin\left(\dfrac{\pi}{2}+\dfrac{n\pi}{2}\right) = \sin(n+1)\dfrac{\pi}{2} = 0$ if n is odd and $(-1)^n$ if n is even. So,

$$\sin 91° = \sum_0^{\infty}(-1)^n\pi^{2n}/(2n)!180^{2n} = 1 - \frac{\pi^2}{2\cdot 180^2} + \cdots = 1 \sim .00015$$

$$= .99985\cdots \sim .9998.$$

5. *Use Taylor's series to prove that*
$\tan^{-1}(x+h) = \tan^{-1}x + (h\sin z)(\sin z)/1 - (h\sin z)^2(\sin 2z)/2 + (h\sin z)^3(\sin 3z)/3 - \ldots,$
where $z = \cot^{-1}x$. **[Delhi 1991]**

By Taylor's theorem

$$f(x+h) = f(x) + hf'(x) + (h^2/2!)f''(x) + \cdots. \qquad \text{(i)}$$

If $z = \cot^{-1}x$, $\cot z = x$, $-\operatorname{cosec}^2 z dz/dx = 1$ or $dz/dx = -\sin^2 z$. Now $f(x) = \tan^{-1}x$

$$\Rightarrow \quad f'(x) = 1/(1+x^2) = 1/(1+\cot^2 z) = \sin^2 z,$$

$$f''(x) = d(\sin^2 z)/dx = \{d(\sin^2 z)/dz\}dz/dx$$

$$= 2\sin z\cos z(-\sin^2 z) = -\sin 2z\sin^2 z$$

and $\quad f'''(x) = df''(x)/dx = -\{d(\sin 2z\sin^2 z)/dz\}dz/dx$

$$= \{2\cos 2z\sin^2 z + \sin 2z\cdot 2\sin z\cos z\}\sin^2 z$$

$$= 2\sin^3 z\sin 3z.$$

Substituting these values of $f(x), f'(x), f''(x)$ and $f'''(x)$ in (i) the required result can be obtained.

6. *Using the power series for both* e^x *and* $\sin x$ *show that* $e^x\sin x = x+x^2+x^3/3-x^5/30+\cdots$.

Since $e^x = \sum x^p/p!$ and $\sin x = \sum(-1)^q x^{2q+1}/(2q+1)!$, where $\sum = \sum_0^{\infty}$, we have

$$e^x\sin x = \sum\sum(-1)^q x^{p+2q+1}/p!(2q+1)!.$$

Taking $n = p+2q+1$, $p = n-2q-1$, $0 \le p < \infty$ and $0 \le q < \infty \Rightarrow 0 \le n-2q-1 < \infty$, whence $0 \le q \le (n-1)/2$ and $1 \le n < \infty$.

$$\therefore \quad e^x\sin x = \sum_{n=1}^{\infty}\sum_{0\le q\le(n-1)/2}(-1)^q x^n/(n-2q-1)!(2q+1)!.$$

So, the coefficient of x^n is $\displaystyle\sum_{0\le q\le(n-1)/2}(-1)^q/(n-2q-1)!(2q+1)!$ or $(1/n!)\sum(-1)^q\,{}^nC_{2q+1}$ for $n = 1,2,3,\ldots$

So, there is no constant term. The coefficient of $x, x^2, x^3, \ldots$ are

$$\sum_0^0(-1)^q\,{}^1C_{2q+1} = {}^1C_1 = 1,$$

$$(1/2!)\sum_0^0(-1)^q\,{}^2C_{2q+1} = (1/2)\,{}^2C_1 = 1,$$

$$(1/3!)\sum_0^1(-1)^q\,{}^3C_{2q+1} = (1/6)[{}^3C_1 - {}^3C_3] = 2/6 = 1/3,$$

$$(1/4!)\sum_0^1(-1)^q\,{}^4C_{2q+1} = (1/24)[{}^4C_1 - {}^4C_3] = 0,$$

$$(1/5!)\sum_0^2(-1)^q\,{}^5C_{2q+1} = (1/120)[{}^5C_1 - {}^5C_3 + {}^5C_5] = -1/30, \ldots.$$

Hence the required result.

7. *Obtain the power series for $(1-x)^{-2}$ by differentiation in $(-1,1)$.*

By Maclaurin's expansion

$$(1-x)^{-1} = \sum x^r \ \forall x \in (-1,1), \text{ where } 0 \le r < \infty.$$

Differentiating

$(1-x)^{-2} = \sum r x^{r-1}$, if we assume $0.x^{-1} = 0$. Otherwise

$$(1-x)^{-2} = \sum_1^\infty r x^{r-1} \ \forall x \in (-1,1).$$

8. *Assuming the validity of expansion of $\sec x$ for $x \ne (2n+1)\frac{\pi}{2}$ for $n \in \mathbb{Z}$, in an ascending power series in x, show that $\sec x = 1 + x^2/2 + 5x^4/24 + \cdots$.*

Let $\sec x = 1/\cos x = \sum a_r x^r$ for $0 \le r < \infty$. (i)

But since $\cos x = \sum(-1)^r x^{2r}/(2r)!$,

$$\sum a_r x^r \cdot \sum(-1)^r x^{2r}/(2r)! \equiv 1$$

or,
$$\sum_{q=0}^\infty\sum_{r=0}^\infty(-1)^r a_q x^{q+2r}/(2r)! \equiv 1. \quad \text{(ii)}$$

Let $n = q + 2r$. Then $q = n - 2r$ and $0 \le q < \infty \Rightarrow 0 \le n - 2r < \infty$. So $r \le n/2$. Hence $0 \le r \le n/2$ and $0 \le n < \infty$.

From (ii) we have

$$\sum_{n=0}^\infty \sum_{0\le r\le n/2} (-1)^r a_{n-2r} x^n/(2r)! \equiv 1.$$

So, the coefficient of x^n is $\sum_{0\le r\le n/2} (-1)^r a_{n-2r}/(2r)! = 0$ for $n \ge 1$ and 1 for $n = 0$. So,

$a_0 = 1$ and for $n = 1, \sum(-1)^r a_{1-2r}/(2r)! = 0$ for $0 \le r \le \frac{1}{2}$, i.e., $a_1 = 0$.

$\sum(-1)^r a_{2-2r}/(2r)! = 0$ for $0 \le r \le 2/2 \Rightarrow a_2 - a_0/2 = 0$. So, $a_2 = a_0/2 = 1/2$.

$\sum(-1)^r a_{3-2r}/(2r)! = 0$ for $0 \le r \le 3/2 \Rightarrow a_3 - a_1/2 = 0 \Rightarrow a_3 = 0$.

$\sum(-1)^r a_{4-2r}/(2r)! = 0$ for $0 \le r \le 4/2 \Rightarrow a_4 - a_2/2 + a_0/24 = 0$,

whence $a_4 = 5/24$ and so on.

Substituting these values of the unknown coefficients we get the required expansion of $\sec x$. This method is known as the **method of unknown coefficients.**

9. *Form a differential equation to get the power series expansion of $e^{ax}\sin bx$ using the method of undetermined coefficients.*

Let $y = e^{ax}\sin bx \equiv \sum a_r x^r,\ 0 \le r < \infty.$ (i)

Then $y_1 = e^{ax}(a\sin bx + b\cos bx) = ay + be^{ax}\cos bx \equiv \sum r a_r x^{r-1}$ (ii)

and $y_2 = ay_1 + be^{ax}(a\cos bx - b\sin bx) = ay_1 + a(y_1 - ay) - b^2 y$

$= 2ay_1 - (a^2 + b^2)y \equiv \sum r(r-1)a_r x^{r-2}.$ (iii)

From (iii) we get the differential equation:

$$y_2 - 2ay_1 + (a^2 + b^2)y = 0. \quad \text{(iv)}$$

Putting the expansions for y, y_1 and y_2 from (i), (ii) and (iii) in (iv)

$$\sum r(r-1)a_r x^{r-2} - 2a\sum r a_r x^{r-1} + (a^2+b^2)\sum a_r x^r \equiv 0$$

or, $$\sum(r+2)(r+1)a_{r+2}x^r - 2a\sum(r+1)a_{r+1}x^r + (a^2+b^2)\sum a_r x^r = 0$$

or, $$\sum[(r+2)(r+1)a_{r+2} - 2a(r+1)a_{r+1} + (a^2+b^2)a_r]x^r \equiv 0.$$

$\therefore$ $$(r+2)(r+1)a_{r+2} - 2a(r+1)a_{r+1} + (a^2+b^2)a_r = 0\ \forall r \in \mathbb{W}. \quad \text{(v)}$$

From (i), $y(0) = 0 \Rightarrow a_0 = 0$. From (ii), $y_1(0) = ay(0) + b = a_1 \Rightarrow a_1 = b$.

From (iii), $y_2(0) = 2ay_1(0) - (a^2+b^2)y(0) = 2a_2$ or $a_2 = ab$.

So, from (v), $3 \cdot 2 \cdot a_3 - 2a \cdot 2a_2 + (a^2+b^2)a_1 = 0 \Rightarrow a_3 = b(3a^2 - b^2)/3!$

and $4 \cdot 3 \cdot a_4 - 2a \cdot 3a_3 + (a^2+b^2)a_2 = 0 \Rightarrow a_4 = 4ab(a^2 - b^2)/4!, \ldots$

$\therefore$ $$e^{ax}\sin bx = bx + abx^2 + b(3a^2 - b^2)x^3/3! + 4ab(a^2 - b^2)x^4/4! + \cdots.$$

1.22 Indeterminate or Undetermined Forms—L'Hospital's Rule*

The symbols

$\frac{0}{0}$, $\frac{\infty}{\infty}$, $0 \times \infty$ or $\infty \times 0$, $\infty - \infty$, 0^0, $1^{\pm\infty}$ and ∞^0, called indeterminate or undetermined forms, do not stand for any numbers. None of them have any numerical values. But each of them is always associated with a pair of functions in limits.

If u and v are two real functions such that $\operatorname{Lt}_{x\to a} u = 0$ and $\operatorname{Lt}_{x\to a} v = 0$, then if

$$f(x) = \frac{u}{v},$$

then $f(x)$ is said to take the **indeterminate form: 0/0 near $x = a$.**

Similarly, if $\operatorname{Lt}_{x\to a} u = \pm\infty$ and $\operatorname{Lt}_{x\to a} v = \pm\infty$, then $f(x) = \dfrac{u}{v}$ takes the **indeterminate form: $\dfrac{\infty}{\infty}$ near $x = a$.**

---

*Also L'Hôpital after Guillaume Francois Antoine de l'Hôpital (1861-1704), French analyst and geometer.

Similarly for all others. Under these circumstances whether the limit of $f(x)$ will exist or not that is uncertain. That is why the forms are called indeterminate forms. The symbols $\infty + \infty$ or $\infty \times \infty$ or ∞^{∞}, etc., do not stand for ordinary real numbers, but they are not indeterminate forms. For, if u and v both tend to $+\infty$ as $x \to a$, then obviously $f(x) = u + v$ will tend to ∞. So the limit (infinite) exists. Similarly, for the others.

Of the fundamental limits discussed in §1.8, the functions $(x^n - a^n)/(x - a)$, $(\sin x)/x$, $\{\ln(1+x)\}/x$ and $(e^x - 1)/x$ are all of the forms $0/0$ near 0, whereas $(1 + 1/x)^x$ is of the form $1^{\pm\infty}$ near $\pm\infty$ and $(1+x)^{1/x}$ is also of the same form near 0. The function $\{\ln(1+x)\} \cdot (1/x)$ and $(e^x - 1) \cdot (1/x)$ are of the form $0 \times \infty$ near 0.

The function $1/x - 1/x$ is of the form $\infty - \infty$ near 0 when $x > 0$ and of the form $-\infty + \infty$ when $x < 0$, but $\text{Lt}(1/x - 1/x)$ exists at $x = 0$ and it is equal to 0, whereas although $2/x - 1/x$ is of the same form but its limit does not exist at $x = 0$.

It might be noted that all these indeterminate forms may ultimately be made to depend upon either $0/0$ or ∞/∞. First of all, if u and v both approach 0 when $x \to a$, then $1/u$ and $1/v$ both approach $\pm\infty$ and since $u/v = (1/v)/(1/u)$, although u/v takes the indeterminate form $0/0$, the same function u/v written in a different form, viz., $(1/v)/(1/u)$ takes the form ∞/∞.

So, so far as the form is concerned, $0/0 = (1/0)/(1/0) = \infty/\infty$.

Similarly, we write

$$0 \times \infty = 0 \times (1/0) = (0 \times 1)/0 = 0/0$$

or, $$0 \times \infty = (1/\infty) \times \infty = \infty/\infty.$$

And $$\infty - \infty = 1/0 - 1/0 = (0 - 0)/0 = 0/0 = (1/0)/(1/0) = \infty/\infty.$$

$$0^0 = e^{0 \ln 0} = e^{0 \times (-\infty)} = e^{-0 \times \infty} = e^{-0/0} = e^{-\infty/\infty},$$

$$1^{\pm\infty} = e^{\pm\infty \ln 1} = e^{\pm\infty \times 0} = e^{\pm 0/0} \text{ or } e^{\pm\infty/\infty},$$

$$\infty^0 = e^{0 \ln \infty} = e^{0 \times \infty} = e^{0/0} = e^{\infty/\infty}.$$

1.22.1 Limit of a Function in the Form: $\frac{0}{0}$

L'Hospital's Rule*. Let $f(x) = u(x)/v(x)$, where

(i) u and v are both continuous in $N(\delta, a)$ for some $\delta > 0$,

(ii) u and v are both differentiable in $N(\delta, \not{a})$,

(iii) $u(a) = v(a) = 0$

and **(iv)** $\underset{x \to a}{\text{Lt}} \dfrac{u'}{v'}$ exists.

Then $$\underset{x \to a}{\text{Lt}} f(x) = \underset{x \to a}{\text{Lt}} \frac{u}{v} = \underset{x \to a}{\text{Lt}} \frac{u'}{v'}.$$

Notes.

(1) Near a, $f(x)$ takes the indeterminate form $0/0$.

(2) a may be ∞ or $-\infty$ also.

*This rule was actually discovered by John Bernoulli and given to L'Hospital in return for salary.

(3) L'Hospital's rule is applicable even if $u(a)$ and/or $v(a)$ are not zero but $\underset{x\to a}{\text{Lt}}\, u = 0 = \underset{x\to a}{\text{Lt}}\, v$.

(4) If $u'(a) = 0 = v'(a)$ or $\underset{x\to a}{\text{Lt}}\, u'(x) = 0 = \underset{x\to a}{\text{Lt}}\, v'(x)$, then

$$\underset{x\to a}{\text{Lt}}\, f(x) = \underset{x\to a}{\text{Lt}}\, \frac{u''(x)}{v''(x)},$$

provided u' and v' satisfy the same condition as u and v in the theorem.

Generalising, if either

$$u^{(r)}(a) = 0 = v^{(r)}(a) \qquad \text{for } r = 1, 2, \ldots, n-1$$

or, $$\underset{x\to a}{\text{Lt}}\, u^{(r)}(x) = 0 = \underset{x\to a}{\text{Lt}}\, v^{(r)}(x) \quad \text{for } r = 1, 2, 3, \ldots, n-1$$

and $u^{(n-1)}$ and $v^{(n-1)}$ satisfy the conditions of the L'Hospital's rule, then

$$\underset{x\to a}{\text{Lt}}\, f(x) = \underset{x\to a}{\text{Lt}}\, \frac{u^n(x)}{v^{(n)}(x)}.$$

Examples 1.13. 1. *Let* $f(x) = u(x)/v(x)\ \forall\, x (\neq 0) \in \mathbb{R}$,
where $u(x) = \sin x\ \forall\, x \in \mathbb{R}$
and $v(x) = x$ *when* $x \neq 0$
$= 1$ *when* $x = 0$.

It may be seen that v is discontinuous at $x = 0$. But still L'Hospital's rule is applicable and

$$\underset{x\to 0}{\text{Lt}}\, f(x) = \underset{x\to 0}{\text{Lt}}\, \frac{u(x)}{v(x)} = \underset{x\to 0}{\text{Lt}}\, \frac{u'(x)}{v'(x)} = \underset{x\to 0}{\text{Lt}}\, \frac{\cos x}{1} = 1.$$

2. *Let* $u(x) = \sin x^3,\ x \neq 0$ *and* $v(x) = x^2,\ x \neq 0$.
Then u and v are both undefined at 0 and they are discontinuous thereat. Consequently, u', v', u'', v'', etc., do not exist at $x = 0$. But still

$$\underset{x\to 0}{\text{Lt}}\, \frac{u(x)}{v(x)} = \underset{x\to 0}{\text{Lt}}\, \frac{u'(x)}{v'(x)} = \underset{x\to 0}{\text{Lt}}\, \frac{3x^2 \cos x^3}{2x} = \frac{3}{2} \underset{x\to 0}{\text{Lt}}\, x \cos x^3 = \frac{3}{2} \times 0 = 0.$$

1.22.2 Limit of a Function in the Form: $\frac{\infty}{\infty}$

L'Hospital's Rule. Let $f(x) = u(x)/v(x)$, where

(i) u and v are continuous in $N(\delta, \not{a})$ for some $\delta > 0$,

(ii) u' and v' exist in $N(\delta, \not{a})$,

(iii) $v'(x) \neq 0$ in $N(\delta, \not{a})$,

(iv) $\underset{x\to a}{\text{Lt}}\, |u(x)| = \underset{x\to a}{\text{Lt}}\, |v(x)| = \infty$

and **(v)** $\underset{x\to a}{\text{Lt}}\, \frac{u'(x)}{v'(x)}$ exists.

Then $\underset{x\to a}{\text{Lt}}\, f(x) = \underset{x\to a}{\text{Lt}}\, \frac{u(x)}{v(x)} = \underset{x\to a}{\text{Lt}}\, \frac{u'(x)}{v'(x)}$.

Here again a may be ∞ or $-\infty$ and the limit $\underset{x\to a}{\text{Lt}}\, \frac{u'(x)}{v'(x)}$ may be finite or infinite.

Examples 1.14. **1.** $\underset{x\to\infty}{\text{Lt}} \dfrac{ax^2+b}{cx^2+d} = \underset{x\to\infty}{\text{Lt}} \dfrac{2ax}{2cx} = \dfrac{a}{c}$ *provided* $c \neq 0$.

If $c = 0$,

$$\underset{x\to\infty}{\text{Lt}} \frac{ax^2+b}{cx^2+d} = \underset{x\to\infty}{\text{Lt}} \frac{ax^2+b}{d},\ d \neq 0$$

$$= \frac{1}{d} \underset{x\to\infty}{\text{Lt}} (ax^2+b) = +\infty \text{ or } -\infty \text{ according as } ad > \text{ or } < 0.$$

2. $\underset{x\to\pi/2}{\text{Lt}} \dfrac{\tan 3x}{\tan x}$, *form:* $\dfrac{\infty}{\infty}$

$$= \underset{x\to\pi/2}{\text{Lt}} \frac{3\sec^2 3x}{\sec^2 x}, \text{ form} : \frac{\infty}{\infty}$$

$$= 3 \underset{x\to\pi/2}{\text{Lt}} \frac{\cos^2 x}{\cos^2 3x}, \text{ form: } \frac{0}{0}$$

$$= 3 \underset{\theta\to 0}{\text{Lt}} \frac{\sin^2\theta}{\sin^2 3\theta}, \text{ form: } \frac{0}{0}, \text{ where } \theta = \pi/2 + x$$

$$= 3 \underset{\theta\to 0}{\text{Lt}} \frac{2\sin\theta\cos\theta}{6\sin 3\theta\cos 3\theta}$$

$$= \underset{\theta\to 0}{\text{Lt}} \frac{\sin\theta}{\sin 3\theta} \cdot \underset{\theta\to 0}{\text{Lt}} \frac{\cos\theta}{\cos 3\theta}$$

$$= \underset{\theta\to 0}{\text{Lt}} \frac{\cos\theta}{3\cos 3\theta} \cdot 1 = \frac{1}{3}.$$

3. *To evaluate* $\underset{x\to 1}{\text{Lt}} \dfrac{2x^3-3x^2+1}{3x^5-5x^3+2}$

we find that, if $u = 2x^3 - 3x^2 + 1$ and $v = 3x^5 - 5x^3 + 2$,

then $u(1) = 2 - 3 + 1 = 0$ and $v = 3 - 5 + 2 = 0$.

So, $f(x) = \dfrac{2x^3-3x^2+1}{3x^5-5x^3+2} = \dfrac{u(x)}{v(x)}$ is of the form: $\dfrac{0}{0}$ in a neighbourhood of 1.

Using l'Hospital's rule,

$$\underset{x\to 1}{\text{Lt}} f(x) = \underset{x\to 1}{\text{Lt}} \frac{u'}{v'} = \underset{x\to 1}{\text{Lt}} \frac{6x^2-6x}{15x^4-15x^2}$$

$$= \frac{2}{5} \underset{x\to 1}{\text{Lt}} \frac{x-1}{x^3-x}, \text{ form: } \frac{0}{0}$$

$$= \frac{2}{5} \underset{x\to 1}{\text{Lt}} \frac{1}{3x^2-1} = \frac{1}{5}.$$

4. *Evaluate the following limit, if it exists:* $\underset{x\to\infty}{\text{Lt}} \dfrac{x^2}{e^x}$.

If $u = x^2$ and $v = e^x$, u/v is of the form: ∞/∞ in a nbd of ∞. So using l'Hospital's rule,

$$l = \underset{x\to\infty}{\text{Lt}} \frac{x^2}{e^x} = \underset{x\to\infty}{\text{Lt}} \frac{2x}{e^x}, \text{ form: } \frac{\infty}{\infty}$$

$$= 2 \underset{x\to\infty}{\text{Lt}} \frac{1}{e^x} = 0.$$

It can be similarly shown that $\underset{x\to\infty}{\text{Lt}}\ x^p/e^{\alpha x} = 0\ \forall\, p$ and $\alpha (> 0) \in \mathbb{R}$.

5. *Ascertain the existence of limit of the following function using l'Hospital's rule:*

$$\underset{x\to 0}{\text{Lt}} \frac{\ln(1-x^2)}{\ln\cos^2 x}.$$

Taking $u = \ln(1-x^2)$ and $v = \ln\cos x$, we find that $u(0) = 0$ and $v(0) = 0$. So,

$$f(x) = \frac{\ln(1-x^2)}{\ln\cos x} \text{ is of the form: } \frac{0}{0} \text{ near } 0.$$

So, using l'Hospital's rule

$$\underset{x\to 0}{\text{Lt}} f(x) = \underset{x\to 0}{\text{Lt}} \frac{u'(x)}{v'(x)} = \underset{x\to 0}{\text{Lt}} \frac{\frac{-2x}{1-x^2}}{-\tan x}$$

$$= 2 \underset{x\to 0}{\text{Lt}} \frac{x}{\tan x} \cdot \underset{x\to 0}{\text{Lt}} \frac{1}{1-x^2}$$

$$= 2 \underset{x\to 0}{\text{Lt}} \frac{x}{\tan x} \cdot 1, \text{ form: } \frac{0}{0}$$

$$= 2 \underset{x\to 0}{\text{Lt}} \frac{1}{\sec^2 x} = 2.$$

6. $\underset{x\to 0}{\text{Lt}} \left(\frac{1}{x} - \frac{1}{\sin x}\right)$, *form:* $\pm(\infty - \infty)$

$$= \underset{x\to 0}{\text{Lt}} \frac{\sin x - x}{x\sin x}, \quad \text{form: } \frac{0}{0}$$

$$= \underset{x\to 0}{\text{Lt}} \frac{\cos x - 1}{\sin x + x\cos x}, \text{ form:} \frac{0}{0}$$

$$= \underset{x\to 0}{\text{Lt}} \frac{-\sin x}{2\cos x - x\sin x} = 0.$$

7. *Evaluate* $\underset{x\to 0}{\text{Lt}}\, x\ln x$, *form:* $0 \times (-\infty)$.

For if $u = x$ and $v = \ln x$ and $f(x) = uv$, then $X = \text{Dom} f = R^+$. So $x \to 0+$. Also $u \to 0$ and $v \to -\infty$. $f(x)$ in this case can be converted into either of the forms: $\frac{0}{0}$ or $\frac{\infty}{\infty}$.

$$f(x) = uv = \frac{u}{1/v}, \quad \text{form: } \frac{0}{0}$$

$$= \frac{v}{1/u}, \quad \text{form: } \frac{\infty}{\infty}.$$

But since the derivative of $1/v$, i.e., of $1/\ln x$ will be complicated we take

$$f(x) = \frac{v}{1/u} = \frac{\ln x}{1/x}.$$

So, $$\underset{x\to 0}{\text{Lt}} f(x) = \underset{x\to 0}{\text{Lt}} \frac{\ln x}{1/x}, \text{ form: } \infty/\infty$$

$$= \underset{x\to 0}{\text{Lt}} \frac{1/x}{-1/x^2} = -\underset{x\to 0}{\text{Lt}}\, x = -0 = 0.$$

8. *Use l'Hospital's rule to find the limit:* $\underset{x\to 0}{\text{Lt}} (x^2)^{x^2}$.

Let $u = v = x^2$ and $f(x) = u^v = (x^2)^{x^2}$.

Obviously, Dom$f = \mathbb{R}_0$. Here since $u \to 0$ and $v \to 0$ as $x \to 0$, $f(x)$ is the form: 0^0 near 0.

If $x \neq 0$, $f(x) = u^v = e^{v \ln u} = e^{x^2 \ln x^2} = e^{g(x)}$, where $g(x) = x^2 \ln x^2$ is of the form: $0 \times (-\infty)$ near 0.

$$g(x) = \frac{\ln x^2}{x^{-2}}, \text{ form: } \frac{-\infty}{\infty}.$$

$$\therefore \quad \underset{x\to 0}{\text{Lt}}\, g(x) = \underset{x\to 0}{\text{Lt}} \frac{2/x}{-2x^{-3}} = -\underset{x\to 0}{\text{Lt}}\, x^2 = 0.$$

$$\therefore \quad \underset{x\to 0}{\text{Lt}}\, f(x) = \underset{x\to 0}{\text{Lt}}\, e^{g(x)} = e^{\underset{x\to 0}{\text{Lt}}\, g(x)} = e^0 = 1.$$

9. *Evaluate* $\underset{x\to 0}{\text{Lt}} \left\{\frac{1}{3}(a^x + b^x + c^x)\right\}^{3/x}$

If $f(x) = \left\{\frac{1}{3}\{(a^x + b^x + c^x)\right\}^{3/x}$, taking $u = \frac{1}{3}(a^x + b^x + c^x)$ and $v = 3/x$ we find that $f(x) = u^v$, where $u \to 1$ as $x \to 0$ and $v \to \pm\infty$ as $x \to 0$.

So, $f(x)$ is of the form: $1^{\pm\infty}$.

We express $f(x)$ in the exponential form $e^{v \ln u} = e^{g(x)}$, where

$$g(x) = v \ln u = \frac{3}{x} \ln \left\{\frac{1}{3}(a^x + b^x + c^x)\right\} = 3 \cdot \frac{U}{V},$$

where $U = \ln\left\{\frac{1}{3}(a^x + b^x + c^x)\right\}$ and $V = x$.

Then $g(x) = \frac{3U}{V}$ is of the form: $\frac{0}{0}$ near 0.

$$\therefore \quad \underset{x\to 0}{\text{Lt}}\, g(x) = 3 \underset{x\to 0}{\text{Lt}} \frac{U'}{V'} = 3 \cdot \underset{x\to 0}{\text{Lt}} \frac{(a^x \ln a + b^x \ln b + c^x \ln c)}{(a^x + b^x + c^x) \cdot 1}$$

$$= \frac{3(\ln a + \ln b + \ln c)}{3} = \ln(abc).$$

$$\therefore \quad \underset{x\to 0}{\text{Lt}}\, f(x) = \underset{x\to 0}{\text{Lt}}\, e^{g(x)} = e^{\underset{x\to 0}{\text{Lt}}\, g(x)} = e^{\ln(abc)} = abc.$$

Exercises 1.5

1. Verify Maclaurin's infinite series expansion of each of the following functions on the indicated intervals:

(i) $\cos^2 x = 1 + \sum_1^\infty (-1)^n 2^{2n-1} x^{2n}/(2n)!$ on $\mathbb{R}$;

(ii) $\sin^2 x = \sum_1^\infty (-1)^{n-1} 2^{2n-1} x^{2n}/(2n)!$ on $\mathbb{R}$;

(iii) $\ln(1 + 2x) = \sum_1^\infty (-1)^{n-1} 2^n x^n/n$ on $(-1/2, 1/2]$;

(iv) $e^x \sin x = x + x^2 + x^3/3 + x^5/30 - \cdots$ on $\mathbb{R}$;

(v) $x(e^x - 1)^{-1} = 1 - x/2 + x^2/12 - x^4/720 + \cdots$ on $\mathbb{R} - \{0\}$;

(vi) $e^{\sin x} = 1 + x + x^2/2 - x^4/8 - \cdots$ on $\mathbb{R}$;

(vii) $\ln \sec x = x^2/2! + 2x^4/4! + 16x^6/6! + \cdots$ on $(-\pi/2, \pi/2)$;

(viii) $\sin^{-1} x = x + \sum_{n=1}^{\infty} \prod_{r=1}^{n} \{(2r-1)/(2r)\} x^{2n+1}/(2n+1)$ on $(-\pi/2, \pi/2]$;

(ix) $\ln(1 - x + x^2) = -x + x^2/2 + 2x^3/3 + x^4/4 - \cdots$ on $((1-\sqrt{5})/2,\ (1+\sqrt{5})/2]$;

(x) $\ln(1 + \sin x) = x - x^2/2 + x^3/6 - \cdots$ on $(-\pi/2, \pi/2]$.

2. Show, by Taylor's expansion that

(i) $1/x = 1/2 \left[1 + \sum_{1}^{\infty} (-1)^r (x-2)^r / 2^r\right]$ for $0 < x < 4$;

(ii) $\ln x = \sum_{1}^{\infty} (-1)^{r-1} (x-1)^r / r$;

(iii) $e^x = e^2 \sum_{0}^{\infty} (x-2)^n / n!$.

3. Use l'Hospital's rule to evaluate the following limits:

(i) $\lim_{x \to 0} [(x - \sin x)/\{x(1 - \cos x)\}]$;

(ii) $\lim_{x \to 0} \{(2 \sin x - \sin 2x)/\sin^3 x\}$;

(iii) $\lim_{x \to \pi/2} [(a^{\sin x} - a)/\ln \sin x]$;

(iv) $\lim_{x \to 0} [(\ln x^2)/(\ln \cot^2 x)]$;

(v) $\lim_{x \to 0} [\{(1+x)^{1/x} - e + ex/2\}/x^2]$.

4. Use l'Hospital's rule to evaluate the following:

(i) $\lim_{x \to 0} [(\ln \sin x)/\cot x]$;

(ii) $\lim_{x \to 0} (\ln \cot 2x)/(\ln \cot x)]$;

(iii) $\lim_{x \to \infty} [(x + \ln x)/(x \ln x)]$;

(iv) $\lim_{x \to 0} [(\ln x^2)/(\ln \cot^2 x)]$.

5. Evaluate the following limits:

(i) $\lim_{x \to 0} [1/x - \{\ln(1+x)\}/x^2]$;

(ii) $\lim_{x \to 0} (1/\sin^2 x - 1/x^2)$;

(iii) $\lim_{x \to 0} \left(\dfrac{\tan x \tan^{-1} x}{x^6} - \dfrac{1}{x^4}\right)$;

(iv) $\lim_{x \to 0} [(e^x + e^{-x} - 2)/x^4 - 2(1 - \cos x)/x^4]$;

(v) $\lim_{x \to \infty} \{x - x^2 \ln(1 + 1/x)\}$.

6. Evaluate the following limits:

(i) $\underset{x\to\infty}{\text{Lt}}\,(1/x)^{1/x}$; **(ii)** $\underset{x\to\pi/2}{\text{Lt}}\,(\cos x)^{2\cos^2 x}$; **(iii)** $\underset{x\to\pi/2}{\text{Lt}}\,(1-\sin x)^{1-\sin x}$.

7. Prove that:

(i) $\underset{x\to 0}{\text{Lt}}\,[(\tan x)/x]^{1/x} = 1$;

[**Hint.** If $u \to 1$, $v \to \pm\infty$ and $(u-1)v \to l$ as $x \to a$, then $u^v \to e^l$ as $x \to a$.]

(ii) $\underset{x\to 1}{\text{Lt}}\,x^{1/(x-1)} = e$;

(iii) $\underset{x\to 1}{\text{Lt}}\,x^{\tan(\pi x/2)} = e^{-2/\pi}$;

(iv) $\underset{x\to 0}{\text{Lt}}\,(1+\sin x)^{\cot x} = e$;

[**Hint.** If $u \to 0$ and $v \to \pm\infty$ and $uv \to l$ as $x \to a$, then $(1+u)^v \to e^l$ as $x \to a$.]

(v) $\underset{x\to\infty}{\text{Lt}}\left\{(1/n)\sum_{1}^{n} a_r^{1/x}\right\}^{nx} = \prod_{1}^{n} a_r$;

(vi) $\underset{x\to 0}{\text{Lt}}\left\{(1/n)\sum_{1}^{n} r^x\right\}^{n/x} = n!$.

[**Hint.** If $u = (1/n)\sum_{1}^{n} r^x$ and $v = n/x$, then $u \to 1$, $v \to \pm\infty$ as $x \to 0$.

$(u-1)v = \sum_{1}^{n}(r^x - 1)/x \to \sum_{1}^{n} \ln r = \ln \prod r = \ln n!$ as $x \to 0$.]

8. Find a and b if $\underset{x\to 0}{\text{Lt}}\,[(a\sin 2x - b\sin x)/x^3] = 1$.

9. Find a, b and c such that $\underset{x\to 0}{\text{Lt}}\,[(ae^x - b\cos x + ce^{-x})/(x\sin x)] = 2$.

Answers

3. (i) 1/3; **(ii)** 1; **(iii)** $a\ln a$; **(iv)** $-e/2$;
(v) $11e/24$. **4. (i)** 0; **(ii)** 1; **(iii)** 0;
(iv) -1. **5. (i)** 1/2; **(ii)** 1/3; **(iii)** 2/9;
(iv) 1/6; **(v)** 1/2. **6. (i)** 1; **(ii)** 1;
(iii) 1. **8.** $a = -1$, $b = -2$. **9.** $a = 1$, $b = 2$, $c = 1$.

2 Integral Calculus

Definite Integral of a Function of One Variable

2.1 Definition of Definite Integral

Let

$$f : X \to \mathbb{R}$$

be a real function and $I = [a, b] \subseteq X$, where a and $b(> a)$ be two ordinary real numbers, so that they are finite. Let $m \leq f(x) \leq M \ \forall x \in I$, so that f is bounded. Let $\{x_i\}_{i=0}^{n}$ be a strictly monotonic increasing sequence with $x_0 = a$ and $x_n = b$. Then we get a set $\{I_i \text{ for } i = 1, 2, \ldots, n\}$ of subintervals $I_i = [x_{i-1}, x_i)$ or $(x_{i-1}, x_i]$ which are pairwise disjoint and $\bigcup_{i=1}^{n} I_i = I$. Such a set of subintervals is called a **partition** of the interval I.

Let $\xi_i \in I_i$ and $\delta x_i = |I_i| = x_i - x_{i-1}$ for $i = 1, 2, 3, \ldots, n$ and $\Delta = \sup\{\delta x_i\}$, called the **norm** of the partition of I. Now, we form the sum

$$S_n = \sum_{1}^{n} f(\xi_i)\delta x_i.$$

Let $\underset{\Delta \to 0}{\text{Lt}}\, S_n$ exist and be independent of the nature of partition of I and of the choice of the point ξ_i in I_i. Then this limit is defined to be the **definite integral** of f w.r.t. x and it is denoted by the symbol

$$\int_a^b f dx \text{ or } \int_a^b f(x)dx.$$

In the above $n \to \infty$ when $\Delta \to 0$ and hence $\displaystyle\int_a^b f(x)dx = \underset{n \to \infty}{\text{Lt}}\, S_n$.

If the integral exists, it is independent of the nature of partition of I. We may then take $\delta x_i = h$, a constant and $\xi_i = x_{i-1}$ or x_i for $i = 1, 2, \ldots$.

Then $\delta x_i = x_i - x_{i-1} = h \Rightarrow \displaystyle\sum_{i=1}^{r}(x_i - x_{i-1}) = \sum_{i=1}^{r} h$

$\Rightarrow \qquad x_r - x_0 = rh \Rightarrow x_r = a + rh, \quad \because x_0 = a.$

So, $\qquad S_n = h \displaystyle\sum_{r=1}^{n} f(a + rh).$

Since $\qquad \Delta = h$ and $b = x_n = a + nh$, $h = \dfrac{b-a}{n}$.

So, $\qquad \displaystyle\int_a^b f(x)dx = \underset{h \to 0}{\text{Lt}}\, S_n = \underset{n \to \infty}{\text{Lt}}\, \frac{b-a}{n} \sum_{1}^{n} f(a + rh).$

The integral defined above is called a **proper integral**. In this case both the interval I and range of f are bounded, i.e., $\forall\, x,\ a \le x \le b$ and $m \le f(x) \le M$, where a, b, m and M are finite real numbers.

2.2 Fundamental Theorem of Integral Calculus—the Newton-Leibnitz Formula

If $f : I \to \mathbb{R}$ is integrable and possesses a primitive F such that DF or $F'(x) = f(x)\,\forall\, x \in I$ then

$$\int_a^b f(x)dx = F(x)|_a^b = F(b) - F(a).$$

2.3 Change of Variable in a Proper Integral

Let **(i)** $f : I \to \mathbb{R}$ be integrable w.r.t. x in $I = [a, b]$,

(ii) $x = g(t)$ define x as a function of t, differentiable in $[\alpha, \beta]$, where $a = g(\alpha)$ and $b = g(\beta)$ and $g'(t) \neq 0$ anywhere in (α, β),

and **(iii)** $f \circ g(t)$ and $g'(t)$ are both integrable w.r.t. t in $[\alpha, \beta]$, then

$$\int_a^b f(x)dx = \int_\alpha^\beta f \circ g(t)g'(t)dt \text{ or } \int_\alpha^\beta f(g(t))g'(t)dt.$$

2.4 Integration by Parts in a Definite Integral

If u, v and u' are all integrable w.r.t. x in $[a, b]$, then

$$\int_a^b uvdx = [uv_1]_a^b - \int_a^b u^{(1)}v_1dx,$$

where $u^{(1)} = Du$ or du/dx and $v_1 = \int vdx$.

2.4.1 Repeated Integration by Parts

Let $u^{(r)} = D^r u = \dfrac{d^r u}{dx^r}$ and $v_r = D^{-r}v = D^{-1}v_{r-1} = \int v_{r-1}dx$ for $r = 1, 2, \ldots$, **with** $u^{(0)} = u$ and $v_0 = v$ and $u^{(r)}$ and v_r are integrable in $[a, b]$, then

$$\int_a^b uvdx = \left[\sum_{r=0}^{n-1}(-1)^r u^{(r)} v_{r+1}\right]_a^b + (-1)^n \int_a^b u^{(n)} v_n dx.$$

In the expanded form

$$\int_a^b uvdx = [uv_1 - u^{(1)}v_2 + u^{(2)}v_3 + \ldots]_a^b + (-1)^n \int_a^b u^{(n)}v_n dx.$$

Note that the first term in the sum:

$$uv_1 - u^{(1)}v_2 + v^{(2)}v_3 + \ldots$$

is uv_1 and the successive terms are obtained from it by taking the product of the derivative of the first factor and integral of the second factor of the previous term and changing the sign of the previous term.

Since the first term is uv_1, the second term is $-\frac{du}{dx} \times \int v_1 dx$, i.e., $-u^{(1)}v_2$, the third term is $+\frac{du^{(1)}}{dx} \int v_2 dx$ or $+u^{(2)}v_3$ and so on.

It is a very useful formula and the series can be extended to an infinite number of terms if it is convergent.

Example.

$$\int_0^1 x^3 e^{2x} dx = \left[x^3 \cdot \frac{e^{2x}}{2} - (3x^2)\frac{e^{2x}}{2^2} + (6x) \cdot \frac{e^{2x}}{2^3} - 6 \cdot \frac{e^{2x}}{2^4}\right]_0^1 = e^2\left(\frac{1}{2} - \frac{3}{4} + \frac{6}{8} - \frac{6}{16}\right) + \frac{6}{16}.$$

2.5 Improper Integral

The integral $\int_a^b f(x)dx$ is **improper** if either a or b or both are infinitely large or $f(x)$ becomes infinitely large at some points in $[a, b]$.

When $f(x)$ is finite,

(i) $\int_a^{\infty} f(x)dx = \underset{b\to\infty}{\text{Lt}} \int_a^b f(x)dx$, when a is finite,

(ii) $\int_{-\infty}^{b} f(x)dx = \underset{a\to-\infty}{\text{Lt}} \int_a^b f(x)dx$, when b is finite

and **(iii)** $\int_{-\infty}^{\infty} f(x)dx = \underset{a\to-\infty}{\text{Lt}} \int_a^c f(x)dx + \underset{b\to\infty}{\text{Lt}} \int_c^b f(x)dx \; \forall\, c \in (a, b)$, c being finite.

When a and b are finite and $a < b$,

(iv) $\int_a^b f(x)dx = \underset{\alpha\to a+}{\text{Lt}} \int_\alpha^b f(x)dx$, where $a < \alpha < b$ and a is the only point of infinite discontinuity of f.

(v) $\int_a^b f(x)dx = \underset{\beta\to b-}{\text{Lt}} \int_a^\beta f(x)dx$, where $a < \beta < b$ and b is the only point of infinite discontinuity of f.

(vi) $\int_a^b f(x)dx = \underset{\gamma\to c-}{\text{Lt}} \int_a^\gamma f(x)dx + \underset{\gamma\to c+}{\text{Lt}} \int_\gamma^b f(x)dx$ if c is the only point of infinite discontinuity in (a, b).

2.6 Reduction Formulae

Let $I(n) = \int f(x, n)dx$, where $n \in \mathbb{Z}$. Then a relation of the form:

$I(n) = \phi(x) + \psi(x)I(n-1)$

or, $I(n) = \phi(x) + \psi(x)I(n-2)$, when $n > 0$

or, $I(n) = \phi(x) + \psi(x)I(n+1)$

or, $I(n) = \phi(x) + \psi(x)I(n+2)$, when $n < 0$

is called a reduction formula.

If $f(x)$ depends upon two integer variables, say m and n, and

$$I(m,n) = \int f(x,m,n)dx$$

then a relation of the form

$$I(m,n) = \phi(x,m,n) + \psi(x,m,n)I(p,q),$$

where either p or q or both numerically less than m and n respectively, is also called a reduction formula.

With the help of these formulae the value of n can be successively reduced numerically and ultimately $I(n)$ can be made to depend on $I(1)$ or $I(0)$ which may be integrated directly and the value of $I(n)$ can be found out for some definite value of n.

Example. 2.1. *Obtain a reduction formula for* $I(n)$, *where*

$$I(n) = \int x^n e^{-x} dx, \text{ where } n \in \mathbb{N}.$$

Integrating by parts,

$$I(n) = x^n(-e^{-x}) - \int nx^{n-1}(-e^{-x})dx = -x^n e^{-x} + nI(n-1). \tag{1}$$

If n is a negative integer

$$I(n) = \int e^{-x}x^n dx = e^{-x}\cdot\frac{x^{n+1}}{n+1} - \int -e^{-x}\cdot\frac{x^{n+1}}{n+1}dx, n \neq -1$$

$$= \frac{x^{n+1}}{n+1}e^{-x} + \frac{1}{n+1}I(n+1), \tag{2}$$

which is the required reduction formula.

To evaluate $I(-3)$ we find from (2)

$$I(-3) = \frac{x^{-2}}{-2}e^{-x} - \frac{1}{2}I(-2),$$

$$I(-2) = \frac{-x^{-1}}{-1}e^{-x} - I(-1),$$

and $\quad I(-1) = \int e^{-x}x^{-1}dx = \int \frac{e^{-x}}{x}dx$, which cannot be evaluated in the closed form.

But $\quad I(3) = -x^3e^{-x} + 3I(2)$, using (1),

$$I(2) = -x^2e^{-x} + 2I(1),$$

$$I(1) = -xe^{-x} + I(0)$$

and $\quad I(0) = \int x^0 e^{-x}dx = \int e^{-x}dx = -e^{-x}.$

So, $I(3)$ can ultimately be evaluated in the closed form.

2.6.1 Reduction Formulae for Some Particular Integrals

I. If $C(n) = \int \cos^n dx \; \forall n \in \mathbb{N}$, then $C(n) = \frac{n-1}{n}C(n-2) + \frac{1}{n}\sin x\cos^{n-1}x$.

$$C(n) = \int \cos^{n-1}x\cos x dx$$

$$= \cos^{n-1}x\sin x - \int (n-1)\cos^{n-2}x(-\sin x)\cdot\sin x dx$$

$$= \cos^{n-1} x \sin x + (n-1) \int \cos^{n-2} x(1 - \cos^2 x)dx$$

$$= \cos^{n-1} x \sin x + (n-1)\{C(n-2) - C(n)\}$$

$\therefore$ $\quad nC(n) = (n-1)C(n-2) + \sin x \cos^{n-1} x$

or, $\quad C(n) = \dfrac{n-1}{n}C(n-2) + \dfrac{1}{n} \sin x \cos^{n-1} x.$

II. If $S(n) = \int \sin^n x dx \ \forall n \in \mathbb{N}$, then $S(n) = \dfrac{n-1}{n}S(n-2) - \dfrac{1}{n}\sin^{n-1} x \cos x$.

III. If $I(n) = \int_0^{\pi/2} \cos^n x dx$, then $I(n) = \int_0^{\pi/2} \sin^n x dx$ and

$$I(n) = \frac{n-1}{n}I(n-2) \ \forall n(\geq 2) \in \mathbb{N},$$

$$= \prod_{r=1}^{(n-1)/2} \left(\frac{2r}{2r+1}\right) \text{ if } n \text{ is odd,}$$

$$= \frac{\pi}{2} \prod_{r=1}^{n/2} \frac{2r-1}{2r} \text{ if } n \text{ is even.}$$

By a property of the definite integral

$$\int_0^{\pi/2} \cos^n x dx = \int_0^{\pi/2} \cos^n(\pi/2 - x)dx = \int_0^{\pi/2} \sin^n x dx.$$

By (I), $\quad I(n) = \int_0^{\pi/2} \cos^n x dx = [(n-1)/n]I(n-2) + [(1/n) \sin x \cos^{n-1} x]_0^{\pi/2}.$

or, $$I(n) = [(n-1)/n]I(n-2), \text{ if } n \geq 2. \tag{2.1}$$

Replacing n by $2r + 1$

$$\frac{I(2r+1)}{I(2r-1)} = \frac{2r}{2r+1} \text{ for } r(\geq 1) \in \mathbb{N},$$

since $n \geq 2 \Rightarrow 2r + 1 \geq 2 \Rightarrow r \geq \dfrac{1}{2} \Rightarrow r(\geq 1) \in \mathbb{N}$.

So, $$\prod_1^n [I(2r+1)/I(2r-1)] = \prod_1^n \{2r/(2r+1)\}.$$

or, $$I(2n+1)/I(1) = \prod_1^n \{2r/(2r+1)\}.$$

But $$I(1) = \int_0^{\pi/2} \cos x dx = \sin x|_0^{\pi/2} = 1.$$

$\therefore$ $$I(2n+1) = \prod_1^n \{2r/(2r+1)\}.$$

Replacing n by $(n-1)/2$

$$I(n) = \prod_1^{(n-1)/2} \{2r/(2r+1)\} \text{ when } n \text{ is odd.}$$

Replacing n by $2r$ in (2.1)

$$I(2r) = \{(2r-1)/2r\}I(2r-2)\ \forall r(\geq 1) \in \mathbb{N}.$$

$$\therefore \quad \prod_{1}^{n}\{I(2r)/I(2r-2)\} = \prod_{1}^{n}\{(2r-1)/2r\}$$

or,
$$I(2n)/I(0) = \prod_{1}^{n}\{(2r-1)/2r\}$$

But
$$I(0) = \int_0^{\pi/2} dx = \frac{\pi}{2}.$$

$$\therefore \quad I(2n) = (\pi/2)\prod_{1}^{n}\{(2r-1)/2r\}.$$

Replacing n by $n/2$

$$I(n) = (\pi/2)\prod_{1}^{n/2}\{(2r-1)/2r\} \text{ when } n \text{ is even.}$$

Note. $I_n = \frac{p}{q}$ multiplied by 1 or $\frac{\pi}{2}$ according as n is odd or even, where the denominator q is the product of all the odd positive integers or all the even positive integers $\leq n$ according as n itself is odd or even and p is the product of all other positive integers $< n$.

IV. If $I(n) = \int \tan^n x dx$, $n \in \mathbb{N}$, then

$$\boldsymbol{I(n) = -I(n-2) + \{1/(n-1)\}\tan^{n-1} x, \ n(\geq 2) \in \mathbb{N}.}$$

$$I(n) = \int \tan^{n-2} x \tan^2 x dx = \int \tan^{n-2} x(\sec^2 x - 1)dx.$$

$$= \int \tan^{n-2} x \sec^2 x dx - I(n-2) = -I(n-2) + \{1/(n-1)\}\tan^{n-1} x. \tag{2.2}$$

Similarly, if $I(n) = \int \cot^n x dx$, it can be shown that $I(n) = -I(n-2) - \{1/(n-1)\}\cot^{n-1} x$.

$$\boldsymbol{J_n = \int_0^{\pi/4} \tan^n x dx = \int_{\pi/4}^{\pi/2} \cot^n x dx = -J_{n-2} + 1/(n-1).}$$

Putting $x = \pi/2 - \theta$, $dx = -d\theta$ and

x	$\pi/4$	$\pi/2$
θ	$\pi/4$	0

.

$$\therefore \int_{\pi/4}^{\pi/2} \cot^n x dx = \int_{\pi/4}^{0} \cot^n(\pi/2 - \theta)(-d\theta) = \int_0^{\pi/4} \tan^n \theta d\theta$$

$$= -J_{n-2} + \{1/(n-1)\}[\tan^{n-1}\theta]_0^{\pi/4} \text{ by (2.2)}$$

$$= -J_{n-2} + 1/(n-1).$$

V. If $I(n) = \int \sec^n x dx\ \forall n \in \mathbb{N}$, then

$$\boldsymbol{I(n) = \frac{n-2}{n-1}I(n-2) + \frac{1}{n-1}\sec^{n-2} x \tan x.}$$

$$I(n) = \int \sec^{n-2} x \sec^2 x dx$$

$$= \sec^{n-2} x \tan x - \int (n-2) \sec^{n-3} x \sec x \tan x \tan x dx$$

$$= \sec^{n-2} x \tan x - (n-2) \int \sec^{n-2} x (\sec^2 x - 1) dx$$

$$= -(n-2)I(n) + (n-2)I(n-2) + \sec^{n-2} x \tan x.$$

$\therefore \quad (n-1)I(n) = (n-2)I(n-2) + \sec^{n-2} x \tan x,$

and $\quad I(n) = \{(n-2)/(n-1)\}I(n-2) + \{1/(n-1)\} \sec^{n-2} x \tan x.$

Similarly, if $I(n) = \int \operatorname{cosec}^n x dx$,

$$I(n) = \frac{n-2}{n-1} I(n-2) - \frac{1}{n-1} \operatorname{cosec}^{n-2} x \cot x.$$

VI. If $I(m,n) = \int \sin^m x \cos^n x dx$, where $m, n \in \mathbb{N}$, then

$$I(m,n) = \frac{m-1}{m+n} I(m-2,n) - \frac{1}{m+n} \sin^{m-1} x \cos^{n+1} x$$
$$= \frac{n-1}{m+n} I(m,n-2) + \frac{1}{m+n} \sin^{m+1} x \cos^{n-1} x.$$

$$I(m,n) = \int \sin^{m-1} x (\cos^n x \sin x) dx$$

$$= \sin^{m-1} x[-\{1/(n+1)\} \cos^{n+1} x] - \int (m-1) \sin^{m-2} x \cdot \cos x \{-1/(n+1)\} \cos^{n+1} x dx$$

$$= -\{1/(n+1)\} \sin^{m-1} x \cos^{n+1} x + \{(m-1)/(n+1)\} \int \sin^{m-2} x \cos^n x (1 - \sin^2 x) dx$$

$$= \{(m-1)/(n+1)\}\{I(m-2,\ n) - I(m,n)\} - \{1/(n+1)\} \sin^{m-1} x \cos^{n+1} x.$$

Transposing and simplifying

$$I(m,n) = \{(m-1)/(m+n)\}I(m-2,\ n) - \{1/(m+n)\} \sin^{m-1} x \cos^{n+1} x.$$

Taking

$$I(m,n) = \int \cos^{n-1} x (\sin^m x \cos x) dx$$

and integrating by parts and proceeding in the same way, it can be shown that

$$I(m,n) = \{(n-1)/(m+n)\}I(m,\ n-2) + \{1/(m+n)\} \sin^{m+1} x \cos^{n-1} x.$$

VII. $$I(m,n) = \int_0^{\pi/2} \sin^m x \cos^n x dx = \int_0^{\pi/2} \cos^m x \sin^n x dx$$
$$= [(m-1)/(m+n)]I(m-2,\ n) = [(n-1)/(m+n)]I(m,\ n-2)$$
$$= \frac{1}{n+1} \prod_{1}^{(m-1)/2} \frac{2r}{n+2r+1}, \text{ if } m \text{ is odd}$$

$$= \frac{1}{m+1} \prod_{1}^{(n-1)/2} \frac{2r}{m+2r+1}, \text{ if } n \text{ is odd}$$

$$= \left[\prod_{1}^{m/2} (2r-1) \prod_{1}^{n/2} (2r-1) \Big/ \prod_{1}^{(m+n)/2} (2r) \right] \frac{\pi}{2},$$

if both m and n are even.

$$I(m,n) = \int_0^{\pi/2} \sin^m x \cos^n x dx = \int_0^{\pi/2} \sin^m(\pi/2 - x) \cos^n(\pi/2 - x) dx$$

$$= \int_0^{\pi/2} \sin^n x \cos^m x dx = I(n,m).$$

Thus $I(m,n)$ is symmetric with respect to m and n. Also by VI

$$I(m,n) = \{(m-1)/(m+n)\} I(m-2,n) - \{1/(m+n)\}[\sin^{m-1} x \cos^{n+1} x]_0^{\pi/2}$$

$$= \{(m-1)/(m+n)\} I(m-2,n).$$

Similarly, $I(n,m) = \{(n-1)/(m+n)\} I(n-2,m)$

$$= \{(n-1)/(m+n)\} I(m,n-2).$$

So, $$I(m,n) = \{(m-1)/(m+n)\} I(m-2,n) = \{(n-1)/(m+n)\} I(m,n-2)). \quad (2.3)$$

Thus any one of the two indices m and n can be reduced, keeping the other unchanged.

If m is odd, we take $m = 2p+1$, $p \in \mathbb{W}$. Then

$$I(2p+1,n) = \{2p/(2p+n+1)\} I(2p-1,n).$$

$$\therefore \quad \prod_{r=1}^{p} [I(2r+1,n)/I(2r-1,n)] = \prod_{r=1}^{p} [2r/(2r+n+1)]$$

or, $$I(2p+1,n)/I(1,n) = \prod_{1}^{p} [2r/(2r+n+1)].$$

But $$I(1,n) = \int_0^{\pi/2} \sin x \cos^n x dx$$

$$= -[(\cos^{n+1} x)/(n+1)]_0^{\pi/2} = 1/(n+1).$$

Since $m = 2p+1 \Rightarrow p = (m-1)/2$,

$$\therefore \quad I(m,n) = \{1/(n+1)\} \prod_{1}^{(m-1)/2} [2r/(2r+n+1)] \text{ if } m \text{ is odd.}$$

Interchanging m and n

$$I(m,n) = I(n,m) = \{1/(m+1)\} \prod_{1}^{(n-1)/2} [2r/(2r+m+1)] \text{ if } n \text{ is odd.}$$

When m and n are both even, being $2p$ and $2q$ respectively, we have from (2.3)

$$I(2r,n) = [(2r-1)/(2r+n)] I(2r-2,n).$$

$$\therefore \qquad \prod_1^p [I(2r,n)/I(2r-2,n)] = \prod_1^p [(2r-1)/(2r+n)]$$

or, $$I(2p,n)/I(0,n) = \prod_1^p [(2r-1)/(2r+n)].$$

But by (III),

$$I(0,n) = \int_0^{\pi/2} \cos^n x dx = (\pi/2) \prod_1^{n/2} [(2r-1)/2r)] \text{ if } n \text{ is even.}$$

$$\therefore \qquad I(2p,n) = (\pi/2) \prod_1^{n/2} [(2r-1)/2r] \prod_1^p (2r-1)/(2r+n)].$$

$\because$ $p = m/2$ we have

$$I(m,n) = (\pi/2) \left[\prod_1^{n/2} (2r-1) \prod_1^{m/2} (2r-1) \right] \Big/ \left[\prod_1^{n/2} (2r) \prod_1^p (2r+n) \right]$$

$$= (\pi/2) \left[\prod_1^{m/2} (2r-1) \prod_1^{n/2} (2r-1) \right] \Big/ \prod_1^{(m+n)/2} (2r).$$

Writing the factors in the reversed order

$$I(m,n) = \frac{(m-1)(m-3)\cdots 4 \cdot 2 \cdot 1}{(m+n)(m+n-2)\cdots(n+1)} \text{ if } m \text{ is odd}$$

$$= \frac{(n-1)(n-3)\cdots 4 \cdot 2 \cdot 1}{(m+n)(m+n-2)\cdots(m+1)} \text{ if } n \text{ is odd}$$

$$= \frac{\{(m-1)(m-3)\cdots 5 \cdot 3 \cdot 1\}\{(n-1)(n-3)\cdots 5 \cdot 3 \cdot 1\}}{(m+n)(m+n-2)\cdots 6 \cdot 4 \cdot 2} \frac{\pi}{2},$$

if m and n are both even.

Combining all of them we can form a simple rule to evaluate the integral:

$I(m,n) = \int_0^{\pi/2} \sin^m x \cos^n x dx = \frac{\{(m-1)(m-3)\cdots\}\{(n-1)(n-3)\cdots\}}{(m+n)(m+n-2)\cdots}$ multiplied by 1 or $\pi/2$ according as only one or both of m and n are even, where two sets of factors in the numerator start with $m-1$ and $n-1$ and one set of factors in the denominator starts with $m+n$ and each of these sets of factors continue diminishing by 2 at a time and end with either 1 or 2 according as the first factor of the set odd or even.

For example,

$$I(2,3) = \frac{(1)(2)}{5 \cdot 3 \cdot 1}.1,$$

$$I(5,6) = \frac{(4 \cdot 2)(5 \cdot 3 \cdot 1)}{11 \cdot 9 \cdot 7 \cdot 5 \cdot 3 \cdot 1} \cdot 1,$$

$$I(4,6) = \frac{(3 \cdot 1)(5 \cdot 3 \cdot 1)}{10 \cdot 8 \cdot 6 \cdot 4 \cdot 2} \cdot \frac{\pi}{2}, \text{ etc.}$$

Note. If m and n are both even, it can be easily proved that

$$I(m,n) = \frac{m!n!}{(m/2)!(n/2)!\{(m+n)/2\}!}\frac{\pi}{2^{m+n+1}}.$$

Examples 2.2. 1. *Evaluate:*

(i) $\int \sin^5 x dx$, **(ii)** $\int \sin^6 x dx$, **(iii)** $\int \cos^4 x dx$, **(iv)** $\int \cos^5 x dx$.

(i) If $S(n) = \int \sin^n x dx$, we know that

$$S(n) = \frac{n-1}{n} S(n-2) - \frac{1}{n}\sin^{n-1} x \cos x.$$

$$\therefore \quad S(5) = \frac{5-1}{5} S(3) - \frac{1}{5}\sin^4 x \cos x$$

$$= \frac{4}{5}\left[\frac{3-1}{3} S(1) - \frac{1}{3}\sin^2 x \cos x\right] - \frac{1}{5}\sin^4 x \cos x$$

and $S(1) = \int \sin x dx = -\cos x$.

So
$$S(5) = \frac{4}{5}\left[\frac{2}{3}\cdot(-\cos x) - \frac{1}{3}\sin^2 x \cos x\right] - \frac{1}{5}\sin^4 x \cos x$$

$$= -\left[\frac{4\cdot 2}{5\cdot 3} + \frac{4\cdot 1}{5\cdot 3}\sin^2 x + \frac{1}{5}\sin^4 x\right]\cos x.$$

(ii)
$$S(6) = \frac{6-1}{6} S(4) - \frac{1}{6}\sin^5 x \cos x$$

$$= \frac{5}{6}\left[\frac{3}{4}\cdot S(2) - \frac{1}{4}\sin^3 x \cos x\right] - \frac{1}{6}\sin^5 x \cos x$$

$$= \frac{5\cdot 3}{6\cdot 4}\left[\frac{1}{2} S(0) - \frac{1}{2}\sin x \cos x\right] - \left(\frac{5\cdot 1}{6\cdot 4}\sin^3 x + \frac{1}{6}\sin^5 x\right)\cos x$$

$$= \frac{5\cdot 3\cdot 1}{6\cdot 4\cdot 2} x - \left[\frac{5\cdot 3\cdot 1}{6\cdot 4\cdot 2} + \frac{5\cdot 1}{6\cdot 4}\sin^2 x + \frac{1}{6}\sin^4 x\right]\sin x \cos x.$$

(iii) If $C(n) = \int \cos^n x dx$, then

$$C(n) = \frac{n-1}{n} C(n-2) + \frac{1}{n}\cos^{n-1} x \sin x.$$

$$\therefore \quad C(4) = \frac{3}{4} C(2) + \frac{1}{4}\cos^3 x \sin x$$

$$= \frac{3}{4}\left[\frac{1}{2} C(0) + \frac{1}{2}\cos x \sin x\right] + \frac{1}{4}\cos^3 x \sin x$$

and $C(0) = \int dx = x$.

$$\therefore \quad C(4) = \frac{3\cdot 1}{4\cdot 2} x + \left(\frac{3\cdot 1}{4\cdot 2} + \frac{1}{4}\cos^3 x\right)\sin x \cos x.$$

(iv) $C(5) = \frac{4}{5}C(3) + \frac{1}{5}\cos^4 x \sin x$

$$= \frac{4}{5}\left[\frac{2}{3}C(1) + \frac{1}{3}\cos^2 x \sin x\right] + \frac{1}{5}\cos^4 x \sin x$$

$$= \left(\frac{4\cdot 2}{5\cdot 3} + \frac{4\cdot 1}{5\cdot 3}\cos^2 x + \frac{1}{5}\cos^4 x\right)\sin x.$$

2. *Evaluate:*

(i) $\int_0^{\pi/2} \sin^7 x dx$, **(ii)** $\int_0^{\pi/2} \sin^8 x dx$, **(iii)** $\int_0^{\pi/2} \cos^5 x dx$, **(iv)** $\int_0^{\pi/2} \cos^6 x dx$.

$$I(n) = \int_0^{\pi/2} \sin^n x dx = \int_0^{\pi/2} \cos^n x dx$$

$$= \prod_{r=1}^{(n-1)/2} \left(\frac{2r}{2r+1}\right) \text{ if } n \text{ is odd}$$

$$= \frac{\pi}{2}\prod_{r=1}^{n/2} \frac{2r-1}{2r} \text{ if } n \text{ is even.}$$

So, **(i)** $I(7) = \prod_{r=1}^{(7-1)/2} \frac{2r}{2r+1} = \prod_{r=1}^{3} \frac{2r}{2r+1} = \frac{2}{3}\cdot\frac{4}{5}\cdot\frac{6}{7},$

(ii) $I(8) = \prod_{r=1}^{4} \frac{2r-1}{2r} = \frac{1\cdot 3\cdot 5\cdot 7}{2\cdot 4\cdot 6\cdot 8}\frac{\pi}{2},$

(iii) $I(5) = \prod_{r=1}^{2} \frac{2r}{2r+1} = \frac{2\cdot 4}{3\cdot 5},$

(iv) $I(6) = \prod_{r=1}^{3} \frac{2r-1}{2r} = \frac{1\cdot 3\cdot 5}{2\cdot 4\cdot 6}\frac{\pi}{2}.$

3. *Integrate the following functions with respect to x, using the reduction formula VII:*

(i) $\sin^2 x \cos^3 x$, **(ii)** $\sin^3 x \cos^2 x$, **(iii)** $\sin^3 x \cos^5 x$, **(iv)** $\sin^2 x \cos^4 x$.

We know that

$$I(m,n) = \int \sin^m x \cos^n x dx = \cos^n x dx = \frac{m-1}{m+n}I(m-2,n) - \frac{1}{m+n}\sin^{m-1} x \cos^{n+1} x$$

or, $$= \frac{n-1}{m+n}I(m,n-2) + \frac{1}{m+n}\sin^{m+1} x \cos^{n-1} x.$$

$\therefore$ **(i)** $I(2,3) = \frac{1}{5}I(0,3) - \frac{1}{5}\sin x \cos^4 x,$

$$I(0,3) = \frac{2}{3}I(0,1) + \frac{1}{3}\cos^2 x \sin x$$

$$= \frac{2}{3}\sin x + \frac{1}{3}\cos^2 x \sin x.$$

$\therefore \quad I(2,3) = \frac{1}{5}\left[\frac{2}{3} + \frac{1}{3}\cos^2 x\right]\sin x - \frac{1}{5}\cos^4 x \sin x$

$$= \left[\frac{1 \cdot 2}{5 \cdot 3} + \frac{1}{5 \cdot 3}\cos^2 x - \frac{1}{5}\cos^4 x\right]\sin x.$$

(ii) $I(3,2) = \frac{2}{5}I(1,2) - \frac{1}{5}\sin^2 x \cos^3 x$

$I(1,2) = \frac{1}{3}I(1,0) + \frac{1}{3}\sin^2 x \cos x.$

$\therefore \quad I(3,2) = \frac{2}{5}\left[\frac{1}{3}(-\cos x) + \frac{1}{3}\sin^2 x \cos x\right] - \frac{1}{5}\sin^2 x \cos^3 x$

$$= -\frac{2}{15}\cos x + \frac{2}{15}\sin^2 x \cos x - \frac{1}{5}\sin^2 x \cos^3 x$$

$$= -\frac{2}{15}\cos^3 x - \frac{1}{5}\sin^2 x \cos^3 x$$

$$= -\frac{1}{15}(2 + 3\sin^2 x)\cos^3 x.$$

(iii) $I(3,5) = \frac{2}{8}I(1,5) - \frac{1}{8}\sin^2 x \cos^6 x.$

Putting $\cos x = t, -\sin x dx = dt.$

$\therefore \quad I(1,5) = \int \sin x \cos^5 x dx = -\int t^5 dt = -\frac{t^6}{6} = -\frac{\cos^6 x}{6}.$

$\therefore \quad I(3,5) = -\frac{2}{8}\cdot\frac{1}{6}\cos^6 x - \frac{1}{8}\sin^2 x \cos^6 x$

$$= -\frac{1}{24}(1 + 3\sin^2 x)\cos^6 x.$$

Alternatively, taking $\cos x = t, \ -\sin x dx = dt.$

$\therefore \quad I(3,5) = \int \sin^3 x \cos^5 x dx = -\int (1-t^2)t^5 dt = -\int (t^5 - t^7)dt$

$$= -\frac{t^6}{6} + \frac{t^8}{8} = -\frac{\cos^6 x}{6} + \frac{\cos^8 x}{8} = -\left(\frac{1}{6} - \frac{1}{8}\cos^2 x\right)\cos^6 x.$$

(iv) $I(2,4) = \frac{1}{6}I(0,4) - \frac{1}{6}\sin x \cos^5 x$

$$= \frac{1}{6}\left[\frac{3}{4}I(0,2) + \frac{1}{4}\sin x \cos^3 x\right] - \frac{1}{6}\sin x \cos^5 x$$

$$= \frac{1}{8}I(0,2) + \left(\frac{1}{24} - \frac{1}{6}\cos^2 x\right)\sin x \cos^3 x$$

$$= \frac{1}{8}\left[\frac{1}{2}I(0,0) + \frac{1}{2}\sin x \cos x\right] + \left(\frac{1}{24} - \frac{1}{6}\cos^2 x\right)\sin x \cos^3 x$$

$$= \frac{1}{16}x + \left(\frac{1}{16} + \frac{1}{24}\cos^2 x - \frac{1}{6}\cos^4 x\right)\sin x \cos x.$$

4. *Use the reduction formula III to evaluate* $\int_0^1 x^n(1-x^2)^{-1/2}dx$.

Putting $x = \sin\theta$, $dx = \cos\theta d\theta$ and knowing that $\theta = 0$ when $x = 0$ and $\theta = \pi/2$ when $x = 1$, the given integral becomes

$$I(n) = \int_0^{\pi/2} \sin^n\theta d\theta = \prod_{r=1}^{(n-1)/2} \frac{2r}{2r+1} \text{ or } \frac{\pi}{2}\prod_{r=1}^{n/2}\frac{2r-1}{2r},$$

according as n is odd or even.

5. *Evaluate* $\int_0^\infty dx/(1+x^2)^n$.

Let $x = \tan\theta$, $-\pi/2 < \theta < \pi/2$. Then $dx = \sec^2\theta d\theta$ and

x	0	∞
θ	0	$\pi/2$

So the given integral is

$$I(n) = \int_0^{\pi/2}\frac{\sec^2\theta d\theta}{\sec^{2n}\theta} = \int_0^{\pi/2}\cos^{2n-2}\theta$$

$$= \frac{\pi}{2}\prod_{r=1}^{n-1}\frac{2r-1}{2r} \text{ if } n > 1$$

$$= \frac{\pi}{2} \text{ if } n = 1.$$

6. *Evaluate* $\int_0^a x^4\sqrt{(a^2-x^2)}dx,\ a > 0$.

Putting $x = a\sin\theta$, the given integral becomes

$$\int_0^{\pi/2} a^4\sin^4\theta\cdot a^2\cos^2\theta\cdot d\theta = a^6 I(4,2)$$

$$= a^6\cdot\frac{3\cdot 1\cdot 1}{6\cdot 4\cdot 2}\frac{\pi}{2} = \frac{a^6}{32}\pi, \text{ applying the formula.}$$

7. *Use the formula for* $\int_0^{\pi/2}\sin^m\theta\cos^n\theta d\theta$ *to evaluate the following:*

(i) $\int_0^\infty x^n(1+x^2)^{-m}dx$, **(ii)** $\int_0^{2a} x^m(2ax-x^2)^{1/2}dx$.

The formula for $I(m,n) = \int_0^{\pi/2}\sin^m x\cos^n x = dx$ is $I(m,n) = \frac{[(m-1)(m-3)\cdots][(n-1)(n-3)\cdots]}{(m+n)(m+n-2)\cdots}$ multiplied by 1 or $\pi/2$ according as only one of m and n or both are even.

(i) Putting $x = \tan\theta$

$$\int_0^\infty x^n(1+x^2)^{-m}dx = \int_0^{\pi/2}\tan^n\theta\sec^{-2m}\theta\sec^2\theta d\theta$$

$$= \int_0^{\pi/2}\sin^n\theta\cos^{2m-n-2}\theta d\theta$$

$$= I(n, 2m-n-2)$$

$$= \frac{[(n-1)(n-3)\cdots][(2m-n-3)(2m-n-5)\cdots]}{(2m-2)(2m-4)\cdots}\cdots \quad (1)$$

multiplied by 1 or $\pi/2$ according as only one of n and $2m-n-2$ is even or both are even. But $2m-n-2$ is odd or even according as n is odd or even. So the value of the integral is the R.H.S. of (1) according as n is odd or even.

(ii) Since the given integral can be written as

$$I = \int_0^{2a} x^m [a^2 - (a-x)^2]^1 \, dx$$

we put $a - x = a\cos\theta$ or, $x = a(1-\cos\theta)$.

Then $dx = a\sin\theta d\theta$. The scheme of change of limits is $\begin{array}{c|c|c} x & 0 & 2a \\ \hline \theta & 0 & \pi \end{array}$.

$$\therefore \quad I = \int_0^{\pi} \{a(1-\cos\theta)\}^m \cdot a\sin\theta \cdot a\sin\theta d\theta$$

$$= a^{m+2}2^m \int_0^{\pi} \sin^{2m}\theta/2 \cdot 4 \cdot \sin^2\theta/2 \cos^2\theta/2 d\theta$$

$$= (2a)^{m+2} \int_0^{\pi} \sin^{2m+2}\theta/2 \cos^2\theta/2 d\theta$$

Putting $x = \theta/2$, $d\theta = 2dx$.

$$\therefore \quad I = (2a)^{m+2} \cdot 2 \int_0^{\pi/2} \sin^{2m+2} x \cos^2 x dx$$

$$= 2\cdot(2a)^{(m+2)} I(2m+2, 2) = 2^{m+3} \cdot a^{m+2} \frac{(2m+2)!2!}{(m+1)!1!(m+2)!} \frac{\pi}{2^{m+5}}$$

$$= \frac{\pi}{2} \cdot \frac{a^{m+2}}{2^m} \cdot \frac{(2m+2)!}{(m+1)!(m+2)!}.$$

Exercises 2.1

1. If $I(n) = \int x^n e^{ax} dx$, show that $I(n) = -(n/a)I(n-1) + x^n e^{ax}/a$.

Derive $I(m,n) = -[n/(m+1)]I(m,n-1) + x^{m+1}/(m+n)$, where

$$I(m,n) = \int x^m (\ln x)^n dx.$$

[**Hint.** Put $z = \ln x$.]

2. If $I(n) = \int_0^a x^n e^{-x} dx$, show that

$$I(n) = (n+a)I(n-1) - a(n-1)I(n-2).$$

3. If $I(n) = \int_0^{\pi/2} \sin^{2n+1}\theta d\theta$, where n is a positive integer, show that

$$I(n) = 2nI(n-1)/(2n+1).$$

Use this to evaluate $\int_0^{\pi/2} \sin^7\theta d\theta$. **[W.B. Sem. Exam. 2001]**

4. Show that $\int_0^{\pi/2} \cos^9 \theta = \int_0^1 x^9(1-x^2)^{-1/2}dx = 128/315.$

5. Evaluate $\int_{-\pi/2}^{\pi/2} (1-\sin\theta)^3 d\theta.$

6. Show that $\int_0^1 x^6 \sqrt{(1-x^2)}dx = 5\pi/384.$

7. Evaluate $\int_0^{2\pi} \sin^3(\theta/4)\cos^4(\theta/4)d\theta.$

8. Evaluate $\int_0^4 x^3\sqrt{(4x-x^2)}dx.$

9. Evaluate $\int_0^2 x^5 dx/(8-x^2)^3.$

10. If $I(n) = \int \operatorname{cosec}^n x dx$, show that

$$I(n) = [(n-2)/(n-1)]I(n-2) - (\operatorname{cosec}^{n-2} x \cot x)/(n-1).$$

Evaluate $\int \operatorname{cosec}^4 x dx.$

11. If $I(n) = \int e^{ax} \sin^n x dx$, show that

$$I(n) = [n(n-1)/(n^2+a^2)]I(n-2) + e^{ax}\sin^{n-1} x(a\sin x - n\cos x)/(n^2+a^2).$$

Evaluate $\int_0^{\pi/2} e^x \sin^3 x dx.$

12. If $I(n) = \int_0^{\pi/2} x^n \sin x dx$, show that

$$I(n) + n(n-1)I(n-2) = n(\pi/2)^{n-1}.$$

13. If $I(n) = \int_0^{\pi/2} x\sin^n x dx$, where $n(\geq 2) \in \mathbb{N}$, show that

$$I(n) = [(n-1)/n]I(n-2) + 1/n^2.$$

[**Hint.** $I(n) = \int_0^{\pi/2} uvdx$, where $u = x\sin^{n-1} x$ and $v = \sin x$.]

Answers

3. $16/35$. **5.** $5\pi/2$. **7.** $8/35$. **8.** 28π.

9. $(\ln 2 - 1/2)/2$. **10.** $-\cot x(2+\operatorname{cosec}^2 x)/3$. **11.** $(4e^{\pi/2}+3)/10$.

2.7 Geometrical Application of Definite Integral

2.7.1 Arc Length of a Plane Curve—Rectification

Let Γ be a curve with its end points A and B lying in the xy-plane, having $y = f(x)$ as its equation, where $a \leq x \leq b$. The abscissae of A and B are respectively a and b. Assuming the derivative $f'(x)$ or y_1 continuous in $[a, b]$, the length of the curve Γ is given by

$$l = \int_a^b \sqrt{1 + y_1^2} dx.$$

If the equation of Γ is of the form $x = g(y)$ for $c \leq y \leq d$ and x_1 or $g'(y)$ is continuous, then

$$l = \int_c^d \sqrt{1 + x_1^2} dy.$$

Examples 2.3. **1.** *Determine the length of the circumference of a circle of radius r.*

The equation to such a circle may be taken as

$$x^2 + y^2 = r^2.$$

The entire circumference consists of two semi-circular arcs whose equations are

$$f(x) = y = +\sqrt{r^2 - x^2} \quad -r \leq x \leq r$$

and $$g(x) = y = -\sqrt{r^2 - x^2} \quad -r \leq x \leq r.$$

But the lengths of these arcs are equal. So the required arc length, i.e., the length of the circumference is given by

$$c = 2\int_{-r}^{r} \sqrt{1 - y_1^2} dx = 2\int_{-r}^{r} \sqrt{1 + \frac{x^2}{r^2 - x^2}} dx$$

$$= 2 \cdot 2r \int_0^r dx/\sqrt{r^2 - x^2} = 4r \sin^{-1} \frac{x}{r}\Big|_0^r = 2\pi r.$$

2. *Find the length of the arc of the parabola $y^2 = 4ax$ measured from its vertex to an end of its latus rectum.*

Taking the equation to the parabola as $x = \frac{1}{4a}y^2$, and knowing that the ordinates of the vertex and the end L of the latus rectum are 0 and $2a$, the required arc length is

$$l = \int_0^{2a} \sqrt{1 + x_1^2} dy = \int_0^{2a} \sqrt{1 + (y/2a)^2} dy.$$

If $\quad y = 2at,$

$$l = 2a \int_0^1 \sqrt{1 + t^2} dt = 2a \left[\frac{t\sqrt{1 + t^2}}{2} + \frac{1}{2} \ln(t + \sqrt{1 + t^2}) \right]_0^1$$

$$= a[\sqrt{2} + \ln(\sqrt{2} + 1)].$$

3. *Compute the length of the astroid given by*

$$x^{2/3} + y^{2/3} = a^{2/3}.$$ **[WBUT Sem. 2003]**

Since the equation to the curve is $(x^{1/3})^2 + (y^{1/3})^2 = (a^{1/3})^2$, which contains even powers of both $x^{1/3}$ and $y^{1/3}$, the curve is symmetrical w.r.t. both the coordinate axes. Hence the required length is $4\,\widehat{BA}$

or, $\quad 4\int_0^a \sqrt{1+y_1^2}\,dx.$

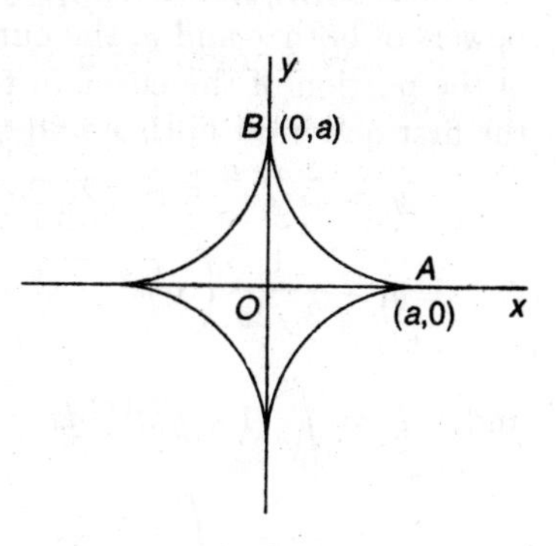

Fig. 2.1

Taking $x = a\cos^3\theta$ and $y = a\sin^3\theta$

$$y_1 = \frac{dy}{dx} = \frac{\dot{y}}{\dot{x}} = \frac{.3a\sin^2\theta\cos\theta}{-3a\cos^2\theta\sin\theta} = -\tan\theta$$

and $\quad dx = -3a\cos^2\theta\sin\theta d\theta.$

Also when $x = 0$, $\theta = \pi/2$ and when $x = a$, $\theta = 0$. So the required length

$$= 4\int_{\pi/2}^{0} \sqrt{1+\tan^2\theta}\cdot(-3a\cos^2\theta\sin\theta)d\theta$$

$$= 12a\int_0^{\pi/2} \sin\theta\cos\theta d\theta$$

$$= 12a\left[\frac{\sin^2\theta}{2}\right]_0^{\pi/2} = 6a.$$

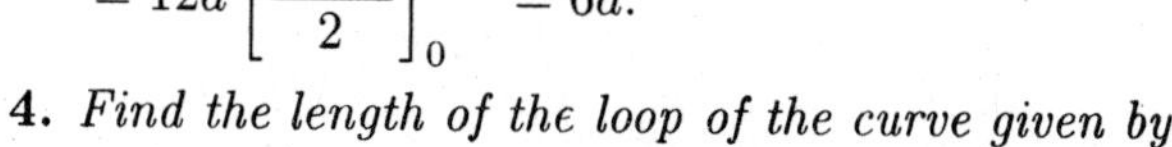

4. *Find the length of the loop of the curve given by*

$$3ay^2 = x(x-a)^2,\ a>0.$$

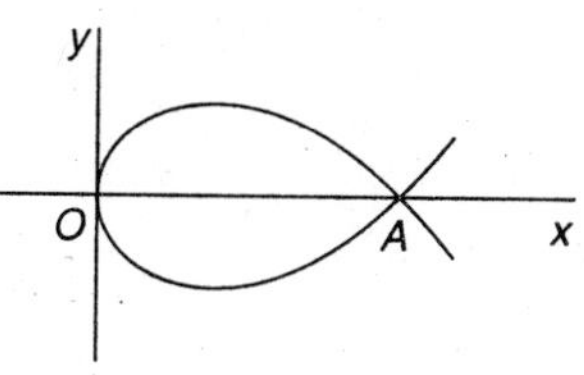

Fig. 2.2

Since y occurs in even powers, the curve is symmetrical in the x-axis. Since $y = 0$ when x is both 0 and a, the curve meets the x-axis in two points $O(0,0)$ and $A(a,0)$. So there is a loop which lies between $x = 0$ and $x = a$.

For the portion of the loop above the x-axis, $y \geq 0$ and hence

$$y = \frac{1}{\sqrt{3a}}|x-a|\sqrt{x},\ x \geq 0$$

is the equation of the portion of the curve lying above the x-axis.

Since for the loop, $0 \leq x \leq a$, $x - a \leq 0$ and hence $|x-a| = a - x$, so,

$$y = \frac{(a-x)\sqrt{x}}{\sqrt{3a}} \text{ and } y_1 = \frac{1}{2\sqrt{3a}}(ax^{-1/2} - 3x^{1/2}).$$

The length of the loop

$$= 2\int_0^a \sqrt{1+y_1^2}\,dx = 2\int_0^a \sqrt{1+\frac{(a-3x)^2}{12ax}}\,dx$$

$$= \frac{1}{\sqrt{3a}}\int_0^a (ax^{-1/2} + 3x^{1/2})dx$$

$$= \frac{1}{\sqrt{3a}}\left[2ax^{1/2} + 2x^{3/2}\right]_0^a = 4a/\sqrt{3}.$$

5. *Show that the whole length of the curve*

$$x^2(a^2 - x^2) = 8a^2y^2$$

is $\pi a\sqrt{2}$.

Since the equation is a polynomial one in both the variables x and y containing only even powers of both x and y, the curve is symmetrical in both the coordinate axes. If l is the length of the portion of the curve in the first quadrant, the whole length of the curve is $4l$. Since, in the first quadrant both x and y are positive,

$$y = (x\sqrt{a^2 - x^2})/2\sqrt{2}a, \ 0 \le x \le a.$$

$$\therefore \quad y_1 = \frac{1}{2\sqrt{2}a}\left(\sqrt{a^2 - x^2} - \frac{x^2}{\sqrt{a^2 - x^2}}\right) = \frac{a^2 - 2x^2}{2\sqrt{2}a\sqrt{a^2 - x^2}}$$

and
$$l = \int_0^a (1 + y_1^2)^{1/2} dx = \int_0^a \left\{1 + \frac{(a^2 - 2x^2)^2}{8a^2(a^2 - x^2)}\right\}^{1/2} dx = \frac{1}{2\sqrt{2}a}\int_0^a \frac{3a^2 - 2x^2}{\sqrt{a^2 - x^2}} dx$$

$$= \frac{1}{2\sqrt{2}a}\int_0^a (a^2/\sqrt{a^2 - x^2} + 2\sqrt{a^2 - x^2})dx$$

$$= \frac{1}{2\sqrt{2}a}\left[a^2 \sin^{-1}\frac{x}{a} + \left\{x\sqrt{a^2 - x^2} + a^2 \sin^{-1}\frac{x}{a}\right\}\right]_0^a$$

$$= \frac{1}{2\sqrt{2}a}\left[x\sqrt{a^2 - x^2} + 2a^2 \sin^{-1}(x/a)\right]_0^a$$

$$= \frac{a\pi}{2\sqrt{2}}.$$

So the whole length $= 4l = \pi a\sqrt{2}$.

6. *Find the length of the arc cut off by $y = mx$ from the curve $y^3 = ax^2$.*

The straight line $y = mx$ cuts the curve: $y^3 = ax^2$ in points whose abscissae are given by

$$m^3x^3 = ax^2 \quad \text{or,} \ x = 0 \quad \text{or,} \ x = a/m^3.$$

$$\because \quad y = a^{1/3}x^{2/3}, \qquad y_1 = \frac{2}{3}\frac{a^{1/3}}{x^{1/3}}.$$

So, the required length of the curve is

$$l = \int_0^{a/m^3} \sqrt{1 + \frac{4a^{2/3}}{9x^{2/3}}} dx = \frac{1}{3}\int_0^{a/m^3} \frac{(9x^{2/3} + 4a^{2/3})^{1/2}}{x^{1/3}} dx.$$

Let $t = (9x^{2/3} + 4a^{2/3})^{1/2}$. Then $t^2 = 9x^{2/3} + 4a^{2/3}$ and

$$2tdt = 6x^{-1/3}dx \text{ or, } dx/x^{1/3} = tdt/3.$$

When $x = 0$, $t = 2a^{1/3}$ and when $x = a/m^3$,

$$t = (9a^{2/3}/m^2 + 4a^{2/3})^{1/2} = a^{1/3}(9 + 4m^2)^{1/2}/m = \tau, \text{ say.}$$

$$\therefore \quad l = \frac{1}{9}\int_{2a^{1/3}}^{\tau} t^2 dt = \frac{1}{27}t^3\Big|_{2a^{1/3}}^{\tau} = \frac{1}{27}[\tau^3 - 8a]$$

$$= \frac{1}{27}\left[\frac{a}{m^3}(9 + 4m^2)^{3/2} - 8a\right] = a[(9 + 4m^2)^{3/2} - 8m^3]/27m^3.$$

Exercises 2.2

1. Find the length of the arc of the parabola $x^2 = 4ay$, $(a > 0)$, cut off by its latus rectum.

2. Show that the length of the arc of the curve $y = \ln \sec x$ from $x = 0$ to $x = \pi/3$ is $\ln(2 + \sqrt{3})$.

3. Show that the length s of the curve given by $x^{2/3} + y^{2/3} = a^{2/3}$, where $a > 0$, measured from $(0, a)$ to the point (x, y) is given by $s = (3/2)(ax^2)^{1/3}$.

4. Find the length of the arc of the curve $y = \ln[(e^x - 1)/(e^x + 1)]$ from $x = 1$ to $x = 2$.

5. Show that the length of the arc of the semicubical parabola $ay^2 = x^3$ from the vertex to $(5a, 5a\sqrt{5})$ is $335a/27$, where $a > 0$.

6. Find the perimeter of the loop of the curve $ay^2 = x^2(a - x)$, $a > 0$.

7. Show that the length of the curve $y^2 = (2x - 1)^3$, cut off by the line $x = 4$ is 37.85 units approximately.

8. Find the length of the arc of the curve $x = t^2 \cos t$ and $y = t^2 \sin t$ from the origin to the point t.

9. Show that the whole length of the curve $x = a\cos^3\theta$, $y = a\sin^3\theta$, where $0 \le \theta < 2\pi$ is $6a$.

10. Prove that the loop of the curve $x = t^2$, $y = t - t^3/3$ is of length $4\sqrt{3}$ units.

Answers

1. $2[\sqrt{2} + \ln(1 + \sqrt{2})]a$. **4.** $\ln(e + 1/e)$. **6.** $4a/\sqrt{3}$. **8.** $[(4 + t^2)^{3/2} - 8]/3$.

2.7.2 Solid of Revolution—Its Volume

Let R be a region of finite area lying in the plane Π and Λ be a straight line lying therein, on one side of R. The solid generated by revolving R completely once about Λ is called the **solid of revolution**, and Λ, the **axis of revolution**.

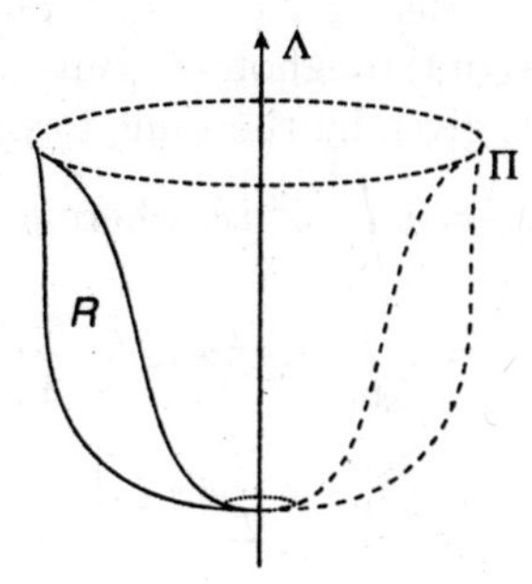

Fig. 2.3

Volume of this solid is called the **volume** and the surface area of the solid is called the **surface area of revolution**.

Case I. Volume of Revolution about the x-axis. Here the plane Π is the xy-Cartesian plane in which the region R lies and the axis Λ, of revolution is the x-axis.

(a) The boundary of R consists of the four lines:
$x = a$, $x = b$, $y = 0$ and $y = f(x) \ge 0 \; \forall x \in I = [a, b]$.

If $P(x, y)$ be any arbitrary point on the boundary $y = f(x)$ and PN is the ordinate of P, then since $y \ge 0$, $PN = y$. An elementary volume $dV = \pi y^2 dx$ and hence the required volume is given by

$$V = \pi \int_a^b y^2 dx, \text{ where } y = f(x).$$

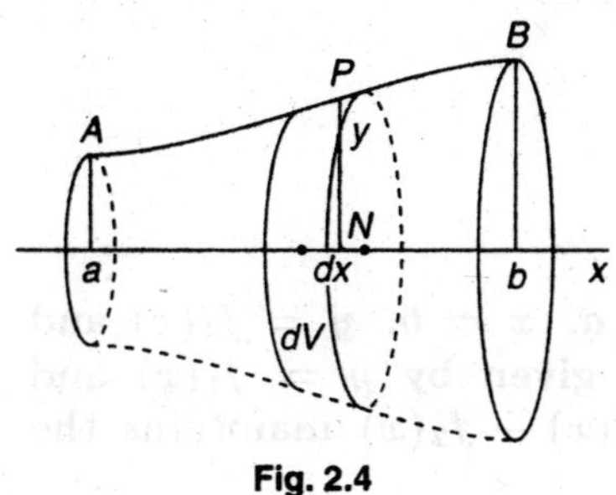

Fig. 2.4

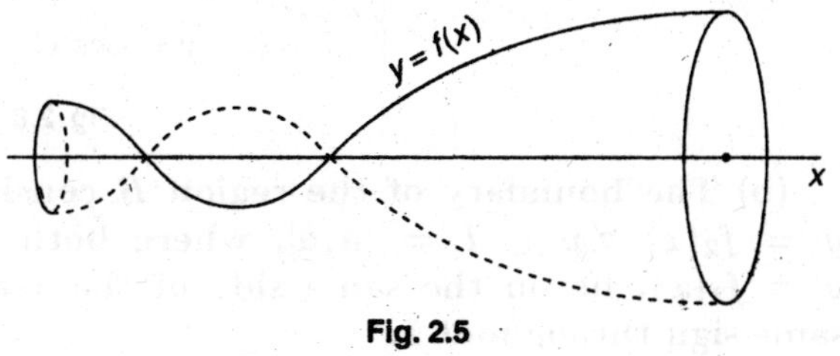

Fig. 2.5

Note. 1. The above result for V is true even if $f(x) \not\geq 0 \; \forall x \in I$. For y or $f(x)$ appears as a squared quantity under the integral sign.

Examples 2.4. 1. *Find the volume of the solid obtained by revolving, about the x-axis, the region bounded by $x = 0$, $x = 1$, $y = 0$ and $y = x \; \forall x \in I = [0, 1]$.*

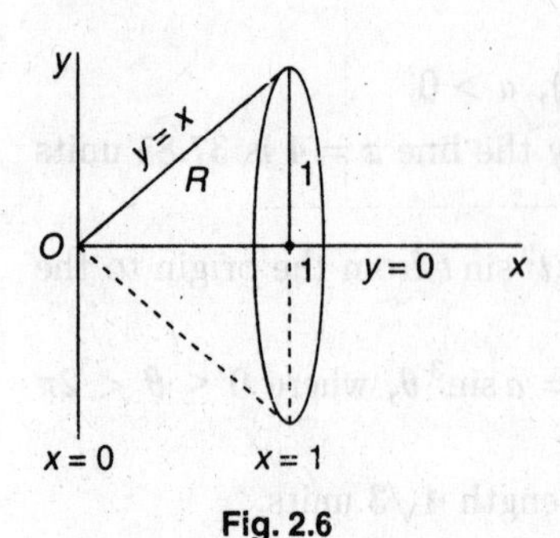

Fig. 2.6

If R is the region bounded by the given lines, obviously R lies on one side of the x-axis. So using the formula

$$V = \pi \int_0^1 y^2 dx = \pi \int_0^1 x^2 dx = \frac{\pi}{3}.$$

Here the solid is a right circular cone with the x-axis as its axis and since $y = 1$ when $x = 1$, the radius of the base and the altitude of the cone are both 1.

So using the formula, from mensuration,

$$V = \frac{1}{3}\pi r^2 h = \frac{1}{3}\pi \cdot 1^2 \cdot 1 = \frac{\pi}{3}.$$

2. *Find the volume of the solid obtained by revolving the region R about the x-axis, where the boundary of R consists of*

$$x = 0, \; x = \pi, \; y = 0, \; y = \cos x \; \forall x \in I = [0, \pi].$$

Here $f(x) = \cos x$ does not maintain its sign throughout I. But even then the volume is given by the same formula.

$$V = \pi \int_0^{\pi} y^2 dx, \text{ where } y = \cos x$$

$$= \pi \int_0^{\pi} \cos^2 x dx = \frac{\pi}{2} \int_0^{\pi} (1 + \cos 2x) dx$$

$$= \frac{\pi}{2} \left[x + \frac{1}{2} \sin 2x \right]_0^{\pi} = \frac{\pi^2}{2}.$$

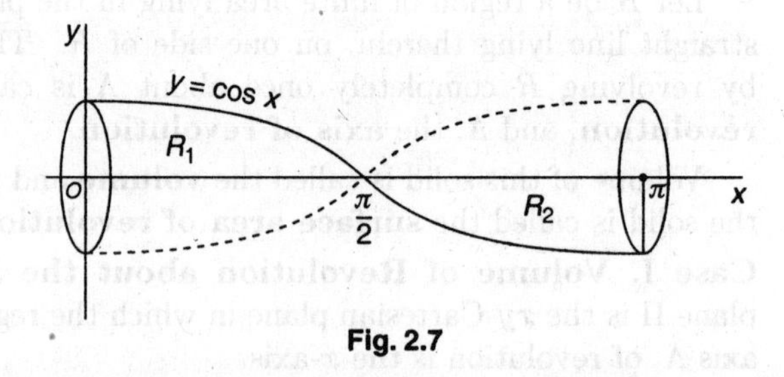

Fig. 2.7

Note. If in place of $\cos x$, y is taken as $|\cos x|$, the volume V will remain the same as is obvious from the figure 2.8. Moreover, $y^2 = |\cos x|^2 = \cos^2 x$. Hence the result.

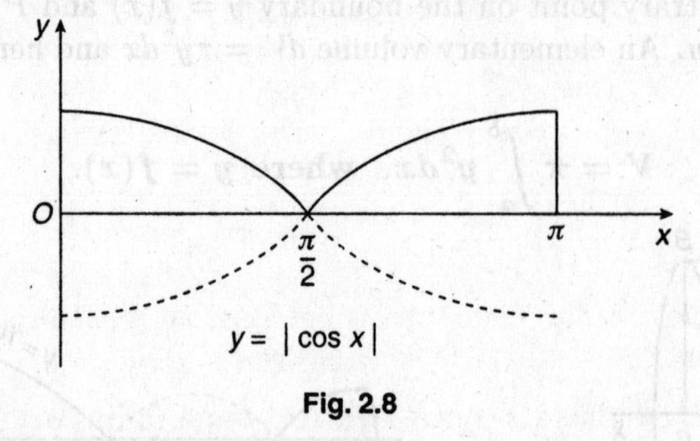

Fig. 2.8

(b) The boundary of the region R consists of $x = a$, $x = b$, $y = f_1(x)$ and $y = f_2(x) \; \forall x \in I = [a, b]$, where both the curves given by $y = f_1(x)$ and $y = f_2(x)$, lie on the same side of the x-axis and $f_1(x) - f_2(x)$ maintains the same sign throughout I.

Then $V = |V_1 - V_2|$, **where** $V_i = \pi \int_a^b f_i^2(x)dx, \ i = 1, 2.$

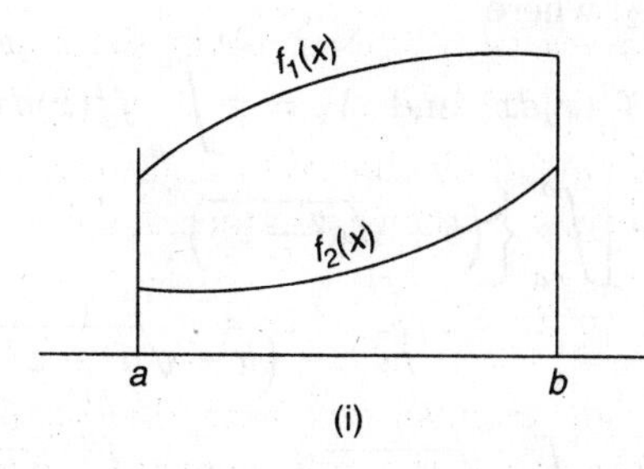

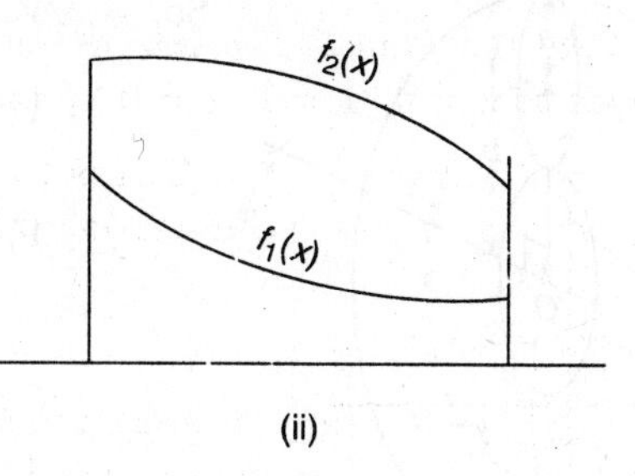

Fig. 2.9

3. *Let R be region of the trapezium OABC (Fig. 2.10), where O is the origin, A is* $\left(\frac{1}{2}, 0\right)$ *and the equations of the line segments OC and AB are*

$$y = x, \qquad 0 \le x \le 1$$

and $$y = x - 1/2, \quad \frac{1}{2} \le x \le 1.$$

Calculate the volume of revolution when R is revolved about the x-axis.

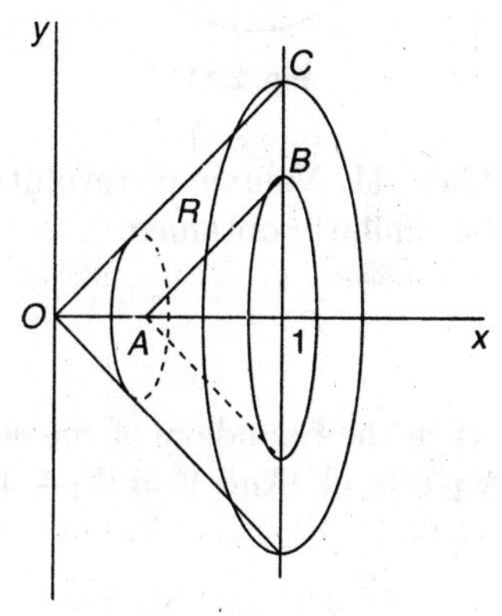

Fig. 2.10

Let $f_1(x) = x, \qquad \text{for } 0 \le x \le 1$

and $f_2(x) = x - 1/2, \text{ for } 1/2 \le x \le 1$

$= 0, \qquad \text{for } 0 \le x < \frac{1}{2}.$

Then the boundary of R will consist of $x = 0, \ x = 1$ and $y = f_1(x)$ and $y = f_2(x)$ for $0 \le x \le 1$.

Here, $V_1 = \pi \int_0^1 f_1^2(x)dx = \pi \int_0^1 x^2 dx = \frac{\pi}{3}$ and

$$V_2 = \pi \int_0^1 f_2^2(x)dx = \pi \left[\int_0^{1/2} f_2^2(x)dx + \int_{1/2}^1 f_2^2(x)dx\right]$$

$$= \pi \int_{\frac{1}{2}}^1 \left(x - \frac{1}{2}\right)^2 dx = \pi \int_0^{\frac{1}{2}} y^2 dy, \ y = x - \frac{1}{2}.$$

$$= \frac{\pi}{3} y^3 \Big|_0^{1/2} = \frac{\pi}{24}.$$

$$\therefore \qquad V = |V_1 - V_2| = \left|\frac{\pi}{3} - \frac{\pi}{24}\right| = \frac{7\pi}{24}.$$

4. *Find the volume of the anchor ring (torus) obtained by revolving the region bounded by the circle:* $x^2 + (y - b)^2 = a^2$*, where* $0 < a < b$*, about the x-axis.*

The circle $x^2 + (y - b)^2 = a^2$ has its centre at $C(0, b)$ and radius a. The circumference can be divided into two parts, the upper and the lower parts, whose equations are

$$y = b + \sqrt{a^2 - x^2} = f_1(x), \ -a \le x \le b$$

and $$y = b - \sqrt{a^2 - x^2} = f_2(x), \ -a \le x \le b.$$

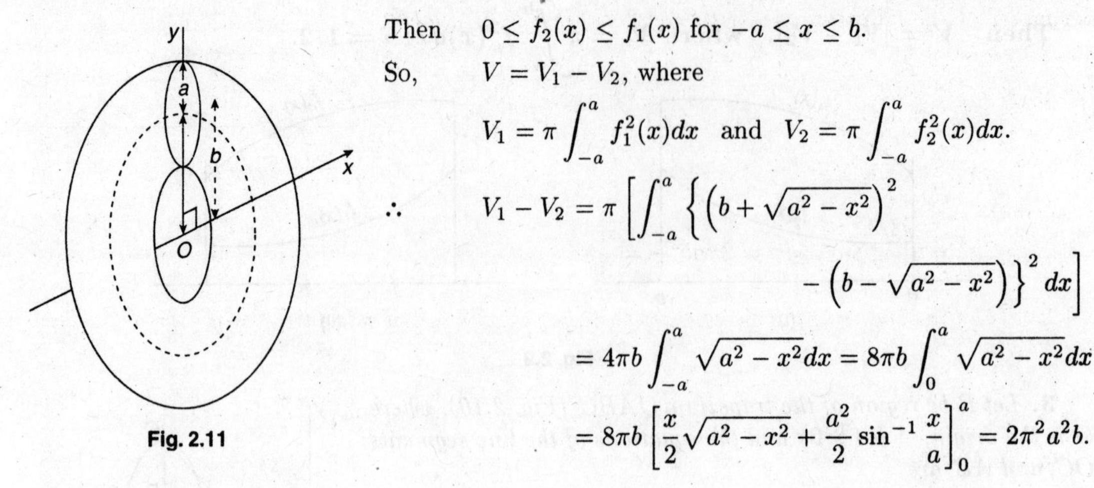

Fig. 2.11

Then $0 \le f_2(x) \le f_1(x)$ for $-a \le x \le b$.

So, $V = V_1 - V_2$, where

$$V_1 = \pi \int_{-a}^{a} f_1^2(x)dx \quad \text{and} \quad V_2 = \pi \int_{-a}^{a} f_2^2(x)dx.$$

$$\therefore \quad V_1 - V_2 = \pi \left[\int_{-a}^{a} \left\{ \left(b + \sqrt{a^2 - x^2}\right)^2 - \left(b - \sqrt{a^2 - x^2}\right) \right\}^2 dx \right]$$

$$= 4\pi b \int_{-a}^{a} \sqrt{a^2 - x^2}dx = 8\pi b \int_{0}^{a} \sqrt{a^2 - x^2}dx$$

$$= 8\pi b \left[\frac{x}{2}\sqrt{a^2 - x^2} + \frac{a^2}{2} \sin^{-1} \frac{x}{a} \right]_0^a = 2\pi^2 a^2 b.$$

Case II. Volume of revolution of a solid obtained by revolving a region about the y-axis can be similarly obtained:

$$V = \pi \int_{c}^{d} x^2 dy,$$

when the boundary of the region consists of the lines $y = c,\ y = d,\ x = 0$ and $x = f(y)$ $\forall\, y \in [c, d]$. And $V = |V_1 - V_2|$, where

$$V_i = \pi \int_{c}^{d} f_i^2(y)dy \text{ for } i = 1, 2,$$

when the region has the boundary consisting of $y = c,\ y = d,\ x = f_1(y)$ and $x = f_2(y)$ $\forall\, y \in [c, d]$, where $f_1(y)$ and $f_2(y)$ are of the same sign, and $f_1(y) - f_2(y)$ always maintains the same sign for all y in $[c, d]$.

5. *Find the volume of the solid generated by revolving, about the y-axis, the region whose boundary is given by*

$$y = -\pi/4,\ y = \pi/4,\ y = \sin^{-1} x.$$

Here $y = \sin^{-1} x \Rightarrow x = \sin y,\ -\pi/4 \le y \le \pi/4$.

$$\therefore \quad V = \pi \int_{-\pi/4}^{\pi/4} x^2 dy = \pi \int_{-\pi/4}^{\pi/4} \sin^2 y dy$$

$$= \frac{\pi}{2} \int_{-\pi/4}^{\pi/4} (1 - \cos 2y)dy$$

$$= \frac{\pi}{2} \left[y - \frac{1}{2} \sin 2y \right]_{-\pi/4}^{\pi/4} = \frac{\pi}{2} \left(\frac{\pi}{2} - 1 \right).$$

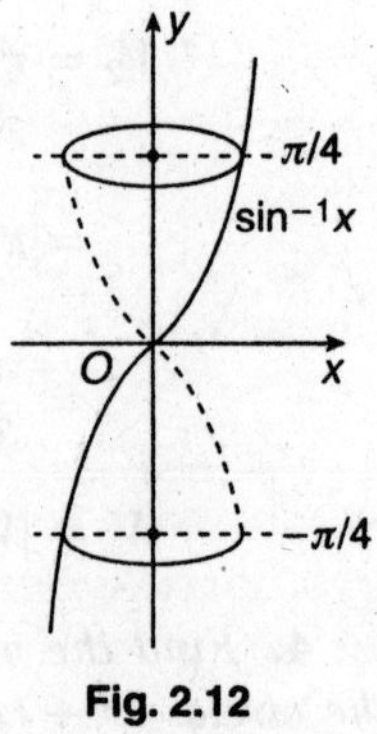

Fig. 2.12

6. *The region, bounded by the straight lines*

$$x = -a,\ x = a$$

and the curve given by $x = a\cos\phi,\ y = b\sin\phi\ \forall\, \phi \in [0, \pi]$, *is rotated about the x-axis. Find the volume of revolution.*

Here the region R is bounded by the entire upper perimeter of the ellipse $x = a\cos\phi,\ y = b\sin\phi,\ 0 \le \phi \le \pi$ and the x-axis.

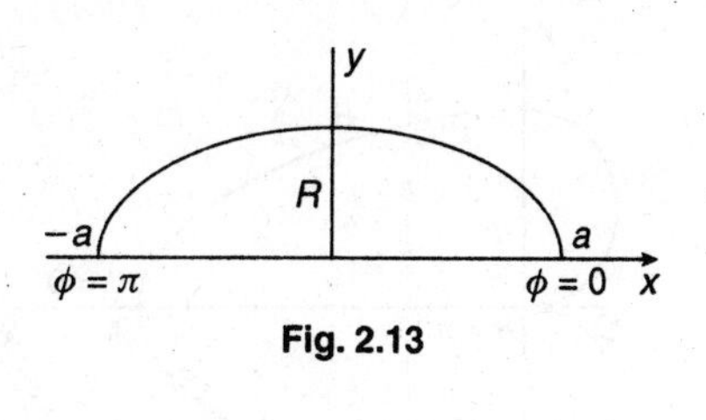

Fig. 2.13

$$\text{So, } V = \pi\int_{-a}^{a} y^2 dx = \pi\int_{\pi}^{0} b^2\sin^2\phi(-a\sin\phi)d\phi$$

$$= \pi ab^2\int_0^{\pi}\sin^3\phi d\phi$$

$$= 2\pi ab^2\int_0^{\pi/2}\sin^3\phi d\phi = 2\pi ab^2\frac{2}{3} = \frac{4}{3}\pi ab^2.$$

7. *Find the volume of the spindle generated by revolving the astroid $x^{2/3} + y^{2/3} = a^{2/3}$ about the y-axis.*

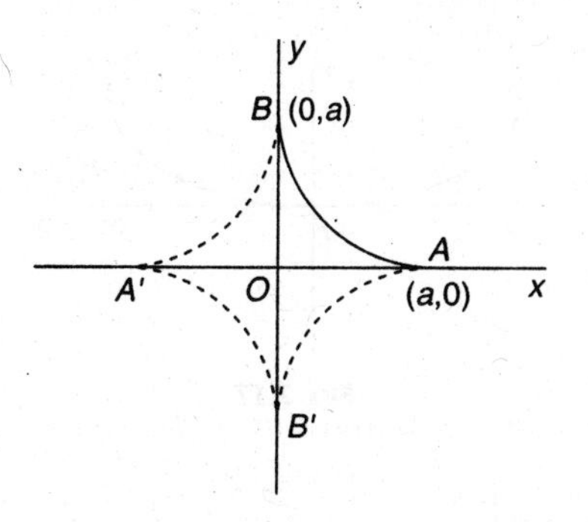

Fig. 2.14

The equation to the astroid is

$$(x^2)^{1/3} + (y^2)^{1/3} = a^{2/3}.$$

Since it contains only even powers of x and y, it is symmetrical with the origin. So the volume of revolution is

$$V = 2\pi\int_0^a x^2 dy.$$

We find that the astroid can be represented parametrically by $x = a\cos^3\theta$ and $y = a\sin^3\theta$, where $0 \le \theta \le \pi/2$ for the arc AB in the figure.

$$\therefore \quad V = 2\pi\int_0^{\pi/2} a^2\cos^6\theta\cdot 3a\sin^2\theta\cos\theta d\theta$$

$$= 6\pi a^3\int_0^{\pi/2}\sin^2\theta\cdot\cos^7\theta d\theta = 6\pi a^3\cdot\frac{1}{9}\cdot\frac{6\cdot 4\cdot 2}{7\cdot 5\cdot 3} = \frac{32\pi a^3}{105}.$$

Case III. The volume of the solid of revolution of a region R when the **boundary is given by polar equations of the type**

$$\boldsymbol{\theta = \alpha,\ \theta = \beta \text{ and } r = f(\theta),}$$

where $0 \le \alpha \le \theta \le \beta \le 2\pi$.

$$\textbf{Then } V = \frac{2}{3}\pi\int_{\alpha}^{\beta} r^3\sin\theta d\theta \textbf{ or, } \frac{2}{3}\pi\int_{\alpha}^{\beta} r^3\cos\theta d\theta,$$

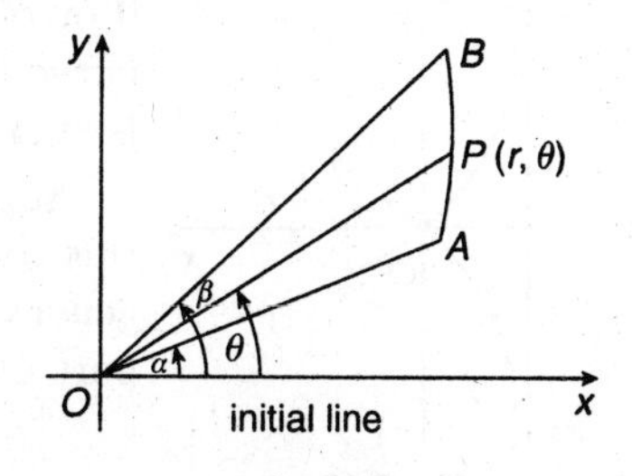

Fig. 2.15

according as the axis of revolution is the initial line ($\theta = 0$ or the x-axis) or a line perpendicular to the initial line ($\theta = \pi/2$ or the y-axis) through the pole.

8. *Find the volume of the solid obtained by revolving the cardioid $r = a(1 + \cos\theta)$ about the initial line.*

The same volume will be obtained if we revolve the arc ABO given by $r = a(1 + \cos\theta)$, $0 \le \theta \le \pi$.

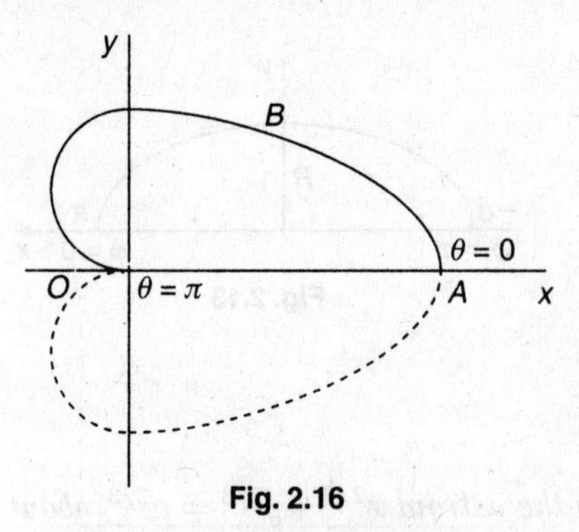

Fig. 2.16

So, the required volume is given by

$$V = \frac{2}{3}\pi \int_0^{\pi} r^3 \sin\theta d\theta$$

$$= \frac{2}{3}\pi \cdot \int_0^{\pi} a^3(1+\cos\theta)^3 \sin\theta d\theta$$

$$= \frac{2}{3}\pi a^3 \int_2^0 t^3(-dt), \text{ where } t = 1+\cos\theta$$

$$= \frac{8\pi a^3}{3}.$$

9. *Find the volume of the reel generated by the revolution of the cycloid $x = a(\theta+\sin\theta)$, $y = a(1-\cos\theta)$ about the x-axis, where $-\pi \le \theta \le \pi$.*

The arc of the cycloid given by $x = a(\theta+\sin\theta)$, $y = a(1-\cos\theta)$, where $-\pi \le \theta \le \pi$ has been shown to be AOB in the figure.

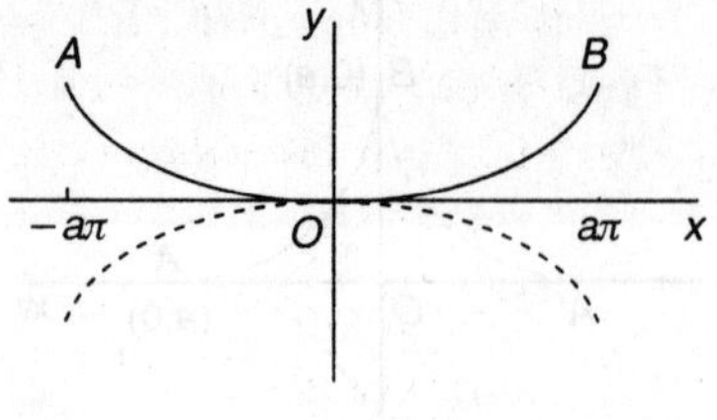

Fig. 2.17

The required volume is

$$\pi \int_{-a\pi}^{a\pi} y^2 dx = \pi \int_{-\pi}^{\pi} a^2(1-\cos\theta)^2 \cdot a(1+\cos\theta)d\theta$$

$$= 2\pi a^3 \int_0^{\pi} \sin^2\theta(1-\cos\theta)d\theta$$

$$= 2\pi a^3 \left[\int_0^{\pi} \frac{1}{2}(1-\cos 2\theta)d\theta - \int_0^{\pi} \sin^2\theta\cos\theta d\theta\right]$$

$$= 2\pi a^3 \left[\frac{1}{2}\left(1-\frac{1}{2}\sin 2\theta\right) - \frac{1}{3}\sin^3\theta\right]_0^{\pi} = \pi^2 a^3.$$

10. *Find the volume of the solid formed by the revolution of the loop of the curve $y^2 = x^2(a-x)/(a+x)$ about the x-axis.*

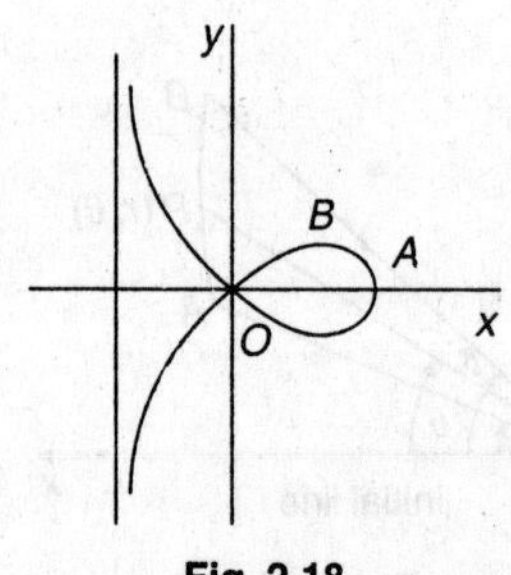

Fig. 2.18

Since y^2 and x^2 are both non-negative for real points $(a-x)/(a+x) \ge 0$ or regarding $a > 0$, $-a < x \le a$. As $x \to -a$, $y \to \pm\infty$, so that the curve has an asymptote: $x = -a$. Since y is in even powers, the curve is symmetrical in the x-axis. Also $y = 0$ when $x = 0$ or $x = a$.

So, the curve has a loop lying between $x = 0$ and $x = a$. Since this loop is symmetrical in the x-axis, the solid of revolution can be generated by revolving only the upper half OBA of the loop. For the part OBA,

$$y = x\sqrt{\frac{a-x}{a+x}}, 0 \le x \le a.$$

So, the required volume

$$V = \pi \int_0^a y^2 dx = \pi \int_0^a \frac{x^2(a-x)}{a+x} dx$$

$$= \pi \int_0^a \left(-x^2 + 2ax - 2a^2 + \frac{2a^3}{a+x}\right) dx$$

$$= \pi\left[-\frac{x^3}{3} + ax^2 - 2a^2x + 2a^3\ln(a+x)\right]_0^a$$

$$= \pi a^3\left(2\ln 2 - \frac{4}{3}\right).$$

Case IV. Volume of revolution about any axis. If the region R bounded by the line segments $AA', A'B', B'B$ and the arc Γ with end points A and B be rotated about the axis Λ, through A' and B' then volume of revolution is given by

$$V = \pi\int_{OA'}^{OB'} PN^2 d(ON).$$

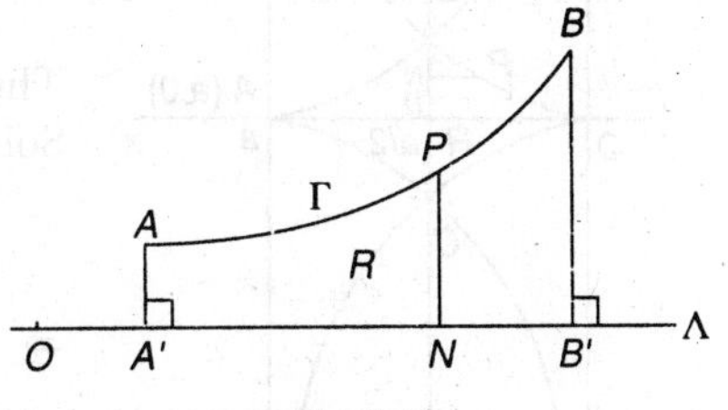

Fig. 2.19

If Λ is the x-axis, $PN = y$, $ON = x$, and if it is the y-axis, $PN = x$, $ON = y$.

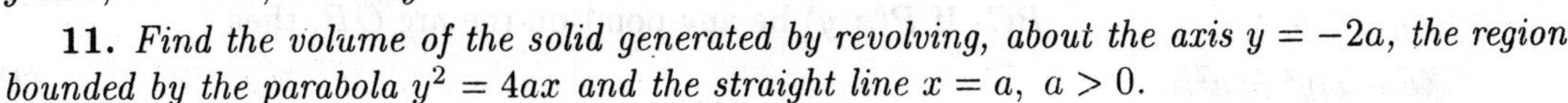

11. *Find the volume of the solid generated by revolving, about the axis $y = -2a$, the region bounded by the parabola $y^2 = 4ax$ and the straight line $x = a$, $a > 0$.*

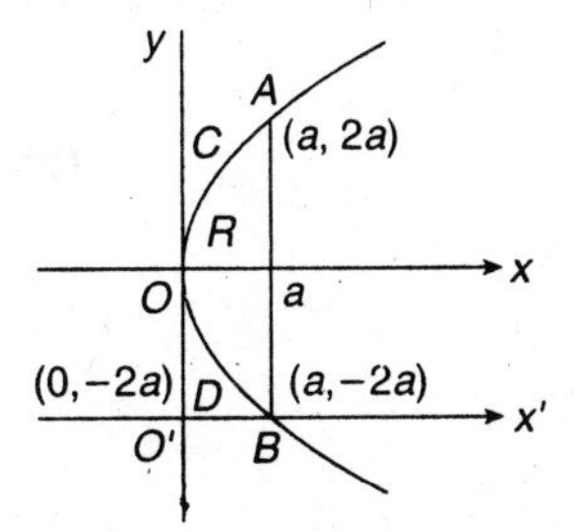

Fig. 2.20

The region R shown in the figure is bounded by the arc AOB of the parabola and the chord AB.

Since the axis of revolution is $y = -2a$, i.e., the line $O'x'$, we shift the origin to $O'(0, -2a)$. Then the equation to the arc will be obtained from $y^2 = 4ax$, $0 \le x \le a$, $-2a \le y \le 2a$ by replacing x by $x + 0$ and y by $y - 2a$. So the new equation to the arc is

$$(y - 2a)^2 = 4ax,$$

which should now be revolved about the new x-axis $O'x'$.

Since the curve is quadratic w.r.t. y,

$$y = 2a \pm 2\sqrt{ax}.$$

$y = f_1(x) = 2a + 2\sqrt{ax}$ for $0 \le x \le a$ is the equation to the upper part OCA and $y = f_2(x) = 2a - 2\sqrt{ax}$ for $0 \le x \le 0$ is the equation to the lower part ODB of the parabola. Also

$$0 \le f_2(x) \le f_1(x)\ \forall x \in [0, a].$$

So, the required volume

$$V = V_1 - V_2 = \pi\int_0^a \left[f_1^2(x) - f_2^2(x)\right] dx$$

$$= \pi\int_0^a 16a\sqrt{ax}dv = 16a^{3/2}\pi\int_0^a \sqrt{x}dx = 16a^{3/2}\pi\frac{2}{3}x^{3/2}\Big|_0^a$$

$$= \frac{32}{3}\pi a^3.$$

12. *Show that the volume, obtained by revolving about $x = a/2$, the region enclosed between the curves $xy^2 = a^2(a-x)$, $(a-x)y^2 = a^2x$, $a > 0$, is $\pi a^3(4-\pi)/4$.*

Let the given curves be Γ_1 and Γ_2, whose equations are $xy^2 = a^2(a-x)$ and $(a-x)y^2 = a^2x$.

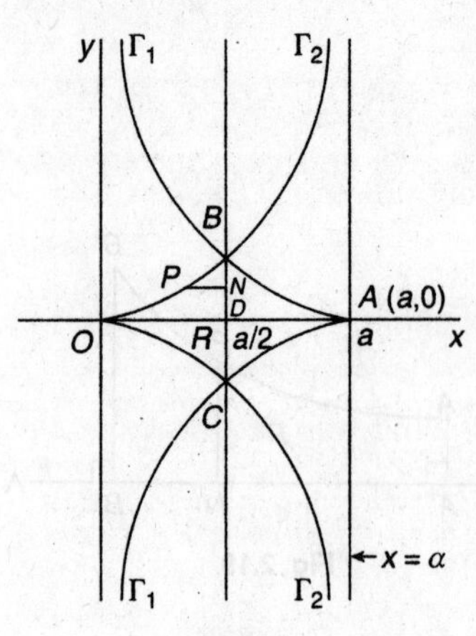

Fig. 2.21

Since for Γ_1, $y^2 = a^2\frac{(a-x)}{x}$ and $y \to \pm\infty$ as $x \to 0$, $x = 0$ is an asymptote of Γ_1. Also $y = 0$ when $x = a$. So $A(a, 0)$ is a point on Γ_1. If $x > a$, $y^2 < 0$. So there is no point on the R.H.S. of the line $x = a$. This curve is symmetric in the x-axis, since y appears in even degrees.

Similarly, Γ_2 has an asymptote $x = a$ and O is its vertex. The region R is bounded by the arcs is OC, CA, AB and OB. Solving the two equations of curves for x,

$$\frac{a^2(a-x)}{x} = \frac{a^2x}{a-x} \Rightarrow (a-x)^2 = x^2$$

$$\Rightarrow \quad a^2 - 2ax = 0 \Rightarrow x = a/2.$$

The equation to the line BC is $x = a/2$.

The volume generated by revolving the region about $x = a/2$ is twice the volume obtained by revolving the region ODB about BC. If $P(x, y)$ be any point on the arc OB, then

$$(a-x)y^2 = a^2x \tag{1}$$

and $PN = a/2 - x$, where $x = ay^2/(a^2 + y^2)$, obtained from (1).

$$\text{So, } PN = \frac{a}{2} - \frac{ay^2}{a^2+y^2} = \frac{a}{2}\,\frac{a^2-y^2}{a^2+y^2}. \tag{2}$$

Also $DN = y$.

$\therefore$ The required volume is given by

$$V = 2\pi \int_0^{DB} PN^2 d(DN).$$

Since, when $x = a/2$, $y = \pm a$, as calculated from (1), $DB = a$.

$$\therefore \quad V = 2\pi \int_0^a \frac{a^2}{4}\left(\frac{a^2-y^2}{a^2+y^2}\right)^2 dy$$

$$= \frac{\pi a^2}{2}\int_0^{\pi/4} \cos^2 2\theta \cdot a\sec^2\theta d\theta, \text{ where } y = a\tan\theta$$

$$= \frac{\pi a^3}{2}\int_0^{\pi/4} (4\cos^2\theta - 4 + \sec^2\theta)d\theta$$

$$= \frac{\pi a^3}{2}\int_0^{\pi/4} (2\cos 2\theta - 2 + \sec^2\theta)d\theta$$

$$= \frac{\pi a^3}{2}[\sin 2\theta - 2\theta + \tan\theta]_0^{\pi/4} = \frac{\pi a^3}{4}(4-\pi).$$

Exercises 2.3

1. Find the volume generated by revolving the portion of the parabola $y^2 = 4ax$, $a > 0$, cut off by its latus rectum, about its axis.

2. The loop of the curve $2ay^2 = x(x-a)^2$, $a > 0$, revolves about the x-axis. Show that the volume of the solid thus generated is $\pi a^3/24$.

3. Find the volume generated by revolving, about the x-axis, the region bounded by the curve $\sqrt{x} + \sqrt{y} = \sqrt{a}$ and the lines $x = 0$ and $y = 0$.

4. Show that the volume generated by the revolution, about the x-axis, of the loop of the curve $y^2 = x^2(a + x)(a - x)$ is $2\pi a^3(\ln 2 - 2/3)$.

5. Find the volume of the solid generated by rotating completely, about the x-axis, the region enclosed in the lines $y^2 = x^3 + 5x$, $x = 2$ and $x = 4$.

6. Show that the volume generated by the portion of the arc of $y = \sqrt{(1 + x^2)}$ lying between $x = 0$ and $x = 4$ as it revolves about the x-axis is $76\pi/3$.

7. Show that the volume of the solid formed by revolving the region, bounded by the arc of the curve $y = e^x \sin x$ between $x = 0$ and $x = \pi$, about the x-axis, is $\pi(e^{2\pi} - 1)/8$.

8. Find the volume of the reel formed by the revolution of the region bounded by an arch of the cycloid $x = a(\theta + \sin\theta)$, $y = a(1 + \cos\theta)$ about the x-axis.

9. Show that the volume of the solid generated by the revolution of the arch of the cycloid $x = a(\theta + \sin\theta)$ and $y = a(1 - \cos\theta)$ about the y-axis, is $\pi a^3(3\pi^2/2 - 8/3)$.

10. Find the volume of the solid generated by the revolution of a loop of the lemniscate $r^2 = a^2 \cos 2\theta$ about the initial line.

11. Prove that the volume generated by the revolution, about the initial line, of the region bounded by the limacon $r = a + b\cos\theta$, $a > b$, and the initial line, is $4\pi a(a^2 + b^2)/3$.

Answers

1. $2\pi a^3$. **3.** $\pi a^3/12$. **5.** 90π. **8.** $5\pi^2 a^3$.

10. $(\pi a^3/4)[(1/\sqrt{2})\ln(\sqrt{2} + 1) - 1/3]$.

2.7.3 Surface Area of Revolution of an Arc

Case I. The axis of revolution is the x-axis.

Let the arc Γ with distinct end points A, B be given in the form:

$$y = f(x),\ a \le x \le b,$$

where a and b are the abscissae of A and B.

We may take, without any loss of generality,

$$f(x) \ge 0\ \forall x \in I = [a, b].$$

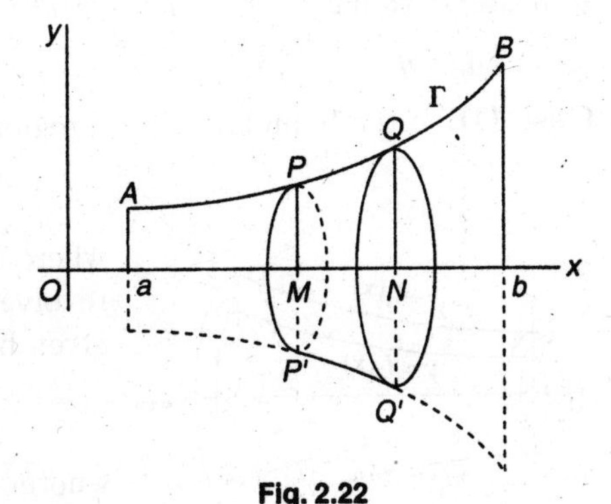

Fig. 2.22

If $f(x) \le 0$ or it does not maintain its sign, we shall take $y = |f(x)| \ge 0$, for in course of revolution $f(x)$ becomes $-f(x)$ and y becomes positive.

If P and Q are two consecutive points on Γ with the arc length $PQ = \delta s$, then for small δs, δs = chord length PQ.

On making one complete revolution the chord PQ generates a curved surface of the frustum of a cone. If δS is the area of this surface, then we know from mensuration

$$\delta S = \pi(r_1 + r_2)l = \pi(PM + QN)PQ.$$

If $y = PM$, $QN = y + \delta y$ and $PQ = \delta s$.

$\therefore$ $\delta S = \pi(2y + \delta y)\delta s = 2\pi y \delta s$,

neglecting higher order small quantities.

But $\delta s^2 = \delta x^2 + \delta y^2$.

$$\therefore \quad \frac{\delta S}{\delta x} = 2\pi y \frac{\delta s}{\delta x} = 2\pi y \sqrt{1 + (\delta y/\delta x)^2}.$$

In the limit $dS/dx = 2\pi y\, ds/dx = 2\pi y \sqrt{1 + y_1^2}$.

$\therefore$ **S, the surface area of revolution of the arc Γ when it makes one complete revolution about the x-axis is given by $S = 2\pi \int_{s_0}^{s_1} y ds = 2\pi \int_a^b y\sqrt{1 + y_1^2} dx$, where $y = f(x)$.**

Since $S = 2\pi \int_a^b y \frac{ds}{dx} dx$, when $a \le x \le b$, $x = r\cos\theta$, $y = r\sin\theta$, where $\alpha \le \theta \le \beta$ gives

$$S = 2\pi \int_\alpha^\beta r\sin\theta \cdot \frac{ds}{d\theta} d\theta,$$

where the equation to the arc Γ is given, in polar coordinates, by

$$r = f(\theta)\ \forall \theta \in [\alpha, \beta].$$

Case II. If the y-axis is the axis of revolution,

$$S = 2\pi \int_c^d x \frac{ds}{dy} dy$$

when Γ is given by $x = x(y)$, $c \le y \le d$

and $$S = 2\pi \int_\alpha^\beta r\cos\theta \frac{ds}{d\theta} d\theta,$$

when Γ is given by $r = f(\theta)$, $\alpha \le \theta \le \beta$

and $x = r\cos\theta$ and $y = r\sin\theta$.

In polar coordinates $\delta s^2 = dr^2 + r^2 d\theta^2$

or, $ds/d\theta = \sqrt{(dr/d\theta)^2 + r^2} = \sqrt{r^2 + r_1^2}$.

Case III. If the boundary Γ of a region consists of the lines

$$x = a,\ x = b,\ y = f_1(x) \text{ and } y = f_2(x),$$

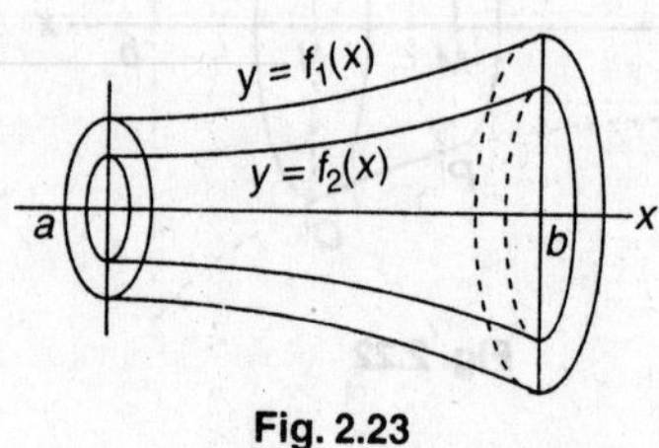

Fig. 2.23

where $0 \le f_2(x) \le f_1(x)\ \forall x \in [a, b]$ and this boundary is revolved about the x-axis, the surface area of revolution is given by

$$S = S_1 + S_2 + S_3 + S_4,$$

where $S_i = 2\pi \int_a^b f_i(x) \frac{ds}{dx} dx$ for $i = 1, 2$

$$S_3 = \pi\{f_1^2(a) - f_2^2(a)\},\ S_4 = \pi\{f_1^2(b) - f_2^2(b)\},$$

S_3 and S_4 are the areas of the two annular bases.

If $f_1(a) = f_2(a)$ and $f_1(b) = f_2(b)$, $S_3 = S_4 = 0$ and $S = S_1 + S_2$, the sum of the two curved surfaces, one inner and the other outer.

Examples 2.5. **1.** *Find the area of the surface formed by the revolution, about its axis, of the arc of the parabola $y^2 = 4ax$, between the vertex and an end of its latus rectum.*

There are two arcs OL and OL' which are symmetrical in the x-axis, the axis of the parabola. The same surface will be obtained by revolving the arc OL or OL'. We, therefore, choose OL. For this arc

$$y = 2\sqrt{ax},\ 0 \le x \le a.$$

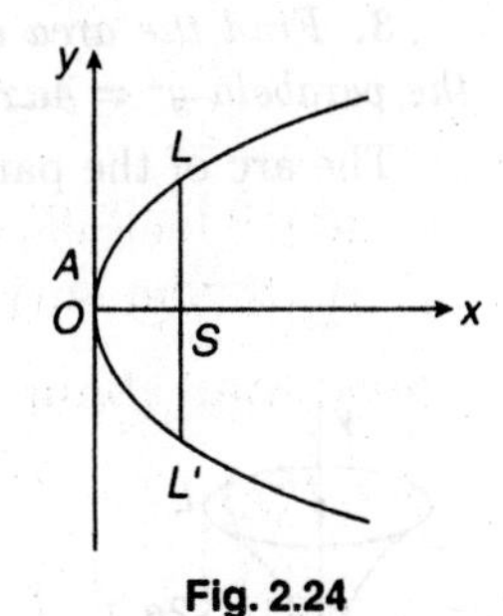

Fig. 2.24

So, $$S = 2\pi \int_0^a y\frac{ds}{dx}dx = 2\pi \int_0^a 2\sqrt{ax}\sqrt{1+y_1^2}dx$$

Now $$y_1 = \frac{\sqrt{a}}{\sqrt{x}}.$$

$$\therefore \quad S = 4\pi\sqrt{a}\int_0^a \sqrt{x}\cdot\sqrt{1+\frac{a}{x}}dx$$

$$= 4\pi\sqrt{a}\int_0^a \sqrt{x+a}dx$$

$$= \frac{8}{3}\pi a^2(2\sqrt{2}-1).$$

2. *Find the surface of revolution of the ellipse $x^2/a^2 + y^2/b^2 = 1$, when revolved about the y-axis.*

$$S = 2\pi \int_{-b}^{b} x(ds/dy)dy,$$

where $$x = \frac{a}{b}\sqrt{b^2-y^2}.$$

Now $$x_1 = -\frac{ay}{b\sqrt{b^2-y^2}}$$

and $$\frac{ds}{dy} = \sqrt{1+x_1^2} = \sqrt{1+\frac{a^2y^2}{b^2(b^2-y^2)}} = \sqrt{\frac{b^4+(a^2-b^2)y^2}{b^2(b^2-y^2)}}.$$

$$\therefore \quad S = \frac{2\pi a}{b^2}\int_{-b}^{b}\sqrt{b^4+(a^2-b^2)y^2}dy.$$

Assuming a and b to be the length of semi-major and semi-minor axes respectively, and e as the eccentricity, $b^2 = a^2(1-e^2)$ or $a^2 - b^2 = a^2e^2$,

$$S = \frac{4\pi a}{b^2}\int_0^b \sqrt{b^4+a^2e^2y^2}dy = \frac{4\pi a^2 e}{b^2}\int_0^b \sqrt{y^2+(b^2/ae)^2}dy$$

$$= \frac{4\pi a^2 e}{b^2}\left[\frac{y}{2}\sqrt{y^2+(b^2/ae)^2} + \frac{b^4}{2a^2e^2}\ln\left(y+\sqrt{y^2+(b^2/ae)^2}\right)\right]$$

$$= \frac{2\pi a^2 e}{b^2}\left[b\sqrt{b^2+(b^2/ae)^2} + \frac{b^4}{a^2e^2}\ln\frac{b+\sqrt{b^2+(b^2/ae)^2}}{b^2/ae}\right]$$

$$= 2\pi a^2 e\left[\frac{\sqrt{a^2e^2+b^2}}{ae} + \frac{b^2}{a^2e^2}\ln\frac{ae+\sqrt{a^2e^2+b^2}}{b}\right]$$

$$= \frac{2\pi}{e}\left[a^2e + b^2 \ln \frac{a(e+1)}{b}\right], \because a^2e^2 + b^2 = a^2$$

$$= 2\pi \left[a^2 + \frac{1}{2}\frac{b^2}{e} \ln \frac{1+e}{1-e}\right], \quad \because \quad b = a\sqrt{1-e^2}.$$

3. *Find the area of the surface generated by the revolution, about the y-axis, of the arc, of the parabola $y^2 = 4ax$, between the ends of its latus rectum.*

The arc of the parabola LOL' is given by

$$y^2 = 4ax, \ 0 \le x \le a, \ -2a \le y \le 2a.$$

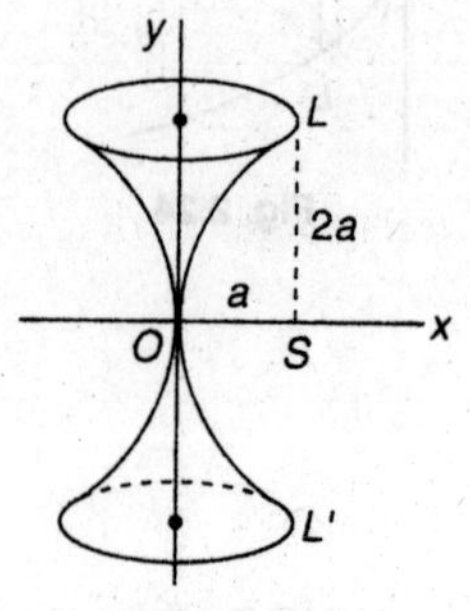

Fig. 2.25

The required surface area is given by

$$S = 2\pi \int_{-2a}^{2a} x\sqrt{1 + x_1^2 dy}.$$

Here $x = \frac{1}{4a}y^2$. So $x_1 = \frac{1}{2a}y$.

$$\therefore \quad S = 2\pi \int_{-2a}^{2a} \frac{1}{4a}y^2\sqrt{1 + \frac{y^2}{4a^2}dy} = \frac{\pi}{a}\int_0^{2a} y^2\sqrt{1+\frac{y^2}{4a^2}dy}$$

$$= 8\pi a^2 \int_0^{\pi/4} \tan^2\theta \sec^3\theta d\theta$$

$$= 8\pi a^2[I(5) - I(3)], \text{ where } I(n) = \int_0^{\pi/4} \sec^n\theta d\theta.$$

But $$I(n) = \frac{n-2}{n-1}I(n-2) + \frac{1}{n-1}\left[\tan^2\theta \ \sec^{n-2}\theta\right]_0^{\pi/4}$$

$$= \frac{n-2}{n-1}I(n-2) + \frac{2^{(n-2)/2}}{n-1}.$$

$$\therefore \quad I(5) - I(3) = \frac{5-2}{5-1}I(3) - I(3) + \frac{2^{3/2}}{4} = \left(\frac{3}{4} - 1\right)I(3) + \frac{\sqrt{2}}{2}$$

$$= -\frac{1}{4}\left[\frac{3-2}{3-1}I(1) + \frac{2^{1/2}}{2}\right] + \frac{\sqrt{2}}{2}$$

$$= -\frac{1}{8}\left[\ln(\sec\theta + \tan\theta)\right]_0^{\pi/4} + \frac{3}{8}\sqrt{2}$$

$$= \frac{1}{8}[3\sqrt{2} - \ln(\sqrt{2}+1)].$$

$$\therefore \quad S = \pi a^2[3\sqrt{2} - \ln(\sqrt{2}+1)].$$

4. *Find the area of the surface formed by revolving the cardioid $r = a(1+\cos\theta)$ about the initial line.* **[WBUT Sem. 2004]**

The formula $S = 2\pi \int_{s_0}^{s_1} yds$, becomes

$$S = 2\pi \int_0^{\pi} r\sin\theta\sqrt{r^2 + r_1^2 d\theta}$$

$$= 2\pi a \int_0^{\pi} (1+\cos\theta)\sin\theta\sqrt{\{a^2(1+\cos\theta)^2 + a^2\sin^2\theta\}d\theta}$$

$$= 2\sqrt{2}\pi a^2 \int_0^{\pi} (1+\cos\theta)^{3/2} \sin\theta d\theta$$

$$= 2\sqrt{2}\pi a^2 \left[-\frac{2}{5}(1+\cos\theta)^{5/2}\right]_0^{\pi} = \frac{32}{5}\pi a^2.$$

5. *Find the surface of the solid generated by the revolution of the lemniscate $r^2 = a^2\cos 2\theta$ about the initial line.*

For real points, $0 \le \cos 2\theta \le 1$. So, $-\pi/4 \le \theta \le \pi/4$. $r \le a$, assuming $a > 0$.

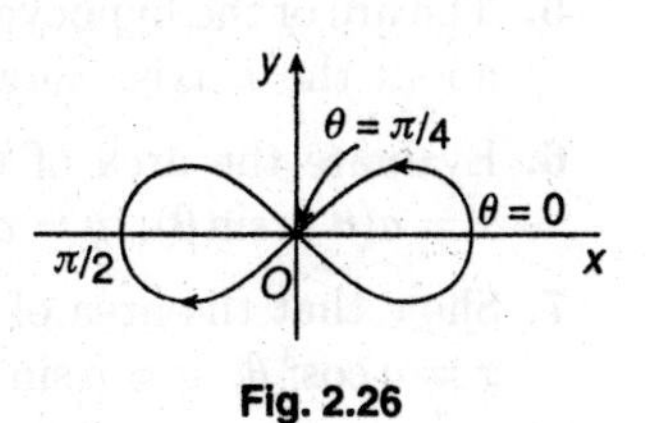

Fig. 2.26

Taking derivatives w.r.t. θ,

$$2rr_1 = -2a^2 \sin 2\theta, \text{ whence } r_1^2 = \frac{a^4 \sin^2 2\theta}{r^2} = \frac{a^2\sin^2 2\theta}{\cos 2\theta}.$$

$$\frac{ds}{d\theta} = \sqrt{r^2 + r_1^2} = \sqrt{a^2\cos 2\theta + \frac{a^2\sin^2 2\theta}{\cos 2\theta}} = a/\sqrt{\cos 2\theta}.$$

Now $\quad S = 2\pi \int_0^{\pi/2} y\frac{ds}{d\theta} d\theta = 2\pi \cdot 2\int_0^{\pi/4} r\sin\theta \frac{ds}{d\theta} d\theta$

$$= 4\pi \int_0^{\pi/4} a\sqrt{\cos 2\theta}\cdot \sin\theta \cdot \frac{a}{\sqrt{\cos 2\theta}} d\theta = 4\pi a^2 \int_0^{\pi/4} \sin\theta d\theta$$

$$= -\, 4\pi a^2 \cos\theta \Big|_0^{\pi/4} = 4\pi a^2\left(1 - \frac{1}{\sqrt{2}}\right) = 2\sqrt{2}\pi a^2(\sqrt{2}-1).$$

6. *Find the surface of the solid generated by the revolution of the astroid $x = a\cos^3\theta$, $y = a\sin^3\theta$ about the y-axis.*

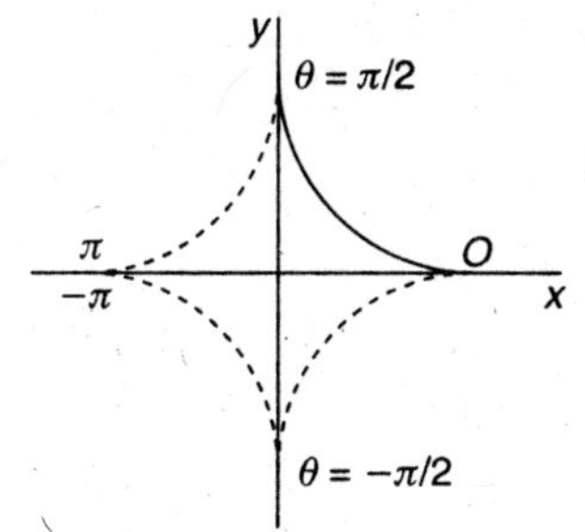

Fig. 2.27

The astroid $x = a\cos^3\theta,\ y = a\sin^3\theta$, where $-\pi < \theta \le \pi$.

From the symmetry of the curve, the required surface will be given by

$$S = 2\cdot 2\pi \int_0^{\pi/2} x\frac{ds}{d\theta} d\theta.$$

But $\quad ds^2 = dx^2 + dy^2 \Rightarrow \frac{ds}{d\theta} = \sqrt{\dot{x}^2 + \dot{y}^2}.$

Now $\quad \dot{x} = -3a\cos^2\theta\sin\theta,\ \dot{y} = 3a\sin^2\theta\cos\theta.$

$\therefore \quad \dot{x}^2 + \dot{y}^2 = 9a^2\sin^2\theta\cos^2\theta$ and $\sqrt{\dot{x}^2+\dot{y}^2} = 3a\sin\theta\cos\theta.$

$$\therefore \quad S = 4\pi \int_0^{\pi/2} a\cos^3\theta \cdot 3a\sin\theta\cos\theta d\theta$$

$$= 12\pi a^2 \int_0^{\pi/2} \sin\theta\cos^4\theta d\theta = 12\pi a^2 \left[-\frac{\cos^5\theta}{5}\right]_0^{\pi/2}$$

$$= \frac{12}{5}\pi a^2.$$

Exercises 2.4

1. Find the surface of the solid generated by revolving, about the x-axis, the region bounded by the arc of the parabola $y^2 = 4ax$, and its latus rectum.

2. An arc of a circle of radius a revolves about its chord. If the length of the arc is $2a\theta$, $(0 < \theta < \pi/2)$, show that the area of the surface generated is $2\pi a^2(\sin\theta - \theta\cos\theta)$.

3. Show that the surface formed by revolving the arc of the curve $y = \sin x$ from $x = 0$ to $x = \pi$ about the x-axis is of area $\pi^2[\sqrt{2} + \ln(1 + \sqrt{2})]$.

4. Find the area of the surface generated by the revolution of the entire ellipse $x^2/a^2 + y^2/b^2 = 1$, $(0 < b < a)$, about the x-axis.

5. The arc of the hypocycloid $x^{2/3} + y^{2/3} = a^{2/3}$, $(a > 0)$, lying in the first quadrant, revolves about the x-axis. Show that the area of the surface thus generated is $6\pi a^2/5$.

6. Evaluate the area of the surface generated by the revolution of an arch of the cycloid $x = a(\theta - \sin\theta)$, $y = a(1 - \cos\theta)$ about the x-axis.

7. Show that the area of the surface of the solid generated by the revolution of the astroid $x = a\cos^3\theta$, $y = a\sin^3\theta$, about the y-axis, is $12\pi a^2/5$.

8. Find the area of the surface obtained by revolving the cardioid $r = a(1 - \cos\theta)$ about the initial line.

9. Show that the area of the surface generated by the revolution of the curve $r = 2a\cos\theta$ about the initial line, is $4\pi a$.

10. Compute the area of the surface formed by revolving the lemniscate $r = a\sqrt{\cos 2\theta}$ about the polar axis.

Answers

1. $8\pi a^2(2\sqrt{2} - 1)/3$. **4.** $2\pi ab\left[b/a + a(a^2 - b^2)^{-1/2}\sin^{-1}\left(\sqrt{a^2 - b^2}/a\right)\right]$. **6.** $64\pi a^2/3$.

8. $32\pi a^2/5$. **10.** $2\pi a^2(2 - \sqrt{2})$.

3 Sequence and Infinite Series

3.1 Sequence

Literally, a sequence implies a succession. Earlier, in mathematics, by a sequence we meant a successive appearance of a few numbers. From that point of view a sequence is regarded as an ordered set numbers. Thus $(a_1, a_2, a_3, \ldots)$ is a sequence of the elements $a_1, a_2, a_3, \ldots$. It is denoted by $\{a_n\}$ or (a_n) or $< a_n >$. A sequence will be abbreviated as sqce.

A sequence is best defined to be a function of an integral variable. The domain of a sequence is either $\mathbb{J}_n = \{1, 2, 3, \ldots, n\}$, the set of first n natural numbers or $\mathbb{N}$, the entire set of natural numbers. A sequence is said to be finite or infinite according as its domain is finite or infinite. Thus, if $f : \mathbb{J}_n \to \mathbb{R}$, then the function f is a finite sequence and if $f : \mathbb{N} \to \mathbb{R}$, then f is an infinite sequence. The f-image of the natural number n is $f(n)$, which is denoted by symbols like $f_n, a_n, b_n, c_n, u_n, v_n, w_n, s_n$, etc., where n has been used as a subscript or the lower index. So a sequence can be very well denoted simply by f or g or, ϕ or ψ. A finite sequence $(a_1, a_2, \ldots, a_n)$ is denoted by $\{a_r\}_{r=1}^{n}$. Since a sequence is nothing but a function, it can be graphically represented by a set of points in the xy-plane, where the natural numbers are plotted on the x-axis and the values of a_n on the y-axis.

3.1.1 Elementary Sequences—Arithmetic, Geometric and Harmonic Sequences

A sequence is said to be arithmetic, geometric or harmonic according as the difference, quotient or the difference of the reciprocals of every pair of consecutive terms of the sequence is constant.

So $\{a_n\}$ is arithmetic if $a_{r+1} - a_r = d$, a constant, for $r = 1, 2, 3 \ldots$, called the **common difference**.

$$\therefore \quad \sum_{r=1}^{n-1}(a_{r+1} - a_r) = \sum_{1}^{n-1} dr \Rightarrow a_n - a_1 = (n-1)d \text{ or, } a_n = a + (n-1)d,$$

where $a = a_1$. Thus the terms of the arithmetic sequence are

$$a,\ a + d,\ a + 2d, \ldots,\ a + \overline{n-1}d, \ldots,$$

which are said to be in arithmetic progression (A.P.).

If $\{g_n\}$ is geometric, then $g_{i+1}/g_i = r$, a constant for $i = 1, 2, 3, \ldots$, called the **common ratio**.

$$\therefore \quad \prod_{i=1}^{n-1}(g_{i+1}/g_i) = \prod_{i=1}^{n-1} r \text{ or, } g_n/g_1 = r^{n-1}$$

or, $\quad g_n = gr^{n-1}$, where $g = g_1$.

The terms of the geometric sequence are $g, gr, gr^2, \ldots, gr^{n-1}, \ldots$, which are said to be in geometric progression (G.P.).

Similarly, $\{h_n\}$ is harmonic if $h_{r+1}^{-1} - h_r^{-1} = \delta$, for $r = 1, 2, 3, \ldots$, a constant. Then

$$\sum_{r-1}^{n-1} (h_{r+1}^{-1} - h_r^{-1}) = \sum_{1}^{n-1} \delta \Rightarrow h_n^{-1} - h_1^{-1} = (n-1)\delta \Rightarrow h_n^{-1} = h^{-1} + \overline{n-1}\delta, \text{ where } h = h_1.$$

$$\therefore \quad h_n = 1/(h^{-1} + \overline{n-1}\delta).$$

The terms of the harmonic sequence are

$$h = 1/h^{-1},\ 1/(h^{-1}+\delta),\ 1/(h^{-1}+2\delta), \ldots, 1/(h^{-1}+\overline{n-1}\delta), \ldots,$$

which are said to be in harmonic progression (H.P.). It is obvious that these terms are the reciprocals of the terms of an arithmetic sequence.

Standard Elementary Sequences: The standard arithmetic geometric and harmonic sequences are respectively

$$\{n\},\ \{\rho^{n-1}\} \text{ and } \{1/n\}$$

or, $(1, 2, 3, 4, \ldots, n, \ldots)$,

$(1, \rho, \rho^2, \rho^3, \ldots, \rho^{n-1}, \ldots)$

and $(1, 1/2, 1/3, \ldots, 1/n, \ldots)$.

We shall be considering real sequences only.

3.2 Boundedness of a Sequence

A sqce $\{a_n\}$ is said to be bounded if there exist real numbers m and M such that $m \le a_n \le M$ for all $n \in \mathbb{N}$.

The arithmetic sequence $\{n\}$ is not bounded, for it has no upper bound.

The geometric sequence $\{\rho^{n-1}\}$ is bounded if $-1 \le \rho \le 1$ and $\rho \ne 0$.

The harmonic sequence $\{1/n\}$ is bounded for,

$$0 < \frac{1}{n} < 1\ \forall n \in \mathbb{N}.$$

3.3 Monotonicity

A sqce $\{a_n\}$ is said to be **monotonic increasing** or **ascending** ($\nearrow$ or $\uparrow$) if

$$a_{n+1} \ge a_n\ \forall n \in \mathbb{N}.$$

It is said to be **strictly increasing** if $a_{n+1} > a_n\ \forall n \in \mathbb{N}$. $\{a_n\}$ is said to be **monotonic decreasing** or **descending** ($\searrow$ or $\downarrow$) if

$$a_{n+1} \le a_n\ \forall n \in \mathbb{N}.$$

It is strictly decreasing if $a_{n+1} < a_n\ \forall n \in \mathbb{N}$.

3.4 Convergence of a Sequence—Limit

A sqce $\{a_n\}$ is said to be **convergent** if $\exists$ a number a satisfying the following condition:

$$\forall \epsilon > 0\, \exists \nu \in \mathbb{N} : n \ge \nu \Rightarrow |a_n - a| < \epsilon.$$

The number a is called **the limit** of the sqce $\{a_n\}$. Symbolically,

$$\underset{n\to\infty}{\text{Lt}}\ a_n = a \text{ or simply } \lim a_n = a \text{ or Lt}\, a_n = a \text{ or } a_n \to a \text{ as } n \to \infty.$$

Roughly speaking, a is the limit of $\{a_n\}$, if for **all** the large values of the integral variable n, the terms a_n of the sequence are either very very close to a or exactly equal to a. Hence $\text{Lt}\, a_n = \text{Lt}\, a_{m+n}$, for any finite $m \in \mathbb{N}$.

Examples 3.1. **1.** *If* $a_n = 1/n\ \forall n \in \mathbb{N}$, $\text{Lt}\, a_n = \underset{n\to\infty}{\text{Lt}}\ 1/n = 0$.

2. *If* $a_n = 1^n\ \forall n \in \mathbb{N}$, $\text{Lt}\, a_n = \underset{n\to\infty}{\text{Lt}}\ 1^n = 1$.

In the Ex. 1 a_n is very very close to 0 when n is very large. In the Ex. 2 all the terms of the sequence are equal to 1.

3. $\text{Lt}\{(n+1)/n\} = \text{Lt}(1+1/n) = 1$.

But if $a_n = \{(n+1)/n\}^n\ \forall n \in \mathbb{N}$, its limit is not easy to find out. Although all the terms of the sequence are rational numbers, its limit can be proved to be not rational. Because of this sort of difficulty in finding the limits of many such sequences, the concept of convergence has been introduced by Cauchy.

3.4.1 Cauchy's Principle of Convergence

A sqce $\{a_n\}$ is convergent iff $\forall \epsilon > 0\, \exists \nu \in \mathbb{N}$: m and n both $\geq \nu \Rightarrow |a_m - a_n| < \epsilon$.

It can be proved that **every convergent sqce possesses a limit and vice versa.** Moreover, if $\{a_n\} \nearrow$ and is bounded above or if $\{a_n\} \searrow$ and is bounded below it is convergent. In other words, **every monotonic bounded sequence is convergent.**

A sqce if not convergent is said to be **divergent.** So, if $\{a_n\}$ does not possess a limit, it is divergent. It is said to **diverge properly** if $a_n \to \infty$ or $-\infty$ as $n \to \infty$, and it is said to **diverge improperly** otherwise. An improperly divergent sqce is said to be **oscillatory**.

Examples 3.2. **1.** *If* $a_n = (-1)^n$, *then the terms of the sequence are* $-1, 1, -1, 1, \ldots$.

So $\{a_n\}$ is oscillatory, oscillating between finite limits -1 and 1.

2. *If* $a_n = \{(-1)^n + 1\}n/2$, *then the terms of the sqce* $\{a_n\}$ *are* $0, 2, 0, 4, 0, 6, \ldots$, *so that* $\{a_n\}$ *oscillates between* 0 *and* ∞.

3. *If* $a_n = (-1)^n n$, *then* $\{a_n\}$ *oscillates between* $-\infty$ *and* ∞.

4. *If* $a_n = n$, $\{a_n\}$ *is the natural sqce of natural numbers and it is properly divergent, diverging to* ∞.

5. *If* $a_n = -n$, $\{a_n\}$ *is properly divergent, diverging to* $-\infty$.

3.4.2 Algebraic Properties of Convergence

If the two sequences $\{u_n\}$ and $\{v_n\}$ are both convergent and $\text{Lt}\, u_n = u$ and $\text{Lt}\, v_n = v$, then

(i) $\{u_n + v_n\}$ and $\{u_n v_n\}$ are both convergent and $\text{Lt}(u_n + v_n) = u + v$, i.e., $\text{Lt}\, u_n + \text{Lt}\, v_n$ and $\text{Lt}(u_n v_n) = uv$, i.e., $(\text{Lt}\, u_n)(\text{Lt}\, v_n)$.

Taking $v_n = k$, a constant $\forall n \in \mathbb{N}$, $\text{Lt}\, v_n = k$ and from $\text{Lt}(u_n v_n) = uv$ we get $\text{Lt}(ku_n) = ku$. So taking $k = -1, \text{Lt}(-u_n) = -1u = -u$ and hence $\text{Lt}(u_n - v_n) = \text{Lt}[u_n + (-v_n)] = \text{Lt}\, u_n + \text{Lt}(-v_n) = u - v$.

(ii) If $\text{Lt}\, v_n = v (\neq 0)$, then $\{v_n^{-1}\}$ or $\{1/v_n\}$ is convergent and $\text{Lt}(1/v_n) = 1/v$. Hence $\{u_n/v_n\}$ is convergent and $\text{Lt}(u_n/v_n) = \text{Lt}[u_n \cdot (1/v_n)] = (\text{Lt}\, u_n)(\text{Lt}\, 1/v_n) = u \cdot 1/v = u/v$.

Generalising, if $\{u_n\}$, $\{u_n\}$ and $\{w_n\}$ are all convergent and u, v and w are their limits, then $\{u_n + v_n + w_n\}$ and $\{u_n v_n w_n\}$ are both convergent and $u + v + w$ and uvw are their limits.

Next, if $u_n > 0 \; \forall n \in \mathbb{N}$ and $\{u_n\}$ and $\{v_n\}$ are both convergent having u and v as their limits, then

$$\text{Lt}\, u_n^{v_n} = (\text{Lt}\, u_n)^{\text{Lt}\, v_n} = u^v.$$

An Important Sequence—Definition of e

It is mentioned earlier that sometimes it becomes difficult to find the limit of a sequence. The sequence $\{e_n\}$, where $e_n = [(n+1)/n]^n$ is such a sequence. But it can be proved that $\{e_n\}$ is $\nearrow$ and bounded above. Hence it is convergent and has a limit. This limit is denoted by the symbol e, i.e., $\text{Lt}\, e_n = e$.

$$\underset{n \to \infty}{\text{Lt}} [(n+1)/n]^n = \text{Lt}(1 + 1/n)^n = e.$$

It is also mentioned that e is not rational, although it is the limit of a sequence of rational numbers. The number e is not even algebraic, it is transcendental. e is also defined as the sum of the infinite series $\sum_0^\infty 1/n!$ so that

$$e = \sum_0^\infty 1/n! = 1/0! + 1/1! + 1/2! + \cdots.$$

It lies between 2 and 3, i.e., $2 < e < 3$ and $e \approx 2.7$.

Examples 3.3. **1.** *Prove that* $\text{Lt}[n^2 + 2n + 3)/(2n^2 + 3n + 4)] = 1/2$.

If $a_n = (n^2 + 2n + 3)/(2n^2 + 3n + 4) = (1 + 2/n + 3/n^2)/(2 + 3/n + 4/n^2)$, taking $u_n = 1 + 2/n + 3/n^2$ and $v_n = 2 + 3/n + 4/n^2$ we find that $\text{Lt}\, u_n = 1$ and $\text{Lt}\, v_n = 2$.

$\therefore \quad \text{Lt}\, a_n = \text{Lt}(u_n/v_n) = (\text{Lt}\, u_n)/(\text{Lt}\, v_n) = 1/2.$

2. *Prove that* $\text{Lt}(\sqrt{(n+1)} - \sqrt{n}) = 0$.

Let $\quad a_n = \sqrt{(n+1)} - \sqrt{n}$. Then

$$a_n = [\sqrt{(n+1)} - \sqrt{n}][\sqrt{(n+1)} + \sqrt{n}]/[\sqrt{(n+1)} + \sqrt{n}] = 1/[\sqrt{(n+1)} + \sqrt{n}]$$

$$= (1/\sqrt{n}) \cdot [1/\{\sqrt{(1 + 1/n)} + 1\}] = u_n \cdot v_n, \text{ say.}$$

Since $\quad \text{Lt}\, u_n = \text{Lt}\, 1/\sqrt{n} = 0$ and $\text{Lt}\, v_n = \text{Lt}[1/\{(1 + 1/n) + 1\}]$

$$= 1/\text{Lt}\{(1 + 1/n) + 1\} = 1/2.$$

$\therefore \quad \text{Lt}\, a_n = \text{Lt}(u_n v_n) = (\text{Lt}\, u_n)(\text{Lt}\, v_n) = 0 \cdot 1/2 = 0.$

3. $\text{Lt}\, r^n = 0$ *if* $|r| < 1$, *i.e.*, $-1 < r < 1$.

Case (i) If $r = 0$, $\text{Lt}\, r^n = \text{Lt}\, 0 = 0$.

Case (ii) If $0 < r < 1, \exists h > 0 : r = 1/(1 + h)$. Then $r^n = 1/(1 + h)^n$. Since $(1 + h)^n = 1 + nh + \ldots > nh$

$$r^n = 1/(1 + h)^n < 1/nh \to 0 \text{ as } n \to \infty.$$

Case (iii) If $-1 < r < 0$, $0 < -r < 1$ and $\text{Lt}(-r)^n = 0$ by Case (ii). So, $\text{Lt}\, r^n = 0$.

Case (iv) If $r = 1$, $\text{Lt}\, r^n = \text{Lt}\, 1^n = 1$.

Case (v) If $r > 1$, $r = 1 + h$, $h > 0$. Then $r^n = (1 + h)^n > nh \to \infty$ as $n \to \infty$, so that $\{r^n\}$ diverges to ∞ when $r > 1$.

Case (vi) If $r = -1$, $r^n = 1$ or -1 according as n is even or odd.

Hence $\text{Lt}\, r^n = 1$ or -1 according as $n \to \infty$ through even or odd values. Since there are two different limiting points, $\{r^n\}$ does not converge. It oscillates between -1 and 1.

Case (vii) If $r < -1$, $-r > 1$. So, $\exists h > 0 : -r = 1 + h$ and $(-r)^n = (1+h)^n > nh$.

If n is even $(-r)^n = r^n > nh \to \infty$ as $n \to \infty$.

If n is odd $(-r)^n = -r^n > nh \Rightarrow r^n < -nh \to -\infty$.

Thus $\{r^n\}$ approaches different limits when $n \to \infty$ through different sets of integral values it does not converge. It oscillates between $-\infty$ and ∞ when $r < -1$.

Thus $\{r^n\}$ converges only when $-1 < r \leq 1$.

4. $\text{Lt}\, a^{1/n} = 1$, *where* $a > 0$. *For* $a^{1/n} = e^{(1/n)\ln a} \to e^0 = 1$ *as* $n \to \infty$.

5. $\text{Lt}[n \ln\{(n+1)/n\}] = 1$.

For, $\text{Lt}(1+1/n)^n = e \Rightarrow \text{Lt}\ln(1+1/n)^n = \ln e = 1$.

Cor. 1. $\lim(1+x_n)^{1/x_n} = e$ if $\text{Lt}\, x_n = 0$

2. $\lim(1/x_n)\ln(1+x_n) = 1$, if $\text{Lt}\, x_n = 0$.

6. $\text{Lt}\, n^{1/n} = 1,\ n \in \mathbb{N}$. *For, if* $n > 1$, $n^{1/n} > 1^{1/n} = 1$.

Taking $n^{1/n} = 1 + u_n$, $n = (1+u_n)^n = 1 + nu_n + \{n(n-1)/2\}u_n^2 + \cdots$ by binomial expansion. So,

$$n > \{n(n-1)/2\}u_n^2 \quad \text{or} \quad u_n^2 < 2/(n-1) \to 0 \text{ as } n \to \infty.$$

So, $\text{Lt}\, u_n = 0$. That is $\text{Lt}\, n^{1/n} = 1$.

3.5 Series

Given a sqce $\{u_n\}$, the expression

$$u_1 + u_2 + u_3 + \cdots \text{ or briefly } \sum u_n,$$

whether terminating or non-terminating is called a **series**. The series is finite or infinite according as it is terminating or non-terminating. We shall be discussing infinite series only. So, henceforth, by a series we shall mean an infinite series denoted by $\sum_1^\infty u_n$ or simply by $\sum u_n$.

For the series $\sum u_n$, u_1, u_2, u_3, etc., are called the first, second, third, etc., terms. The nth term is u_n. The sum of the first n terms, denoted usually by S_n, such that $S_n = \sum_1^n u_r$. Then obviously,

$$S_1 = u_1,\ S_2 = u_1 + u_2,\ S_3 = u_1 + u_2 + u_3, \ldots,\ S_n = u_1 + u_2 + u_3 + \ldots + u_n.$$

Thus we get a sequence $\{S_n\}$, called the **sequence of partial sums** of the series $\sum u_n$.

3.5.1 Convergent, Divergent and Oscillatory Series

A series $\sum u_n$ is said to be convergent, divergent or oscillatory according as the sqce $\{S_n\}$ of partial sums is convergent, divergent or oscillatory. If $\{S_n\}$ converges, it will have a limit. This limit is called the **sum** of the series $\sum u_n$. Thus, if $S = \underset{n\to\infty}{\text{Lt}} S_n$, then S is the sum of the series $\sum u_n$ and we write

$$\sum_1^\infty u_n = \underset{n\to\infty}{\text{Lt}} \sum_1^n u_r = S$$

and say that the series $\sum u_n$ converges to S. The series is said to diverge to ∞ or $-\infty$ according as $\text{Lt}\, S_n$ is ∞ or $-\infty$.

Examples 3.4. **1.** *The geometric series* $1 + 1/2 + 1/2^2 + \ldots$ *or* $\sum 1/2^{n-1}$ *is convergent and has the sum 2. For, here*

$$S_n = \sum_{r=1}^{n} 1/2^{r-1} = (1 - 1/2^n)/(1 - 1/2) = 2(1 - 1/2^n)$$

and $\quad \underset{n \to \infty}{\text{Lt}} S_n = \underset{n \to \infty}{\text{Lt}} 2(1 - 1/2^n) = 2.$

2. *If* $u_n = 2^{n-1}$, $\sum u_n = \sum 2^{n-1}$ *is an infinite series which is divergent. For, here*

$$\sum S_n = \sum_{r=1}^{n} 2^{r-1} = (2^n - 1)/(2 - 1) = 2^n - 1 \text{ and } \underset{n \to \infty}{\text{Lt}} S_n = \underset{n \to \infty}{\text{Lt}} (2^n - 1) = \infty.$$

3. *The series* $1 - 1 + 1 - 1 + \ldots$ *is oscillatory.*

For, here $u_r = (-1)^{r-1} \forall r \in \mathbb{N}$ and

$$S_n = \sum_{1}^{n} u_r = \sum_{1}^{n} (-1)^{r-1} = \frac{(-1)^n - 1}{-1 - 1} = \frac{1}{2}\{1 - (-1)^n\} = \frac{1}{2} + \frac{1}{2}(-1)^{n+1}.$$

But $\{(-1)^{n+1}\}$ is oscillatory. So $\underset{n \to \infty}{\text{Lt}} S_n$ does not exist and it is neither ∞ nor $-\infty$. Hence the result. But

$$S_{2n} = 0 \quad \text{and} \quad S_{2n+1} = 1.$$

The sum is said to oscillate between 0 and 1.

3.5.2 Convergence of Arithmetic, Geometric, Harmonic and Arithmetico-geometric Series

Arithmetic Series: If $u_n = a + (n - 1)d$, then $\sum u_n$ is an arithmetic series. We know that

$$S_n = \sum_{1}^{n} u_r = \sum \{a + (r - 1)d\} = \frac{n}{2}[2a + (n - 1)d] = \frac{n}{2}[(2a - d) + nd]$$

$\to \infty$ or $-\infty$ according as $d >$ or < 0. If $d = 0$, $S_n = na$ which diverges to ∞ or $-\infty$ according as $a >$ or < 0. If a and d are both zero, then $S_n = 0$ and the series then converges and has the sum zero also.

Geometric Series: The most general geometric series is $\sum u_n$, where $u_n = gx^{n-1}$, where $g \neq 0$ and $x \neq 0$.

$$S_n = \sum_{1}^{n} u_r = g \sum_{1}^{n} x^{r-1} = g\frac{1 - x^n}{1 - x}, \text{ if } x \neq 1$$

$$= ng, \text{ if } x = 1.$$

If $|x| < 1$, $\text{Lt}_{n \to \infty} x^n = 0$. So $\underset{n \to \infty}{\text{Lt}} S_n = \dfrac{g}{1 - x}$ and the geometric series $\sum u_n$ converges and has the sum $g/(1 - x)$.

If $x = 1$, $S_n = ng \to \infty$ or $-\infty$ according as $g >$ or < 0. So the series diverges.

If $x = -1$, $S_n = \frac{g}{2}\{1 - (-1)^n\}$. Since $\{(-1)^n\}$ is oscillatory, the series is also oscillatory.

If $x > 1$, $S_n = g(x^n - 1)/(x - 1)$ and $x - 1 > 0$ and $x^n \to \infty$ as $n \to \infty$. So, $S_n \to \infty$ or $-\infty$ according as $g >$ or < 0. In this case, therefore, the series diverges.

If $x < -1$, $x - 1 < -1 - 1$, i.e., $-2 < 0$ and x^n oscillates between $-\infty$ and ∞. So, the series is also oscillatory.

Summarising, **we find that the geometric converges only when the common ratio lies between -1 and 1.**

Harmonic Series: Every harmonic series is divergent. In particular, the elementary harmonic series, viz.

$$1 + 1/2 + 1/3 + \cdots \quad \text{or} \quad \sum 1/n$$

diverges to ∞. Similarly, $1 + 1/3 + 1/5 + \cdots$ and $1/2 + 1/5 + 1/7 + \cdots$ are divergent.

Arithmetico-geometric Series: If $\{a_n\}$ is arithmetic and $\{g_n\}$ is geometric, then the series $\sum a_n g_n$ is said to be arithmetico-geometric. For example, if $a_n = n$ and $g_n = x^{n-1}$, $x \neq 0$, then $\{a_n\}$ is arithmetic, whereas $\{g_n\}$ is geometric. So, the series $\sum a_n g_n$, i.e., $t \sum nx^{n-1}$ or $1 + 2x + 3x^2 + \cdots + nx^{n-1} + \cdots$ is an arithmetico-geometric series. Since

$$S_n = 1 + 2x + 3x^2 + 4x^3 + \cdots + nx^{n-1}$$

$$xS_n = x + 2x^2 + 3x^3 + \cdots + (n-1)x^{n-1} + nx^n.$$

On subtraction

$$(1 - x)S_n = 1 + x + x^2 + \cdots + x^{n-1} - nx^n = \frac{1 - x^n}{1 - x} - nx^n.$$

If $-1 < x < 1$, we have already seen that $(1 - x^n)/(1 - x) \to 1/(1 - x)$ as $n \to \infty$. It can be proved that $nx^n \to 0$ as $n \to \infty$ if $-1 < x < 1$. Hence $(1 - x)S_n \to 1/(1 - x)$ or $S_n \to 1/(1 - x)^2$ as $n \to \infty$. Thus

$$1 + 2x + 3x^2 + \cdots + nx^{n-1} = 1/(1 - x)^2$$

or, $\sum nx^{n-1} = 1/(1 - x)^2$.

3.5.3 Sum and Difference of Two Convergent Series and the Product of a Convergent Series by a Constant

If $\sum u_n$ and $\sum v_n$ are both convergent having their sums U and V respectively, then

$$\sum (u_n + v_n) \quad \text{and} \quad \sum (u_n - v_n),$$

called the sum and the difference of the two series, converge and have $U + V$ and $U - V$ as their sums and we write

$$\sum u_n + \sum v_n = \sum (u_n + v_n) = U + V$$

and $\sum u_n - \sum v_n = \sum (u_n - v_n) = U - V.$

Also, if k is any constant

$$k \sum u_n = \sum k u_n = kU.$$

Cor. If the series $\sum u_n$ and $\sum v_n$ are both convergent having sums U and V, and k_1 and k_2 are constants, then

$$\sum (k_1 u_n + k_2 v_n) = k_1 \sum u_n + k_2 \sum v_n = k_1 U + k_2 V.$$

3.5.4 Introduction and Removal of Terms

If $\sum_{n=1}^{\infty} u_n$ be any series, then $\sum_{n=1}^{\infty} u_{n+m}$, where $m \in \mathbb{N}$ is the series obtained from the former by removing its first m terms. Similarly, the series $\sum_{n=1}^{\infty} u_{n-m}$ is the series obtained from $\sum u_n$ by introducing, in the beginning the m terms $u_{1-m}, u_{2-m}, \ldots, u_{-1}$ and u_0. Also the first series, viz., $u_1 + u_2 + u_3 + \cdots$ may be thought to have obtained from the second series $u_{m+1} + u_{m+2} + u_{m+3} + \cdots$ by adding the m terms, viz., $u_1, u_2, \ldots, u_m$ in the beginning.

Theorem 3.1. The two series $\sum u_n$ and $\sum u_{n+m}$, where m is a fixed natural number, either both converge or both diverge.

For the proof see §3.6, Appendix A.

3.5.5 Cauchy's Principle of Convergence of a Series

A necessary and sufficient condition for the convergence of a series $\sum u_n$, according to Cauchy, is that

$$\forall \epsilon > 0\ \exists \nu \in \mathbb{N} : m, n > \nu \Rightarrow |S_m - S_n| < \epsilon$$

$$\text{or, } \forall \epsilon > 0\ \exists \nu \in \mathbb{N} : n > \nu \Rightarrow |S_{n+p} - S_n| < \epsilon,\ \forall p \in \mathbb{W}. \tag{3.1}$$

Here $S_{n+p} - S_n = u_{n+1} + u_{n+2} + \cdots + u_{n+r}$. It is called the pth partial sum of the series $\sum u_n$ after the first n terms. If $\forall p \in \mathbb{N}$, this partial sum tends to zero as $n \to \infty$, then the series $\sum u_n$ converges and has a sum.

Theorem 3.2. A necessary condition for the convergence of the series $\sum u_n$ is that $u_n \to 0$ as $n \to \infty$.

Proof. By Cauchy's principle, if $\sum u_n$ converges, $S_{n+p} - S_n \to 0$ as $n \to \infty$ for every $p \in \mathbb{N}$. Putting $p = 1$, $S_{n+1} - S_n = u_{n+1}$ must tend to zero as $n \to \infty$. So $u_n \to 0$ as $n \to \infty$.

Note. The converse is not necessarily true. For, if $u_n = \frac{1}{n}, u_n \to 0$ as $n \to \infty$. But $\sum u_n = \sum 1/n$, being harmonic is divergent.

Examples 3.5. 1. *Show that the following series are not convergent:*

(i) $\sum \sqrt[3]{[(2n^2 - 1)/(n^2 + n + 1)]}$;

(ii) $\sum [(n+1)^2/n^2 - (n+1)/n]^{-n}$; **[WBUT '01]**

(iii) $\sum n/(1 + 2^{-n})$.

(i) Here $u_n = \sqrt[3]{[(2 - 1/n^2)/(1 + 1/n + 1/n^2)]} \to \sqrt[3]{2}$ as $n \to \infty$. Since $u_n \not\to 0$ as $n \to \infty$, $\sum u_n$ is not convergent.

(ii) Here $u_n = \left[\frac{n+1}{n}\left(\frac{n+1}{n} - 1\right)\right]^{-n} = \left(\frac{n+1}{n} \cdot \frac{1}{n}\right)^{-n} = \frac{n^n}{(1 + 1/n)^n}$.

But $(1 + 1/n)^n \to e$ and $n^n \to \infty$ as $n \to \infty$. Hence $u_n \to \infty$. Since $u_n \not\to 0$ as $n \to \infty$, the series does not converge.

(iii) Here, if $u_n = n/(1 + 2^{-n})$, then as $1 + 2^{-n} \to 1$, but $n \to \infty$ as $n \to \infty$, so u_n also $\to \infty$ as $n \to \infty$. Hence $\sum u_n$ cannot converge.

3.5.6 Series of Positive Terms

If $u_n > 0\ \forall n \in \mathbb{N}$, then $\sum u_n$ is called a series of positive terms or simply a positive series.

Theorem 3.3. A series $\sum u_n$, where $u_n > 0 \ \forall n \in \mathbb{N}$, either converges or diverges to ∞.

Proof. If $S_n = \sum_1^n u_r$, then since $S_{n+1} - S_n = u_{n+1} > 0$, $\{S_n\} \uparrow$. So $\{S_n\}$ is convergent if it is bounded above, otherwise it is divergent.

Examples 3.6. 1. *If* $u_n = 1/2^{n-1} \ \forall n \in \mathbb{N}$, $u_n > 0 \ \forall n \in \mathbb{N}$ *and* $S_n = \sum_1^n u_r = \sum_1^n 1/2^{r-1} = 2(1 - 1/2^n) < 2 \ \forall n \in \mathbb{N}$.

So, $\{S_n\}$ is bounded above and hence it is convergent. Since

$$\operatorname*{Lt}_{n\to\infty} S_n = \operatorname*{Lt}_{n\to\infty} 2(1 - 1/2^n) = 2, \sum u_n = 2.$$

2. *If* $u_n = 1/(n-1)! \ \forall n \in \mathbb{N}$, $u_n > 0 \ \forall n \in \mathbb{N}$.

$S_n = 1 + \sum_2^n 1/(r-1)!$.

If $r \geq 2, (r-1)! = 1 \cdot 2 \cdots (r-1) \geq 2 \cdot 2 \cdots 2 = 2^{r-2}$.

$\therefore \quad 1/(r-1)! < 1/2^{r-2}$ if $r > 2$.

So, if $n > 2$,

$$S_n \leq 1 + \sum_2^n 1/2^{r-2} = 1 + 2(1 - 1/2^{n-1}) < 3.$$

Thus $\{S_n\}$ is bounded above. Hence $\sum 1/(n-1)!$ is convergent.

3.5.7 Introduction and Removal of Brackets

The terms of a given series $\sum u_n$ may be grouped and a new series $\sum v_n$ may be formed, where

$$v_1 = u_1 + u_2 + \cdots + u_{m_1}, \ v_2 = u_{m_1+1} + u_{m_1+2} + \cdots + u_{m_2}, \cdots$$

Then the series $\sum v_n$ is said to be formed from $\sum u_n$ by introducing brackets:

$$\sum v_n = v_1 + v_2 + \cdots = (u_1 + \cdots + u_{m_1}) + (u_{m_1+1} \cdots + u_{m_2}) + \cdots$$

So, the series $\sum u_n$ may be regarded as obtained from $\sum v_n$ by removing the brackets.

For example, by grouping the terms of the series

$$1 + 2 + 3 + 4 + \cdots$$

we get a series

$(1) + (2 + 3) + (4 + 5 + 6) + \cdots$

or, $\quad 1 + 5 + 15 + \cdots$

Here $v_1 = 1, v_2 = 2 + 3 = 5, \ v_3 = 4 + 5 + 6 = 15, \ldots$

Theorem 3.4. Let $\sum u_n$ be a positive series and $\sum v_n$ is obtained from $\sum u_n$ by grouping its terms, so that each group contains a finite number of terms. Then $\sum u_n$ and $\sum v_n$ either both converge or both diverge.

3.5.8 Rearrangement of Terms of a Series

If $\sum u_n$ and $\sum v_n$ are such that every term of $\sum u_n$ is a term of $\sum v_n$ and vice versa, then one of them is regarded as obtained from the other by rearrangement of its terms.

For example, given the series $1 + 2 + 3 + 4 + \cdots$, the series $2 + 1 + 4 + 3 + \cdots$ is obtained by rearranging the terms of the first series.

Theorem 3.5. If $\sum u_n$ is a convergent series of positive terms and $\sum u'_n$ is obtained from $\sum u_n$ after rearranging its terms, then $\sum u'_n$ is also convergent having the same sum as that of $\sum u_n$.

For the proof see Appendix A, §3.6.

3.5.9 Tests of Convergence of Series of Positive Terms

As per definition, a series is convergent iff the sequence of its partial sums is convergent. Since it is not always possible to express S_n, the nth partial sum, as a convenient function of n, the limit of S_n cannot be easily found out. Other methods, not based on S_n, are used. These methods are known as 'tests of convergence'.

3.5.10 Comparison Test

Theorem 3.6. Let $\sum u_n$ and $\sum v_n$ be both series of positive terms. If either (a) $u_n \le kv_n \; \forall n \ge m$, where $m(\in \mathbb{N})$ is fixed and $k(> 0)$ is also fixed or, (a′) $\operatorname*{Lt}_{n\to\infty} (u_n/v_n) = l$, then $\sum u_n$ is convergent, whenever $\sum v_n$ is convergent.

If either (b) $u_n \ge kv_n \; \forall n \ge m$, where as before m and k are fixed or, (b′) $\operatorname*{Lt}_{n\to\infty} (u_n/v_n) = l'$, a positive number or ∞, then $\sum u_n$ is divergent, whenever $\sum v_n$ is divergent.

The series $\sum v_n$, with the help of which we test the convergence or divergence of $\sum u_n$ is called a **test series.**

For the proof see Appendix A, §3.6.

3.5.11 *p*-series—An Important Test Series

Theorem 3.7. The series $\sum 1/n^p$ is convergent iff $p > 1$.

For the proof see Appendix A, §3.6.

Examples 3.7. 1. *Test the series* $1/1 \cdot 2 \cdot 3 + 3/2 \cdot 3 \cdot 5 + 5/3 \cdot 4 \cdot 5 + \cdots$ *for convergence.*

Let u_n be the nth term of the given series. Then

$$u_n = (2n - 1)/n(n + 1)(n + 2),$$

whose numerator is a polynomial of degree one in n and the denominator is that of degree 3. The degree of the denominator minus the degree of the numerator $= 3 - 1 = 2$. So we take $v_n = 1/n^2$. Then

$$\operatorname*{Lt}_{n\to\infty} (u_n/v_n) = \operatorname*{Lt}_{n\to\infty} \frac{n(2n-1)}{(n+1)(n+2)} = \operatorname*{Lt}_{n\to\infty} \frac{2 - 1/n}{(1+1/n)(1+2/n)} = 2.$$

But $\sum v_n = \sum 1/n^2$ is convergent. So, $\sum u_n$ is also convergent.

2. *Test the convergence of the series:*

$$(1+2)/2^3 + (1+2+3)/3^3 + (1+2+3+4)/4^3 + \cdots$$

Here $u_{n-1} = (1+2+3+\cdots+n)/n^3$ for $n = 2, 3, 4, \ldots$

$= (n+1)/2n^2$

So, $u_n = (n+2)/2(n+1)^2$.

Degree of denominator − Degree of numerator $= 2 - 1 = 1$.

So, we take $v_n = 1/n$ for comparison.

$$\operatorname*{Lt}_{n\to\infty} (u_n/v_n) = \frac{1}{2} \operatorname*{Lt}_{n\to\infty} \frac{n(n+2)}{(n+1)^2} = \frac{1}{2} \operatorname*{Lt}_{n\to\infty} \frac{1+2/n}{1+1/n)^2} = \frac{1}{2}.$$

Since $\sum v_n$, being a harmonic series, is divergent, so the given series is also divergent.

3. *Test the series* $\sum u_n$ *for convergence, where* $u_n = \sqrt{n^4+1} - \sqrt{n^4-1}$.

Here $u_n = \dfrac{2}{\sqrt{n^4+1}+\sqrt{n^4-1}} = \dfrac{2}{n^2\sqrt{1+1/n^4}+n^2\sqrt{1-1/n^4}}$

$= \dfrac{2}{n^2}.w_n,$ where $w_n = \dfrac{1}{\sqrt{(1+1/n^4)}+\sqrt{(1-1/n^4)}} \to 1$ as $n \to \infty$.

So, we choose $v_n = 1/n^2$ for the test series.

$$u_n/v_n = 2w_n \to 2 \text{ as } n \to \infty.$$

Since $\sum v_n$, i.e., $\sum 1/n^2$ is convergent, the given series is also convergent.

4. *Prove that the series* $2/1 + 3/4 + 4/9 + \cdots$ *is divergent.*

Here $u_n = (n+1)/n^2 = 1/n + 1/n^2 \to 0$ as $n \to \infty$. So we cannot make any definite conclusion regarding the convergence of the series. Since u_n is a rational function of n and the degree of the denominator minus the degree of the numerator is 1, we take $v_n = 1/n^1$ or $1/n$.

Since $u_n/v_n = 1 + 1/n > 1\ \forall n \in \mathbb{N}$ and the harmonic series $\sum v_n$ divergent so the given series is divergent.

5. *Show that the series* $\sum u_n$ *is divergent, where*

$$u_n = \sqrt[3]{(2n^2-1)}/\sqrt[4]{(3n^3+2n+5)}.$$

Here $u_n = \left[n^{2/3}\sqrt[3]{(2-1/n^2)}\right] \div \left[n^{3/4}\sqrt[4]{(3+2/n^2+5/n^3)}\right]$

$= \dfrac{1}{n^{1/12}}w_n,$ where $w_n = \dfrac{\sqrt[3]{(2-1/n^2)}}{\sqrt[4]{(3+2/n^2+5/n^3)}} \to \dfrac{\sqrt[3]{2}}{\sqrt[4]{3}}$ as $n \to \infty$.

So taking $v_n = 1/n^{1/12}$, $u_n/v_n = w_n \to \sqrt[3]{2}/\sqrt[4]{3}$.

But $\sum v_n = \sum 1/n^p$ is divergent, since $p = 1/12 < 1$. Hence $\sum u_n$ is divergent.

6. *For what values of* x *is the series* $1/2^{\alpha+x} + 2^\alpha/3^{\alpha+x} + 3^\alpha/4^{\alpha+x} + \cdots$ *convergent?*

Here $u_n = n^\alpha/(n+1)^{\alpha+x} = \left(\dfrac{n}{n+1}\right)^\alpha \dfrac{1}{(n+1)^x}$.

We take $v_n = 1/n^x$. Then $u_n/v_n = \left(\dfrac{n}{n+1}\right)^{\alpha+x} \to 1$ as $n \to \infty$.

But $\sum v_n$ converges only when $x > 1$. So the given series also converges when $x > 1$.

Exercises 3.1

1. Prove that $\sum u_n$ is convergent where u_n is

   (i) $(n+1)/(n^3+2)$, (ii) $(n^2+n-1)/n^4$, (iii) $(n^2+n-1)^{1/2}/(n^5-2)^{1/2}$,

   (iv) $(1/n)\sin(1/n)$, (v) $\tan(1/n^2)$.

2. Test the convergence of the following series:

   (a) $1/1\cdot 2^2 + 1/2\cdot 3^2 + 1/3\cdot 4^2 + \cdots$,

   (b) $\sum[\sqrt{(n^3+1)} - \sqrt{(n^3-1)}]$,

   (c) $\sum(n^3+1)/(2n^4+3n-1)$,

   (d) $\sum\sqrt{[n/(n+1)^3]}$,

   (e) $\sum\sqrt{[(2^n-1)/(3^n-1)]}$.

3. Test the following series for convergence or divergence mentioning the range of values of x for convergence:

   (a) $\sum x^{n-1}/(1+x^n)$, (b) $\sum 1/(x^n+x^{-n})$, (c) $\sum a^n/(x^n+a^n)$.

4. Test the convergency of the series $\sum\{\sin(1/n)\}/\sqrt{n}$. **[WBUT '01]**

5. Test the series $\sum\sqrt{n}/(an^{3/2}+b)$, where $a>0$. **[WBUT '04]**

6. Test the convergence of the series $\sum u_n$, where $u_n=(n^3+1)^{1/3}-n$.

Answers

2.(a) conv.; **(b)** conv.; **(c)** div.; **(d)** div.;
(e) conv. **3.** conv. **(a)** $x<1$; **(b)** $x\neq 1$; **(c)** $x>a$.
4. conv. **5.** div. **6.** conv.

3.5.12 D'Alembert's Ratio Test

Theorem 3.8. A series $\sum u_n$ of positive terms is convergent if either (a) $u_{n+1}/u_n \le k < 1\ \forall n \ge m$, where m and k are fixed or, (a′) $\operatorname{Lt}_{n\to\infty}(u_{n+1}/u_n) = l < 1$; and the series is divergent if either (b) $u_{n+1}/u_n \ge 1\ \forall n \ge m$ or (b′) $\operatorname{Lt}_{n\to\infty}(u_{n+1}/u_n) = l > 1$.

For the proof see Appendix A, §3.6.

Note. If $\operatorname{Lt}_{n\to\infty}(u_{n+1}/u_n)=1$, then the ratio test fails and further investigation is required. For example, if $u_n = 1/n^p$, the convergence of $\sum u_n$ depends upon the value of p. But $\operatorname{Lt}_{n\to\infty}(u_{n+1}/u_n) = \operatorname{Lt}_{n\to\infty}\{n^p/(n+1)^p\} = \operatorname{Lt}_{n\to\infty}\{1/(1+1/n)^p\} = 1$ for all p. So the ratio test gives no information regarding the convergence of the series when the limit is 1.

Examples 3.8. 1. *Show that the series $\sum(2n-1)/n!$ is convergent.*

Here $u_n = (2n-1)/n! = 2/(n-1)! - 1/n! \to 0$ as $n\to\infty$.

So, no immediate conclusion can be made. But in (2), Ex. 3.6 we have seen that $\sum v_n = \sum 1/(n-1)!$ is convergent. So, $\sum w_n = \sum 1/n!$ is also convergent. Hence $\sum u_n$, i.e., $2\sum v_n - \sum w_n$ is also convergent.

The convergence of $\sum u_n$ can be easily proved by the ratio test.

$\because \quad u_{n+1}/u_n = [(2n+1)/(n+1)!]/[(2n-1)/n!] = (2n+1)/(2n-1)(n+1)$,

$\therefore \quad u_{n+1}/u_n \to 0$ as $n \to \infty$. Since here $l = 0 < 1$, $\sum u_n$ is convergent.

2. *Test the convergence of* $1 + \sum (n^n/n!)$.

Here $u_1 = 1,\ u_2 = 1^1/1! = 1,\ u_3 = 2^2/2!, \ldots$

If $n > 1$, $u_n = (n-1)^{n-1}/(n-1)!$.

$$\therefore \quad u_{n+1}/u_n = \frac{n^n}{n!}\,\frac{(n-1)!}{(n-1)^{n-1}} = \frac{n^{n-1}}{(n-1)^{n-1}} = \frac{1}{(1-1/n)^{n-1}}$$

$$= \left(\frac{n}{n-1}\right)^{n-1} = \left(1 + \frac{1}{n-1}\right)^{n-1} \to e(>1) \text{ as } n \to \infty,$$

for $(1+1/n)^n \to e \Rightarrow [1 + 1/(n-1)]^{n-1} \to e$ also as $n \to \infty$. Since $l = e > 1$, the series is divergent.

3. *Find whether the series:* $1 + 1/2^2 + 2^2/3^3 + 3^3/4^4 + \cdots$ *is convergent or divergent.*

Let $u_n = n^n/(n+1)^{n+1}\ \forall n \in \mathbb{N}$. Then

$$u_n = \frac{1}{n+1} \cdot \frac{n^n}{(n+1)^n} = \frac{1}{n+1} \cdot \frac{1}{(1+1/n)^n} \to 0 \cdot \frac{1}{e} = 0 \text{ as } n \to \infty.$$

So, no immediate conclusion can be made regarding the convergence of the series. But taking $v_n = 1/n$ we find that $u_n/v_n = n^n/(n+1)^n = 1/(1+1/n)^n \to 1/e$ as $n \to \infty$. But $\sum v_n$, i.e., $\sum 1/n$ diverges. Hence $\sum u_n$ also diverges and $1 + \sum u_n$, i.e., the given series diverges.

Here the ratio test fails. For,

$$\frac{u_{n+1}}{u_n} = \frac{(n+1)^{n+1}}{(n+2)^{n+2}} \cdot \frac{(n+1)^{n+1}}{n^n} = \left(\frac{n+1}{n+2}\right)^{n+1} \frac{1}{n+2} \cdot (n+1) \cdot \left(\frac{n+1}{n}\right)^n$$

$$= \frac{1}{\{1 + 1/(n+1)\}^{n+1}} \cdot \frac{n+1}{n+2}(1+1/n)^n \to \frac{1}{e} \cdot 1 \cdot e, \text{ i.e., } 1 \text{ as } n \to \infty.$$

4. *For which positive values of x is the series* $\sum x^{n-1}/[(3n-2)(3n-1)3n]$ *convergent?*

Taking $\quad u_n = x^{n-1}/[(3n-2)(3n-1)3n]$,

$$u_{n+1}/u_n = (3n-1)3n(3n+1)x/[(3n-2)(3n-1)3n]$$

$$= (3 - 1/n)(3 + 1/n)x/[(3 - 2/n)(3 - 1/n)] \to x \text{ as } n \to \infty.$$

So, by the ratio test $\sum u_n$ converges when $0 \le x < 1$ and diverges when $x > 1$.

5. *Test the series* $1 + 2^2 \cdot x/3^2 + 2^2 \cdot 4^2 \cdot x^2/3^2 \cdot 5^2 + 2^2 \cdot 4^2 \cdot 6^2 \cdot x^3/3^2 \cdot 5^2 \cdot 7^2 + \cdots$ *for convergence, where* $x \ne 1$. **[WBUT '04]**

Taking $u_0 = 1,\ u_n = 2^2 \cdot 4^2 \cdots (2n)^2 x^n/3^2 \cdot 5^2 \cdots (2n+1)^2$,

$$u_{n+1} = 2^2 \cdot 4^2 \cdots (2n+2)^2 x^{n+1}/3^2 \cdot 5^2 \cdots (2n+3)^2.$$

So, $\quad u_{n+1}/u_n = (2n+2)^2 x/(2n+3)^2 = (2 + 2/n)^2 x/(2 + 3/n)^2 \to x$ as $n \to \infty$.

Hence by the ratio test $\sum u_n$ converges if $0 \le x < 1$ and diverges if $x > 1$.

6. *Show that the series:* $x/(a + \sqrt{1}) + x^2/(a + \sqrt{2}) + x^3/(a + \sqrt{3}) + \cdots$ *converges when* $x < 1$*, x being positive.*

If $u_n = x^n/(a+\sqrt{n})$,

$$\frac{u_{n+1}}{u_n} = \frac{x(a+\sqrt{n})}{a+\sqrt{(n+1)}} = \frac{a/\sqrt{n}+1}{a/\sqrt{n}+\sqrt{(1+1/n)}}x \to x \text{ as } n \to \infty.$$

So, $\sum u_n$ converges if $x < 1$ and diverges if $x > 1$.

When $x = 1$, $u_n = \dfrac{1}{a+\sqrt{n}}$. Taking $v = \dfrac{1}{\sqrt{n}}$,

$$u_n/v_n = \sqrt{n}/(a+\sqrt{n}) = 1/(a/\sqrt{n}+1) \to 1 \text{ as } n \to \infty.$$

But $\sum v_n$ diverges, because $p = 1/2$. So, $\sum u_n$ also diverges.

7. *Test the convergence of the series*

$$\left(\frac{1}{3}\right)^2 + \left(\frac{1\cdot 2}{3\cdot 5}\right)^2 + \left(\frac{1\cdot 2\cdot 3}{3\cdot 5\cdot 7}\right)^2 + \cdots$$ **[WBUT '02]**

Here $u_n = \left\{\dfrac{1\cdot 2\cdot 3\cdots n}{3\cdot 5\cdot 7\cdots(2n+1)}\right\}^2 \forall n \in \mathbb{N}$.

$u_{n+1}/u_n = \left(\dfrac{n+1}{2n+3}\right)^2 = \left(\dfrac{1+1/n}{2+3/n}\right)^2 \to \dfrac{1}{4}$ as $n \to \infty$ and since $1/4 < 1$, by the ratio test the series converges.

Exercises 3.2

1. Discuss the convergence of the following series, where the nth term is—

(i) $n^2/3^n$, **(ii)** $n!/n^n$, **(iii)** $n!2^n/n^n$ **[WBUT '03]**,

(iv) $(n!)^2/\prod_{1}^{2n-1}(2r-1)$, **(v)** $\prod_{1}^{n}[r/(2r+1)]^2$, **(vi)** $n^2/2^n + 1/n^2$.

2. Test, for convergence, the following series:

(i) $1/(1+2) + 2/(1+2^2) + 3/(1+2^3) + \cdots$;

(ii) $1!2/1^1 + 2!2^2/2^2 + 3!2^3/3^3 + \cdots$;

(iii) $2/1 + 2\cdot 5\cdot 8/1\cdot 5\cdot 9 + 2\cdot 5\cdot 8\cdot 11/1\cdot 5\cdot 9\cdot 13 + \cdots$;

(iv) $1 + (\sqrt{2}-1)/1! + (\sqrt{2}-1)^2/2! + (\sqrt{2}-1)^3/3! + \cdots$. **[WBUT '01]**

3. If $x > 0$, test the following series for convergence mentioning the range of values of x for which it is convergent:

(i) $\sum nx^n$; **(ii)** $\sum x^n/n$; **(iii)** $\sum x^n/(n^2+1)$; **[WBUT '03]**

(iv) $\sum x^{2n-2}/(n+1)\sqrt{n}$; **(v)** $\sum(2^{n+1}-2)\cdot x^n/(2^n+1)$.

4. Discuss the convergence of the following series. Assuming $x > 0$, find the range of values of x for which the series converges:

(i) $\sum(\ln x)^{-n}$; **(ii)** $\sum n/2^n$; **(iii)** $\sum\{n/(n+1)\}^{n^2}$;

(iv) $\sum\{n/(2n+1)\}^n$; **(v)** $\sum 2^{-n-(-1)^n}$; **(vi)** $\sum\{n/(n+1)\}^n x^n$;

(vii) $\sum\{(n+1)x/n\}^n/n$.

Answers

1. **(i)** conv.; **(ii)** conv.; **(iii)** conv.; **(iv)** div.;
(v) conv.; **(vi)** conv.

2. **(i)** conv.; **(ii)** conv.; **(iii)** conv.; **(iv)** conv.

3. conv. for **(i)** $x < 1$; **(ii)** $x < 1$; **(iii)** $x \le 1$;
(iv) $|x| \le 1$; **(v)** $x < 1$.

4. **(i)** conv.; **(ii)** conv.; **(iii)** conv.; **(iv)** conv.;
(v) conv.; **(vi)** conv. for $x < 1$; **(vii)** conv. for $x < 1$.

3.5.13 Cauchy's Root Test

Theorem 3.9. A series of positive terms $\sum u_n$ is convergent if either (a) $u_n^{1/n} \le k < 1\ \forall n \ge m$, k and n being fixed or, (a′) $\underset{n\to\infty}{\text{Lt}}\ u_n^{1/n} = l$, where $l < 1$; and divergent if either (b) $u_n^{1/n} \ge 1$ for an infinite number of values of n or, (b′) $\underset{n\to\infty}{\text{Lt}}\ u_n^{1/n} = l$, where $l > 1$.

For the proof see §3.6, Appendix A.

Examples 3.9. **1.** *Test the convergence of the series:*

$$\frac{1}{2} + \frac{1}{3} + \frac{1}{2^2} + \frac{1}{3^2} + \frac{1}{2^3} + \frac{1}{3^3} + \cdots$$

Here $u_{2n-1} = \dfrac{1}{2^n}$, $u_{2n} = \dfrac{1}{3^n}$ and $u_{2n+1} = \dfrac{1}{2^{n+1}}$.

Now $u_{2n}/u_{2n-1} = (2/3)^n \to 0$ as $n \to \infty$ and $u_{2n+1}/u_{2n} = (1/2)(3/2)^n \to \infty$ as $n \to \infty$.

So, the limit of u_{n+1}/u_n as $n \to \infty$ does not exist. Therefore, no conclusion can be made from the ratio test. Applying the root test we find that

$$u_{2n-1}^{1/(2n-1)} = (1/2^n)^{1/(2n-1)} = (1/2)^{n/(2n-1)} = (1/2)^{1/(2-1/n)}.$$

Since $2 - 1/n < 2$, $1/(2 - 1/n) > 1/2$.

But $\quad \dfrac{1}{2} < 1$ and $1/(2 - 1/n) > \dfrac{1}{2} \Rightarrow u_{2n-1}^{1/(2n-1)} < 1/\sqrt{2}\ \forall n \in \mathbb{N}$.

Also $\quad u_{2n}^{1/2n} = \left(\dfrac{1}{3^n}\right)^{1/2n} = \dfrac{1}{\sqrt{3}}\ \forall n \in \mathbb{N}$.

Since $u_n^{1/n} \le k < 1$, where $k = 1/\sqrt{2}$ or $1/\sqrt{3}$, $\forall n \in \mathbb{N}$, so, by root test the series converges.

Alternatively, grouping the terms in pairs

$$1/2^n + 1/3^n \le 2 \cdot 1/2^n = 1/2^{n-1} = v_n, \text{ say.}$$

Then $\sum v_n$ converges. Hence by Theorem 3.4 the given series is convergent.

2. *Test the series $\sum u_n$ for convergence where $u_n = (2^n + 1)/(3^n + 2)$.*

Here $\quad u_n = \dfrac{(2/3)^n + (1/3)^n}{1 + 2/3^n} \to 0$ as $n \to \infty$.

So, no immediate conclusion regarding convergence.

But $\quad \dfrac{u_{n+1}}{u_n} = \dfrac{2^{n+1}+1}{3^{n+1}+2} \cdot \dfrac{3^n+2}{2^n+1} = \dfrac{2+1/2^n}{1+1/2^n} \cdot \dfrac{1+2/3^n}{3+2/3^n} \to \dfrac{2}{3} < 1$ as $n \to \infty$.

So, by the ratio test $\sum u_n$ converges.

3.5.14 Alternating Series—Its Convergence

If $u_n > 0\ \forall n \in \mathbb{N}$, then the series $\sum(-1)^{n-1}u_n$ or $\sum(-1)^n u_n$ is called an alternating series. The signs of the terms alternate in such a series.

Leibnitz's Test: Theorem 3.10. If $\{u_n\}$ is a monotonically decreasing null sequence, then the series $\sum(-1)^{n-1}u_n$ converges.

For the proof see Appendix A, §3.6.

Example. *If $p > 0$, $1 - 1/2^p + 1/3^p - 1/4^p + \cdots$ is convergent for taking $u_n = 1/n^p$, we find that $u_{n+1} - u_n = 1/(n+1)^p - 1/n^p = -\{(n+1)^p - n^p\}/\{n(n+1)\}^p < 0$. So $\{u_n\}$ is $\downarrow$ and since $\operatorname{Lt}_{n\to\infty} u_n = \operatorname{Lt}_{n\to\infty} (1/n^p) = 0$, $\{u_n\}$ is null. Hence by Leibnitz's test the given series is convergent.*

3.5.15 Absolute and Conditional Convergence

Let $\sum u_n$ be any series, not necessarily of positive terms. Then $\sum |u_n|$ is a series of non-negative terms. The series $\sum u_n$ is said to be **absolutely convergent** if $\sum |u_n|$ is convergent.

Since $u_n \leq |u_n|\ \forall n \in \mathbb{N}$, we find that $\sum u_n$ converges whenever $\sum |u_n|$ is convergent. Thus **every absolutely convergent series is convergent.** But it may happen that $\sum u_n$ is convergent, but $\sum |u_n|$ is not convergent, that is to say, $\sum u_n$ is not absolutely convergent, then the series is said to be **conditionally convergent or semi-convergent.**

Examples 3.10. 1. *$\sum(-1)^{n-1}/2^n$ is absolutely convergent. For, if $u_n = (-1)^{n-1}/2^n$, $|u_n| = 1/2^n$ and $\sum |u_n|$ or $\sum 1/2^n$, being a geometric series with the common ratio $1/2 (< 1)$ is convergent.*

2. *If $u_n = (-1)^{n-1}/n$, $\sum u_n$ is semi-convergent. For, $|u_n| = 1/n$ which tends to zero as n tends to ∞ and it is monotonically decreasing. So, by Leibnitz's test $\sum u_n$ is convergent. But $\sum |u_n| = \sum 1/n$, which is divergent.*

3. *If $a \geq 0$, the series $\sum(-1)^{n-1}/(n+a)^p$ is (i) absolutely convergent if $p > 1$ and (ii) conditionally convergent if $p \leq 1$.*

Let $u_n = 1/(n+a)^p = \left|(-1)^{n-1}/(n+a)^p\right|$. Then $\{u_n\}$ is $\downarrow$ and null if $p > 0$. So the given series is convergent when $p > 0$.

Taking $v_n = 1/n^p$ we find that $u_n/v_n = n^p/(n+a)^p = 1/(1+a/n)^p \to 1$ as $n \to \infty$. So u_n and $\sum v_n$ either both converge or both diverge. But $\sum v_n$ converges only when $p > 1$ and diverges when $p \leq 1$. So $\sum u_n$ converges when $p > 1$ and diverges when $p \leq 1$. Therefore, the given series is (i) absolutely convergent when $p > 1$ and (ii) conditionally convergent when $p \leq 1$.

Exercises 3.3

1. Test the convergence of the following alternating series:

(i) $\sum(-1)^{n-1}/(2n-1)!$; **(ii)** $\sum(-1)^{n-1}(n+1)/n$; **(iii)** $\sum(-1)^{n-1}n/(n^2+1)$;

(iv) $\sum(-1)^{n-1}n/(2n+3)$; (v) $\sum(\cos n\pi)/n(n+1)$.

2. Find the range of values of x for the absolute convergence of $\sum n(-x)^{n-1}$.

3. If $0 < x < 1$, show that $\sum(-1)^{n-1}x^n/(1+x^n)$ is convergent.

4. Examine the following series for absolute convergence:

(i) $\sum(-1)^n(n+1)/n^3$; (ii) $\sum(-1)^{n-1}n^2/n!$; (iii) $\sum(-1)^n(n!)^2/(2n)!$.

5. State Leibnitz's test for the convergence of an alternating series. Prove that the series $\sum(-1)^n x^n/n$ is absolutely convergent when $|x| < 1$ and conditionally convergent when $x = 1$. **[WBUT '01]**

Answers

1. **(i)** conv.; **(ii)** div.; **(iii)** conv.; **(iv)** div.; **(v)** conv. **2.** Absolutely convergent for $|x| < 1$.

4. **(i)** abs. conv.; **(ii)** div.; **(iii)** abs. conv.

3.6 Appendix A

Proofs of Some Theorems in Chapter 3

1. Theorem 3.1. Since m is fixed, $S_m = \sum_1^m u_r$ is a constant.

$$\text{Now } S_{m+n} = \sum_1^{m+n} u_r = \sum_1^m u_r + \sum_1^n u_{m+r} = S_m + T_n, \tag{i}$$

where T_n is the sum of the first n terms of the series $\sum_{r=1}^{\infty} u_{m+r}$.

$$\text{But } \operatorname*{Lt}_{n\to\infty} S_n = S \iff \operatorname*{Lt}_{n\to\infty} S_{m+n} = S. \text{ So, } \operatorname*{Lt}_{n\to\infty} T_n = \operatorname*{Lt}_{n\to\infty} S_{m+n} - S_m = S - S_m. \tag{ii}$$

So, $\{T_n\}$ converges, i.e., $\sum_1^{\infty} u_{m+n}$ converges. Thus $\sum u_n$ and $\sum u_{m+n}$ converge simultaneously.

Again, from (i) since $T_n = S_{m+n} - S_m$, if $\{S_{m+n}\}$ i.e., $\{S_n\}$ diverges $\{T_n\}$ diverges. So $\sum u_{m+n}$ diverges if $\sum u_n$ diverges. Thus the series $\sum u_n$ and $\sum u_{m+n}$ either both converge or both diverge. In other words, the nature of a series with regard to convergence does not alter by the introduction of removal of a finite number of terms of the series. But from (ii) we observe that the sum may alter by the introduction or removal of terms.

2. Theorem 3.5. Let $T_n = \sum_{i=1}^{n} u_{r_i} = u_{r_1} + u_{r_2} + \cdots + u_{r_n} = u'_1 + u'_2 + \cdots + u'_n$.

Here $r_1, r_2, \ldots, r_n$ are all natural numbers. Let $m = \max\{r_1, r_2, \ldots, r_n\}$. Then obviously $T_n \le S_m < S$, where $S = \sum u_n$. So $\{T_n\}$ is bounded above. But $\{T_n\} \uparrow$. Hence $\{T_n\}$ and consequently $\sum u_{r_n}$ converges and if T is its sum, $T \le S$. Similarly, the original series $\sum u_n$ may be regarded as one obtained from $\sum u_{r_n}$ by rearranging its terms. So $S \le T$. Hence $T = S$. Similarly, it is not difficult to prove that $\sum u_{r_n}$ diverges when $\sum u_n$ diverges.

3. Theorem 3.6. **(a)** Let $u_n \le kv_n \, \forall n \in \mathbb{N}$ and let $\sum v_n$ be convergent and V be its sum. Now $u_n \le kv_n \, \forall n \in \mathbb{N} \Rightarrow \sum_1^n u_r \le k \sum_1^n v_r \le kV$. (i)

If $S_n = \sum_1^n u_r$, (i) shows that $\{S_n\}$ is bounded above. Since $\{S_n\} \uparrow$, $\sum u_n$ converges.

If $u_n \leq kv_n$ not for all $n \in \mathbb{N}$ but for $n \geq m$, where m is a fixed natural number, then since the removal of a finite number of terms of a series does not alter the convergence, $\sum_{n=0}^{\infty} v_{m+n}$ converges whenever $\sum v_n$ converges. But

$$u_n \leq kv_n \; \forall n \geq m \Rightarrow \sum_1^n u_{m+r} \leq k \sum_1^n v_{m+r} \; \forall n \geq 1.$$

So, $\sum u_{m+n}$ is convergent. Introducing the first m terms in the beginning, we get $\sum_1^{\infty} u_n$, which is therefore convergent. Thus $\sum u_n$ converges whenever $\sum v_n$ converges.

(a′) When $\underset{n\to\infty}{\text{Lt}} (u_n/v_n) = l$, $\forall \epsilon > 0 \, \exists m \in \mathbb{N}$: $|u_n/v_n - l| < \epsilon \; \forall n \geq m$

or, $l - \epsilon \leq u_n/v_n < l + \epsilon \; \forall n \geq m$.

Taking $k = l + \epsilon$ we get $u_n/v_n < k \; \forall n \geq m$, a condition which is the same as in (a).

(b) Next, let $u_n \geq kv_n \; \forall n \in \mathbb{N}$. Then

$$S_n = \sum_1^n u_r \geq k \sum_1^n v_r = kS'_n, \text{ say.}$$

If $\sum v_n$ diverges, it diverges to ∞. If $u_n \geq kv_n$ not for all $n \in \mathbb{N}$ but for $n \geq m$, a fixed number, then since by the removal of a finite number of terms the convergence or divergence is not affected $\sum u_n$ will diverge if $\sum v_n$ diverges. So, $\sum_{n=1}^{\infty} u_{m+n}$ will be divergent. Introducing the first m terms in the beginning, the series $\sum u_n$ will also be divergent.

(b′) When $\underset{n\to\infty}{\text{Lt}} (u_n/v_n) = l'$ and $l' > 0$, choosing $\epsilon (> 0)$, small enough, $l' - \epsilon$ may be made positive. We shall then get $m \in \mathbb{N}$ such that $l' - \epsilon < u_n/v_n < l' + \epsilon$ for $n \geq m$. This implies that $u_n/v_n > k (= l' - \epsilon) > 0$ for $n \geq m$, a condition, which is the same as in (b).

4. Theorem 3.7

Case I. Let $p > 1$. Grouping the terms of the series $\sum u_n$, where $u_n = 1/n^p$, we get a new series $\sum v_n$, where

$$v_1 = 1, \; v_2 = 1/2^p + 1/3^p, \; v_3 = 1/4^p + \cdots + 1/7^p, \ldots, v_n = 1/(2^{n-1})^p + \ldots + 1/(2^n - 1)^p,$$

which contains 2^{n-1} terms.

Since each of the terms of $v_n \leq 1/(2^{n-1})^p$, the first of its terms,

$$v_n \leq 2^{n-1} \cdot 1/(2^{n-1})^p = (1/2^{p-1})^{n-1} = w_n, \text{ say.}$$

Taking $x = 1/2^{p-1}$, we find that $0 < x < 1$, since $p > 1$.

So $\sum w_n = \sum x^{n-1}$ is a convergent geometric series. Since $v_n \leq w_n$, $\sum v_n$ is also convergent. Since the grouping of terms of a positive series does not alter the convergence of the sum $\sum u_n$, i.e., $\sum 1/n^p$ is convergent.

Case II. Let $p = 1$. Then $u_n = 1/n$ and $\sum u_n$ is a standard harmonic series. Again, we group the terms and form a new series $\sum v_n$ where

$$v_1 = 1 + 1/2, \; v_2 = 1/3 + 1/4, \; v_3 = 1/5 + \cdots + 1/8, \ldots, v_n = 1/(2^{n-1} + 1) + \cdots + 1/2^n$$

for $n \geq 2$. There are 2^{n-1} terms in v_n and each of its terms $\geq 1/2^n$, the last of its terms. So, $v_n \geq 2^{n-1} \cdot 1/2^n = 1/2, v_1 \geq 1/2$ also. So, $v_n \not\to 0$ as $n \to \infty$. Consequently, $\sum v_n$ is divergent and hence $\sum 1/r$ is divergent.

Case III. Let $p < 1$. Then $n^p < n$ and hence $1/n^p > 1/n$. Since $\sum 1/n$ is divergent, by comparison test $\sum 1/n^p$ is also divergent.

5. Theorem 3.8

(a) Without any loss of generality we may assume

$$u_{n+1}/u_n \leq k \; \forall n \in \mathbb{N}, \quad \text{where } k < 1.$$

Then $\prod_{1}^{n-1} (u_{r+1}/u_r) \leq \prod_{1}^{n-1} k$ or $u_n/u_1 \leq k^{n-1}$ or $u_n < u_1 k^{n-1}$.

Since $0 < k < 1$, $\sum k^{n-1}$ is a convergent geometric series. So by comparison test $\sum u_n$ is convergent.

If $u_{n+1}/u_n \leq k$ not for all $n \in \mathbb{N}$, but for $n \geq m$, then proceeding in the same way we shall get $\sum u_{m+n}$ to be convergent. Hence $\sum u_n$ will also be convergent.

(a′) If $\underset{n\to\infty}{\text{Lt}} (u_{n+1}/u_n) = l < 1$, then $\forall \epsilon > 0 \; \exists m \in \mathbb{N}$:

$$l - \epsilon > u_{n+1}/u_n < l + \epsilon \; \forall n \geq m.$$

$\epsilon(> 0)$ may be chosen small enough so that $l + \epsilon < 1$.

Taking $k = l + \epsilon$, we find that $u_{n+1}/u_n < k < 1 \; \forall n \geq m$. So, by (a) $\sum u_n$ is convergent.

(b) Let $u_{n+1}/u_n \geq 1 \; \forall n \epsilon \mathbb{N}$. Then

$$u_n/u_1 = \prod_{r=1}^{n-1} (u_{r+1}/u_r) \geq 1 \Rightarrow u_n > v_1.$$

So, $u_n \not\to 0$ as $n \to \infty$. Hence $\sum u_n$ diverges.

The proof can be modified in case $u_{n+1}/u_n \geq 1$, not for all $n \in \mathbb{N}$, but for $n \geq m$.

(b′) Let $\underset{n\to\infty}{\text{Lt}} (u_{n+1}/u_n) = l > 1$. Then $\epsilon(> 0)$ can be chosen so that $l - \epsilon > 1$. Then $l - \epsilon < u_{n+1}/u_n < l + \epsilon \Rightarrow u_{n+1}/u_n > k(= l - \epsilon) > 1$. Hence, by (b), $\sum u_n$ diverges.

6. Theorem 3.9

(a) Let $u_n^{1/n} \leq k < 1 \; \forall n \in \mathbb{N}$. Then $u_n \leq k^n \; \forall n \in \mathbb{N}$, where $0 \leq k < 1$. But $\sum k^n$ is, then, a convergent geometric series. Hence by comparison test $\sum u_n$ is convergent.

If $u_n^{1/n} \leq k < 1 \; \forall n \geq m$, a fixed number, the proof can be modified.

(a′) Let $\underset{n\to\infty}{\text{Lt}} u_n^{1/n} = l < 1$. Then $\epsilon(> 0)$ may be chosen so that $l + \epsilon < 1$. Then for such a choice of $\epsilon, \exists m$:

$$l - \epsilon < u_n^{1/n} < l + \epsilon \; \forall n \geq m.$$

So, by (a), $\sum u_n$ is convergent.

(b) Let $u_n^{1/n} \geq 1$, not for all $n \geq m$, but for an infinite number of integral values of n. Then $u_n \geq 1$ for all these infinite number of values of n. So $u_n \not\to 0$ as $n \to \infty$. So $\sum u_n$ cannot converge. It must diverge.

Note. If $l = 1$, the test fails and then further investigation is required.

7. Theorem 3.10

Let $S_n = \sum_{r=1}^{n} (-1)^{r-1} u_r$. Then $S_{2n} = \sum_{1}^{2n} (-1)^{r-1} u_r$ and $S_{2n+2} - S_{2n} = u_{2n+1} - u_{2n+2} \geq 0 \; \forall n \in \mathbb{N}$, since $\{u_n\} \downarrow$. So $\{S_{2n}\} \uparrow$. Also

$$S_{2n} = u_1 - (u_2 - u_3) - \cdots - (u_{2n-2} - u_{2n-1}) - u_{2n} < u_1 \; \forall n \in \mathbb{N}.$$

So, $\{S_{2n}\}$ is bounded above. Hence $\{S_{2n}\}$ is convergent.

Next, $S_{2n+1} - S_{2n-1} = -u_{2n} + u_{2n+1} < 0 \; \forall \, n \in \mathbb{N}$. So, $\{S_{2n-1}\} \downarrow$. Also

$$S_{2n-1} = (u_1 - u_2) + \cdots + (u_{2n-3} - u_{2n-2}) + u_{2n-1} > u_1 - u_2 \; \forall \, n \in \mathbb{N}.$$

So, $\{S_{2n-1}\}$ is bounded below. Hence it is convergent. Since both $\{S_{2n}\}$ and $\{S_{2n-1}\}$ converge, so $\{S_n\}$ also converges. Moreover,

$$\underset{n\to\infty}{\text{Lt}} (S_{2n} - S_{2n-1}) = \underset{n\to\infty}{\text{Lt}} u_{2n} = 0, \text{ since } \{u_n\} \text{ is null.}$$

So, $\underset{n\to\infty}{\text{Lt}} S_{2n} = \underset{n\to\infty}{\text{Lt}} S_{2n-1}$ and the common value is the sum of the given series.

4 Differential Calculus

Function of Several Variables

4.1 Introduction

In the case of a function of one real variable the domain $X \subseteq \mathbb{R}$. In the case of a function of two real variables the domain is a subset of $\mathbb{R}^2 = \mathbb{R} \times \mathbb{R} = \{(x, y) : x, y \in \mathbb{R}\}$ viz., the entire xy-plane. Let $E \subseteq \mathbb{R}^2$. Then a function $f : E \to Z$, where $Z \subseteq \mathbb{R}$, is, by definition, a subset of $E \times Z$, so that $\forall\,(x, y) \in E,\ E!z \in Z : ((x, y), z) \in f$. $((x, y), z)$ is denoted by (x, y, z). Thus, f is a subset of ordered triples $(x, y, z) \in X \times Y \times Z$, if $E = X \times Y$. The number z is called the **f-image of (x, y)** and we write $z = f(x, y)$. The point (x, y) is called a **pre-image** of z. Here $\text{Dom} f = E \subseteq \mathbb{R}^2$ and the range of f, i.e., $\text{Ran} f$ or, $f(E) \subseteq \mathbb{R}$.

To give a geometrical representation of a function of two real variables x and y we take three mutually perpendicular straight lines intersecting at a point O called the **origin**. These three lines will serve as the coordinate lines: $x-, y-$ and $z-$axes, each of which serves as a number line.

The xy-plane is $\mathbb{R}^2$ and z-axis represents $\mathbb{R}$ (Fig. 4.1). If $E \subseteq \mathbb{R}^2$, then geometrically E is a part of the xy-plane. The function $f : E \to \mathbb{R}$ with E as its domain gives, for every point $P(x, y) \in E$ its f-image Q representing a real number z on the z-axis such that $Q = f(P)$ or, $z = f(x, y)$.

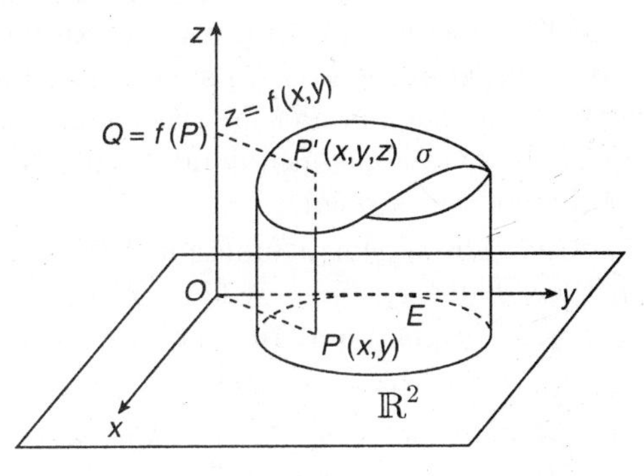

Fig. 4.1

So $f = \{(x, y, z) : z = f(x, y)\}$, which represents geometrically the set of all those points $P'(x, y, z)$ in the 3-D space $\mathbb{R}^3$ for which $z = f(x, y)$. In the figure E has been shown as a region in the xy plane, so that f is a surface σ over E. The domain E could be just a point or a line in the xy plane. Then geometrically f would represent a point or a space curve.

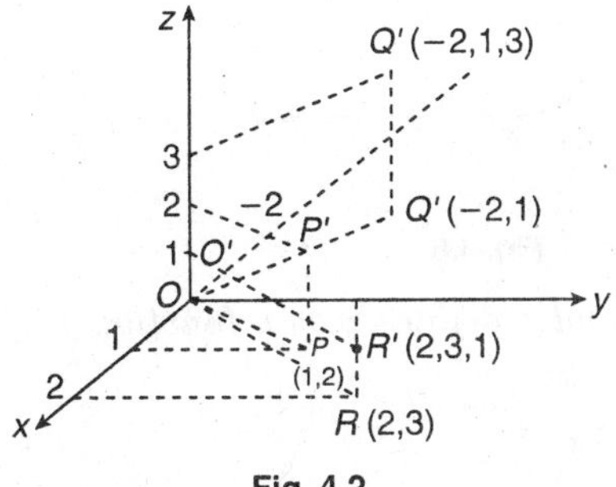

Fig. 4.2

Examples 4.1. 1. *Let f be a function mapping E into $\mathbb{R}$, where $E = \{O, P, Q, R\}, O, P, Q$ and R being respectively* $(0, 0)$, $(1, 2)$, $(-2, 1)$ *and* $(2, 3)$. *Also let* $f(O) = 1$, $f(P) = 2$, $f(Q) = 3$ *and* $f(R) = 1$. *Then* $f = \{(0, 0, 1), (1, 2, 2), (-2, 1, 3), (2, 3, 1)\}$.

$\text{Dom} f = E$, $\text{Ran} f = \{1, 2, 3\}$, which represents the points $1, 2$ and 3 on the z-axis. These are the f-images of the four points O, P, Q and R, the images of O and R being the same. Geometrically, $f = \{O', P', Q', R'\}$, a set of four points distributed in the xyz-space.

2. *Let $E = \mathbb{R}^2$, the entire xy-plane and the function $f : E \to \mathbb{R}$ be such that $z = f(x,y)$, where $f(x,y) = x + y$. In this case $z = x + y \; \forall (x,y) \in \mathbb{R}^2$. So this equation represents a plane surface passing through the origin (Fig. 4.3). Here the range of the function is the entire z-axis. Because $\forall z \in \mathbb{R}$, x and y can be found out from the equation $z = x + y$.*

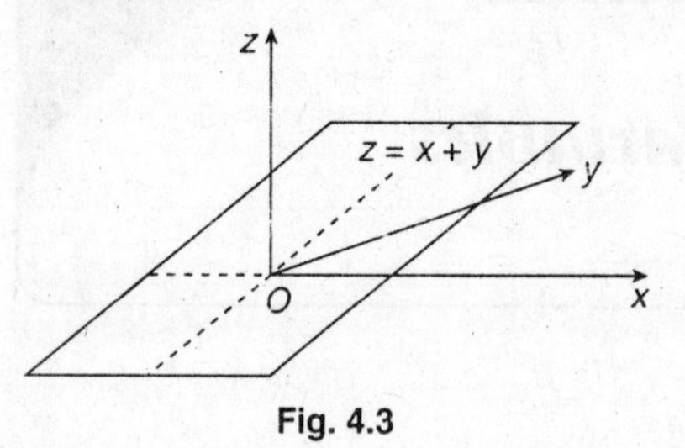

Fig. 4.3

If f is a function of three real variables x, y and z its domain $D \subseteq \mathbb{R}^3 = \{(x,y,z) : x,y,z, \in \mathbb{R}\}$ and range is a subset of $\mathbb{R}$. For every real function we take $\mathbb{R}$ as its codomain. Thus, if the function

$$f : D \to \mathbb{R}$$

then $f \subseteq D \times \mathbb{R}$ and $\forall (x,y,z) \in D$ there exists a unique $u \in \mathbb{R}$ such that $((x,y,z),u)$ or $(x,y,z,u) \in f$ so that u is the f-image of $P(x,y,z)$ and we write

$$u = f(x,y,z) \text{ or, } f(P).$$

Here P or, (x,y,z) is a pre-image of u.

It is not possible to give a proper geometrical representation of a function of 3 or more variables. But a physical interpretation is possible. For example, if u is the density of a non-homogeneous solid body placed in the (x,y,z) space, then the space D occupied by the body is the domain and for every point $P(x,y,z)$ in the domain the density u of the body is unique. Similarly u could be the temperature of the body at a point.

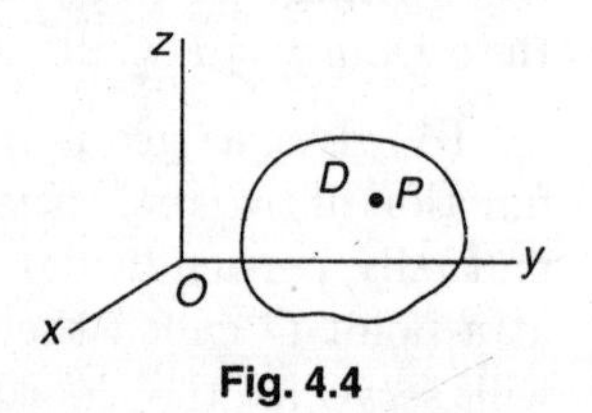

Fig. 4.4

Generalising, for a function f of n real variables its domain D consists of points having n coordinates x_i, for $i = 1, \ldots, n$ and $f \subseteq D \times \mathbb{R}$. Every point of f has $n + 1$ coordinates $x_1, x_2, \ldots, x_n$ and u so that $(x_1, x_2, \ldots, x_n, u) \in f$ and $u = f(x_1, x_2 \ldots, x_n)$.

3. *If $z = f(x,y) = \sqrt{1 - x^2 - y^2}$, then for f to be real $x^2 + y^2 \leq 1$ so that the domain $D = \{(x,y) : x^2 + y^2 \leq 1\}$, which is a circular region in the (x,y) plane. Since $z \geq 0$ and $z^2 = 1 - x^2 - y^2$, i.e., $x^2 + y^2 + z^2 = 1$, $f = \{(x,y,z) : x^2 + y^2 + z^2 = 1 \;\&\; z \geq 0\}$ so that geometrically f is a hemispherical surface lying over the circular region D.*

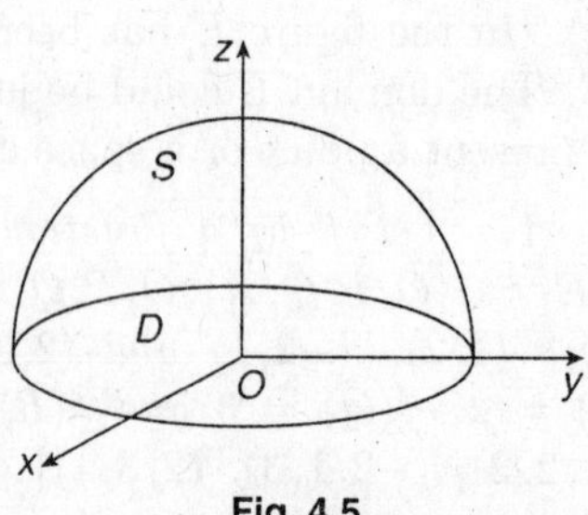

Fig. 4.5

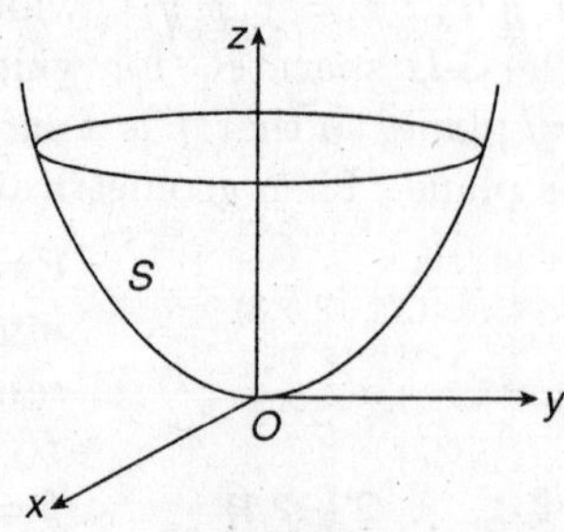

Fig. 4.6

4. *If $z = x^2 + y^2 \; \forall (x,y) \in \mathbb{R}^2$, then the equation in x, y and z defines z as a function f of x and y, whose domain is $\mathbb{R}$ and range:*

$$f(\mathbb{R}^2) = \{z \in \mathbb{R} : z \geq 0\}.$$

Here $z = x^2 + y^2$ represents a surface which is a paraboloid of revolution obtained by revolving the parabola, given by $z = y^2$, $x = 0$ lying in the (y,z) plane, about the z-axis.

4.2 Neighbourhood of a Point in $\mathbb{R}^2$

Let P be any point in $\mathbb{R}^2$, i.e., in the (x, y) plane. Then by the delta neighbourhood of P we mean the set $N(\delta, P)$ of all the points Q in the (x, y)-plane such that $PQ < \delta$, δ being a positive number. Set theoretically

$$N(\delta, P) = \{Q \in \mathbb{R}^2 : PQ < \delta\}.$$

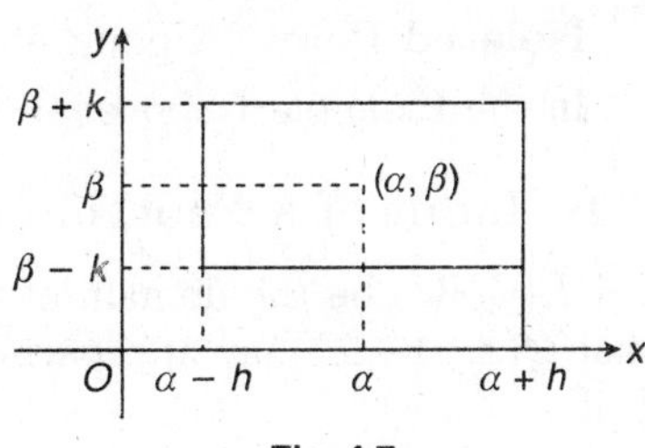

Fig. 4.7

Geometrically, $N(\delta, P)$ is a circular disc with centre P and radius δ.

If P is the origin $O(0,0)$ and $\delta = 1$, then

$$N(1, 0) = \{(x, y) \in \mathbb{R}^2 : \sqrt{x^2 + y^2} < 1 \text{ or, } x^2 + y^2 < 1\},$$

which is obviously a circular region of radius 1 and centre 0.

In place of taking a neighbourhood (nbd) in the form of circular region we may take a rectangular region. Thus, if P is (α, β), then

$$N = \{(x, y) : |x - \alpha| < h \text{ and } |y - \beta| < k\}$$

is also a nbd of P, with P at the centre of the rectangular region.

Deleted δ-nbd of P is the set of points given by $N(\delta, \not{P}) = N(\delta, P) - \{P\}$.

4.3 Limit Points and Isolated Points of a Set $E \subseteq \mathbb{R}^2$

Let $E \subseteq \mathbb{R}^2$ and P is a given point (α, β) of $\mathbb{R}^2$. Then P is called a **limit point** of E if, $\forall \delta > 0$, $N(\delta, \not{P})$ contains at least one point of E. In this case we write $(x, y) \to (\alpha, \beta)$, where $(x, y) \in E$. Here the point P may or may not belong to E.

Examples 4.2. **1.** *Let* $E = \left\{\left(\frac{1}{2^n}, \left(\frac{-1}{3}\right)^n\right) : n \in \mathbb{N}\right\}$

$$= \left\{\left(\frac{1}{2}, -\frac{1}{3}\right), \left(\frac{1}{2^2}, \frac{1}{3^2}\right), \left(\frac{1}{2^3}, -\frac{1}{3^3}\right), \ldots\right\}.$$

Let P be $(0,0)$. Then although $P \notin E$, P is a limit point of E. For if $(x, y) \in E$, $\exists n \in N$: $x = 1/2^n,\ y = (-1/3)^n$.

Now, $\forall \delta > 0,\ N(\delta, P) = \left\{(x, y) \in \mathbb{R}^2 : \sqrt{(x-0)^2 + (y-0)^2} < \delta\right\}$.

Thus, $\forall (x, y) \in N(\delta, P),\ x^2 + y^2 < \delta^2$.

Now, $(1/2^n)^2 + \{(-1/3)^n\}^2 < \delta^2 \Rightarrow 1/4^n + 1/9^n < \delta^2$ *(1)*

$\because$ $1/4^n + 1/9^n < 2 \cdot 1/4^n$,

(1) will hold if $2/2^{2n} < \delta^2$ *or,* $2^{2n} > 2/\delta^2$

or, $2n \ln 2 > \ln(2/\delta^2)$ *or,* $n > \dfrac{\ln(2/\delta^2)}{\ln 4}$.

If $m = \left[\dfrac{\ln 2/\delta^2}{\ln 4}\right]$, *then when $n > m + 1$ the point $(1/2^n, (-1/3)^n)$ of E will ie in $N(\delta, \not{P})$.*

Apart from $(0,0)$ there is no limit point of E.

2. *If $E = \{(x,y) \in \mathbb{R}^2 : |x-1| \leq 2 \,\&\, |y-2| \leq 3\}$, then E is a rectangle and all its points are limit points.*

Isolated Point: A point of E is an isolated point if it is not a limit point of E.

In the Example 1 above, all the points of E are isolated points.

4.4 Limit of a Function of Two Variables

Let $E(\subseteq \mathbb{R}^2)$ be the domain of the function $f : E \to \mathbb{R}$. Let $P(\alpha, \beta)$ be a limit point of E, so that $Q(x,y) \in E$ may approach P or $Q \to P$. A number $l(\in \mathbb{R})$ is said to be the limit of f at P if

$$\forall \epsilon > 0 \,\exists\, \delta > 0 : f(Q) \in N(\epsilon, l) \text{ whenever } Q \in E \cap N(\delta, \not{P}).$$

Symbolically, $\underset{Q \to P}{\text{Lt}} f(Q) = l$ or, $\underset{(x,y)\to(\alpha,\beta)}{\text{Lt}} f(x,y) = l$.

Roughly speaking, l is the limit of f at P if $\forall Q(\in E)$ very very close to P (but not the same as P) $f(Q)$ is either very nearly equal to or exactly equal to $f(P)$.

Notes. 1. In the above definition the deleted δ-nbd of P may be replaced by any other nbd, viz., a square nbd:

$$N = \{(x,y) \in \mathbb{R}^2 : 0 < |x - \alpha| < \delta \,\&\, 0 < |y - \beta| < \delta\}.$$

2. It should be noted that P may or may not belong to E, the domain of f and the limit l may or may not belong to the range $f(E)$ of f.

3. If the limit of f exists at P, then $Q(\in E)$ may approach P through any system of points in E. In particular Q may approach P along any continuous (or discontinuous) curve given by $y = \phi(x)$, where $x \to \alpha$, $y \to \beta$ and the limit of f must be the same for all such curves lying in E.

If the limits of f when $Q(\in E)$ approaches P along two different paths are found to be different, then the limit of f cannot exist at P. The limit of f, if it exists, is unique.

Theorems on Limits. Let $u = f(x,y)$ and $v = g(x,y)$ and $P(\alpha,\beta) \in \text{Dom}\, f \cap \text{Dom}\, g = E$, say and $R(x,y)$ is a variable point in E. Let

$$\underset{R\to P}{\text{Lt}}\, u = l \text{ and } \underset{R\to P}{\text{Lt}}\, v = m.$$

Then **(i)** $\underset{R\to P}{\text{Lt}}(u \pm v) = l \pm m = \underset{R\to P}{\text{Lt}}\, u \pm \underset{R\to P}{\text{Lt}}\, v$,

(ii) $\underset{R\to P}{\text{Lt}}(uv) = lm = \left(\underset{R\to P}{\text{Lt}}\, u\right)\left(\underset{R\to P}{\text{Lt}}\, v\right)$

and **(iii)** $\underset{R\to P}{\text{Lt}}(u/v) = l/m = \left(\underset{R\to P}{\text{Lt}}\, u\right) / \left(\underset{R\to P}{\text{Lt}}\, v\right)$,

provided $m = \underset{R\to P}{\text{Lt}}\, v \neq 0$ only in the last case.

Converses of these theorems are not necessarily true.

Example 4.3. 1. *Show that the function defined by*

$$f(x,y) = 2xy/(x^2 + y^2)$$

has no limit at the origin.

Here $E = \text{Dom}\, f = R^2 - \{0,0\} = \mathbb{R}^2 - \{O\}$. But O is a limit point of E. Let us choose the path Γ_1 defined by

$$\Gamma_1 = \{(x,y) \in E : y = x\}.$$

Then if $R = (x, y) \in \Gamma_1$, $R = (x, y)$ and R can approach O along Γ_1. Then

$$f(x,y) = f(x,x) = 2x \cdot x(x^2 + x^2) = 2x^2/(2x^2) \to 1 \text{ as } R \to 0.$$

Choosing a different path Γ_2 defined by

$$\Gamma_2 = \{(x,y) \in E;\ y = -x\}$$

a point R on it will have its coordinates $(x, -x)$ and

$$f(R) = f(x,-x) = 2x(-x)/(x^2 + x^2) - 2x^2/(2x^2) \to -1$$

as $R \to 0$. Thus we get two different limits of f at the same point, viz., the origin O. So the $f(x, y)$ has no limit at O.

4.4.1 Repeated Limits

Let $F(a, b)$ be a lt. pt. of the domain D of a function f of two real variables x and y. Then a point $P(x, y) \in D$ may approach F. Then keeping x fixed the variable y may approach b. $\forall x$, $\lim\limits_{y \to b} f(x, y)$, if it exists, is obviously a function ϕ of x. In other words $\phi(x) = \lim f(x, y)$ as $y \to b$.

Let $X = \text{Dom}\,\phi$. Then a must be a limit point of X. So, $\lim \phi(x)$ may exist when $x \to a$. Let $\lambda = \lim \phi(x)$ as $x \to a$. Then λ is a repeated limit of f at (a, b) and we write

$$\lambda = \lim_{x \to a} \lim_{y \to b} f(x,y).$$

Similarly, $\lim\limits_{y \to b} \lim\limits_{x \to a} f(x, y)$ may exist and let if be λ'.

It may be mentioned that λ and λ' are not necessarily equal. It can be shown that they are equal when the so called simultaneous limit, viz., $\lim\limits_{(x,y) \to (a,b)} f(x, y)$ exists. But converse is not necessarily true.

Examples 4.4. 1. *Let* $f(x, y) = (x - y)/(x + y)$. *Then* $\text{Dom}\, f = \mathbb{R}^2 - \Lambda$, *where* Λ *is the straight line whose equation is* $x + y = 0$ *or*

$$\Lambda = \{(x,y) : x + y = 0\}.$$

Although $O \in \text{Dom}\, f$ but $O \in \Lambda$ and it is a lt. pt. of $\text{Dom}\, f$.

Now $\quad \mathop{\text{Lt}}\limits_{x \to 0} f(x, y) = \mathop{\text{Lt}}\limits_{x \to 0} [(x - y)/(x + y)] = -y/y = -1$ if $y \neq 0$.

Similarly, $\quad \mathop{\text{Lt}}\limits_{y \to 0} f(x, y) = x/x = 1$ if $x \neq 0$.

$\therefore \quad \mathop{\text{Lt}}\limits_{y \to 0} \mathop{\text{Lt}}\limits_{x \to 0} f(x, y) = \mathop{\text{Lt}}\limits_{y \to 0} (-y/y) = -1 = \lambda'$,

whereas $\quad \mathop{\text{Lt}}\limits_{x \to 0} \mathop{\text{Lt}}\limits_{y \to 0} f(x, y) = \mathop{\text{Lt}}\limits_{x \to 0} (x/x) = 1 = \lambda$.

Thus $\quad \lambda \neq \lambda'$.

That the simultaneous limit of f at O does not exist, can be shown by following straight paths given by $y = mx$ towards the origin, m taking different values.

2. *Show that* $\underset{(x,y)\to(0,0)}{\text{Lt}}[(x+y)\{y+(x+y)^2\}/\{y-(x+y)^2\}]$ *does not exist.*

For proving the non-existence of the limit we try to find different paths of approach to the origin $O(0,0)$, so that the limits are different along different paths. Let

$$\Gamma_m = \{(x,y) : (x+y)[y+(x+y)^2] = m[y-(x+y)^2]\}.$$

The number m exists, since for no point (x,y) belonging to the domain of the function, $y-(x+y)^2 = 0$. There are points on Γ_m which belong to the domain of the functions and which can approach the origin. For all such points $f(x,y) = m$, so that $\underset{(x,y)\to(0,0)}{\text{Lt}} f(x,y) = m$, which is different for different values of m. Hence the limit does not exist.

3. *Show that the limit of* $f(x,y)$, *as* $(x,y)\to(0,0)$ *exists and it is zero, where*

$$f(x,y) = x^2y^2/(x^2+y^2).$$

Obviously, $E = \text{Dom } f = \mathbb{R}^2 - \{(0,0)\}$. So O is limit point of E. Now $\forall\, x, y \in \mathbb{R}$, $x^2 \le x^2+y^2$ and $y^2 \le x^2+y^2$.

$\therefore$ $x^2y^2 \le (x^2+y^2)^2$ and hence $x^2y^2/(x^2+y^2) \le x^2+y^2 < \delta^2 = \epsilon$.

So, $\forall \epsilon > 0\, \exists\, \delta = \sqrt{\epsilon} : |f(x,y) - 0| < \epsilon$, whenever $(x,y) \in N(\delta, 0)$. Hence Lt $f(x,y) = 0$ at O.

4. *Show that* $\lim[xy(x^2-y^2)/(x^2+y^2)] = 0$ *at* O.

$\forall\,(x,y) \in \text{Dom } f$, $|x| \le \sqrt{(x^2+y^2)}$, $|y| \le \sqrt{(x^2+y^2)}$

and $-y^2 \le y^2 \Rightarrow |x||y|$, i.e., $|xy| \le (x^2+y^2)$ and $x^2 - y^2 \le x^2+y^2$.

$\therefore$ $|xy(x^2-y^2)/(x^2/y^2) - 0| \le x^2+y^2 < \delta^2 = \epsilon$ if $\sqrt{(x^2+y^2)} < \delta$. Hence the result.

5. *For the function* $f(x,y) = x^2y^2/(x^2+y^2)$, *we have already seen that the simultaneous limit exists at* O *and the limit is zero. Now we shall see that the two repeated limits exist and both of them are zero. In fact,*

$$\lim_{x\to 0} f(x,y) = y^2 \underset{x\to 0}{\text{Lt}} \{x^2/(x^2+y^2)\} = y^2.0 = 0$$

$\lim_{y\to 0} f(x,y) = 0$ *also, by symmetry.*

$$\therefore \quad \lim_{y\to 0}\lim_{x\to 0} f(x,y) = \lim_{x\to 0}\lim_{y\to 0} f(x,y) = 0.$$

4.5 Continuity

A function $f : D \to \mathbb{R}$ is said to be continuous at a point $P(\alpha,\beta) \in D$ if $\forall \epsilon > 0 \exists \delta$:

$$(x,y) \in N(\delta, P) \cap D \Rightarrow f(x,y) \in N(\epsilon, f(\alpha,\beta)).$$

f is said to be continuous in D if it is continuous at every point of D. Moreover, if P is a lt. pt. of D, for continuity of f at P, the simultaneous limit of f at P must be equal to $f(P)$.

Like limits, the sum, difference and product of continuous functions are continuous in their common domain and the quotient of two continuous functions is also continuous at all the points common to their domains except the vanishing points of the denominator therein.

It is to be noted that if $f(x,y)$ is continuous at (α,β), then $f(x,\beta)$ as a function of the single variable x is also continuous at $x = \alpha$. Similarly, $f(\alpha, y)$ is continuous at $y = \beta$. Also, if $P(\alpha,\beta)$ is a limit point of the domain D of a function f, then for continuity of f, P must belong to D, the limit of f must exist at P and the limit must be equal to $f(\alpha,\beta)$.

Examples 4.5. **1.** *Show that the following function is continuous at* $(0,0)$*:*

$$f(x,y) = xy\frac{x^2-y^2}{x^2+y^2} \quad \textit{when } (x,y) \neq (0,0),$$
$$= 0 \quad \textit{when } (x,y) = (0,0).$$

Taking $x = r\cos\theta,\ y = r\sin\theta$, where $r \neq 0$, we get a point P within the domain of the function at a distance r from $O(0,0)$. For this point P,

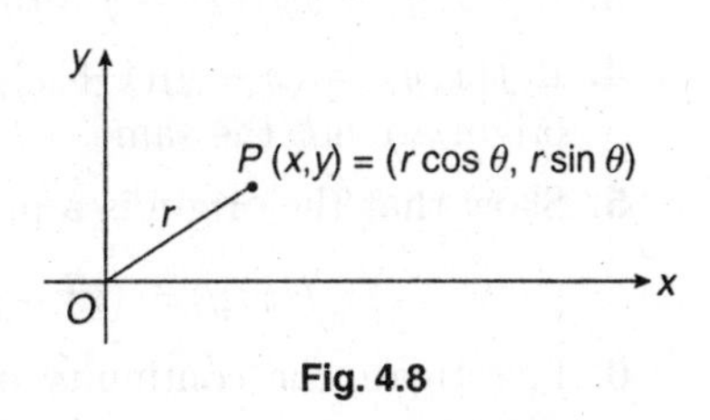

Fig. 4.8

$$f(P) = f(r\cos\theta, r\sin\theta)$$
$$= r^2\cos\theta\sin\theta\frac{\cos^2\theta-\sin^2\theta}{\cos^2+\sin^2\theta}$$
$$= \frac{1}{4}r^2\sin 4\theta.$$

f will be continuous at O if $\forall\epsilon > 0\,\exists\,\delta > 0$:

$$|f(x,y) - f(0,0)| < \epsilon \text{ whenever } OP < \delta. \tag{1}$$

Now $$|f(x,y) - f(0,0)| = \left|\frac{1}{4}r^2\sin 4\theta - 0\right| = \frac{1}{4}r^2|\sin 4\theta| \leq \frac{1}{4}r^2 < \epsilon,$$

whenever $r^2 < 4\epsilon$ or, $OP^2 < 4\epsilon$, or, $OP < \sqrt{4\epsilon} = 2\sqrt{\epsilon}$.

So taking $\delta = 2\sqrt{\epsilon}$ we find the condition (1) satisfied. Hence f is continuous at 0.

2. *Given* $f(x,y) = x^2/(x^2+y^2-x)$ *when* $(x,y) \neq (0,0)$ *and* $f(0,0) = 0$, *show that although* $f(x,y) \to 0$ *when* $(x,y) \to (0,0)$ *along any straight line whatever but still it is discontinuous at the origin.*

Taking $y = mx$, a straight line with slope m a point $P(x,y)$ may approach $O(0,0)$ along this line. For such a point

$$f(P)f(x,y) = f(x,mx) = x^2/(x^2+m^2x^2-x)$$
$$= x/(x+mx-1) \to 0 \text{ as } (x,mx) \to (0,0).$$

Since m is arbitrary, we find that $f(x,y) \to 0$ whenever $(x,y) \to (0,0)$ along a straight line.

But if $P(x,y)$ is a point on the parabola $y^2 = x$, P can approach 0 along this parabola. But in that case

$$f(x,y) = f(y^2,y) = y^4/(y^4+y^2-y^2) = 1$$

since $y \neq 0$. So,

$$f(x,y) \to 1 \text{ as } (y^2,y) \to (0,0).$$

Since $f(x,y) \not\to$ a definite limit as $(x,y) \to (0,0)$ along different paths, $\underset{(x,y)\to(0,0)}{\text{Lt}} f(x,y)$ does not exist. And hence f is discontinuous at O.

The limit points, limit and continuity for a function of more than two variables can be discussed exactly in the same way as in the case of two variables.

Exercises 4.1

1. Show that the limits of the following functions do not exist at the origin.

(i) $x^2y/(x^4+y^2)$, **(ii)** $x^2y^2/[x^2y^2+(x^2-y^2)^2]$, **(iii)** $(x^3+y^3)/(x-y)$.

2. Show that

(i) $\lim\limits_{(x,y)\to(0,0)} [\{x\sin(x^2+y^2)\}/(x^2+y^2)] = 0,$

(ii) $\lim\limits_{(x,y)\to(2,1)} [\{\sin^{-1}(xy-2)\}/\{\tan^{-1}(3xy-6)\}] = 1/3.$

3. If $f(x,y) = xy/(x^2+y^2)$ then show that the two repeated limits at the origin are equal.

4. If $f(x,y) = (x-y)(1+x)/\{(x+y)(1+y)\}$, show that the two repeated limits at the origin are not the same.

5. Show that the origin is a point of discontinuity of f, where

$$f(x,y) = (x^2-y^2)/(x^2+y^2) \text{ for } (x,y) \neq (0,0) \text{ and } f(0,0) = 0.$$

6. Investigate the continuity of f at $(1,2)$, where $f(x,y) = x^2 + 2y$ if $(x,y) \neq (1,2)$ and $f(1,2) = 0$.

Answer

6. Discontinuous.

4.6 Partial Derivatives and Related Problems

Given a function f of two variables x and y and a point (α,β) in its domain, the limit:

$$\operatorname*{Lt}_{x\to\alpha} \frac{f(x,\beta)-f(\alpha,\beta)}{x-\alpha},$$

is said to be the **first partial derivative of f with respect to x at $P(\alpha,\beta)$**. It is denoted by symbols like.

$$f_x(\alpha,\beta), \quad \frac{\partial f(\alpha,\beta)}{\partial x} \quad \text{or,} \quad D_x f(\alpha,\beta).$$

If $P(x,y)$ is assumed to be a fixed point, then $\operatorname*{Lt}\limits_{X\to x} \dfrac{f(X,y)-f(x,y)}{X-x} = f_x(x,y)$ or, $\partial f(x,y)\partial y$ or $D_x f(x,y)$. Obviously, this limit is the same as

$$\operatorname*{Lt}_{\delta x\to 0} \frac{f(x+\delta x,y)-f(x,y)}{\delta x}.$$

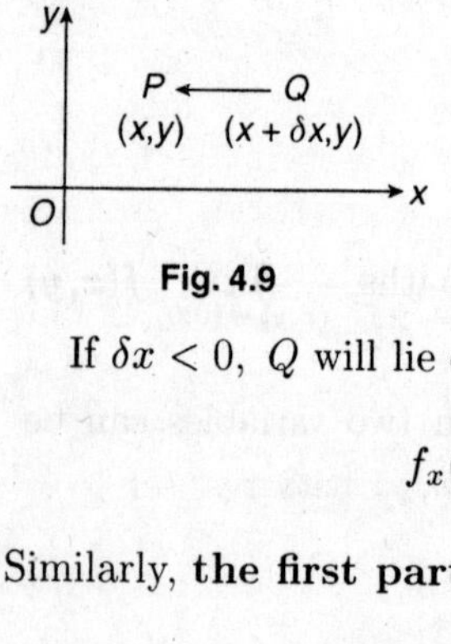

Fig. 4.9

Here the points $P(x,y)$ and $Q(x+\delta x,y)$ have their ordinates the same (Fig. 4.9). If $\delta x > 0$, Q will lie on the R.H.S. of P on a straight line parallel to the x-axis.

So, $f_x(x,y) = \operatorname*{Lt}\limits_{Q\to P} \dfrac{f(Q)-f(P)}{PQ}$, when Q approaches P along a line parallel to the x-axis.

If $\delta x < 0$, Q will lie on the L.H.S. of P and since $PQ > 0$, $\delta x = -PQ$. And

$$f_x(x,y) = \operatorname*{Lt}_{Q\to P} \frac{f(Q)-f(P)}{-PQ} = \operatorname*{Lt}_{Q\to P} \frac{f(P)-f(Q)}{PQ}.$$

Similarly, **the first partial derivative of f with respect to y** at (x,y) is defined by

$$f_y(x,y) = \operatorname*{Lt}_{Y\to y} \frac{f(x,Y)-f(x,y)}{Y-y} = \operatorname*{Lt}_{\delta y\to 0} \frac{f(x,y+\delta y)-f(x,y)}{\delta y}.$$

If $\delta y > 0$ and P and Q are respectively (x, y) and $(x, y + \delta y)$ then

$$f_y(x, y) = \operatorname*{Lt}_{Q \to P} \frac{f(Q) - f(P)}{PQ}.$$

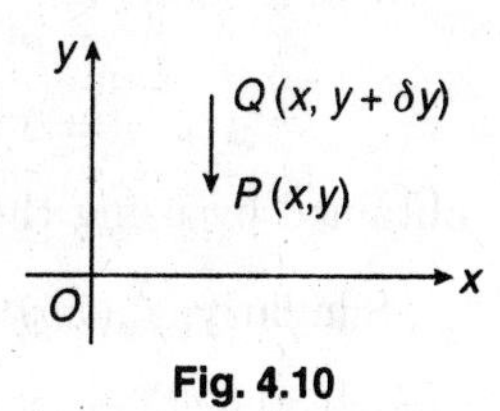

Fig. 4.10

Here $Q \to P$ along a straight line parallel to the y-axis. To evaluate f_x at a point $P(x, y)$ we find that y remains constant. Similarly, to evaluate f_y, by using limits, x remains constant. So, some times using the formulae for derivatives of a function of a single variable, we evaluate f_x and f_y regarding y and x respectively as constants.

Examples 4.6. 1. *If* $f(x, y) = (x^3 + y^3)/(x - y)$ *when* $x \neq y$,
$= 0$, *when* $x = y$,

show that the function is discontinuous at $O(0, 0)$, *although the 1st partial derivatives* $f_x(0, 0)$ *and* $f_y(0, 0)$ *both exist.*

By definition, $f_x(0,0) = \operatorname*{Lt}_{x \to 0} \dfrac{(0 + x, 0) - f(0, 0)}{x}$

$$= \operatorname*{Lt}_{x \to 0} \frac{x^3/x - 0}{x} = \operatorname*{Lt}_{x \to 0} x = 0$$

and $$f_y(0,0) = \operatorname*{Lt}_{x \to 0} \frac{f(0, y) - f(0, 0)}{y} = \operatorname*{Lt}_{y \to 0} \frac{-y^2}{y} = 0.$$

Thus both the partial derivatives exist and are 0 at the origin.

If $\Gamma_m = \{(x, y) : x^3 + y^3 = m(x - y)\}$ then if $R \in \Gamma_m$, R can approach 0. When $R \to 0$, $f(x, y) \to m$, which is different for different values of m. So $\lim f(x, y)$ does not exist at 0 and hence it is discontinuous thereat.

2. *If* $f(x, y) = ax^2 + 2hxy + by^2 + 2gx + 2fy + c \; \forall (x, y) \in \mathbb{R}^2$, *then show that* $f_x(x, y) = 2(ax + hy + g)$ *and* $f_y(x, y) = 2(hx + by + f)$.

Using definition,

$$f_x(x, y) = \operatorname*{Lt}_{\delta x \to 0} \frac{f(x + \delta x, y) - f(x, y)}{\delta x}.$$

Since $$f(x + \delta x, y) - f(x, y) = a(x + \delta x)^2 + 2h(x + \delta x)y + by^2 + 2g(x + \delta x) + 2fy + c$$
$$-\{ax^2 + 2hxy + by^2 + 2gx + 2fy + c\}$$
$$= a\{(x + \delta x)^2 - x^2\} + 2hy\delta x + 2g\delta x,$$

so, $$f_x(x, y) = \operatorname*{Lt}_{\delta x \to 0} [a(2x + \delta x) + 2hy + 2g] = 2(ax + hy + g).$$

Similarly, since $f(x, y + \delta y) - f(x, y) = 2hx\delta y + b(2y + \delta y)\delta y + 2f\delta y$

so, $$f_y(x, y) = \operatorname*{Lt}_{\delta y \to 0} \frac{f(x, y + \delta y) - f(x, y)}{\delta y} = \operatorname*{Lt}_{\delta y \to 0} [2hx + b(2y + \delta y) + 2f]$$
$$= 2(hx + by + f).$$

Alternatively, since $f_x(x, y) = \frac{\partial}{\partial x} f(x, y)$ and it is the derivative of $f(x, y)$ with respect to x regarding **y constant**, so

$$f_x(x, y) = \frac{\partial}{\partial x}(ax^2 + 2hxy + by^2 + 2gx + 2fy + c)$$
$$= a\frac{\partial}{\partial x}x^2 + 2hy\frac{\partial}{\partial x}x + b\frac{\partial}{\partial x}y^2 + 2g\frac{\partial}{\partial x}x + 2f\frac{\partial}{\partial x}y + \frac{\partial}{\partial x}c$$

$$= a \cdot 2x + 2hy \cdot 1 + 0 + 2g \cdot 1 + 0 + 0$$
$$= 2(axy + hy + g),$$

obtained by using the formulae for derivatives of a function of a single variable.

Similarly, $f_y(x,y) = 2hx\frac{\partial}{\partial y}y + b\frac{\partial}{\partial y}y^2 + 2f\frac{\partial}{\partial y}y = 2(hx + by + f).$

4.6.1 Geometrical Interpretation of First Partial Derivatives

By definition,

$$f_x(a,b) = \underset{\delta x \to 0}{\text{Lt}} \frac{f(a+\delta x, b) - f(a,b)}{\delta x}.$$

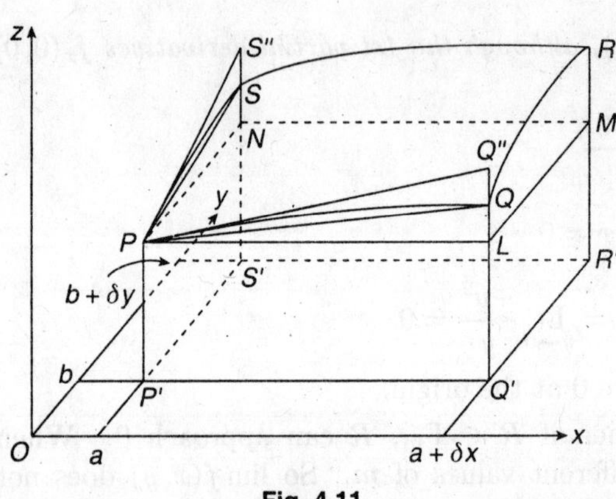

Fig. 4.11

Let $z = f(x,y)$ represent a surface of which $PQRS$ is a part bounded by the four planes

$$x = a, \; x = a + \delta x,$$
$$y = b \text{ and } y = b + \delta y.$$

Let $z_1 = f(a,b), \; z_2 = f(a+\delta x, \; b)$

$$z_3 = f(a+\delta x, b+\delta y) \text{ and } z_4 = f(a, b+\delta y).$$

Then the coordinates of P, Q, R, S are respectively (a, b, z_1), $(a+\delta x, b, z_2)$, $(a+\delta x, b+\delta y, z_3)$ and $(a, b+\delta y, z_4)$.

Let $PLMN$ be the plane parallel to the (x,y)-plane cutting the lines parallel to the z-axis through P, Q, R and S in P, L, M and N respectively. We have then

$QL = z_2 - z_1 = f(a+\delta x, b) - f(a,b),$

$PL = \delta x$, assumed positive.

Now, $\delta x \to 0$ as $Q \to P$.

$$\therefore \quad f_x(a,b) = \underset{Q \to P}{\text{Lt}} \frac{QL}{PL} = \underset{\alpha \to \phi}{\text{Lt}} \tan\alpha = \tan\phi = \tan QPQ'',$$

where $\tan\alpha$ may be regarded as the gradient of chord PQ of the curve of intersection of the surface by the plane through P parallel to the (z,x) plane. So $f_x(a,b)$ is $\tan\phi$, the gradient of the tangent to this curve at P relative to the x-axis.

Again, $SN = z_4 - z_1 = f(a, b+\delta y) - f(a,b)$ and $PN = \delta y$, assumed positive. Also $\delta y \to 0$ when $S \to P$. So,

$$f_y(a,b) = \underset{\delta y \to 0}{\text{Lt}} \frac{f(a, b+\delta y) - f(a,b)}{\delta y} = \underset{S \to P}{\text{Lt}} \frac{SN}{PN} = \underset{\beta \to \psi}{\text{Lt}} \tan\beta = \tan\psi = \tan NPS'',$$

where $\tan\beta$ is the gradient of the chord PN of the curve of intersection of the surface by the plane through P parallel to the (y,z)-plane. So $f_y(a,b)$ is $\tan\psi$, the gradient of the tangent to this curve at P relative to the y-axis.

Note. By the **gradient** $\tan\phi$ of the tangent to the curve PQ, we mean the slope at P, of the surface in the direction of the x-axis, assuming the (x,y)-plane horizontal and the positive z-axis vertically upward. Similarly, $\tan\psi$ is the slope, at P, of the surface in the direction of the y-axis. In fact, the slope of the surface, at P, in any direction can be found out.

$f_x(a,b)$, the first partial derivative of f with respect to x at $P(a,b)$ can be regarded as the rate of change of $u(= f(x,y))$ with respect to x, at P, when y is held fixed, viz., b. Similarly, $f_y(a,b)$ is the rate of change of u, at P, with respect to y when $x = a$. For example, if u is the area of a triangle whose base is x and altitude y then

$$u = \frac{1}{2}xy.$$

$u_x(2,1)$ is the rate of change of the area with respect to the base at $(2,1)$ when $y = 1$. Since $u_x = \frac{1}{2}y, u_x(2,1) = \frac{1}{2}$.

Similarly, $u_y(2,1)$ is the rate of change of the area with respect to the altitude of $(2,1)$ when $x = 2$. Now, $u_y = \frac{1}{2}x$. So $u_y(2,1) = \frac{1}{2} \cdot 2 = 1$.

4.7 First Partial Derivatives of a Function of Three Variables

If $u = f(x,y,z)$, a function of three real variables x, y and z, then

$$u_x = \frac{\partial u}{\delta x} = f_x = \operatorname*{Lt}_{\delta x \to 0} \frac{f(x+\delta x, y, z) - f(x,y,z)}{\delta x},$$

so that the 1st partial derivative of u w.r.t.x can be obtained by using, whenever possible, the formula for ordinary derivative du/dx of u w.r.t. x regarding two other variables, viz., y and z constants. Similarly,

$$u_y = \frac{\partial u}{\partial y} = \operatorname*{Lt}_{\delta y \to 0} \frac{f(x, y+\delta y, z - f(x,y,z)}{\delta y}$$

and $$u_z = \frac{\partial u}{\partial z} = \operatorname*{Lt}_{\delta z \to 0} \frac{f(x, y, z+\delta y) - f(x,y,z)}{\delta z}.$$

4.8 Differentiability and Differentials

A function $f : D \to \mathbb{R}^2$ is said to be **differentiable at a point $P(a,b) \in D$** if the difference

$$\delta f = f(a+\delta x, b+\delta y) - f(a,b),$$

where $Q(a+\delta x, b+\delta y)$ is another point in D, can be expressed as

$$A\delta x + B\delta y + \phi\delta x + \psi\delta y,$$

where A and B are independent of δx and δy, but ϕ and ψ are functions of δx and δy such that ϕ and ψ both tend to 0 as $(\delta x, \delta y) \to (0,0)$. Thus,

$$\delta f = A\delta x + B\delta y + \phi\delta x + \phi\delta y, \tag{1}$$

where $\phi \to 0$ and $\psi \to 0$ when $(\delta x, \delta y) \to (0,0)$. The part $A\delta x + B\delta y$ in the R.H.S. of (1), called the **principal part** of δf, is defined to be the **differential** of f and it is denoted by df. Thus,

$$df = A\delta x + B\delta y.$$

Under certain conditions, it can be shown, by taking limits, that

$$A = f_x(a,b) \text{ and } B = f_y(a,b), \text{ so that}$$

$$df = f_x(a,b)\delta x + f_y(a,b)\delta y. \tag{2}$$

To find the differential dx of x we take $f(x,y) = x$.

Then $f_x = 1,\ f_y = 0$ and $df = dx$.

So $dx = 1\delta x + 0\delta y$.

Thus $\delta x = dx$ and similarly $\delta y = dy$.

So from (2) we have

$$df = f_x(a,b)dx + f_y(a,b)dy.$$

Replacing (a,b) by (x,y) and $f_x(x,y)$ and $f_y(x,y)$ by f_x and f_y respectively we have

$$df = f_x dx + fydy. \tag{3}$$

The R.H.S. of (3) is called the **differential** of f at (x,y) and it is denoted by df, the L.H.S. of (3).

It can be shown that if

$$u = f(x,y) \text{ and } v = g(x,y),$$

$$d(u \pm v) = du \pm dv,$$

$$d(uv) = udv + vdu,$$

and $\quad d(u/v) = (vdu + udv)/v^2.$

If $u = f(x,y,z)$, then $du = f_x dx + f_y dx + f_z dz$ or $u_x dx + u_y dy + u_z dz$.

4.9 Total Differential Coefficient

If x and y are both functions of a single variable t, then we know that

$$dx = (dx/dt)dt \quad \text{or} \quad \dot{x}dt \text{ and } dy = (dy/dt)dt \quad \text{or} \quad \dot{y}dt.$$

If now $z = f(x,y)$, where $x = x(t)$ and $y = y(t)$, then $z = f(x(t),\ y(t)) = \phi(t)$ $\forall\, t \in \text{Dom}\, x \cap \text{Dom}\, y$, giving z as a function of single variable t. Then

$$dz = (dz/dt)dt. \tag{1}$$

But z as a function of the two variables x and y.

$$dz = f_x dx + f_y dy = z_x dx + z_y dy = z_x \dot{x} dt + z_y \dot{y} dt$$

or $$dz = [(\partial z/\partial x)(dx/dt) + (\partial z/\partial y)(dy/dt)]dt. \tag{2}$$

From (1) and (2),

$$\dot{z} = dz/dt = (\partial z/\partial x)(dx/dt) + (\partial z/\partial y)(dy/dt) \text{ or } z_x\dot{x} + z_y\dot{y}.$$

dz/dt is called **the total differential coefficient** or the **total derivative** of z w.r.t. t.

If u is a function of three variables x, y and z each of which is a function of the same single variable t, then the total derivative of u w.r.t. t is given by

$$\dot{u} = u_x\dot{x} + u_y\dot{y} + u_z\dot{z}.$$

4.9.1 An Important Case

If $\quad u = f(x,y)$, where $y = y(x)$ or $\phi(x)$, then

$u = f(x, y(x))$, so that in this case u is ultimately a function of x. Then

$du = (du/dx)dx$, where $du = u_x dx + u_y dy$.

But $\quad dy = \phi'(x)dx$ or $(dy/dx)dx$. So,

$du = u_x dx + u_y y' dx$, whence $du/dx = u_x + u_y y'$.

If $\quad u = 0,\ du/dx = 0$ also, and we get

$$u_x + u_y y' = 0 \text{ or, } \boldsymbol{y' = -u_x/u_y} \textbf{ or, } \boldsymbol{-f_x/f_y},$$

which gives the formula for the derivative of an implicit function defined by $f(x, y) = 0$.

Examples 4.7. **1.** *If $f(x, y) = 0$, where $f(x, y) = x^3 + y^3 - 3axy$, assuming that this equation defines y as a real function of x in some domain X, evaluate dy/dx.*

Here $f_x = 3x^2 - 3ay$ and $f_y = 3y^2 - 3ax$.

So, $\quad dy/dx = -f_x/f_y = -(x^2 - ay)/(y^2 - ax)$, provided $y^2 - ax \neq 0$.

2. *Let the equation: $xy \sin x + \cos y = 0$ define y as a real function of x in some nbd of $(0, \pi/2)$. Find dy/dx at $(0, \pi/2)$.*

Taking $f(x, y) = xy \sin y + \cos y$,

$$f_x = y(\sin x + x \cos x) \text{ and } f_y = x \sin x - \sin y.$$

$$\therefore \quad dy/dx = -f_x/f_y = -y(\sin x + x \cos x)/(x \sin x - \sin y)$$

$$= 0 \text{ at } (0, \pi/2).$$

3. *Find dy/dx at $(1, 1)$ when $2xy - \ln(xy) = 2$.*

Let $\quad f(x, y) = 2xy - \ln(xy) - 2 = 2xy - \ln x - \ln y - 2$.

Then $\quad f_x = 2y - 1/x$ and $f_y = 2x - 1/y$.

$$\therefore \quad dy/dx = -(2y - 1/x)/(2x - 1/y) = -1 \text{ at } (1, 1).$$

4.10 Homogeneous Function

A function $u = f(x, y)$ of two real variables is said to be homogeneous of degree $n(\in \mathbb{R})$ in x and y if either

(i) $u = x^n \phi(y/x)$ or, $y^n \psi(x/y)$ or (ii) $f(tx, ty) = t^n f(x, y)$ or, $t^n u$.

Similarly, if $u = f(x, y, z)$, a function of three real variables x, y and z then it is homogeneous of degree n in x, y and z if, either

(i) $u = x^n \phi(y/x, z/x)$ or, $y^n \psi(x/y, z/y)$ or, $z^n \chi(x/z, y/z)$,

(ii) $f(tx, ty, tz) = t^n f(x, y, z)$ or, $t^n u$.

Examples 4.8. **1.** *If $u = ax^2 + 2hxy + by^2$, in which all the terms are of the same degree, viz., 2, then u is a homogeneous function of degree 2 in x and y.*

For,

$$u = x^2(a + 2hy/x + b(y/x)^2) = x^2 \phi(y/x).$$

If $v = y/x$, ϕ is a function of single variable v and

$$\phi(v) = a + 2hv + bv^2.$$

2. *If $u = \sqrt{ax} + \sqrt{by} + \sqrt{cy}$, then u is a homogeneous function of degree $1/2$ in x, y and z.*

For

$$u = y^{\frac{1}{2}} \left[\sqrt{ax/y} + \sqrt{b} + \sqrt{cz/y}\right].$$

3. *If $u = e^{x/y}$ or, $e^{y/x}$, then u is a homogeneous function of degree 0 in x and y.*

For, $e^{tx/ty}$ or $e^{ty/tx} = t^0 e^{x/y}$ or $e^{y/x} = t^0 u$.

4. *If $u = \ln x - \ln y$, then u is homogeneous of degree 0.*

For, $\ln(tx) - \ln(ty) = \ln t + \ln x - \ln t - \ln y = \ln x - \ln y = t^0 u$.

Alternatively, $u = \ln(x/y) = y^0 \ln(x/y)$.

5. *If $u = x^2 \tan^{-1} \dfrac{x}{y}$, then u is obviously homogeneous of degree 2. Also,*

$$u = x^2 \tan^{-1}[1/(y/x)] \quad \textit{or,} \quad x^2 \cot^{-1}(y/x),$$

or $$(tx)^2 \tan^{-1} \frac{tx}{ty} = t^2 \left(x \tan^{-1} \frac{x}{y} \right) = t^2 u.$$

4.10.1 Euler's Theorem on Homogeneous Functions

If u is a homogeneous function of degree n in x and y, possessing the partial derivatives u_x and u_y, then $xu_x + yu_y = nu$.

Proof. Let $u = f(x, y) = x^n \phi(y/x) = x^n \phi(v)$, where $v = y/x$. Then

$$u_x = nx^{n-1}\phi(v) + x^n \phi'(v) v_x.$$

But $$v = y/x \Rightarrow v_x = -y/x^2.$$

$\therefore$ $$u_x = nx^{n-1}\phi(v) - x^{n-2} y \phi'(v).$$

Similarly, $$u_y = x^n \phi'(v) v_y = x^n \phi'(v)(1/x) = x^{n-1}\phi'(v).$$

$\therefore$ $$xu_x + yu_y = nx^n \phi(v) - x^{n-1} y\phi'(v) + x^{n-1}\phi'(v) = nu.$$

Alternatively, if $f(x, y)$ is a homogeneous function of degree n in x and y, then

$$f(tx, ty) = t^n f(x, y) \; \forall \, t \in \mathbb{R} \text{ except 0 when } n \text{ is negative.}$$

Let $$F(x, y, t) = t^{-n} f(tx, ty) = f(x, y). \tag{1}$$

Taking $$u = tx \text{ and } v = ty,$$

$$F(x, y, t) = t^{-n} f(u, v).$$

$\therefore$ $$\begin{aligned} F_t &= -nt^{-n-1} f(u, v) + t^{-n}(f_u u_t + f_v v_t) \\ &= -nt^{-n-1} f(u, v) + t^{-n}(x f_u + y f_v) \end{aligned}$$

and $$\begin{aligned} t^{n+1} F_t &= -nf(u, v) + (tx f_u + ty f_v) \\ &= -nf(u, v) + u f_u + v f_v. \end{aligned} \tag{2}$$

But from (1) it is obvious that F is independent of t.

$\therefore$ $$F_t = 0$$

and hence $u f_u + v f_v = nf(u, v) \; \forall \, t \in \mathbb{R}$.

Taking $t = 1$ we get $u = x$ and $v = y$.

$\therefore$ $$x f_x + y f_y = nf(x, y) \quad \text{or,} \quad xu_x + yu_y = nu.$$

Note. Although a proof has been given for a homogeneous function of two variables, the theorem is true for any finite number of variables. The converse of this theorem is also true. So **a necessary and sufficient condition for a function u of m variable $x_i, i = 1, \ldots, m$ to be homogeneous of degree n in all the variables x_i is that**

$$\sum_{i=1}^{m} x_i u_{x_i} = nu.$$

4.10.2 Euler's Theorem on a Composite Function of a Homogeneous Function

Let $u = \phi(v)$, where ϕ is a derivable function possessing a unique inverse: ϕ^{-1} and let

$$v = \phi^{-1}(u) = \psi(u).$$

Then if v is a homogeneous function of degree n in the two variables x and y possessing the derivatives v_x and v_y, then

$$xu_x + yu_y = n\psi(u)/\psi'(u).$$

Proof. Invertibility and derivability of ϕ ensure that of ψ. Since $v = \psi(u)$, and $\psi'(u)$ exists, and $\neq 0$.

$v_x = \psi'(u)u_x$ and $v_y = \psi'(u)u_y$.

$\therefore \quad xv_x + yv_y = nv = n\psi(u)$

or, $\quad \psi'(u)(xu_x + yu_y) = n\psi(u)$

or, $\quad xu_x + yu_y = n\psi(u)\psi'(u)$.

Cor. If v is a homogeneous function of three variables x, y and z, we get also

$$xu_x + yu_y + zu_z = n\psi(u)/\psi'(u).$$

Examples 4.9. 1. *Verify Euler's Theorem for the following functions:*

(i) $z = ax^2 + 2hxy + by^2$, (ii) $u = (x^{1/4} + y^{1/4})/(x^{1/5} + y^{1/5})$,

(iii) $z = \sin^{-1}[(\sqrt{x} - \sqrt{y})/(\sqrt{x} + \sqrt{y})]$.

(i) $z = x^2[bv^2 + 2hv + a]$, where $v = y/x$, implies that z is a homogeneous function of degree 2 in x and y.

So, by Euler's theorem, $xz_x + yz_y = 2z$.

Now $\quad z_x = 2x[bv^2 + 2hv + a] + x^2[b2v \cdot v_x + 2hv_x]$

$\quad = 2x^{-1}z + x^2[-2bvy/x^2 - 2hy/x^2]$

$\quad = 2x^{-1}z - 2y[bv + h]$

and $\quad z_y = x^2[2bvv_y + 2hv_y] = x^2[2bv/x + 2h/x] = 2x[bv + h]$.

So, $\quad xz_x + yz_y = 2z$.

(ii) $u = x^{1/4-1/5}(1 + v^{1/4})/(1 + v^{1/5})$

$\quad = x^{1/20}(1 + v^{1/4})/(1 + v^{1/5})$.

So, u is a homogeneous function of degree 1/20 in x and y. By Euler's theorem, we must have

$xu_x + yu_y = u/20$.

Now $\quad u_x = \dfrac{1}{20}x^{-19/20}\dfrac{1 - v^{1/4}}{1 + v^{1/5}} + x^{1/20}\dfrac{[(1/4)v^{-3/4}(1 + v^{1/5}) - (1/5)v^{-4/5}(1 - v^{1/4})]v_x}{(1 - v^{1/5})^2}$

$$xu_x = \frac{1}{20}x^{1/20}\frac{1 - v^{1/4}}{1 + v^{1/5}} - \frac{x^{1/20}x^{-1}y}{20}\,\frac{5v^{-3/4}(1 + v^{1/5}) - 4v^{-4/5}(1 - v^{1/4})}{(1 - v^{1/5})^2}$$

$$u_y = x^{1/20}\frac{5v^{-3/4}(1+v^{1/5}) - 4v^{-4/5}(1-v^{1/4})}{(1-v^{1/5})^2}\cdot v_y$$

$$yu_y = x^{1/20}x^{-1}y\frac{5v^{-3/4}(1+v^{1/5}) - 4v^{-4/5}(1-v^{1/4})}{(1-v^{1/5})^2}.$$

$$\therefore \quad xu_x + yu_y = \frac{1}{20}x^{1/20}\frac{1-v^{1/4}}{1+v^{1/5}} = \frac{1}{20}u.$$

Thus Euler's theorem is verified.

(iii) $u = \sin z = x^0(1-\sqrt{v})/(1+\sqrt{v}) = x^0\phi(v)$, where $\phi(v) = (1-\sqrt{v})(1+\sqrt{v})$ and v being y/x.

$\therefore u$ is a homogeneous function of degree 0 in x and y.

Now $\quad u_x = (\cos z)z_x = \phi'(v)v_x = \phi'(v)\dfrac{-y}{x^2}$

and $\quad u_y = (\cos z)z_y = \phi'(v)v_y = \phi'(v)\dfrac{1}{x}$.

So, $\quad xu_x + yu_y = x\phi'(v)\dfrac{-y}{x^2} + y\phi'(v)\dfrac{1}{x} = 0.$

$\therefore \quad x(\cos y)z_x + y\cdot(\cos z)z_y = 0, \text{ or } xz_x + yz_y = 0.$

2. *Using Euler's theorem on the composite function of a homogeneous function verify that if*

(i) $u = \sin^{-1}[(x+y)/(\sqrt{x}+\sqrt{y})]$, *then* $xu_x + yu_y = (1/2)\tan u$,

(ii) $u = \tan^{-1}[(x^3+y^3)/(x-y)]$, *then* $xu_x + yu_y = \sin 2u$,

(iii) $u = \ln(x^2+y^2+z^2) - \ln\left(\sqrt{yz}+\sqrt{zx}+\sqrt{xy}\right)$, *then* $xu_x + yu_y + zu_z = 1$.

(i) If $v = (x+y)/(\sqrt{x}+\sqrt{y})$, then v is a homogeneous function of degree 1/2 in x and y. So

$$xv_x + yv_y = (1/2)v. \tag{1}$$

But $\quad v = \sin u$. So, $v_x = (\cos u)u_x$ and $v_y = (\cos u)u_y$.

$\therefore$ from (1)

$$x(u_x \cos u) + y(u_y \cos u) = (1/2)\sin u,$$

whence $\quad xu_x + yu_y = (1/2)\tan u$.

If we want to directly apply the formula, here

$$\phi = \sin^{-1} \Rightarrow \psi = \phi^{-1} = \sin, \psi' = \cos.$$

$$\therefore \quad xu_x + yu_y = \frac{1}{2}\psi(u)/\psi'(u) = (1/2)\sin u/\cos u = (1/2)\tan u.$$

(ii) Here $v = (x^3+y^3)/(x-y)$, a homogeneous function of degree 2 in x and y. So,

$$xv_x + yv_y = 2v. \tag{2}$$

But $\quad v = \tan u$. So, $v_x = u_x \sec^2 u$ and $v_y = u_y \sec^2 u$.

$\therefore$ from (2)

$$xu_x \sec^2 u + yu_y \sec^2 u = 2\tan u,$$

whence $\quad xu_x + yu_y = 2\tan u \cos^2 u = \sin 2u$.

(iii) Here $u = \ln\left(\sum x^2/\sum \sqrt{yz}\right) = \ln v$ say, so that $v = e^u$. Since v is homogeneous of degree 1,

$$xv_x + yv_y + zv_z = 1v.$$

But $\quad v_x = e^u u_x,\ v_y = e^u u_y$ and $v_z = e^u u_z$.

$\therefore \quad xe^u u_x + ye^u u_y + ze^u u_z = e^u$

or, $\quad xu_x + yu_y + zu_z = 1.$

4.11 Partial Derivatives of Composite Functions—Chain Rule

Let $x = x(u, v)$ and $y = y(u, v)$ possess continuous first order partial derivatives x_u, x_v, y_u, y_v in the domain E common to both x and y, and $z = f(x, y)$ possess continuous first order partial derivatives z_x and z_y in E', where E' is the image of E, so that $\forall (u, v)$ in E, (x, y), calculated from $x = x(u, v)$ and $y = y(u, v)$ lies in E'. Then z becomes a composite function $\phi(u, v)$ of u and v. Under the above conditions z possesses continuous first order partial derivatives with respect to u and v and

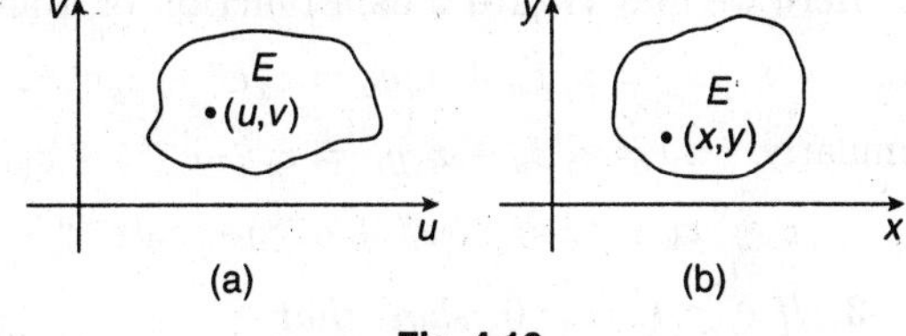

Fig. 4.12

$$z_u = z_x x_u + z_y y_u \text{ and } z_v = z_x x_v + z_y y_v.$$

For, since $z = \phi(u, v)$

$$dz = z_u du + z_v dv. \tag{1}$$

But $\quad z = f(x, y).$

So, $\quad dz = z_x dx + z_y dy. \qquad (2)$

Again, $\quad x = x(u, v)$ and $y = y(u, v)$.

So, $\quad dx = x_u du + x_v dv$ and $dy = y_u du + y_v dv$.

So from (2)

$$\begin{aligned} dz &= z_x(x_u du + x_v dv) + z_y(y_u du + y_v dv) \\ &= (z_x x_u + z_y y_u)du + (z_x x_v + z_y y_v)dv. \end{aligned}$$

Comparing with (1), we get

$$z_u = z_x x_u + z_y y_u \text{ and } z_v = z_x x_v + z_y y_v.$$

These formulae are called **Chain rule** for the calculation of first partial derivatives of a composite function.

Examples 4.10. 1. *Find the total derivative dz/dt when*

$$z = xy^2 + x^2y;\ x = at^2,\ y = 2at.$$

Verify by direct substitution.

Using the formula

$$dz/dt = z_x\dot{x} + z_y\dot{y}$$

we have, since $z_x = y^2 + 2xy$ and $z_y = 2xy + x^2$,

$$\dot{x} = 2at,\ \dot{y} = 2a,$$

so, $dz/dt = (y^2 + 2xy)2at + (2xy + x^2)2a$

$= (4a^2t^2 + 2 \cdot at^2 \cdot 2at)2at + (2 \cdot at^2 \cdot 2at + a^2t^4)2a$

$= 2(5t + 8)a^2t^3.$

Verification:

$$z = xy^2 + x^2y = at^2 \cdot 4a^2t^2 + a^2t^4 2at$$
$$= 4a^3t^4 + 2a^3t^5$$
$$= 2a^3(2t^4 + t^5).$$

$\therefore \quad dz/dt = 2a^3(8t^3 + 5t^4) = 2a^3t^3(5t + 8).$

2. *Given that z is a function of x and y, where $x = e^u + e^{-v}$ and $y = e^{-u} - e^v$, prove that $z_u - z_v = xz_x - yz_y$.*

Here we may regard z as a function of u and v.

So, $\quad z_u = z_x x_u + z_y y_u = z_x e^u - z_y e^{-u}.$

Similarly, $\quad z_v = z_x x_v + z_y y_v = z_x(-e^{-v}) + z_y(-e)^v.$

$\therefore \quad z_u - z_v = z_x(e^u + e^{-v}) - z_y(e^{-u} - e^v) = xz_x - yz_y.$

3. *If $F(p, t, v) = 0$, show that*

$(dp/dt)_v$ *const.* $\times (dv/dp)_t$ *const.* $\times (dt/dv)_p$ *const.* $= -1.$ **[W.B. Sem. Exam. '01]**

If $F(x, y) = 0$, we know that $dy/dx = -F_x/F_y$.

When $v =$ constant, $F(p, t, v) = 0$ defines p as a function of t and hence $dp/dt = -F_t/F_p$, i.e., $(dp/dt)_v$ const. $= -F_t/F_p$.

Similarly, $(dv/dp)_t$ const. $= -F_p/F_v$

and $\quad (dt/dv)_p$ const. $= -F_v/F_t$.

Multiplying we get the required result.

Exercises 4.2

1. Find f_x and f_y at $(2, 1)$, where $f(x, y) = (x + y - 1)/(x - y + 1)$.

2. Show that both $f_x(0, 0)$ and $f_y(0, 0)$ are zero, where

(i) $f(x, y) = \sqrt{|xy|}$,

(ii) $f(x, y) = (x^3 + y^3)/(x - y)$ when $x \neq y$ and zero when $x = y$,

(iii) $f(x, y) = xy/\sqrt{(x^2 + y^2)}$ when $(x, y) \neq (0, 0)$ and $f(0, 0) = 0$.

3. Show that the function f, defined by $f(x, y) = (x^3 - y^3)/(x^2 + y^2)$ when $x^2 + y^2 \neq 0$ and $f(0, 0) = 0$, is continuous and possesses first partial derivatives at the origin.

4. If $u = x\ln(xy)$, where $x^3 + y^3 + 3xy = 1$, find du/dx.

5. Evaluate z_x and z_y, where z is given by

(i) $x^2 + y^2 - e^z = 0$, **(ii)** $\tan^{-1}[(x^2 + y^2)/(x + y)]$ and **(iii)** $x + y + z = \ln z$.

6. If $z = e^{ax+by} f(ax - by)$, where f is differentiable, then show that $az_y + bz_x = 2abz$.

7. If $z = (x^2 + y^2)/(x + y)$, show that $(z_x - z_y)^2 = 4(1 - z_x - z_y)$.

8. **(i)** If $u = \sin^{-1}(x - y)$, where $x = 3t$ and $y = 4t^3$, show that $du/dt = 3/\sqrt{(1 - t^2)}$.
(ii) If $u = \sin(x/y)$, where $x = e^t$ and $y = t^2$, show that $du/dt = \{(t-2)/t^3\}e^t \cos(e^t/t^2)$.

9. If $u = \tan^{-1}(y/x)$, where $x = e^t - e^{-t}$ and $y = e^t + e^{-t}$, find du/dt.

10. If $z = \ln(u^2 + v)$, where $u = e^{x^2+y^2}$ and $v = x^2 + y$, show that $z_x = 2x(2u^2 + 1)/(u^2 + v)$ and $z_y = (4yu^2 + 1)/(u^2 + v)$.

11. If $u = f(r, s)$, where $r = x + at$ and $s = y + bt$, x, y and t being independent variables and a and b are constants, then show that $u_t = au_x + bu_y$.

12. If $u = f(x, y)$, where $x = r\cos\theta$ and $y = r\sin\theta$, prove that $u_x^2 + u_y^2 = u_r^2 + u_\theta^2/r^2$.

13. If $u = e^{r\cos\theta}\cos(r\sin\theta)$ and $v = e^{r\cos\theta}\sin(r\sin\theta)$, prove that $u_\theta = -rv_r$ and $v_\theta = ru_r$.

14. If $u = f(y - z, z - x, x - y)$, prove that $u_x + u_y + u_z = 0$.

15. Verify Euler's Theorem on homogeneous functions, when

(i) $u = \sin[(x^2 + y^2)/(xy)]$,
(ii) $u = (z^2 + xy)/(zx)$,
(iii) $u = (x^{1/4} + y^{1/4})/(x^{1/5} + y^{1/5})$.

16. **(i)** If $\theta = \sin^{-1}\sqrt{[(x^{1/3} + y^{1/3})/(\sqrt{x} + \sqrt{y})]}$, show that $x\theta_x + y\theta_y + (1/12)\tan\theta = 0$,
(ii) If $\theta = \cos^{-1}[(x+y)/(\sqrt{x}+\sqrt{y})]$, verify that $\cos\theta$ is a homogeneous function of degree $1/2$ in x and y. Hence show that $x\theta_x + y\theta_y + (1/2)\cot\theta = 0$.
(iii) If $\theta = \tan^{-1}[(x^3 + y^3)/(x - y)]$, show that $x\theta_x + y\theta_y = \sin 2\theta$.

17. If $u = \sin^{-1}(x/y) + \tan^{-1}(y/x)$, show that $xu_x + yu_y = 0$.

18. If $\ln u = (x^3 + y^3)/(3x + 4y)$, show that $xu_x + yu_y = 2u\ln u$.

19. If $e^z = x^2 + xy + y^2$, show that $xz_x + yz_y = 2$.

20. If $u = x/(y + z) + y/(z + x) + z/(x + y)$, show that $xu_x + yu_y + zu_z = 0$.

21. **(i)** If $u = \sin^{-1}[(x + 2y + 3z)/(x^8 + y^8 + z^8)]$, show that $xu_x + yu_y + zu_z + 7\tan u = 0$.
(ii) If $u = \sin^{-1}[(x^3 + y^3 + z^3)/(ax + by + cz)]$, show that $xu_x + yu_y + zu_z = 2\tan u$.

22. If $x^2/(a^2+u)+y^2/(b^2+u)+z^2/(c^2+u) = 1$, show that $u_x^2+u_y^2+u_z^2 = 2(xu_x+yu_y+zu_z)$.

23. If $u = f\left(\sum x^2\right) g\left(\sum yz\right)$, prove that $\sum(y - z)u_x = 0$.

24. If $f(v^2 - x^2, v^2 - y^2, v^2 - z^2) = 0$, where v is a function of x, y and z, show that $v_x/x + v_y/y + v_z/z = 1/v$.

25. If $u = f(by - cz, cz - ax, ax - by)$, prove that $u_x/a + u_y/b + u_z/c = 0$.

26. If $u = f(\xi, \eta, \zeta)$, where $\xi = y/z$, $\eta = z/x$ and $\zeta = x/y$, show that $\sum xu_x = 0$.

27. If f is a function of x, y and z, where $x = u + v + w$, $y = vw + wu + uv$ and $z = uvw$, show that

$$uf_u + vf_v + wf_w = xf_x + 2yf_y + 3zf_z.$$

28. If $u = f(x^2 + y^2 + z^2)$, where $x = r\cos\theta\cos\phi$, $y = r\cos\theta\sin\phi$ and $z = r\sin\theta$, find u_θ and u_ϕ.

Answers

1. $f_x = 0,\ f_y = 1.$ **4.** $1 + \ln(2ay) - x(x^2+y^2)/\{y(x+y^2)\}.$

5. (i) $2x/(x^2+y^2),\ 2y/(x^2+y^2);$

(ii) $(x^2+2xy-y^2)/[(x^2+y^2)^2+(x+y)^2],\ (y^2+2xy-x^2)/[(x^2+y^2)^2+(x+y)^2];$

(iii) $z_x = z_y = z/(1-z).$ **9.** $-2/(e^{2t}+e^{-2t})$ or $-\text{sech}\, 2t.$

28. $u_\theta = 0 = u_\phi.$

4.12 Higher Order Partial Derivatives and Differentials

Let $u = f(x,y)$, define a function of two real variables x and y, possessing the first order partial derivatives $u_x(x,y)$ and $u_y(x,y)$. Each of these partial derivatives is a function of x and y in some domain and each may possess the first order partial derivatives:

$$\frac{\partial}{\partial x}u_x,\ \frac{\partial}{\partial y}u_x,\ \frac{\partial}{\partial x}u_y \text{ and } \frac{\partial}{\partial y}u_y \text{ for some points } (x,y) \in \mathbb{R}^2.$$

By definition,

$$\frac{\partial}{\partial x}u_x = \underset{\delta x \to 0}{\text{Lt}} \frac{u_x(x+\delta x, y) - u_x(x,y)}{\delta x}.$$

But $u_x(x,y) = \underset{\Delta x \to 0}{\text{Lt}} \dfrac{u(x+\Delta x, y) - u(x,y)}{\Delta x}$

and $u_x(x+\delta x, y) = \underset{\Delta x \to 0}{\text{Lt}} \dfrac{u(x+\delta x+\Delta x, y) - u(x+\delta x, y)}{\Delta x}.$

So, ultimately

$$u_{xx} = \frac{\partial}{\partial x}u_x = \underset{\delta x \to 0}{\text{Lt}} \frac{1}{\delta x}\left[\underset{\Delta x \to 0}{\text{Lt}} \frac{u(x+\delta x+\Delta x, y) - u(x+\delta x, y) - u(x+\delta x, y) + u(x,y)}{\Delta x}\right]$$

or, $$u_{xx} = \underset{\delta x \to 0}{\text{Lt}}\ \underset{\Delta x \to 0}{\text{Lt}} \frac{1}{\delta x \Delta x}[u(x+\delta x+\Delta x, y) - 2u(x+\delta x, y) + u(x,y)],$$

which involves repeated limits.

Similarly,

$$u_{yx} = \frac{\partial}{\partial y}u_x = \underset{\delta y \to 0}{\text{Lt}}\ \underset{\delta x \to 0}{\text{Lt}} \frac{1}{\delta y \delta x}[u(x+\delta x, y+\delta y) - u(x+\delta x, y) - u(x, y+\delta y) + u(x,y)].$$

In this way all the second order partial derivatives involve successive repeated limits and we have

$$u_{x^2} = u_{xx} = \partial u_x/\partial x,\ u_{y^2} = u_{yy} = \partial u_y/\partial y,$$

$$u_{yx} = \partial u_x/\partial y \text{ and } u_{xy} = \partial u_y/\partial x.$$

So, normally $u_{yx} \neq u_{xy}$, although most often these two partial derivatives are equal. In fact

$$u_{yx} = u_{xy}$$

at a point $P(a,b)$, if both the first partial derivatives u_x and u_y exist in a nbd. of P and one of the second order partial derivatives u_{yx} and u_{xy} is continuous at P.

That the second order partial derivatives u_{xy} and u_{yx} are not always equal is shown by the following example:

Example. *If* $u = f(x, y) = xy(x^2 - y^2)/(x^2 + y^2)$ *when* $(x, y) \neq (0, 0)$,

$= 0$ *when* $(x, y) = (0, 0)$,

show that $u_{xy}(O) \neq u_{yx}(O)$, *O being the origin* $(0, 0)$. **[Sem. Exam. 2003]**

If $(x, y) \neq (0, 0)$ and P is (x, y), then $P \neq O$.

$$u_{xy}(O) = \underset{\delta x \to 0}{\text{Lt}} \underset{\delta y \to 0}{\text{Lt}} \frac{1}{\delta x \delta y}[u(0 + \delta x, 0 + \delta y) - u(\delta x, 0) - u(0, \delta y) + u(0, 0)]$$

$$= \underset{x \to 0}{\text{Lt}} \underset{y \to 0}{\text{Lt}} \frac{1}{xy}[u(x, y) - u(x, 0) - u(0, y) + u(0, 0)]$$

$$= \underset{x \to 0}{\text{Lt}} \frac{1}{x} \underset{y \to 0}{\text{Lt}} \frac{1}{y}\left[\frac{xy(x^2 - y^2)}{x^2 + y^2}\right] = \underset{x \to 0}{\text{Lt}}\left[\underset{y \to 0}{\text{Lt}} \frac{(x^2 - y^2)}{x^2 + y^2}\right] = \underset{x \to 0}{\text{Lt}} \frac{x^2}{x^2} = 1.$$

And $$u_{yx}(0) = \underset{y \to 0}{\text{Lt}} \underset{x \to 0}{\text{Lt}} \frac{1}{xy}[u(x, y) - u(0, y) - u(x, 0) - u(0, 0)]$$

$$= \underset{y \to 0}{\text{Lt}} \underset{x \to 0}{\text{Lt}} \frac{1}{xy} \cdot \frac{xy(x^2 - y^2)}{x^2 + y^2} = \underset{y \to 0}{\text{Lt}}\left[\underset{x \to 0}{\text{Lt}} \frac{x^2 - y^2}{x^2 + y^2}\right] = \underset{y \to 0}{\text{Lt}} \frac{y^2}{y^2} = -1.$$

Thus $u_{yx}(0) \neq u_{xy}(0)$.

At any other point

$$u_x = \frac{\partial}{\partial x}\left\{xy\frac{x^2 - y^2}{x^2 + y^2}\right\} = y\left[\frac{x^2 - y^2}{x^2 + y^2} + \frac{zx^2(x^2 + y^2) - (x^2 - y^2)}{(x^2 + y^2)^2}\right]$$

$$= y\left[\frac{(x^2 + y^2)(x^2 - y^2) + 2x^2 \cdot 2y^2}{(x^2 + y^2)^2} = \frac{y(x^4 - y^4 + 4^2xy^2)}{(x^2 + y^2)^2}\right],$$

$$u_y = \frac{\partial}{\partial y}\left\{xy\frac{x^2 - y^2}{x^2 + y^2}\right\} = \frac{x(x^4 - y^4 - 4x^2y^2)}{(x^2 + y^2)^2}$$

and $$u_{yx} = \frac{x^6 + 9x^4y^2 - 9x^2y^4 - y^6}{(x^2 + y^2)^3} = u_{xy}.$$

Since each of the four second order partial derivatives u_{xx}, u_{yy}, u_{xy}, u_{yx} may possess partial derivatives with respect to x and y, we have in all eight third order derivatives of u

$$u_{x^3},\ u_{y^3},\ u_{x^2y},\ u_{xy^2},\ u_{xyx},\ u_{yx^2},\ u_{yxy},\ u_{y^2x}.$$

If a change in the order of derivation gives the same result, we have only four third order derivatives, viz.,

$$u_{x^3},\ u_{y^3},\ u_{x^2y},\ u_{xy^2}.$$

In case $u = f(x, y, z)$ a function of three variables there are three first order partial derivatives

$$u_x,\ u_y \text{ and } u_z,$$

six second order partial derivatives

$$u_{x^2},\ u_{y^2},\ u_{z^2},\ u_{yz},\ u_{zx},\ u_{xy},$$

if $u_{xy} = u_{yx}$, etc., and nine third order derivatives

$$u_{x^3},\ u_{y^3},\ u_{z^3},\ u_{y^2z},\ u_{yz^2},\ u_{z^2x},\ u_{zx^2},\ u_{x^2y},\ u_{xy^2},$$

if a change in the order of derivation does not alter the value.

Higher Order Differentials

If $\quad u = f(x, y)$, we have seen

$$du = f_x dx + f_y dy \quad \text{or,}\ u_x dx + u_y dy = \left(dx\frac{\partial}{\partial x} + dy\frac{\partial}{\partial y}\right) u.$$

Similarly, the second order differential

$$\begin{aligned} d^2u &= d(du) = d[u_x dx + u_y dy] \\ &= (du_x)dx + (du_y)dy, \text{ since } dx \text{ and } dy \text{ are constants} \\ &= (u_{xx}dx + u_{yx}dy)dx + (u_{xy}dx + u_{yy}dy)dy. \end{aligned}$$

or, $\quad \mathbf{d^2u = u_{x^2}dx^2 + 2u_{xy}dxdy + u_{y^2}dy^2}$,

where $dx^2 = (dx)^2$, $dy^2 = (dy)^2$ and $u_{xy} = u_{yx}$.

Since $\quad d^2u = \left(dx^2\dfrac{\partial^2}{\partial x^2} + 2dxdy\dfrac{\partial^2}{\partial x \partial y} + dy^2\dfrac{\partial^2}{\partial y^2}\right) u,$

we have symbolically,

$$d^2u = (dx\partial/\partial x + dy\partial/\partial y)^2 u \quad \text{or} \quad (dxD_x + dyD_y)^2 u.$$

Similarly,

$$\begin{aligned} d^3u &= u_{x^3}dx^3 + 3u_{x^2y}dx^2dy + 3u_{xy^2}dxdy^2 + u_{y^3}dy^3 \\ &= (dx\partial/\partial x + dy\partial/\partial y)^3 u \quad \text{or} \quad (dxD_x + dyD_y)^3 u. \end{aligned}$$

Examples 4.11. 1. *If $z(= f(x, y))$ is a homogeneous function of degree n, then show that its first order partial derivatives are each a homogeneous function of degree $n - 1$.*

Since z is a homogeneous function of degree n,

$\therefore \quad z = x^n \phi(y/x)$.

Then $\quad z_x = nx^{n-1}\phi(u) + x^n\phi'(u)u_x$, where $u = y/x$

$$\begin{aligned} &= nx^{n-1}\phi(u) - x^{n-2}y\phi'(u) \\ &= x^{n-1}\psi(u), \text{ where } \psi(u) = n\phi(u) - u\phi'(u). \end{aligned}$$

Hence $\quad z_x$ is a homogeneous function of degree $n - 1$ in x and y.

Again, $\quad z_y = x^n\phi'(u)u_y = x^{n-1}\phi'(u)$,

which implies that z_u is a homogeneous function of degree $n - 1$ in x and y.

2. *If z is a homogeneous function of degree n in x and y, then prove that $x^2z_{xx} + 2xyz_{xy} + y^2z_{yy} = n(n-1)z$, assuming $z_{xy} = z_{yx}$.*

By Euler's theorem, $\qquad xz_x + yz_y = nz. \qquad (1)$

Since z_x is a homogeneous function of degree $n - 1$, by the same theorem

$$xz_{xx} + yz_{yx} = (n-1)z_x. \qquad (2)$$

Similarly, since z_y is a homogeneous function of degree $n - 1$,

$$xz_{xy} + yz_{yy} = (n-1)z_y. \qquad (3)$$

$x \cdot (2) + y \cdot (3)$ gives

$$x^2 z_{xx} + 2xyz_{xy} + y^2 z_{yy} = (n-1)(xz_x + yz_y)$$
$$= n(n-1)z \text{ by (1).}$$

4.13 Use of Operators $\partial/\partial x$ and $\partial/\partial y$ or D_x and D_y

For convenience, we take $D_x = \partial/\partial x$ and $D_y = \partial/\partial y$.

Then $\quad D_x z = (\partial/\partial x)z = \partial z/\partial x = z_x$ and $D_y z = (\partial/\partial y)z = \partial z/\partial y = z_y$.

If h and k are constants,

$$(hD_x + kD_y)z = hz_x + kz_y.$$

Operating both sides with hD_x,

$$hD_x(hD_x + kD_y)z = hD_x(hz_x + kz_y) = h^2 z_{xx} + hkz_{xy} \tag{1}$$

Similarly, $kD_y(hD_x + kD_y)z = hkz_{yx} + k^2 z_{yy}$. (2)

Adding, $\quad hD_x(hD_x + kD_y) + kD_y(hD_x + kD_y)z = h^2 z_{xx} + 2hkz_{xy} + k^2 z_{yy}$

or, $\quad (hD_x + kD_y)^2 z = (h^2 D_x^2 + 2hkD_xD_y + k^2 D_y^2)z$

$$= h^2 z_{xx} + 2hkz_{xy} + k^2 z_{yy}.$$

Thus we have

$$\boldsymbol{(hD_x + kD_y)^2 \equiv h^2 D_x^2 + 2hkD_xD_y + k^2 D_y^2}$$

or, $\quad (h\partial/\partial x + k\partial/\partial y)^2 \equiv h^2\partial^2/\partial x^2 + 2hk\partial^2/\partial x\partial y + k^2\partial^2/\partial y^2$. (3)

In place of h and k if we take x and y respectively, then we shall get

$$xD_x(xD_x + yD_y)z = xD_x(xz_x + yz_y)$$
$$= x[xz_{xx} + z_x + yz_{xy}]$$
$$= x^2 z_{xx} + xyz_{xy} + xz_x,$$

in place of (1) and

$$yD_y(xD_x + yD_y)z = yD_y(xz_x + yz_y)$$
$$= y[xz_{yx} + yz_{yy} + z_y] = xyz_{xy} + y^2 z_{yy} + yz_y$$

in place of (2). Their sum is

$$xD_x(xD_x + yD_y)z + yD_y(xD_x + yD_y)z = x^2 z_{xx} + 2xyz_{xy} + y^2 z_{yy} + xz_x + yz_y$$

or, $\quad \boldsymbol{(xD_x + yD_y)^2 z} = x^2 z_{xx} + 2xyz_{xy} + y^2 z_{yy} + xz_x + yz_y$

$$= \boldsymbol{(x^2 D_x^2 + 2xyD_xD_y + y^2 D_y^2 + xD_x + yD_y)z}.$$

Thus in this case, it might be carefully noted that

$$(x\partial/\partial x + y\partial/\partial y)^2 = (x\partial/\partial x + y\partial/\partial y)(x\partial/\partial x + y\partial/\partial y) \neq x^2\partial^2/\partial x^2 + 2xy\partial^2/\partial x\partial y + y^2\partial^2/\partial y^2.$$

For convenience the partial derivatives z_{xx}, z_{xy} and z_yy are represented by p, q and r, respectively.

4.14 The Laplacian Operator ∇^2 or Δ

The operator $\partial^2/\partial x^2 + \partial^2/\partial y^2 + \partial^2/\partial z^2$ is denoted by ∇^2, called **'del squared'** or Laplacian. It plays an important role in solving higher physical problems like wave propagation and heat conduction. The partial differential equation

$$(\partial^2/\partial x^2 + \partial^2/\partial y^2 + \partial^2/\partial z^2)V = 0$$

or, $\nabla^2 V = 0$

or, $V_{xx} + V_{yy} + V_{zz} = 0$

is the well-known **Laplacian Equation**. Any homogeneous function of degree n in x, y and z satisfying Laplace's equation is called a **Spherical Harmonic of degree** n.

It is customary to denote $x^2 + y^2 + z^2$ by r^2.

Examples 4.12. **1.** *If* $V = r$, $V_x = x/r$.

For, $r^2 = x^2 + y^2 + z^2 \Rightarrow 2rr_x = 2x$.

So, $r_x = x/r$.

Similarly, $r_y = y/r$ and $r_z = z/r$.

Again $r_{xx} = 1/r - (x/r^2)r_x = 1/r - x^2/r^3$

Similarly, $r_{yy} = 1/r - y^2/r^3$ and $r_{zz} = 1/r - z^2/r^3$.

$\therefore$ $\nabla^2 r = r_{xx} + r_{yy} + r_{zz} = 3/r - (x^2 + y^2 + z^2)/r^3 = 2/r$.

Thus $r = \sqrt{(x^2 + y^2 + z^2)}$, although a homogeneous function of degree 1 in x, y and z, is not a Spherical Harmonic.

2. *If* $u = \phi(x, y)$ *and* $\psi(x, y) = 0$, *prove that* $du/dx = (\phi_x\psi_y - \phi_y\psi_x)/\psi_y$.

$$\frac{du}{dx} = \phi_x + \phi_y\frac{dy}{dx}.$$

$\because$ $\psi(x, y) = 0$, $\therefore$ $dy/dx = -\psi_x/\psi_y$.

$\therefore$ $du/dx = \phi_x - \phi_y\psi_x/\psi_y = (\phi_x\psi_y - \phi_y\psi_x)/\psi_y$.

3. *If u is a homogeneous function of the n-th degree in x, y, z and if $u = f(X, Y, Z)$, where $X = u_x$, $Y = u_y$ and $Z = u_z$, then prove that*

$$Xf_X + Yf_Y + Zf_Z = nu/(n-1).$$

Since u is a homogeneous function of degree n, each of $X(= u_x)$, $Y(= u_y)$ and $Z(= u_z)$ is homogeneous of degree $n - 1$.

$$\therefore \quad xX_x + yX_y + zX_z = (n-1)X, \tag{1}$$

$$xY_x + yY_y + zY_z = (n-1)Y \tag{2}$$

$$\text{and} \quad xZ_x + yZ_y + zZ_z = (n-1)Z. \tag{3}$$

Now $(1)f_X + (2)f_Y + (3)f_Z$ gives

$$(n-1)(Xf_X + Yf_Y + Zf_Z] = x(f_XX_x + f_YY_x + f_ZZ_x) + y(f_XX_y + f_YY_y + f_ZZ_y) + z(f_XX_z + f_YY_z + f_ZZ_z)$$

$$= xu_x + yu_y + zu_z = nu, \quad \because \; u_x = f_XX_x + f_YY_x + f_ZZ_x.$$

$\therefore$ $Xf_X + Yf_Y + Zf_Z = nu/(n-1)$.

4. *If $x = r\cos\theta$ and $y = r\sin\theta$, where $r \geq 0$, show that $rdr = xdx + ydy$ and $r^2 d\theta = xdy - ydx$.*

Since $x^2 + y^2 = r^2$,

$$dr^2 = d(x^2 + y^2) = dx^2 + dy^2$$

or, $\quad 2rdr = 2xdx + 2ydy,$

whence $\quad rdr = xdx + ydy.$

Again, $\quad \tan\theta = y/x \Rightarrow \sec^2\theta d\theta = (xdy - ydx)/x^2.$

$\because \quad x = r\cos\theta,\ r^2 d\theta = xdy - ydx.$

5. *If $A = xy$, $x \geq 0$, $y \geq 0$, explain geometrically the equation $\delta A = \frac{\partial A}{\partial x}\delta x + \frac{\partial A}{\partial y}\delta y$ with reference to the area of a rectangle whose side lengths x and y are allowed to increase to $x + \delta x$, $y + \delta y$ the increments being infinitesimals of the first order.*

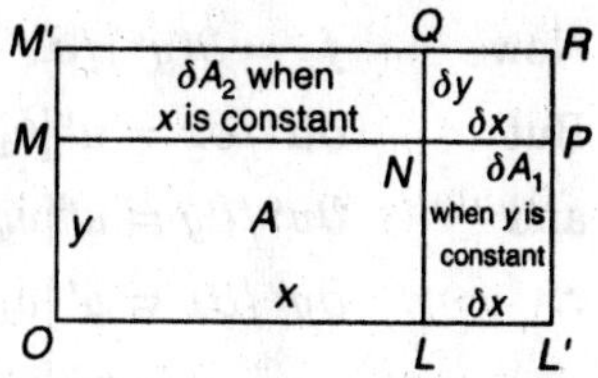

Let A = area of the rectangle $OLNM = OL \cdot OM = xy$.

If y is held fixed and x is allowed to increase to $x + \delta x = OL'$, then the new rectangle is $OL'PM$, whose area is $A + \delta A_1$, so that

$$\delta A_1 = \text{area of } LL'PN \approx \frac{\partial A}{\partial x}\delta x.$$

Fig. 4.13

Similarly, when x is held fixed and y is allowed to increase to $y + \delta y = OM'$, the new rectangle is $OLQM'$, whose area is $A + \delta A_2$, so that δA_2 = area of $MNQM' \approx \frac{\partial A}{\partial y}\delta y$. Area of $NPRQ = \delta x \delta y \approx 0$ to the first order of smallness.

If δA is the actual increment in the area A when both x and y have increased to $x + \delta x$ and $y + \delta y$,

$$A + \delta A = \text{area of } OL'RM'$$

$$= \text{ar}\, OLNM + \text{ar}\, LL'PN + \text{ar}\, MNQM' + \text{ar}\, NPRQ = A + \delta A_1 + \delta A_2 + \delta x \delta y$$

$$= A + \frac{\partial A}{\partial x}\delta x + \frac{\partial A}{\partial y}\delta y + \delta x \delta y$$

$$\therefore \quad \delta A \approx \frac{\partial A}{\partial x}\delta x + \frac{\partial A}{\partial y}\delta y.$$

6. *Given $H = f(y - z, z - x, x - y)$, prove that $H_x + H_y + H_z = 0$.*

Let $\quad u = y - z,\ v = z - x,\ w = x - y$. Then $H = f(u, v, w)$.

But $\quad H_x = f_u u_x + f_v v_x + f_w w_x = -f_v + f_w,$

$$H_y = f_u u_y + f_v v_y + f_w w_y = f_u - f_w$$

and $\quad H_z = f_u u_z + f_v v_z + f_w w_z = -f_u + f_v.$

Adding, $\quad H_x + H_y + H_z = 0.$

7. *If u is a homogeneous function of three independent variables x, y and z and of degree n, then show that*

$$xu_x + yu_y + zu_z = nu.$$

Let $\quad u = f(x, y, z) = x^n \phi(v, w)$, where $v = y/x$ and $w = z/x$.

Now $\quad u_x = nx^{n-1}\phi + x^n(\phi_v v_x + \phi_w w_x)$

$$= nx^{n-1}\phi + x^n[-y\phi_v/x^2 - z\phi_w/x^2]$$

$$= nx^{n-1}\phi - x^{n-2}y\phi_v - x^{n-2}z\phi_w \quad (1)$$

$$u_y = x^n[\phi_v v_y + \phi_w w_y] = x^{n-1}y\phi_w \quad (2)$$

and $\quad u_z = x^n[\phi_v v_z + \phi_w w_z] = x^{n-1}z\phi_w. \quad (3)$

$\therefore \quad x(1) + y(2) + z(3)$ gives the required result.

8. *Find dy/dx when*

(i) $y^{x^y} = \sin x,$ *(ii)* $(\cos x)^y - (\sin y)^x = 0,$ *(iii)* $(\tan x)^y + y^{\cot x} = a.$

(i) Let $f(x,y) = y^{x^y} - \sin x = y^v - \sin x$, where $v = x^y$.

$$y' = dy/dx = -f_x/f_y \text{ when } f(x,y) = 0 \text{ or, } y^{x^y} = \sin x.$$

Now $\quad f_x = \partial(y^v)/\partial x - \partial \sin x/\partial x = \partial y^v/\partial x - \cos x.$

But $\quad \partial u^v/\partial x = u^v[v_x \ln u + (v/u)u_x]$

and $\quad \partial u^v/\partial y = u^v[v_y \ln u + (v/u)u_y].$

$\therefore \quad \partial y^v/\partial x = y^v[v_x \ln y + (v/y)y_x]$

$$= y^v[yx^{y-1}\ln y], \quad \because y_x = 0.$$

Similarly, $\partial y^v/\partial y = y^v[v_y \ln y + v/y\ y_y]$

$$= y^v\left[x^y \ln x \ln y + \frac{x^y}{y}1\right], \quad \because y_y = 1.$$

$\therefore \quad f_x = y^v(yx^{y-1}\ln y) - \cos x = yx^{y-1}\sin x \ln y - \cos x, \quad \because y^v = \sin x$

and $\quad f_y = x^y y^{v-1}(y \ln x \ln y + 1) = x^y y^{-1} \sin x(y \ln x \ln y + 1).$

$$\therefore \quad dy/dx = \frac{y(\cot x - yx^{y-1}\ln y)}{x^y(y\ln x \ln y + 1)} = \frac{y}{x}\cdot\frac{(x\cot x - yx^y \ln y)}{x^y(y \ln x \ln y + 1)}.$$

Since $\quad y^{x^y} = \sin x,\ x^y \ln y = \ln \sin x.$

$$\therefore \quad dy/dx = \frac{y(x\cot x - y\cdot\ln\sin x)}{\frac{x\cdot\ln\sin x}{\ln y}(y\ln x\ln y + 1)} = \frac{y}{x}\ \frac{\ln y(x\cot x - y\ln\sin x)}{\ln\sin x(y\ln x\ln y + 1)}.$$

Alternatively,

$$y^{x^y} = \sin x \Rightarrow x^y \ln y = \ln \sin x.$$

Taking total derivatives with respect to x

$$x^y\left[y'\ln x + \frac{y}{x}\right]\ln y + x^y\frac{1}{y}y' = \cot x.$$

$\therefore \quad x^y y'(\ln x \ln y + 1/y) = \cot x - x^y(y/x)\ln y$

and $\quad y' = (y/x)(x\cot x - yx^y \ln y)/x^y(y\ln x \ln y + 1).$

(ii) $f(x,y) = (\cos x)^y - (\sin y)^x = 0.$

Using the formula: $\dfrac{\partial}{\partial x}(u^v) = u^v\left[v_x \ln u + \dfrac{v}{u}u_x\right],$

$$f_x = (\cos x)^y\left[0\cdot\ln\cos x + \frac{y}{\cos x}(-\sin x)\right] - (\sin y)^x\left[1\cdot\ln\sin y + \frac{x}{\sin y}\cdot 0\right]$$

$$= -y(\cos x)^{y-1}\sin x - (\sin y)^x \ln \sin y.$$

Similarly, $\partial(u^v)/\partial y = u^v[v_y \ln u + (v/u)u_y]$ gives

$$f_y = (\cos x)^y[1 \cdot \ln\cos x + (y/\cos x)\cdot 0] - (\sin y)^x[0 \cdot \ln\sin y + (x/\sin y)\cos y]$$
$$= (\cos x)^y \ln\cos x - x(\sin y)^{x-1}\cos y.$$

$$\therefore \quad dy/dx = -f_x/f_y = \frac{y(\cos x)^{y-1}\sin x + (\sin y)^x \ln\sin y}{(\cos x)^y \ln\cos x - x(\sin y)^{x-1}\cos y}.$$

Since $(\cos x)^y = (\sin y)^x$,

$$dy/dx = \frac{y(\sin y)^x \frac{\sin x}{\cos x} + (\sin y)^x \ln\sin y}{(\sin y)^x \ln\cos x - x(\sin y)^x \frac{\cos y}{\sin y}}$$
$$= \frac{\sin y}{\cos x}\,\frac{(y\sin x + \cos x\ln\sin y)}{(\sin y\ln\cos x - x\cos y)}. \quad (1)$$

Alternatively, since $(\cos x)^y = (\sin y)^x$, taking natural logarithms

$$y\ln\cos x = x\ln\sin y.$$

So, differentiating term by term with respect to x

$$y'\ln\cos x - y\frac{\sin x}{\cos x} = \ln\sin y + x\frac{\cos y}{\sin y}y'.$$

or, $$y'\left(\ln\cos x - x\frac{\cos y}{\sin y}\right) = y\frac{\sin x}{\cos x} + \ln\sin y,$$

whence (1) is easily obtained.

(iii) $(\tan x)^y + y^{\cot x} = a.$

Taking the total derivatives with respect to x term by term

$$(\tan x)^y[y'\ln\tan x + y\cot x\sec^2 x] + y^{\cot x}\left[-\operatorname{cosec}^2 x\ln y + \frac{\cot x}{y}y'\right] = 0$$

or, $$y'[(\tan x)^y \ln\tan x + y^{\cot x - 1}\cot x] = y^{\cot x}\operatorname{cosec}^2 x\ln y - y(\tan x)^y\cot x\sec^2 x.$$

$$\therefore \quad y' = \frac{y^{\cot x}\operatorname{cosec}^2 x\ln y - y(\tan x)^y\cot x\sec^2 x}{(\tan x)^y\ln\tan x + y^{\cot x-1}\cot x}.$$

9. *If $f(x,y)=0$, defines y as a function of x possessing the second derivative y'', show that*

$$y'' = -[f_{xx}f_y^2 - 2f_{xy}f_xf_y + f_{yy}f_x^2] \div f_y^3.$$

We know that $y' = -f_x/f_y = -u/v$, say,

where $u = f_x(x,y)$ and $v = f_y(x,y)$. Then

$$y'' = -(vu_x - uv_x)/v^2.$$

But $$u_x = f_{xx} + f_{xy}y' = f_{xx} - \frac{f_x}{f_y}f_{xy} = \frac{f_{xx}f_y - f_xf_{xy}}{f_y}$$

and $$v_x = f_{xy} + f_{yy}y' = f_{xy} - \frac{f_x}{f_y}f_{yy} = \frac{f_yf_{xy} - f_xf_{yy}}{f_y}.$$

$$\therefore \quad y'' = -\frac{1}{f_y^2}[f_{xx}f_y - f_xf_{xy} - (f_x/f_y)(f_yf_{xy} - f_xf_{yy})]$$
$$= -[f_{xx}f_y^2 - 2f_xf_yf_{xy} + f_{yy}f_x^2] \div f_y^3.$$

10. *Show that, at a point on the surface given by $x^x y^y z^z = c$, where $x = y = z$,*

$$z_{xy} = -\{x \ln(ex)\}^{-1}.$$

The equation $x^x y^y z^z = c$ defines z as a function $f(x, y)$ of two independent variables x and y.

From the equation

$$x \ln x + y \ln y + z \ln z = \ln c, \tag{1}$$

treating z as a function of the two independent variables x and y, differentiating (1) term by term, partially with respect to x, we get

$$1 + \ln x + z_x \ln z + z_x = 0, \tag{2}$$

whence $\quad z_x = -\dfrac{1 + \ln x}{1 + \ln z} = -1$, when $x = y = z$.

Similarly, $\quad z_y = -\dfrac{1 + \ln y}{1 + \ln z} = -1$, when $x = y = z$.

Differentiating (2) partially with respect to y

$$z_{xy} \ln z + z_x \frac{1}{z} z_y + z_{xy} = 0,$$

whence $\quad z_{xy} = -\dfrac{z_x z_y}{z(1 + \ln z)}.$

When $\quad x = y = z,\ z_{xy} = -\dfrac{(-1)(-1)}{x(1 + \ln x)} = -\dfrac{1}{x \ln(ex)} = -\{x \ln(ex)\}^{-1}.$

11. *Find y'' when $x^3 + y^3 - 3axy = 0$.*

Let $\quad f(x, y) = x^3 + y^3 - 3axy.$

If $f(x, y) = 0$ defines y as a function of x possessing the second derivative and if $f_y \neq 0$, then we know that

$$y'' = -[f_{xx} f_y^2 - 2 f_x f_y f_{xy} + f_{yy} f_x^2] \div f_y^3.$$

Here, $f_x = 3x^2 - 3ay,\ f_{xx} = 6x,$

$f_y = 3y^2 - 3ax,\ f_{xy} = -3a,\ f_{yy} = 6y.$

$$\therefore \quad y'' = -[6x \cdot 9(y^2 - ax)^2 - 2 \cdot 3(x^2 - ay)3(y^2 - ax)(-3a) + 6y9(x^2 - ay)^2] \div 27(y^2 - ax)^3$$

$$= -54[xy(x^3 + y^3 - 3axy) + a^3 xy] \div 27(y^2 - ax)^3$$

$$= -2a^3 xy/(y^2 - ax)^3.$$

Exercises 4.3

1. If $u = xf(y/x) + g(y/x)$, then show that

$$x^2 u_{xx} + 2xy u_{xy} + y^2 u_{yy} = 0.$$

2. (i) If H be a homogeneous function of degree n in x and y and if $u = (x^2 + y^2)^{-n/2}$, show that

$$\partial(H u_x)/\partial x + \partial(H u_y)/\partial y = 0.$$

(ii) If H be homogeneous function of degree n in x, y and z and if $u = \left(\sum x^2\right)^{-(n+1)/2}$,

then show that $\sum \partial(H u_x)/\partial x = 0.$

3. If $u = x\phi(x+y) + y\psi(x+y)$, show that $u_{xx} - 2u_{xy} + u_{yy} = 0$, assuming the validity of the change in the order of derivation.
4. If V is a function of the single variable r, where $r^2 = \sum x^2$, show that
$$\sum V_{xx} = d^2V/dr^2 + (2/r)dV/dr.$$
5. If by the substitution: $u = x^2 - y^2,\ v = 2xy,\ f(x,y) = g(u,v)$, show that
$$f_{xx} + f_{yy} = 4(x^2+y^2)(g_{uu} + g_{vv}).$$
6. Given that z is a function of x and y, where $x = u + v$ and $y = uv$, prove that
$$z_{uu} - 2z_{uv} + z_{vv} = (x^2 - 4y)z_{yy} - 2z_y.$$
7. Let $y = f(x,t)$, where f is a differentiable function of two independent variables x and t which are related to two variables u and v by $u = x + ct$ and $v = x - ct$. Prove that the equation
$$y_{xx} - (1/c^2)y_{tt} = 0$$
can be transformed into $y_{uv} = 0$.
8. If $U = \sin^{-1}[(x^{1/3} + y^{1/3})/(x^{1/2} + y^{1/2})]^{1/2}$, show that
$x^2U_{xx} + 2xyU_{xy} + y^2U_{yy} = (1/144)\tan U(13 + \tan^2 U)$. **[W.B. Sem. Exam. '01]**
9. If $U = \sqrt{xy}$, find the value of $\partial^2U/\partial x^2 + \partial^2U/\partial y^2$. **[W.B. Sem. Exam. '01]**
10. If f is a function of two variables x and y possessing continuous partial derivatives, where $x = u^2v$ and $y = uv^2$, then show that
$$2x^2 f_{xx} + 2y^2_{yy} + 5xyf_{xy} = uvf_{uv} - (2/3)(uf_u + vf_v).$$
11. State Euler's theorem on homogeneous functions of two variables. If
$u = \tan^{-1}[(x^3 + y^3)/(x - y)]$, show that $x^2u_{xx} + 2xyu_{xy} + y^2u_{yy} = (1 - 4\sin^2 u)\sin 2u$.
12. If $u = e^{xyz}$, prove that $u_{xyz} = (1 + 3xyz + x^2y^2z^2)u$.
13. A function $f(x,y)$ having continuous second order partial derivatives, when expressed in terms of new variables u and v, defined by $x = (u+v)/2$ and $y = \sqrt{uv}$, is transformed into $g(u,v)$. Show that
$$g_{uv} = (1/4)[f_{xx} + 2(x/y)f_{xy} + f_{yy}].$$
14. If $u(x,y) = \phi(xy) + \sqrt{xy}\,\psi(y/x),\ x \neq 0,\ y \neq 0$, where ϕ and ψ are twice differentiable, prove that $x^2u_{xx} - y^2u_{yy} = 0$.
15. Transform the equation $x^2z_x + y^2z_y = z^2$ by introducing new independent variables u and v and a new dependent variable w, where $u = x, v = 1/y - 1/x$ and $w = 1/z - 1/x$.

Answers

9. $-(x^2+y^2)/(4x^{3/2}y^{3/2})$. 15. $u^2w_u = w^2$.

4.15 Taylor's Theorem for a Function of Two Variables

If a function f of two real variables x and y possesses continuous nth order partial derivatives in a certain nbd $N(P)$ of a point $P(a,b)$ and if $Q(a+\delta x, b+\delta y)$ be another point in $N(P)$, then

$$f(Q) = f(P) + (1/1!)Df(P) + (1/2!)D^2 f(P) + \cdots + (1/n!)D^n f(R)$$

$$= \sum_0^{n-1} (1/r!)D^r f(P) + (1/n!)D^n f(R),$$

where $D^0 f(P) = f(P)$ and $D^r f(P) = (\delta x \partial/\partial x + \delta y \partial/\partial y)^r f(P)$,

for $r = 1, 2, \ldots, n$ and R is $(a+\theta\delta x, b+\theta\delta y)$, θ being a real number lying in $(0,1)$.

In the expanded form

$$f(a+\delta x, b+\delta y) = f(a,b) + (1/1!)(\delta x D_x + \delta y D_y) f(a,b) + (1/2!)(\delta x D_x + \delta y D_y)^2 f(a,b) + \cdots$$
$$+ \{1/(n-1)!\}(\delta x D_x + \delta y D_y)^{n-1} f(a,b) + (1/n!)(\delta x D_x + \delta y D_y)^n f(a+\theta\delta x, b+\theta\delta y),$$

where $0 < \theta < 1$.

4.16 Extrema of a Function of Two Variables

A function $f : E \to \mathbb{R}^2$ is said to have a **local extremum** or a **local extreme** at a point $P \in E$ if $\exists N(\delta, \not{P}) : f(Q) - f(P)$ maintains the same sign $\forall\, Q \in E \cap N(\delta, \not{P})$.

This extremum is a **local maximum** or a **local minimum** according as $f(Q) - f(P) <$ or $> 0\ \forall\, Q \in E \cap N(\delta, \not{P})$.

Examples 4.13. 1. *The function f defined by $f(x,y) = \sqrt{a^2 - x^2 - y^2}$ has a local maximum at $O(0,0)$, where $a > 0$.*

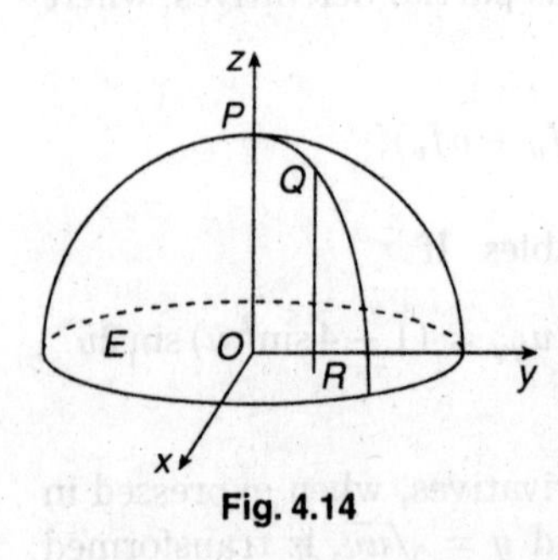

Fig. 4.14

Here $E = \text{Dom}\, f = \{(x,y) \in \mathbb{R}^2 : a^2 - x^2 - y^2 \geq 0\}$. Since $x^2 + y^2 \leq a^2$, E is a circular region with O as its centre and a as its radius. For f to have a local extremum at O, $f(R) - f(O)$ must maintain the same sign $\forall\, R \in E \cap N$, where N is some deleted nbd of O. Taking (x,y) as the coordinates of R

$$f(R) - f(O) = f(x,y) - f(0,0) = \sqrt{a^2 - x^2 - y^2} - a < 0.$$

Since $(\sqrt{a^2 - x^2 - y^2})^2 - a^2 = -(x^2 + y^2) < 0$, if $R \neq O$.

Thus f has a maximum at O.

If $\quad z = f(x,y) = \sqrt{a^2 - x^2 - y^2}$, then $z \geq 0\ \forall\, (x,y) \in E$.

This represents a hemisphere with its base in the xy-plane, supposed to be horizontal. If the positive direction of the z-axis is vertically upwards and P is the topmost point where the value of z or f is a, the value of z at any other point R is QR which is less than OP.

So z or f must have a maximum at O.

2. *If $z = f(x,y) = a + \sqrt{x^2 + y^2}$, then z has a local minimum at $O(0,0)$. For if $P \neq O$, where P is (x,y)*

$$f(P) - f(O) = a + \sqrt{x^2 + y^2} - a = \sqrt{x^2 + y^2} > 0.$$

Global Extremum: Given E as the domain of f, f is said to have a **global extremum** at P if $\forall\, Q(\neq P) \in E$, $f(P) - f(Q)$ always maintains its sign. It is said to be maximum or minimum at P according as $f(P) - f(Q) >$ or < 0.

In the Example 1 above f has a global maximum whereas in the example 2 above f has a global minimum at O.

4.16.1 Necessary Conditions for the Existence of an Extremum of a Function Possessing First Order Partial Derivatives

If a function f of two independent variables x and y has an extremum at a point $P(\alpha, \beta)$ and possesses the first order partial derivatives f_x and f_y at P, then the necessary condition that f possesses an extremum at P is that both f_x and f_y vanish at P, i.e.,

$$f_x(P) = f_y(P) = 0.$$

For, if $f(x, y)$ is extreme at $P(\alpha, \beta)$, $f(x, \beta)$ must be extreme at $x = \alpha$. So $f_x(x, \beta) = 0$ at $x = \alpha$, i.e., $f_x(P) = 0$.

Similarly, $f(\alpha, y)$ must be extreme at $y = \beta$ so that $f_y(P) = 0$. The point P is called a **stationary point** and $f(P)$ is called the **stationary value** of the function f at P.

4.16.2 Sufficient Conditions for a Function to be Extreme

Let a function $f(x, y)$ possess continuous second order partial derivatives. Then a set of sufficient conditions for f to possess an extremum at $P(\alpha, \beta)$ is that

(i) $f_x(P) = 0 = f_y(P)$,

(ii) the second order derivatives are not all zero in a certain nbd of P, and

(iii) $H = f_{xx}(P)f_{yy}(P) - \{f_{xy}(P)\}^2$ or, $\begin{vmatrix} f_{xx}(P) & f_{xy}(P) \\ f_{xy}(P) & f_{yy}(P) \end{vmatrix} > 0$.

This extremum is a maximum or a minimum according as $f_{xx}(P)$ and $f_{yy}(P)$ are both negative or positive.

For, f to be extreme at P, three must exist a nbd N of P such that $\forall Q(\neq P) \in N$, $\delta f (= f(Q) - f(P))$ always maintains a definite sign. Let Q be $(\alpha + \delta x, \beta + \delta y)$. Then

$$\begin{aligned}\delta f &= f(Q) - f(P) = f(\alpha + \delta x, \beta + \delta y) - f(\alpha, \beta) \\ &= df(P) + (1/2)d^2 f(R),\end{aligned}$$

by Taylor's theorem, where $R(\alpha + \theta\delta x, \beta + \theta\delta y)$ is a point of the segment PQ so that $0 < \theta < 1$. At a stationary point P, $f_x(P) = f_y(P) = 0$, so that

$$df(P) = f_x(P)dx + fy(P)dy = 0, \text{ where } \delta x = dx \text{ and } \delta y = dy.$$

$$\begin{aligned}\text{So, } \delta f &= \frac{1}{2}d^2 f(R) = \frac{1}{2}(dxD_x + dyD_y)^2 f(R) \\ &= \frac{1}{2}\left[f_{xx}(R)dx^2 + 2f_{xy}(R)dxdy + f_{yy}(R)dy^2\right]\end{aligned}$$

If dx and dy are very small, because of continuity of f_{xx}, f_{xy} and f_{yy}, their values at R will differ from those at P by a negligibly small amount. So, so far as the sign of δf is concerned we may take

$$\delta f \approx \frac{1}{2}[Adx^2 + 2Bdxdy + Cdy^2], \tag{1}$$

where $A = f_{xx}(P)$, $B = f_{xy}(P)$ and $C = f_{yy}(P)$.

Now the R.H.S. of (1) in the square brackets is a quadratic in dx and dy. It can be written as $(AX^2 + 2BX + C)dy^2$ or Qdy^2 where $X = dx/dy$. dy^2 always positive, if not zero. So the sign of δf ultimately depends upon that of Q. Now we know that Q always maintains the sign of A if the discriminant $D < 0$, i.e.,

$$4(B^2 - AC) < 0 \quad \text{or,} \quad AC - B^2 > 0$$

$$\text{or,} \quad H = \begin{vmatrix} A & B \\ B & C \end{vmatrix} = \begin{vmatrix} f_{xx}(P) & f_{xy}(P) \\ f_{xy}(P) & f_{yy}(P) \end{vmatrix} > 0.$$

For f to be maximum $f(Q) - f(P)$, i.e., $\delta f < 0 \; \forall Q (\neq P) \in N$. So $A < 0$ and for H to be > 0, $B < 0$. Thus for f to be maximum at P.

(i) $f_x(P) = 0 = f_y(P)$,

(ii) A, i.e., $f_{xx}(P) < 0$ and C, i.e., $f_{yy}(P) < 0$

and (iii) H, i.e., $\begin{vmatrix} A & B \\ B & C \end{vmatrix}$ or, $\begin{vmatrix} f_{xx}(P) & f_{xy}(P) \\ f_{xy}(P) & f_{yy}(P) \end{vmatrix} > 0$.

Similarly, for f to be minimum at P, $A > 0$, $C > 0$ and $H > 0$.

Note 1. $Q = AX^2 + 2BX + C$, $D = 4(B^2 - AC) = -4H$.

If $D > 0$, the sign of Q is not definite. Q takes the sign of A when X does not lie between the two real and unequal roots of $Q = 0$ and it takes the opposite sign when X lies between the roots. For different values of X, Q may be $>, =$ or < 0. So, δf and hence $d^2 f$ is **indefinite** (in sign) if $H < 0$ and in this case $f(P)$ cannot be extreme.

If $H = 0$, then $D = 0$ and the two roots of $Q = 0$ coincide. Since Q may be zero, i.e., signless for a real value X it is said to be **semi-definite**. In this case f may or may not have an extremum at P. The case is doubtful and further investigation is necessary.

In terms of differentials of f a set of sufficient conditions for f to be extreme at P is that

(i) $df(P) = 0$ and (ii) $d^2 f(P)$ is definite.

f is maximum or minimum at P according as A and C are both negative or both positive.

Note 2. The expression $H = f_{xx}(x, y) f_{yy}(x, y) - \{f_{xy}(x, y)\}^2$ or, $\begin{vmatrix} f_{xx}(x,y) & f_{xy}(x,y) \\ f_{xy}(x,y) & f_{yy}(x,y) \end{vmatrix}$ is called the **Hessian** of the function at (x, y). In fact, H is a function of the two variables x and y and at the same time it is an algebraic expression involving all the three second order partial derivatives of f. For a function f to be extreme at P, $H(P) > 0$. f can never have an extremum if $H(P) < 0$ and if $H(P) = 0$, the case is doubtful. In this case f may or may not have an extremum and further investigation is necessary.

Examples 4.14. 1. $f(x, y) = 1$, *when neither x nor y is zero,*

$= 0$, *when either x or y is zero.*

Investigate the function for the existence of its extremum at the origin.

Actually, $f(x, y) = 1$, when $xy \neq 0$

$= 0$, when $xy = 0$.

So, $f(O) = f(0, 0) = 0$, $f(0, y) = 0$ and $f(x, 0) = 0$.

If P is any point close to O, then

$$\delta f = f(P) - f(O) = f(P),$$

which may be either 1 or 0. So, δf is not definite.

$\delta f = 0$ if P lies on any of the coordinate axes,

$= 1$ if P does not lie on any of the axes.

So f has no extremum at 0, although we find

$$f_x(0) = \underset{x \to 0}{\text{Lt}} \frac{f(x,0) - f(0,0)}{x} = 0, \quad \because f(x,0) = 0 = f(0,0)$$

and $$f_y(0) = \underset{y \to 0}{\text{Lt}} \frac{f(0,y) - f(0,0)}{y} = 0.$$

2. *Show that u has a local minimum at $P(a,a)$ and $u_{min} = 3a^2$, where*

$$u = xy + a^3(1/x + 1/y).$$

$u_x = y - a^3/x^2$ and $u_y = x - a^3/y^2$.

$\therefore \quad u_x = 0 = u_y \Rightarrow x^2y = xy^2 = a^3$

$x^2y = xy^2 \Rightarrow xy(x - y) = 0 \Rightarrow x = y,$

since $xy = 0$ for $x = 0$ or, $y = 0$. u_x and u_y will not exist.

$x = y$ and $x^2y = a^3 \Rightarrow x = y = a.$

Thus $P(a,a)$ is a stationary point.

$A = u_{xx} = 2a^3/x^3 = 2$ at P

$B = u_{xy} = 1$ and $C = u_{yy} = 2a^3/y^3 = 2$ at P.

$\therefore \quad H = AC - B^2 = 2 \cdot 2 - 1 = 3 > 0$ at P.

And A and C are both positive. So u is minimum at P and $U_{min} = a \cdot a + a^3(1/a + 1/a) = 3a^2$.

3. *Investigate $f(x,y)$ for local extrema, where $f(x,y) = xy(a - x - y)$.*

$f_x = y(a - x - y) - xy$ and $f_y = x(a - x - y) - xy$.

$\therefore \quad f_x = 0 = f_y \Rightarrow y(a - 2x - y) = 0 \qquad$ (i)

and $\quad x(a - x - 2y) = 0. \qquad$ (ii)

From (i) either $\quad x = 0 \qquad$ (iii)

or, $\quad a - x - 2y = 0. \qquad$ (iv)

From (ii) either $\quad y = 0 \qquad$ (v)

or, $\quad a - 2x - y = 0. \qquad$ (vi)

(iii) and (v) give a stationary point $O(0,0)$; (iii) and (vi) give $P(0,a)$.

From (iv) and (v), we get $Q(a,0)$ and from (iv) and (vi), we get $R(a/3, a/3)$. Thus there are four stationary points.

Now $$f_{xx} = \frac{\partial}{\partial x}(ay - 2xy - y^2) = -2y,$$

$$f_{xy} = \frac{\partial}{\partial y}(ay - 2xy - y^2) = a - 2x - 2y$$

and $$f_{yy} = \frac{\partial}{\partial p}(ax - 2xy - x^2) = -2x.$$

At O, $A = f_{xx}(0,0) = 0$, $B = f_{xy}(0,0) = a$, and $C = f_{xy}(0,0) = 0$.

So, $H = AC - B^2 = -a^2 < 0$.

Therefore, f has no extremum at O.

At $P(0,a)$, $A = -2a$, $B = -a$, $C = 0$.

$\therefore$ $H = -a^2 < 0 \Rightarrow f$ has no extremum.

Similarly at $Q(0,0) f$ has no extremum.

At $P(a/3, a/3)$, $A = -2a/3$, $B = -a/3$, $C = -2a/3$.

$\therefore$ $H = 4a^2/9 - a^2/9 = a^2/3 > 0$.

If $a > 0$, $A < 0$, $H > 0$, so f has a maximum at R.

If $a < 0$, $A > 0$, $H > 0$ and hence f has a minimum.

4.16.3 Extrema of a Function of Three Independent Variables

If $u = u(x,y,z)$, where x, y and z are three independent variables, then u is a function of three variables. Suppose, u possesses second order partial derivatives at a point $P(\alpha, \beta, \gamma)$. Then u possesses an extreme value at P if $du = 0$ and d^2u has a definite sign at P. For if $\delta x, \delta y$ and δz (or dx, dy and dz) are arbitrary quantities not all zero, then

$$\delta u = u(\alpha + \delta x, \beta + \delta y, \gamma + \delta z) - u(\alpha, \beta, \gamma)$$

$$= du + \frac{1}{2}!\, d^2u + \frac{1}{3}!\, d^3u(\alpha + \theta\delta x, \beta + \theta\delta y,\ \gamma + \theta\delta z).$$

Since d^3u contains third powers of small quantities δx, δy and δz and $du = 0$, the sign of δu depends essentially on d^2u.

Now $d^2u = (dxD_x + dyD_y + dzD_z)^2 u(\alpha, \beta, \gamma)$

$$= adx^2 + bdy^2 + cdz^2 + 2fdydz + 2gdzdx + 2hdxdy$$

$$= (1/a)[(adx + hdy + gdz)^2 + (ab - h^2)dy^2 + 2(af - gh)dydz + (ac - g^2)dz^2]$$

$$= (1/a)\left[\left(\sum adx\right)^2 + (Cdy^2 - 2Fdydz + Bdz^2)\right], \tag{1}$$

where it is assumed

$$a = u_{xx}(P) \neq 0,\ b = u_{yy}(P),\ c = u_{zz}(P),$$

$$f = u_{yz}(P),\ g = u_{zx}(P),\ h = u_{xy}(P)$$

and B, C and F are the cofactors of b, c and f in the expansion of the determinant

$$H = \begin{vmatrix} a & h & g \\ h & b & f \\ g & f & c \end{vmatrix} = \begin{vmatrix} u_{xx} & u_{xy} & u_{xz} \\ u_{yx} & u_{yy} & u_{yz} \\ u_{zx} & u_{zy} & u_{zz} \end{vmatrix} = abc + 2fgh - af^2 - bg^2 - ch^2,$$

which is the Hessian of u at P.

From (1) it is obvious that the sign of d^2y will be maintained if the sign of

$$Cdy^2 - 2Fdydz + Bdz^2 \tag{2}$$

is always positive, i.e., it is positive definite and it is so if $C > 0$ and the discriminant $4(F^2 - BC) < 0$, i.e.,

$$BC - F^2 > 0. \tag{3}$$

But $BC - F^2 = (ac - g^2)(ab - h^2) - (gh - af)^2 = aH$.

So, (3) is true if a and H have the same sign.

Thus we find that **d^2u is positive definite if a, B, C and H are all positive** and it is **negative definite if $a < 0$, B and C both > 0 and $H < 0$**. It might be noted that a and $C = \begin{vmatrix} a & h \\ h & b \end{vmatrix}$ are the two principal minors of H.

So, u is maximum at P if the principal minors a and C of H and H are alternately negative, positive and negative and it is minimum if they are all positive in addition to the condition that all the three first order partial derivatives are zero at P.

Examples 4.15. 1. *Find the stationary points of the function*

$$u = x^2 + y^2 + z^2 + 2xyz - 4zx - 2yz - 2x - 4y + 4z$$

and examine its extrema at those points.

Here, $u_x = 2x + 2yz - 4z - 2$, $u_y = 2y + 2zx - 2z - 4$

and $\quad u_z = 2z + 2xy - 4x - 2y + 4.$

$\therefore \quad u_x = u_y = u_z = 0 \Rightarrow$

$$yz + x - 2z - 1 = 0, \tag{1}$$

$$zx + y - z - 2 = 0 \tag{2}$$

and
$$xy - 2x - y + z + 2 = 0. \tag{3}$$

(2) + (3) gives

$$x(y + z - 2) = 0. \tag{4}$$

Solution of (1), (2) and (3) is the same as that (1), (2) and (4). But (4) consists of two independent equations

$$x = 0 \tag{5}$$

and
$$y + z - 2 = 0. \tag{6}$$

So, we are required to solve (1), (2) and (5) and (1), (2) and (6).

From (1), (2) and (5)

$x = 0, \; yz + x - 2z - 1 = 0$ and $zx + y - z - 2 = 0,$

or, $\quad x = 0, \; yz - 2z - 1 = 0$ and $y - z - 2 = 0,$

or, $\quad x = 0, \; (y - 2)z - 1 = 0$ and $z = y - 2,$

or, $\quad x = 0, \; z^2 - 1 = 0,$ since $z = y - 2,$

or, $\quad x = 0, \; z = \pm 1, y = 1$ or 3.

So, one set of solutions of (1), (2) and (5) is

$$(0, 1, -1) \text{ and } (0, 3, 1). \tag{7}$$

Next we are to solve (1), (2) and (6). From (6) $z = 2 - y$. Substituting in (1) and (2)

$(y - 2)(2 - y) + x - 1 = 0$ and $(x - 1)(2 - y) + y - 2 = 0,$

i.e., $\quad -y^2 + 4y - 4 + x - 1 = 0$ and $-xy + 2x + 2y - 4 = 0$

or, $\quad x = y^2 - 4y + 5$ and $x(y - 2) - 2y + 4 = 0.$

$\therefore \quad (y^2 - 4y + 5)(y - 2) - 2(y - 2) = 0$

or, $\quad (y^2 - 4y + 3)(y - 2) = 0,$

whence either $y^2 - 4y + 3 = 0$, whose roots are 1 and 3 or, $y = 2$. Thus $y = 1, 2$ and 3.

$$\therefore \quad x = y^2 - 4y + 5 = 2 \text{ and } z = 2 - y = 2 - 1 = 1 \quad \text{when } y = 1$$
$$= 1 \qquad = 0 \qquad \text{when } y = 2$$
$$= 2 \qquad = -1 \qquad \text{when } y = 3.$$

So, we get another set of solutions:

$$(1, 2, 0),\ (2, 1, 1) \text{ and } (2, 3 - 1). \tag{8}$$

(7) and (8) give five stationary points

$$P(0, 1, -1),\ Q(0, 3, 1),\ R(1, 2, 0),\ S(2, 1, 1) \text{ and } T(2, 3, -1).$$

To determine the nature of the function at the stationary points we require the second order partial derivatives.

$$a = u_{xx} = 2 > 0,\ b = u_{yy} = 2,\ c = u_{zz} = 2,$$
$$f = u_{yz} = -2(1 - x),\ g = u_{zx} = -2(2 - y),\ h = u_{xy} = 2z.$$

At $\quad P,\ a = 2,\ b = 2,\ c = 2,\ f = -2,\ g = -2,\ h = -2$

$$C = ab - h^2 = 0 \text{ and } F = gh - af = 8$$
$$H = abc + 2fgh - af^2 - bg^2 - ch^2 = -24.$$

Since $\quad C = 0,\ BC - F^2 = -24 < 0$, so $H < 0$,

so $d^2 f$ is indefinite and there is no extremum at A.

Similarly, it might be shown that, since $h = \pm 2$ at all other points except $R(1, 2, 0)$, where $h = 0$ and hence $C = 0$ and $d^2 f$ is indefinite. At R, $C = 4 > 0$. Thus a, B, C and H are all positive at R. Hence u has a minimum at R and

$$u_{\min} = 1^2 + 2^2 + 0^2 + 2 \cdot 1 \cdot 2 \cdot 0 - 4 \cdot 0 \cdot 1 - 2 \cdot 2 \cdot 0 - 2 \cdot 1 - 4 \cdot 2 - 4 \cdot 0$$
$$= -11.$$

2. *Show that function given by*

$$f(x, y, z) = 3 \ln \sum x^2 - 2 \sum x^3$$

has an extreme, viz., $\ln(3/e^2)$ *at* $(1/\sqrt[3]{3}, 1/\sqrt[3]{3}, 1/\sqrt[3]{3})$.

Here $\quad f_x = 6x/\sum x^2 - 6x^2,$

$$f_y = 6y/\sum x^2 = 6y^2,$$
$$f_z = 6z/\sum x^2 - 6z^2,$$
$$a = f_{xx} = 6(y^2 + z^2 - x^2)/\left(\sum x^2\right)^2 - 12x,$$
$$b = f_{yy} = 6(z^2 + x^2 - y^2)/\left(\sum x^2\right)^2 - 12y,$$
$$c = f_{zz} = 6(x^2 + y^2 - z^2)/\left(\sum x^2\right)^2 - 12z,$$

$$f = f_{yz} = -12yz/\left(\sum x^2\right)^2,$$

$$g = f_{zx} = -12zx/\left(\sum x^2\right)^2,$$

and $$h = f_{xy} = -12xy/\left(\sum x^2\right)^2.$$

So, $f_x = 0 \Rightarrow$ either $x = 0$ (1)

or, $x\sum x^2 = 1.$ (2)

Similarly, $f_y = 0 \Rightarrow$ either $y = 0$ (3)

or, $y\sum x^2 = 1$ (4)

and $f_z = 0 \Rightarrow$ either $z = 0$ (5)

or, $z\sum x^2 = 1.$ (6)

If P is the point $(1/\sqrt[3]{3}, 1/\sqrt[3]{3}, 1/\sqrt[3]{3})$, then $f_x = f_y = f_z = 0$, so that P is a stationary point.

At P $a = b = c = -10/\sqrt[3]{3}$ and $f = g = h = -4/\sqrt[3]{3}$.

So, $H = -54 < 0$ and $C = ab - h^2 = 84/\sqrt[3]{9} = B$.

Since a, B and C, and H are alternately negative, positive and negative, f is maximum at P and

$$f_{\max} = f(P) = 3\ln(3/\sqrt[3]{9}) - 2 = \ln(3/e^2).$$

4.16.4 Constrained Extrema

We have already seen how to find the local extrema of a function. Sometimes, we are required to find the extrema under some given conditions called **Subsidiary conditions**. For example, if we are asked to find the minimum distance of the origin from a given plane, we shall have to consider the function

$$f(x, y, z) = \sqrt{x^2 + y^2 + z^2} \quad \text{or} \quad \sqrt{\sum x^2} \tag{1}$$

representing the distance of the point (x, y, z) from the origin, which much be subjected to a condition of the form

$$ax + by + cz + d = 0, \tag{2}$$

representing a given plane.

The condition (2) is the subsidiary condition. Since the domain of f is the entire three-dimensional Euclidean space, a point $R(x, y, z)$ is free to move anywhere in the space. There is a global minimum 0 of the function although there is no maximum. But because of the subsidiary condition (2) the point R is not now free to move anywhere in space but it should move only on the plane given by (2). We know that the minimum distance of the plane (2) from the origin is is $|d|/\sqrt{a^2 + b^2 + c^2}$, which is therefore the minimum value of f subject to the subsidiary condition (2).

There are two different approaches for determining the stationary values of a function subjected to subsidiary conditions.

(I) Reduction of the number of independent variables with the help of subsidiary conditions.

(II) Lagrange's method of undetermined multipliers.

(I) Case I. $f : E \to \mathbb{R}$ is a function of two variables which is subject to the constraint $g(x,y) = 0 \forall (x,y) \in E$.

Let $z = f(x,y) \forall (x,y) \in E$ (1)

and $g(x,y) = 0 \forall (x,y) \in E_1$ (2)

Solving (2), at least theoretically, for y, we get

$$y = \phi(x). \tag{3}$$

Substituting in (1), we get

$$z = f(x, \phi(x)) = \psi(x), \text{ say,} \tag{4}$$

a function of one independent variable x. The stationary points of $\psi(x)$ will give, with the help of (3), that of f. Also for an extreme of f,

$$dz = 0 \Rightarrow f_x + f_y\phi'(x) = 0, \; y = \phi(x). \tag{5}$$

From (2), $g = 0 \Rightarrow g_x + g_y\phi'(x) = 0$. (6)

Eliminating $\phi'(x)$ from (5) and (6), we get

$$I = \begin{vmatrix} f_x & f_y \\ g_x & g_y \end{vmatrix} = 0. \tag{7}$$

J is called a **functional determinant** or the **Jacobian** (§4.17.2).

Solving (2) and (7) for x and y, we get the stationary points f in $E \cap E_1$.

Examples 4.16. 1. *Find the extrema of* $x^2 + y^2$ *when* $x - y + 2 = 0$.

Here $f(x,y) = x^2 + y^2 \forall (x,y) \in \mathbb{R}$ (i)

and $g(x,y) = x - y + 2 = 0$. (ii)

From (ii) $y = \phi(x) = x + 2$ (iii)

and $\psi(x) = f(x, \phi(x)) = x^2 + (x+2)^2 = 2x^2 + 4x + 4$.

At a stationary point, $\psi'(x) = 4x + 4 = 0$ if $x = -1$.

When $x = -1$, $y = 1$, from (iii).

$\therefore$ $(-1, 1)$ is a stationary point of f, subject to (ii).

$\because$ $\psi''(x) = 4$, $\psi''(1) = 4 > 0$, ψ has a minimum value at -1.

So, f has also a minimum at $(-1, 1)$.

The minimum value of f is $(-1)^2 + 1^2 = 2$.

Since $f(x,y)$ may be interpreted as the square or the distance of (x,y) from the origin, the square of the shortest distance of the origin from the straight line (ii) is 2.

Alternatively,

$$J = 0 \Rightarrow \begin{vmatrix} 2x & 2y \\ 1 & -1 \end{vmatrix} = 0 \Rightarrow x + y = 0. \tag{iv}$$

Solving (iii) and (iv), we get the stationary point $(-1, 1)$.

Case II. Let $f : E \to \mathbb{R}$ be a function of three variables x, y and z and the subsidiary condition be

$$g(x, y, z) = 0. \tag{1}$$

Solving for z from (1), we get

$$z = \phi(x, y). \tag{2}$$

If $\quad u = f(x, y, z)$, we get

$$u = f(x, y, \phi(x, y)) = \psi(x, y), \tag{3}$$

a function of two variables. The stationary values of ψ will, with the help of (2), give that of f.

Also, substituting ϕ for z in f, we get

$$u = f(x, y, \phi). \tag{4}$$

At a stationary point $du = 0$, which gives

$$f_x + f_z\phi_x = 0 \quad \text{and} \quad f_y + f_z\phi_y = 0. \tag{5}$$

Also, $g(x, y, \phi) = 0 \Rightarrow dg = 0$

or $$g_x + g_z\phi_x = 0 \quad \text{and} \quad g_y + g_z\phi_y = 0. \tag{6}$$

Eliminating ϕ_x and ϕ_y from (5) and (6), we get

$$J_1 = \begin{vmatrix} f_x & f_z \\ g_x & g_z \end{vmatrix} = 0 \quad \text{and} \quad J_2 = \begin{vmatrix} f_y & f_z \\ g_y & g_z \end{vmatrix} = 0. \tag{7}$$

Solving the three equation (1) and (7), the stationary values are obtained.

2. *Find the shortest distance of the origin from the plane $ax + by + cz + d = 0$.*

The distance of any point (x, y, z) from the origin O is $\sqrt{x^2 + y^2 + z^2} = \sqrt{\sum x^2}$.

Let $\quad u = f(x, y, z) = \sum x^2. \qquad$ (i)

Then u is the square of the distance of (x, y, z) from O.

Let $ax + by + cz + d = 0$ be abbreviated to

$$\sum ax + d = 0. \tag{ii}$$

Then the minimum value of u subject to the constraint (ii) will give the square of the shortest distance.

From (ii) $z = \phi(x, y) = -\dfrac{1}{c}(ax + by + d). \qquad$ (iii)

$$\therefore \quad \phi_x = -a/c, \; \phi_y = -b/c. \tag{iv}$$

Also, $f_x = 2x, \quad f_y = 2y, \quad f_z = 2z,$

$$g_x = a, \quad g_y = b, \quad g_z = c.$$

$$\therefore \quad J_1 = \begin{vmatrix} 2x & 2z \\ a & c \end{vmatrix} = 0 \Rightarrow cx - az = 0 \tag{v}$$

and $$J_2 = \begin{vmatrix} 2y & 2z \\ b & c \end{vmatrix} = 0 \Rightarrow cy = bz = 0. \tag{vi}$$

$$\therefore \quad \frac{x}{a} = \frac{y}{b} = \frac{z}{c} = \lambda, \text{ say.}$$

Substituting in (ii), $\lambda\sum a^2 + d = 0$ or, $\lambda = -d/\sum a^2$. (vii)

$\therefore \quad x = -ad/\sum a^2,\ y = -bd/\sum a^2$ and $z = -cd/\sum a^2$

are the coordinates of the only stationary point of f subject to (ii).

That f attains a minimum at this stationary point can be shown by considering the 2nd partial derivatives.

$$f_x = 2x \Rightarrow A = f_{xx} = 2,\ f_y = 2y \Rightarrow B = f_{yy} = 2,\ C = f_{xy} = 0$$

and $\quad H = f_{xx}f_{yy} - f^2xy = 4$ at P.

Since f_{xx}, f_{yy} and H are all positive, f must have a minimum at P.

The minimum value of f is $f(P)$ given by

$$\sum\left(-ad/\sum a^2\right)^2 = \frac{d^2}{\left(\sum a^2\right)^2}\sum a^2 = d^2/\sum a^2$$

and therefore the shortest distance is $|d|/\sqrt{\sum a^2}$.

Case III. Let $f : E \to \mathbb{R}$ be a function of three independent and variables x, y and z and let it be subjected to two subsidiary conditions

$$g(x,y,z) = 0 \text{ and } h(x,y,z) = 0 \tag{1}$$

Let $\quad u = f(x,y,z)\,\forall\,(x,y,z) \in E$. (2)

Theoretically, two equations in (1) may be solved for two of the three variables x, y and z. Let y and z be obtained in terms of x:

$$y = \phi_1(x) \text{ and } z = \phi_2(x). \tag{3}$$

Then x is the only independent variable. Substituting in (2), we get

$$u = f(x, \phi_1(x), \phi_2(x) = \psi(x), \tag{4}$$

a function of only one independent variable. We can now search for the stationary values of this function. If α is a stationary point of ψ then a stationary point of f will be $P(\alpha, \beta\gamma)$, where

$$\beta = \phi_1(\alpha) \text{ and } \gamma = \phi_2(\alpha).$$

If ψ becomes extreme at α, f will be extreme at P.

Also making use of differentials,

$$df = f_x + f_y\phi_1' + f_z\phi_2' = 0 \tag{5}$$

and $\quad dg = 0$ and $dh = 0 \Rightarrow g_x + g_y\phi_1' + g_z\phi_2' = 0$ and $h_x + h_y\phi_1' + h_z\phi_2' = 0$ for $i = 1, 2$ (6)

at a stationary point, Eliminating ϕ_1' and ϕ_2' from the three equations in (5) and (6), we get an equation

$$J(x,y,z) = \begin{vmatrix} f_x & f_y & f_z \\ g_x & g_y & g_z \\ h_x & h_y & h_z \end{vmatrix} = 0. \tag{7}$$

Solving the two equations in (1) and the equation (7), the stationary points are obtained.

Note. The function J in (7) is the Jacobian of the transformation transforming (x, y, z) to (f, g, h) (see § 4.17.3).

3. *Find the shortest distance of the origin from the straight line given by*

$$\sum l_i x - p_i = 0,\ i = 1, 2.$$

Let $\quad u = f(x, y, z) = \sum x^2 \quad$ (i)

and $\quad g_i = \sum l_i x - p_i = 0$ for $i = 1, 2. \quad$ (ii)

Then u is the square of the distance between the origin O and a point (x, y, z). So the distance is $\sqrt{u}$.

Since $f_x = 2x,\ f_y = 2y,\ f_z = 2z,\ \partial g_1/\partial x = l_1,\ \partial g_1/\partial y = m_1,\ \partial g_1/\partial z = n_1,\ \partial g_2/\partial x = l_2,\ \partial g_2/\partial y = m_2,\ \partial g_2/\partial z = n_2$, the equation $J = 0$ gives

$$\begin{vmatrix} x & y & z \\ l_1 & m_1 & n_1 \\ l_2 & m_2 & n_2 \end{vmatrix} = 0,$$

which may be written as $lx + my + nz = 0$, (iii)

where $\quad l = m_1 n_2 - m_2 n_1,\ m = n_1 l_2 - n_2 l_1$ and $n = l_1 m_2 - l_2 m_1$.

Solving (ii) and (iii) for x, y and z, we shall get a stationary point P whose coordinates are $\alpha = D_1/D,\ \beta = D_2/D$ and $\gamma = D_3/D$, where

$$D_1 = \begin{vmatrix} p_1 & m_1 & n_1 \\ p_2 & m_2 & n_2 \\ 0 & m & n \end{vmatrix},\ D_2 = \begin{vmatrix} l_1 & p_1 & n_1 \\ l_2 & p_2 & n_2 \\ l & 0 & n \end{vmatrix},\ D_3 = \begin{vmatrix} l_1 & m_1 & p_1 \\ l_2 & m_2 & p_2 \\ l & m & 0 \end{vmatrix}$$

and $\quad D = \begin{vmatrix} l_1 & m_1 & n_1 \\ l_2 & m_2 & n_2 \\ l & m & n \end{vmatrix}.$

As before, since $f_{xx} = f_{yy} = 2$ and $f_{xy} = 0,\ H = 4$, f is minimum at P. The shortest distance is

$$\sqrt{f(\alpha, \beta, \gamma)} = (1/|D|)\sqrt{(D_1^2 + D_2^2 + D_3^2)}.$$

(II) Lagrange's Method of Undetermined Multipliers

Lagrange, a great French mathematician, gave a more systematic and uniform method of finding the stationary points of a function subject to some constraints.

Let the function f of $n+m$ variables x_i for $i = 1, \ldots, n$ and u_j for $j = 1, \ldots, m$ be subjected to m constraints

$$g_j = 0, \tag{1}$$

where g_j are all functions of $m + n$ variables x_i and u_j. Theoretically, these m functions, supposed to be independent, can be solved for the m variables u_j in terms of the remaining n variables and when they are substituted in f, we get a function of n independent variables x_i ultimately. Then the differential df will be, ultimately, a linear combination of n differentials dx_i. Now g_j will be linearly independent if the Jacobian

$$J = \partial(g_1, g_2, \ldots, g_m)/\partial(u_1, u_2, \ldots, u_m) \neq 0.$$

Since there are m constraints, Lagrange has made use of m parameters λ_j and formed a function

$$L = f + \sum_{j=1}^{m} \lambda_j g_j. \tag{2}$$

If E is the common domain of f and g_j for $j = 1, \ldots, m$, then basically

$$L \equiv f, \text{ since } g_j = 0 \text{ for } j = 1, \ldots, m.$$

Now $$dL = df + \sum_1^m \lambda_j dg_j. \tag{3}$$

The m parameters λ_j which appear as multipliers of g_j in (2) are called **Lagrangian multipliers** and the function L is called the **Lagrangian function.**

At a stationary point $dL = 0$

$$\Rightarrow \quad \sum_{l=1}^n L_{x_i} \cdot dx_i + \sum_{j=1}^m L_{u_j} du_j = 0$$

$$\Rightarrow \quad L_{x_i} = 0 \text{ for } i = 1, \ldots, n \text{ and } L_{u_j} = 0 \text{ for } j = 1, \ldots, m,$$

since dx_i and du_j are all independent.

$$\therefore \quad L_{x_i} = \partial f/\partial x_i + \sum_{j=1}^m \lambda_j \partial g_j/\partial x_i = 0 \text{ for } i = 1, \ldots, n \tag{4}$$

and $$L_{u_j} = \partial f/\partial u_j + \sum_{k=1}^m \lambda_k \partial g_k/\partial u_j = 0 \text{ for } j = 1, \ldots m. \tag{5}$$

(1), (4) and (5) give $m + n + m$ or $2m + n$ equations which can be solved for the $n + m$ variables and m parameters.

A set of corresponding $n + m$ values of the variables will give a stationary point.

4. *Use Lagrange's method to find the shortest distance of the origin from the straight line given by*

$$g_j = 0,$$

where $$g_j = \sum l_j x - p_j = 0 \text{ for } j = 1, 2.$$

Let $$f = \sum x^2.$$

Then if $L = f + \lambda_1 g_1 + \lambda_2 g_2$, $dL = 0$ implies

$$L_x = L_y = L_z = 0$$

$$f_x + \lambda_1 \partial g_1/\partial x + \lambda_2 \partial g_2/\partial x = 0, \ldots$$

or, $$2x + \lambda_1 l_1 + \lambda_2 l_2 = 0, \; 2y + \lambda_1 m_1 + \lambda_2 m_2 = 0 \text{ and } 2z + \lambda_1 n_1 + \lambda_2 n_2 = 0.$$

Eliminating λ_1 and λ_2, we get as before $lx + my + nz = 0$,

where $l = m_1 n_2 - m_2 n_1$, $m = n_1 l_2 - n_2 l_1$ and $n = l_1 m_2 - l_2 m_1$.

This equation along with the given constraints $g_j = 0$ will give the stationary point.

5. *Show that the lengths of the semi-axes of the section of the ellipsoid $\sum x^2/a^2 = 1$ by the plane $\sum lx = 0$ are the roots of the equation $\sum a^2 l/(r^2 - a^2) = 0$ in r.*

Let r be the length of a semi-axis of the ellipse Γ, which is the line of intersection of the surface σ, given by

$$\sum x^2/a^2 = 1 \tag{i}$$

and the plane Π given by

$$\sum lx = 0. \qquad \text{(ii)}$$

Now $\quad r = \sqrt{\sum x^2}. \qquad \text{(iii)}$

We are therefore required to show that the extremes of r will be the roots of the quadratic equation in r^2:

$$\sum a^2 l/(r^2 - a^2) = 0. \qquad \text{(iv)}$$

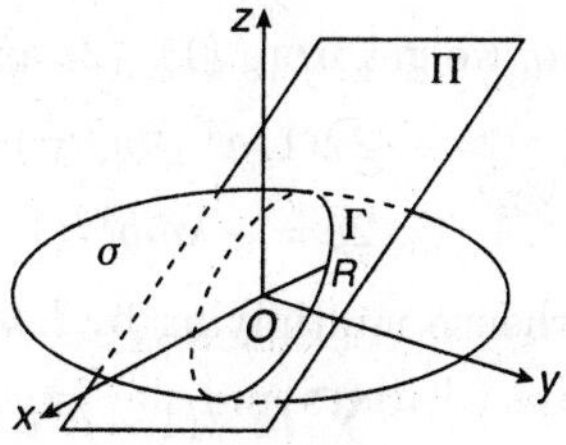

Fig. 4.15

Let us take $f = r^2 = \sum x^2 \qquad \text{(v)}$

for convenience and consider the extrema of f.

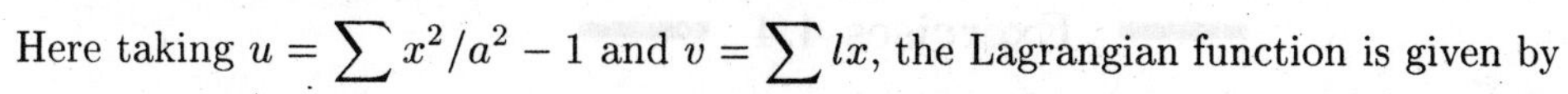

Here taking $u = \sum x^2/a^2 - 1$ and $v = \sum lx$, the Lagrangian function is given by

$$L = f + \lambda u + \mu v,$$

where λ and μ are the Lagrangian multipliers.

$L_x = L_y = L_z = 0$ implies

$$f_x + \lambda u_x + \mu v_x = 0 \quad \text{or,} \quad 2x + 2\lambda x/a^2 + \mu l = 0, \qquad \text{(vi)}$$

$$f_y + \lambda u_y + \mu v_y = 0 \quad \text{or,} \quad 2y + 2\lambda y/b^2 + \mu m = 0 \qquad \text{(vii)}$$

and $\quad f_z + \lambda u_z + \mu v_z = 0 \quad \text{or,} \quad 2z + 2\lambda z/c^2 + \mu n = 0. \qquad \text{(viii)}$

Now $\quad xL_x + yL_y + zL_z = 0$ gives

$$2\sum x^2 + 2\lambda \sum x^2/a^2 + \mu \sum lx = 0.$$

Using (i), (ii) and (v), we get

$$2r^2 + 2\lambda \cdot 1 + \mu \cdot 0 = 0 \text{ or, } \lambda = -r^2.$$

Substituting in (vi), (vii) and (viii), we get

$$x(1 - r^2/a^2) + \mu l = 0,\ y(1 - r^2/b^2) + \mu m = 0 \text{ and } z(1 - r^2/c^2) + \mu n = 0,$$

whence $x/\mu = a^2 l/(r^2 - a^2)$, $y/\mu = b^2 m/(r^2 - b^2)$ and $z/\mu = c^2 n/(r^2 - c^2)$.

Multiplying these by l, m and n respectively and adding, we get the required equation (iv), since by (ii) $\sum lx = 0$.

6. *Find the extreme values of* $u = x^2/a^4 + y^2/b^4 + z^2/c^4$, *when* $lx + my + nz = 0$ *and* $x^2/a^2 + y^2/b^2 + z^2/c^2 = 1$.

Let $\quad u = \sum x^2/a^2,\ g = \sum lx$ and $h = \sum x^2/a^2 - 1$.

Using Lagrange's multipliers λ and μ,

$$du + \lambda dg + \mu dh = 0$$

$$\Rightarrow \quad \sum 2x/a^4 dx + \lambda \sum l dx + 2\mu \sum (x/a^2) dx = 0,$$

whence $\quad 2x/a^4 + \lambda l + 2\mu x/a^2 = 0, \qquad (1)$

$$2y/b^4 + \lambda m + 2\mu y/b^4 = 0 \qquad (2)$$

and $\quad 2z/c^4 + \lambda n + 2\mu z/b^4 = 0. \qquad (3)$

$(1)x + (2)y + (3)z$ gives $2\sum x^2/a^4 + \lambda \sum lx + 2\mu \sum x^2/a^2 = 0$

or, $\quad 2u + 2\mu \cdot 1 = 0$ or, $\mu = -u$, using subsidiary conditions.

So, we get from (1), (2) and (3),

$$2x(1/a^4 - u/a^2) + \lambda l = 0 \quad \text{or,} \quad 2x = -\lambda l a^4/(1 - a^2 u),$$

$$2\dot{y} = -\lambda m b^4/(1 - b^2 u) \quad \text{and} \quad 2z = -\lambda n c^4/(1 - c^2 u),$$

whence multiplying by l, m and n respectively and adding, we get

$$\sum la^4/(1 - a^2 u) = 0,$$

a quadratic equation in u, whose roots give the extreme values of u.

Exercises 4.4

1. Show that $x^4 + x^2 y + y^2$ has a minimum at the origin.
2. Show that $2x^4 - 3x^2 y + y^2$ has no extremum at $(0, 0)$ although both the first-order partial derivatives vanish thereat.
3. Investigate the following functions for extreme values:

   (i) $x^3 - y^3$, **(ii)** $x^2 + y^2 + xy + x + y$,

   (iii) $x^2 + xy + y^2 - 5x - 4y$, **(iv)** $x^4 + y^4 - 6(x^2 + y^2) + 8xy$.
4. If $u = xy + a^3(1/x + 1/y)$, show that u has a minimum at (a, a). Find its minimum value.
5. Find the maximum value of $x^3 y^2 z$ subject to the condition $x + y + z = 1$.
6. Show, by using Lagrange's multipliers that the extreme values of $u = x^2 + y^2 + z^2$ subject to the conditions

   $$ax^2 + by^2 + cz^2 = 1 \quad \text{and} \quad lx + my + nz = 0$$

   are given by the roots of $\sum l^2/(au - 1) = 0$.
7. Given $2^x 3^y 5^z = 1$, find the maximum value of $(x + 1)(y + 1)(z + 1)$.

Answers

3. (i) Stationary point $(0, 0)$, but no extremum; **(ii)** Minimum at $(-2/3, -2/3)$;

(iii) Minimum at $(2, 1)$;

(iv) Stationary points are $(0, 0)$, $(\pm 1, \pm 1)$, $(\pm\sqrt{5}, \mp\sqrt{5})$ with maximum at $(0, 0)$, minimum at $(\pm\sqrt{5}, \mp 5)$, but no extrema at $(\pm 1, \pm 1)$.

5. $2^8 3^9$. **7.** $(\ln 30)^3/(27 \ln 2 \cdot \ln 3 \cdot \ln 5)$.

4.17 Transformation

Let u and v be two functions of the two real variables x and y given by $u = f(x, y)$ and $v = g(x, y)$ and let

$$E = \operatorname{Dom} f \cap \operatorname{Dom} g \subseteq \mathbb{R}^2.$$

Let $E' = \{(u, v) : u = f(x, y)$

and $v = g(x, y)\, \forall (x, y) \in E\}$.

Then the pair of functions u and v (or f and g) is said to define a **transformation*** or **mapping** T which maps E to E'. Symbolically,

$$T : E \to E'.$$

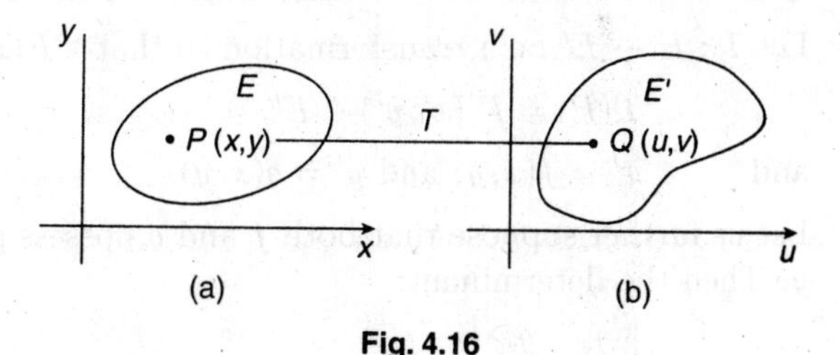

Fig. 4.16

Here E is called the **domain** and E' the **range** of T and $\forall$ point $P(x, y) \in E$ there is a unique ***T*-image** $Q(u, v)$ of P so that

$$Q = T(P)$$

and T is said to map P to Q. Symbolically,

$$T : P \to Q.$$

Just like a function of one variable, the transformation T may be one-one or many one according as distinct elements of E have distinct T-images in E' or not.

4.17.1 Composition or Product of Transformations in $\mathbb{R}^2$

Let T_1 and T_2 be two transformations, where $T_1 : E \to E'$ and $T_2 : E' \to E''$ so that $\forall\, P(x, y) \in E$, $T_1 : P \to P'$, where $P'(x', y') \in E'$ and $\forall P' \in E'$ $T_2 : P' \to P''$, where $P''(x'', y'') \in E''$.

Then $P' = T_1(P)$ and $P'' = T_2(P')$.

$\therefore$ $P'' = T_2(T_1(P)) = T(P)$, say.

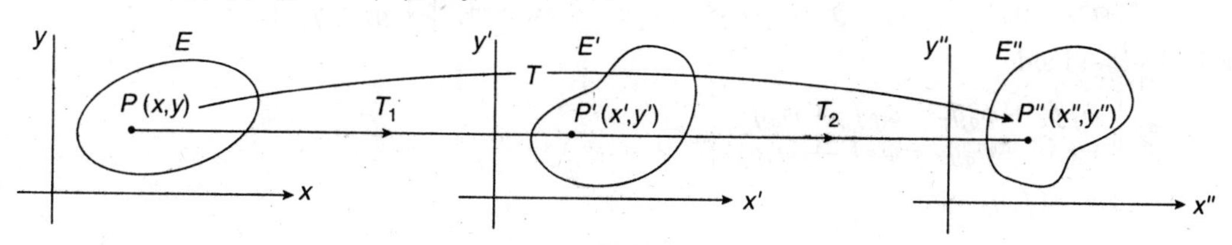

Fig. 4.17

Then T is a transformation mapping E into E'', i.e.,

$$T : E \to E'',$$

such that $\forall\, P(x, y) \in E\, \exists P''(x'', y'')$, where

$$T(P) = P''.$$

This transformation T, called a **composite** or **resultant** or a **product transformation**, which is composed of T_1 and T_2.

Since $T(P) = T_2(T_1(P))$ we may take $T = T_2 \cdot T_1$ or simply $T_2 T_1$.

---

*A transformation is a more general type of a function. Following the previous definitions we find, that a function maps a point in any finite dimensional space into a point in one dimensional space $\mathbb{R}$. Whereas a transformation maps a point in $\mathbb{R}^m$ into a point in $\mathbb{R}^n$.

4.17.2 Jacobian of a Transformation in $\mathbb{R}^2$

Let $T : E \to E'$ be a transformation so that $\forall\, P(x, y) \in E$

$$T(P) = P'(x', y') \in E'$$

and $\quad x' = f(x, y)$ and $y' = g(x, y)$.

Let us further suppose that both f and g possess partial derivatives with respect to both x and y. Then the determinant:

$$\begin{vmatrix} f_x & g_x \\ f_y & g_y \end{vmatrix}$$

is called the **Jacobian of the transformation** T or the **Jacobian of the functions** f **and** g **w.r.t.** x **and** y. It is denoted by $J(T)$ or simply by J.

Thus, $\quad J = \begin{vmatrix} f_x & g_x \\ f_y & g_y \end{vmatrix}$ or, $\begin{vmatrix} f_x & f_y \\ g_x & g_y \end{vmatrix}$ or, $\dfrac{\partial(f, g)}{\partial(x, y)}$ or, $\partial(f, g)/\partial(x, y)$.

As in the previous section if $T = T_2T_1$ then it can be shown that

$$J(T) = J(T_1)J(T_2).$$

Let us take, for convenience, $x' = f(x, y)$, $y' = g(x, y)$ and $x'' = \phi(x', y')$ and $y'' = \psi(x', y')$ and P, P' and P'' are respectively (x, y), (x', y') and (x'', y''). Then

$$T_1 : P \to P',\ T_2 : P' \to P'' \text{ and } T(= T_2T_1) : P \to P''.$$

$\because \quad T_1 : (x, y) \to (f, g)$, $T_2 : (f, g) \to (\phi, \psi)$ and $T : (x, y) \to (\phi, \psi)$,

$$J(T) = \begin{vmatrix} \phi_x & \phi_y \\ \psi_x & \psi_y \end{vmatrix} \text{ and } J(T_1) = \begin{vmatrix} f_x & f_y \\ g_x & g_y \end{vmatrix} \text{ and } J(T_2) = \begin{vmatrix} \phi_f & \phi_g \\ \psi_f & \psi_g \end{vmatrix}.$$

We are required to show that

$$\begin{vmatrix} \phi_x & \phi_y \\ \psi_x & \psi_y \end{vmatrix} = \begin{vmatrix} f_x & f_y \\ g_x & g_y \end{vmatrix} \begin{vmatrix} \phi_f & \phi_g \\ \psi_f & \psi_g \end{vmatrix} = \begin{vmatrix} \phi_f & \phi_g \\ \psi_f & \psi_g \end{vmatrix} \begin{vmatrix} f_x & f_y \\ g_x & g_y \end{vmatrix}.$$

The right-hand side is

$$\begin{vmatrix} \phi_f f_x + \phi_g g_x & \phi_f f_y + \phi_g g_y \\ \psi_f f_x + \psi_g g_x & \psi_f f_y + \psi_g g_y \end{vmatrix}.$$

But $\quad \phi = \phi(f, g)$. So

$$\phi_x = \phi_f f_x + \phi_g g_x \text{ and } \phi_y = \phi_f f_y + \phi_g g_y.$$

Similarly, $\psi_x = \psi_f f_x + \psi_g g_x$ and $\psi_y = \psi_f f_y + \psi_g g_y$.

Hence the required result.

Symbolically,

$$\frac{\partial(\phi, \psi)}{\partial(x, y)} = \frac{\partial(\phi, \psi)}{\partial(f, g)} \cdot \frac{\partial(f, g)}{\partial(x, y)}.$$

Example 4.17. 1. *Calculate the Jacobian of the transformation:*

$$T : (x, y) \to (r, \theta),$$

where $\quad x = r\cos\theta, \quad$ *and* $\quad y = r\sin\theta,$

$$x_r = \cos\theta, \qquad x_\theta = -r\sin\theta,$$

$$y_r = \sin\theta, \qquad y_\theta = r\cos\theta.$$

$$\therefore \quad J(T) = \begin{vmatrix} \cos\theta & -r\sin\theta \\ \sin\theta & r\cos\theta \end{vmatrix} = r \begin{vmatrix} \cos\theta & -\sin\theta \\ \sin\theta & \cos\theta \end{vmatrix} = r.$$

If T^{-1} maps $(r,\theta) \to (x,y)$, then

$$r = \sqrt{x^2+y^2} \text{ and } \theta = \tan^{-1}(y/x) \text{ or, } \pi + \tan^{-1}(y/x)$$

according as $x >$ or, < 0.

$$r_x = \frac{x}{\sqrt{x^2+y^2}}, \; r_y = \frac{y}{\sqrt{x^2+y^2}}, \; \theta_x = \frac{-y/x^2}{1+y^2/x^2} = \frac{-y}{x^2+y^2}$$

and $\quad \theta_y = \dfrac{1/x}{1+y^2/x^2} = \dfrac{x}{x^2+y^2}.$

$$\therefore \quad J(T^{-1}) = \begin{vmatrix} \frac{x}{\sqrt{x^2+y^2}} & \frac{y}{\sqrt{x^2+y^2}} \\ \frac{-y}{x^2+y^2} & \frac{x}{x^2+y^2} \end{vmatrix} = \frac{1}{(x^2+y^2)^{3/2}} \begin{vmatrix} x & y \\ -y & x \end{vmatrix} = \frac{1}{\sqrt{x^2+y^2}} = \frac{1}{r} = \frac{1}{J(T)}.$$

Thus $\quad J(T)J(T^{-1}) = 1 = J(TT^{-1}) = J(T^{-1}T) = J(I),$

where I is the identity transformation.

4.17.3 Transformation in $\mathbb{R}^3$

Let $T : E \to E'$, where E and E' are both subsets of $\mathbb{R}^3$. Then $\forall P(x,y,z) \in E, E!P'(u,v,w) \in E'$ such that

$$P' = T(P).$$

Let $u = u(x,y,z)$, $v = v(x,y,z)$ and $w = w(x,y,z)$, all of which are supposed to be differentiable. Then, the Jacobian of the transformation T is given by

$$J(T) = \partial(u,v,w)/\partial(x,y,z) \doteq \begin{vmatrix} u_x & u_y & u_z \\ v_x & v_y & v_z \\ w_x & w_y & w_z \end{vmatrix}.$$

Here also if $T = T_2T_1$, then $J(T) = J(T_1)J(T_2)$.

If $\quad T_1 : P(x,y,z) \to P'(u,v,w)$

and $\quad T_2 : P'(u,v,w) \to P''(X,Y,Z)$,

then $\quad T : P \to P''$

and $\quad J(T) = J(T_1)J(T_2)$

i.e., $\quad \partial(X,Y,Z)/\partial(x,y,z) = [\partial(X,Y,Z)/\partial(u,v,w)][\partial(u,v,w)/\partial(x,y,z)]$.

This result is equivalent to the chain rule in $\mathbb{R}$:

$dX/dx = (dX/du)(du/dx)$.

Some Transformations in $\mathbb{R}^3$

1. Let $T : (x,y,z) \to (r,\theta,\phi)$,

where $x = r\sin\theta\cos\phi$, $y = r\sin\theta\sin\phi$ and $z = r\cos\theta$.

Then T is the transformation from Cartesian to spherical polar coordinates.

The Jacobian of x,y,z w.r.t. r,θ,ϕ is

$$J = \partial(x,y,z)/\partial(r,\theta,\phi).$$

Let us take $\lambda = \cos\theta$, $\mu = \sin\theta$, $\xi = \cos\phi$, $\eta = \sin\phi$.

Then $\lambda^2 + \mu^2 = \xi^2 + \eta^2 = 1$, $d\lambda/d\theta = -\mu$, $d\mu/d\theta = \lambda$, $d\xi/d\phi = -\eta$ and $d\eta/d\phi = \xi$. So,

$$\dot{J} = \begin{vmatrix} x_r & x_\theta & x_\phi \\ y_r & y_\theta & y_\phi \\ z_r & z_\theta & z_\phi \end{vmatrix} = \begin{vmatrix} \mu\xi & r\lambda\xi & -r\mu\eta \\ \mu\eta & r\lambda\eta & r\mu\xi \\ \lambda & -r\mu & 0 \end{vmatrix}$$

$$= r^2[\lambda^2\mu(\xi^2 + \eta^2) + \mu^3(\xi^2 + \eta^2)] = r^2\mu = r^2 \sin\theta.$$

2. Let $T : (x, y, z) \to (\rho, \phi, z)$,

where $x = \rho\cos\phi$, $y = \rho\sin\phi$ and $z = z$. Then the transformation is from Cartesian to cylindrical polar coordinates.

To get $J(T)$, we take $\xi = \cos\phi$ and $\eta = \sin\phi$. Then

$$\xi^2 + \eta^2 = 1,\ d\xi/d\phi = -\eta \text{ and } d\eta/d\phi = \xi.$$

Now $x_\rho = \xi,\ x_\phi = -\rho\eta,\ x_z = 0,\ y_\rho = \eta,\ y_\phi = \rho\xi,\ y_z = 0$

$z_\rho = z_\phi = 0$ and $z_z = 1$.

$$\therefore \quad J(T) = \begin{vmatrix} \xi & -\rho\eta & 0 \\ \eta & \rho\xi & 0 \\ 0 & 0 & 1 \end{vmatrix} = \rho(\xi^2 + \eta^2) = \rho.$$

3. $T : (\rho, \phi, z) \to (r, \theta, \phi)$,

where $\rho = r\sin\theta$, $\phi = \phi$, $z = r\cos\theta$.

Then T is the transformation from cylindrical to spherical polar coordinates.

Taking $\lambda = \cos\theta$, $\mu = \sin\theta$, $\lambda^2 + \mu^2 = 1$

$\rho = r\mu$, $\phi = \phi$, $z = r\lambda$.

Now $\rho_r = \mu,\ \rho_\theta = r\lambda,\ \rho_\phi = 0,\ \phi_r = \phi_\theta = 0,\ \phi_\phi = 1,$

$z_r = \lambda,\ z_\theta = -r\mu,\ z_\phi = 0$

$$\text{So,} \quad J(T) = \begin{vmatrix} \mu & r\lambda & 0 \\ 0 & 0 & 1 \\ \lambda & -r\mu & 0 \end{vmatrix} = r(\lambda^2 + \mu^2) = r.$$

5 Integral Calculus of a Function of Several Variables

5.1 Introduction

Definite integral of a function of a single variable has been defined in Chapter 2 as the limit of a sum. The function has been defined in a one dimensional space $\mathbb{R}$ or the x-axis. For a function with its domain in $\mathbb{R}^2, \mathbb{R}^3$ or even $\mathbb{R}^n$ for $n = 3, 4$, etc., definite integral may be defined over $\mathbb{R}, \mathbb{R}^2$, etc. In particular for a function $f(x, y)$ defined in $E \subseteq \mathbb{R}^2 = \{(x, y) : x, y \in \mathbb{R}\}$, its definite integral can be defined along a curve in E or over a region contained in E. Similarly, for a function $f(x, y, z)$ defined in a certain domain in $\mathbb{R}^3 = \{(x, y, z) : x, y, z \text{ all } \in \mathbb{R}\}$, a definite integral may be defined along a curve, over a surface or over a region contained in the domain.

5.2 Integral of a Function of Two Variables over an Interval Parallel to a Coordinate Axis—Repeated Integrals

Let $f : E \to \mathbb{R}$ be a function with its domain E, a rectangle with its sides parallel to the coordinate lines, defined by

$$E = \{(x, y) \in \mathbb{R}^2 : a \leq x \leq b \quad \text{and} \quad c \leq y \leq d\}.$$

Then $\forall$ fixed $y \in [c, d]$, the integral

$$\int_a^b f(x, y)dx,$$

when it exists, becomes a function $F(y)$ of y so that

$$F(y) = \int_a^b f(x, y)dx. \tag{1}$$

If $\beta \in [c, d]$, $F(\beta) = \displaystyle\int_a^b f(x, \beta)dx$, which is the integral of $f(x, y)$ along the line segment:

$$AB = \{(x, y) : a \leq x \leq b,\ y = \beta\} \text{ (Fig. 5.1)}.$$

So, taking (X, Y) as the current coordinates of a point in the xy-plane (1) represents the integral of (x, y) w.r.t. x along the line segment

$$PQ = \{(X, Y) \in \mathbb{R}^2 : a \leq X \leq b \quad \text{and} \quad Y = y\}.$$

Similarly, the integral $\int_c^d f(x, y)dy$, when it exists $\forall$ fixed $x \in [a, b]$ defines a function $G(x)$ of x so that

$$G(x) = \int_c^d f(x, y)dy, \tag{2}$$

which is the integral of $f(x, y)$ along the line segment TU defined by

$$TU = \{(X, Y) \in \mathbb{R}^2 : X = x \text{ and } c \leq Y \leq d\}.$$

The integral along an arc lying in E will be defined later.

Repeated Integrals

If $F(y)$ is bounded and integrable in $[c, d]$, then we have

$$I = \int_c^d F(y)dy = \int_c^d \left\{ \int_a^b f(x, y)dx \right\} dy = \int_c^d dy \int_a^b f(x, y)dx,$$

which is denoted by

$$\int_c^d \int_a^d f(x, y)dxdy. \tag{3}$$

In (3), $F(y)$ or, $\int_a^b f(x, y)dx$ is called the **inner integral** and 1 is the **repeated** integral in the sense that $f(x, y)$ is successively integrated, first w.r.t. x and then w.r.t. y.

Similarly, the integral:

$$\int_c^d f(x, y)dy,$$

if it exists $\forall\ x \in [a, b]$, is a function $G(x)$ of the variable x, so that

$$G(x) = \int_c^d f(x, y)dy.$$

If this function is integrable in $[a, b]$, we have

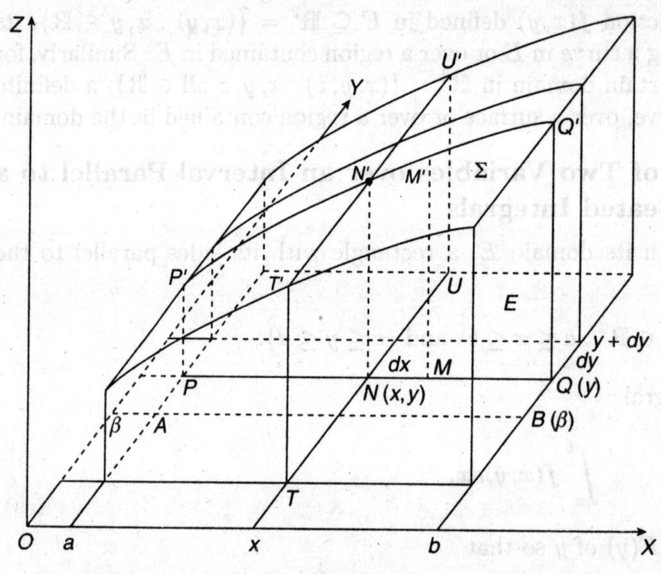

Fig. 5.1

$$J = \int_a^b G(x)dx = \int_a^b \left\{ \int_c^d f(x, y)dy \right\} dx = \int_a^b dx \int_c^d f(x, y)dy$$

$$= \int_a^b \int_c^d f(x, y)dydx, \tag{4}$$

which is also another repeated integral obtained by integrating $f(x, y)$, this time, successively first w.r.t. y and then w.r.t. x. In this case the inner integral is $G(x)$.

Geometrical Interpretations of $F(y)$ and $G(x)$

$F(y) = \int_a^b f(x, y)dx$ is the integral of $f(x, y)$ w.r.t. x along the line segment PQ from P to Q, where $PQ = \{(X, Y) \in \mathbb{R}^2 : a \leq X \leq b,\ Y = y\}$. Now, if $f(x, y) \geq 0,\ z = f(x, y)$ represents a surface $\sum$ over E. If through $N(x, y)$ a line is drawn parallel to the z-axis meeting $\sum$ in N', then $f(x, y) = NM'$. $f(x, y)dx$ is the area of the strip $MNN'M'$ and hence $F(y)$ represents the area of $PQQ'P'$, a portion of the plane $Y = y$ bounded by $x = a,\ x = b,\ z = f(x, y)$ and

$z = 0$. Similarly, $G(x)$ represents the area of $TUU'T'$. Similarly, since $F(y)dy$ represents an elementary volume of a slice of the solid bounded by $x = a$, $x = b$, $z = 0$ and $z = f(x, y)$ between the planes $Y = y$ and $Y = y+dy$, I represents the entire volume of this solid. Similarly, J also represents the same volume.

5.3 Double Integral

5.3.1 Double Integral over a Rectangular Region

Let $f : E \to \mathbb{R}$ be a bounded function of two variables x and y so that $E \subseteq \mathbb{R}^2$ and for two finite numbers m and M, $m \leq f(x, y) = M \forall (x, y) \in E$. Then the integral of f over E is called a **double integral** defined below:

Let $E = R = \{(x, y) \in \mathbb{R}^2 : a \leq x \leq b \text{ and } c \leq y \leq d\}$.

Then R is a rectangular region in the xy-plane (Fig. 5.2).

For a proper double integral of f over R we assume that f is bounded there. To define the double integral of f as the limit of a sum we divide R into a finite number n of sub-rectangles R_i by drawing lines parallel to the coordinate axes. Let P_i be any arbitrary point of R_i of area δA_i. We form the sum:

$$S_n = \sum_{i=1}^{n} f(P_i)\delta A_i.$$

Let us now increase n, the number of parts (R_i) of R, indefinitely so that the diagonal of the greatest of the sub-rectangles tends to zero. If now

$$\underset{n\to\infty}{\text{Lt}}\ S_n$$

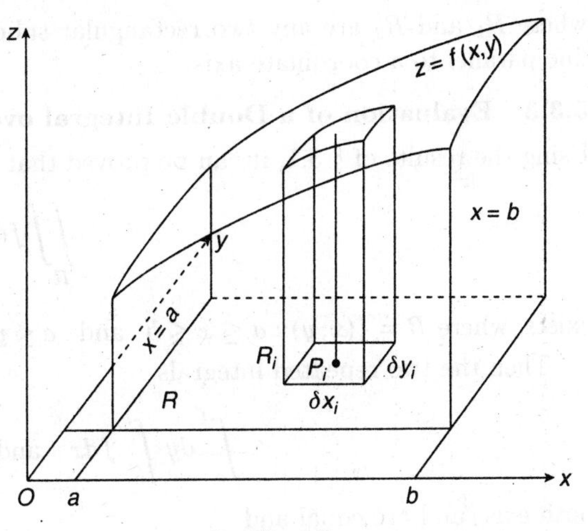

Fig. 5.2

exists and is independent of the choice of P_i in R_i, then this limit, called **the double integral of f over R**, is denoted by

$$f(x, y)dA, \quad \text{or,} \quad \iint_R f(x, y)dxdy$$

or simply by $\int\int_R f dxdy$.

If $f(P) \geq 0 \ \forall P \in R$, then the double integral geometrically represents the **volume of the space enclosed by the surfaces:**

$$\boldsymbol{x = a,\ x = b,\ y = c,\ y = d,\ z = f(x, y),\ \text{and}\ z = 0\ \text{(Fig. 5.2)}.}$$

In particular when $f(P) = 1 \forall P \in R$, the double integral becomes $\iint_R dxdy$, which represents the area $A = (b - a)(d - c)$ of the rectangle.

Such a double integral always exists if f is either continuous in R or it is discontinuous at a finite number of points or on a finite number of continuous lines in R.

5.3.2 Properties of Double Integrals

The double integrals over a rectangular domain R in the xy-plane satisfies the following properties.

(1) $\iint\limits_R kf(x,y)dxdy = k\iint\limits_R f(x,y)dxdy$, k being a constant,

(2) $\iint\limits_R (f \pm g)dxdy = \iint\limits_R fdxd \pm \iint\limits_R gdxdy$,

$$\iint\limits_R \sum f_i \cdot dxdy = \sum \iint\limits_R f_i dxdy$$

and (3) $\iint\limits_R fdxdy = \iint\limits_{R_1} fdxdy + \iint\limits_{R_2} fdxdy$,

where R_1 and R_2 are any two rectangular subdomains obtained by dividing R by a straight line parallel to a coordinate axis.

5.3.3 Evaluation of a Double Integral over a Rectangular Region

Using the results of § 5.1, it can be proved that if the double integral

$$\iint\limits_R fdxdy$$

exists, where $R = \{(x,y) : a \le x \le b \text{ and } c \le y \le d\}$.

Then the two repeated integrals

$$\int_c^d dy \int_a^b fdx \quad \text{and} \quad \int_a^b dx \int_c^d fdy$$

both exist and are equal and

$$\iint\limits_R fdxdy = \int_c^d dy \int_a^b fdx = \int_a^b dx \int_c^d fdy.$$

That this result is true can be explained physically in the following way:

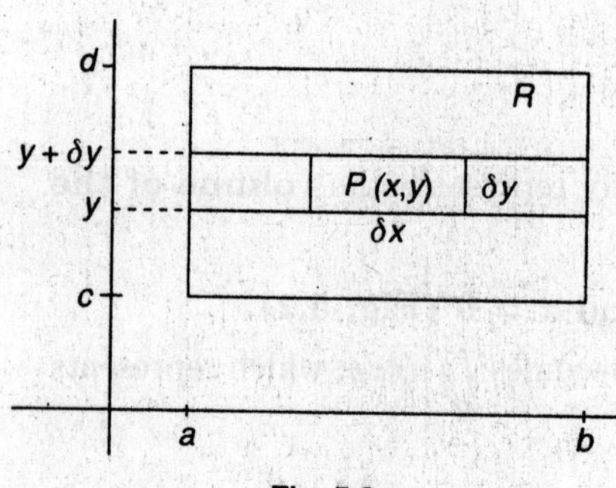

Fig. 5.3

Let R be a rectangular lamina and $f(P)$ the surface density of the lamina at a point $P(x,y)$. Let us take a thin strip of the lamina lying between y and $y + \delta y$. Then the mass of the strip is approximately

$$\sum f(P)\delta x\delta y = \delta y \sum f(P)\delta x \quad \text{(Fig. 5.3)}.$$

In the limit the mass of this strip is $\delta m \delta y$,

where $\delta m = \int_a^b fdx$.

Then $\sum \delta m \delta y$ is the approximate mass of entire rectangle R.

In the limit we have

$$\int_c^d dmdy = \int_c^d dy \int_a^b f dx,$$

the actual mass of the rectangular lamina, which is also the double integral $\iint_R f dxdy$. For, if $\delta A = \delta x \delta y, f(P)\delta A$ is the approximate mass of δA and

$$\sum f(P)\delta A$$

is the sum of the approximate masses of all the rectangular subdomains. In the limit, we get

$$\iint_R f dxdy$$

as the actual mass of the lamina.

Next, if we divide the rectangle into strips parallel to the x-axis, we can similarly show that the mass is equal to the repeated integral

$$\int_a^b dx \int_c^d f dy.$$

Thus physically, we find that

$$\iint_R f dxdy = \int_c^d dy \int_a^b f dx = \int_a^b dx \int_c^d f dy.$$

The integral $\int_c^d dy \int_a^b f dx$ is sometimes expressed as $\int_c^d \left(\int_a^b f dx \right) dy$ or, as $\int_c^d \int_a^b f dxdy$.

Note. The existence of the repeated integrals however, does not entail that of the double integral.

Examples 5.1. 1. *Evaluate* $I = \iint xy(x^2+y^2)dxdy$ *over the rectangle*

$$R = \{(x,y) : 0 \le x \le a \;\; and \;\; 0 \le y \le b\}.$$

Here $f(x,y) = xy(x^2+y^2)$, which, being a polynomial in x and y, is continuous. So I exists and it equals the repeated integral

$$\int_0^a dx \int_0^b xy(x^2+y^2)dxdy = \int_0^a xdx \int_0^b (x^2y+y^3)dy$$

$$= \int_0^a xdx \left[x^2y^2/2 + y^4/4\right]_0^b = \int_0^a x(b^2x^2/2 + b^4/4)dx = a^2b^2(a^2+b^2)/8. \tag{1}$$

The other repeated integral is

$$\int_0^b dy \int_0^a xy(x^2+y^2)dx = \int_0^b \left\{ \int_0^a x(x^2+y^2)dx \right\} ydy$$

$$= \int_0^b \left[x^4/4 + x^2y^2/2\right]_{x=0}^a ydy = (1/4)\int_0^b (a^4 + 2a^2y^2)ydy = a^2b^2(a^2+b^2)/8, \tag{2}$$

which is the same as (1).

2. *Evaluate the two repeated integrals:*

(i) $I = \int_0^1 \int_0^1 \frac{x-y}{(x+y)^3} dydx$ *and* **(ii)** $J = \int_0^1 \int_0^1 \frac{x-y}{(x+y)^3} dxdy$ *and show that they are not equal. What conclusion can you make regarding the double integral*

$$\iint_R \frac{x-y}{(x+y)^3} dxdy,$$

where R is rectangle: $\{(x,y) : 0 \le x \le \text{ and } 0 \le y \le 1\}$*?*

Let $f(x,y) = (x-y)/(x+y)^3$. This function is not bounded over the diagonal $x+y=0$ of the rectangle and so it is violating the condition for the given double integral to be proper.

Let $F(x) = \int_0^1 \frac{x-y}{(x+y)^3} dy$. Then since $\frac{x-y}{(x+y)^3} = \frac{2x}{(x+y)^3} - \frac{1}{(x+y)^2}$,

$F(x) = [-x/(x+y)^2 + 1/(x+y)]_{y=0}^1 = 1/(1+x)^2$ of $x \neq 0$ and

$F(0) \int_0^1 (-y/y^3)dy = -\int_0^1 (1/y^2)dy = [1/y]_0^1$ does not exist.

But $I = \int_0^1 F(x)dx = \int_0^1 \frac{dx}{(1+x)^2} = -\left[\frac{1}{1+x}\right]_0^1 = -\frac{1}{2} + 1 = \frac{1}{2}.$

If $G(y) = \int_0^1 \frac{x-y}{(x+y)^3} dx$, then since $\frac{x-y}{(x+y)^3} = \frac{1}{(x+y)^2} - \frac{2y}{(x+y)^3}$

$\therefore$ $G(y) = \left[-\frac{1}{x+y} + \frac{y}{(x+y)^2}\right]_0^1 = -\frac{1}{(1+y)^2}$. But $G(0)$, like $F(0)$, does not exist.

But $J = \int_0^1 G(y)dy = -\int_0^1 \frac{dy}{(1+y)^2} = \left[\frac{1}{1+y}\right]_0^1 = -\frac{1}{2}.$

Thus $I \neq J$.

This shows that the double integral does not exist. Because if the double integral exists, I and J both must exist and they will be equal.

5.3.4 Double Integral over an Arbitrary Plane Region

In the Art. 5.3.1, we have defined double integral of a function over a rectangular region with sides parallel to the coordinate axes. Next we shall define the same double integral over an arbitrary plane region E in the xy-plane. For the integral to be proper the integrand $f(x,y)$ and the region E both are assumed to be **bounded**, so that there exist numbers m and M such that $m = \le f(x,y) = M \ \forall \ (x,y) \in E$ and there is a rectangle R with sides parallel to the coordinate axes so that $E \subseteq R$ (Fig. 5.4).

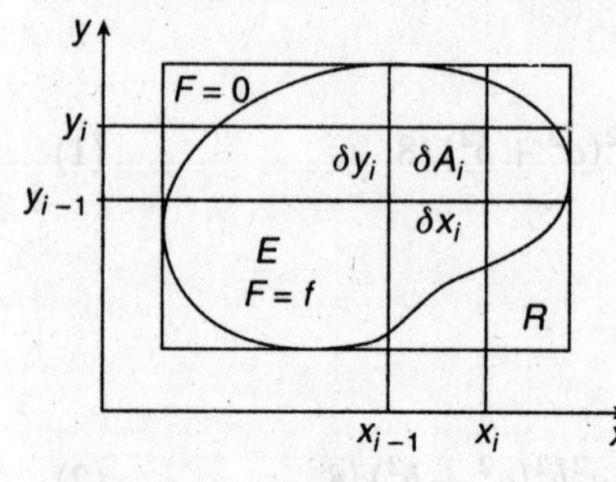

Fig. 5.4

We define the double integral of f over E with the help of the double integral of the function F over R, defined as follows:

$$F(x,y) = f(x,y) \forall (x,y) \in E$$
$$= 0 \ \forall (x,y) \in R - E.$$

By definition,

$$\iint\limits_E f(x,y)dxdy = \iint\limits_R F(x,y)dxdy.$$

Alternatively, E can be divided into a finite number n of subdomains E_i with areas δA_i such that $\bigcup_{i=1}^{n} E_i = E$, where $E_i \cap E_j = \phi$ for $i \neq j$. Taking any arbitrary point $P_i(x_i, y_i)$ in E_i, we form the sum $S_n = \Sigma_1^n f(P_i)\delta A_i$. Now when the partition of E is made finer so that Δ, the greatest of all the δA_i's tends to zero and the limit of S_n when $\Delta \to 0$ exists and is independent of the choice of P_i in E_i, then this limit is defined to be the double integral of $f(x,y)$ w.r.t. x and y over E. Since $n \to \infty$ when $\Delta \to 0$, we have, therefore

$$\underset{n\to\infty}{\text{Lt}}\ S_n = \iint\limits_E f(x,y)dA \quad \text{or,} \quad \iint\limits_E f(x,y)dxdy.$$

Geometrically, if $f(P) \geq 0 \forall\ P \in E$, the double integral represents **the volume of a cylinder** with E as its plane base and the portion of the surface given by $z = f(x,y)$ over the base. In particular, if $f(P) = 1 \forall\ P \in E$, the integral represent the volume of a cylinder with base E and height unity and hence $\iint\limits_E dxdy$ represents, numerically, **the area of the base E.**

This area is also the **area of the plane region bounded by the closed curve**, which is the boundary of E.

A sufficient condition for integrability

If E is bounded by a finite number continuous curves defined by functions like $y = \phi(x)$ or, $x = \psi(y)$ defined in suitable intervals and f is continuous in E, then f is integrable in E.

Following the definition of double integral, it is not always very convenient to evaluate it. Evaluation will be easier if it can be converted into integrals of a function of a single variable.

5.3.5 Evaluation of a Double Integral over an Arbitrary Region

Let E, the domain of integration, be bounded by piecewise continuous curves defined by functions like $y = \phi(x)$ or $x = \psi(y)$ in suitable intervals and $f(x,y)$ be continuous in E. Then f is integrable over E. Now to make the evaluation of the double integral $\iint\limits_E f dxdy$ possible with the help of ordinary integrals we consider a few cases concerning the boundary of E.

Case I. E is quadratic with respect to the y-axis

E is said to be **quadratic** with respect to the y-axis if either a line drawn parallel to the y-axis meets the boundary of E in not more than two points or a few segments of the line form parts of the boundary of E.

The region E in the xy-plane shown in Fig. 5.5 is quadratic with respect to the y-axis.

The lines $x = x_1$ and $x = x_4$ meet the boundary of E in only one point, $x = x_3$ in two distinct points, two segments of the line $x = x_2$ form parts of the boundary and the line $x = x_5$ meets the boundary of E in no point at all. Then the

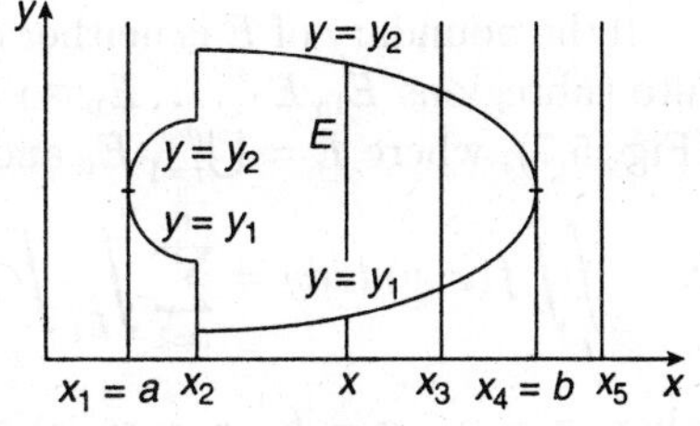

Fig. 5.5

domain E can be expressed as $E = \{(x, y) \in R^2 : a \leq x \leq b,\ y_1(x) \leq y \leq y_2(x) \forall\ x \in [a, b]\}$ and the boundary of such a region can be expressed as $x = a,\ x = b,\ y = y_1(x)$ and $y = y_2(x) \forall\ x \in [a, b]$, where $y_1 \leq y_2 \forall\ x \in [a, b]$, where y_1 and y_2 are piecewise continuous. In the figure above both y_1 and y_2 are discontinuous at x_2. In this case

$$\iint_E f dy dx = \int_a^b \left\{ \int_{y_1}^{y_2} f dy \right\} dx, = \int_a^b \int_{y_1}^{y_2} f dy dx, \tag{1}$$

where $F(x) = \int_{y_1}^{y_2} f(x, y) dy$ is an ordinary integral of a function of the single variable y(x being a parameter) with respect to y in the interval $[y_1, y_2]$. If x is fixed, y_1 and y_2 are fixed. As x varies in $[a, b]$, the end points y_1 and y_2 of the interval $[y_1, y_2]$ also vary. Here $F(x)$ geometrically represents the area under the curve Γ of intersection of the surface $z = f(x, y)$, supposed non-negative, and the vertical plane through x parallel to the yz-plane (Fig. 5.6).

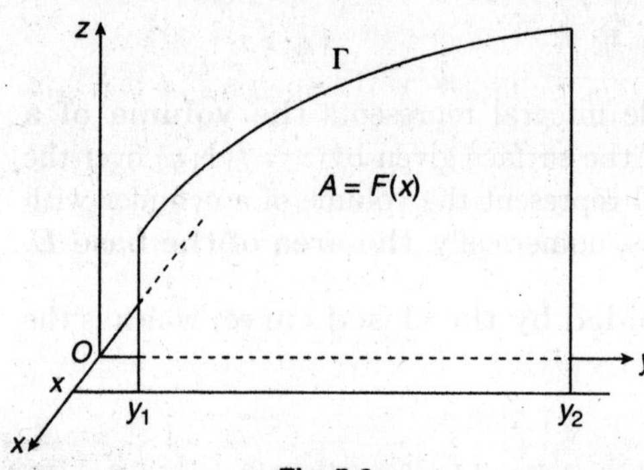

Fig. 5.6

Since the integral $F(x)$ is a function of the single variable x,

$$\iint_E f dx dy = \int_a^b F(x) dx,$$

just an ordinary definite integral of a function of a single variable. The integral on the R.H.S. of (1) is a repeated integral.

Case II. E is quadratic with respect to the x-axis

The boundary of E can be described as

$y = c,\ y = d,\ x = x_1(y),\ x = x_2(y)$, where $x_1 \leq x_2 \forall\ y \in [c, d]$, where x_1 and x_2 are piecewise continuous in $[c, d]$. Then as in Case I,

$$\iint_E f dx dy = \int_c^d \left\{ \int_{x_1}^{x_2} f(x, y) dx \right\} dy = \int_c^d \int_{x_1}^{x_2} f dx dy,$$

where $E = \{(x, y) \in R^2 : c \leq y \leq d,\ x_1(y) \leq x \leq x_2(y) \forall\ y \in [c, d]\}$ and its boundary is given by

$$y = c,\ y = d,\ x = x_1(y),\ x = x_2(y), \text{ where } x_1 \leq x_2 \forall\ y \in [c, d].$$

Case III.

If the boundary of E is neither quadratic in x nor quadratic in y, the region E can be divided into subregions $E_1, E_2, \ldots, E_n$ so that the boundary of each of these regions is quadratic in y (Fig. 5.7), where $E = \bigcup_{i=1}^n E_i$ and $E_i \cap E_j = \phi$ for $i \neq j$.

$$\therefore \quad \iint_E f(x, y) dx dy = \sum_{i=1}^n \int_{E_i} \int f(x, y) dx dy = \sum_{i=1}^n \int_{a_i}^{b_i} \int_{y_i}^{y_i'} f dy dx,$$

where $x = a_i,\ x = b_i,\ y = y_i(x)$ and $y = y_i'(x)$, so that $y_i \leq y_i' \ \forall\ x \in [a_i, b_i]$ for $i = 1, \ldots, n$, constituting the boundary of the region E_i.

E could be partitioned into subregions $E'_1, E'_2, \ldots, E'_m$ so that each of them is quadratic in x and

$$E = \bigcup_{j=1}^{m} E'_j \quad \text{and} \quad E'_j \bigcap_{j \neq k} E'_k = \phi \text{ for } j \neq k.$$

Then $$\iint_E f(x,y)dxdy = \sum_{j=1}^{m} \int_{c_j}^{d_j} \int_{x_j}^{x'_j} fdxdy.$$

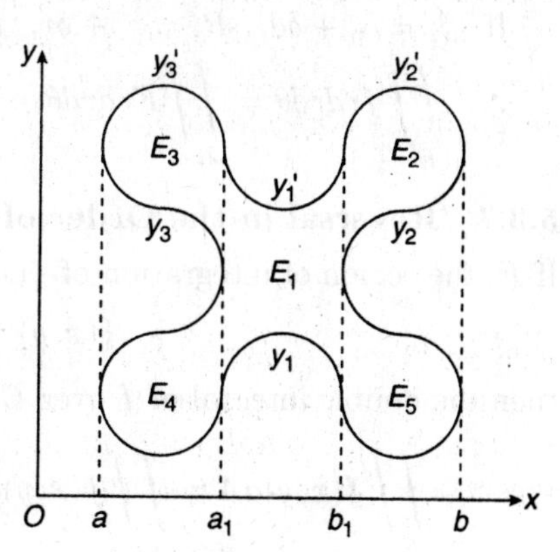

Fig. 5.7

The properties of integrals over any arbitrary plane region like E are the same as that over a rectangular region. Also, if $f(x, y)$ is continuous and non-negative over E, then the integral $\iint_E fdxdy$ represents, geometrically, the volume of the space enclosed by the cylindrical surface over the boundary of E and with generators parallel to the z-axis, bounded by the surfaces: $z = 0$ and $z = f(x, y)$.

Thus, if $f(x, y) \geq 0 \;\forall\; (x, y) \in E$, then

$$V = \iint_E fdxdy,$$

where V is the volume of the cylinder described above. Putting $f = 1$, we get

$$A = \iint_E dxdy,$$

which represents the area of the plane region E.

5.3.6 Polar Partitioning of the Domain of Integration

Instead of enclosing E in a rectangular region, it could be enclosed in any type of suitable region R, which could be partitioned into subregions in some other way. If δA_i is the area of the subregion E_i, then in place of $\iint_E fdxdy$ the symbol $\iint_E fdA$ could be used. In rectangular partitions $dA = dxdy$. If the function f is defined in polar coordinates r, θ and $z = f(r, \theta) \forall\; (r, \theta) \in E$, then E can be enclosed in a region R whose boundary consist of the straight lines $\theta = \alpha$ and $\theta = \beta$, and the arcs $r = \rho$ and $r = \rho'$.

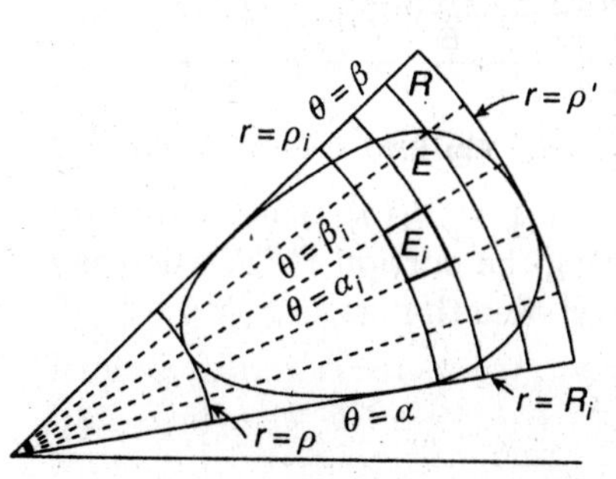

Fig. 5.8

We define a function $F(r, \theta)$ such that

$$F(r, \theta) = f(r, \theta) \forall\; (r, \theta) \in E = 0 \text{ or, if } (r, \theta) \notin E.$$

Then R could be divided into subregions E_i bounded by

$$\theta = \alpha_i,\; \theta = \beta_i,\; r = \rho_i \quad \text{and} \quad r = R_i.$$

Let δA_i be the area of E_i (Fig. 5.8). Then

$$\iint_E fdA = \iint_R FdA.$$

If $\beta_i = \alpha_i + \delta\theta_i,\ R_i = r_i + \delta r_i$, then $\delta A_i \approx r_i \delta r_i \delta\theta_i$.

$$\therefore \quad \iint_E f r dr d\theta = \iint_R F r dr d\theta.$$

5.3.7 Reversal in the Order of Integration

If E, the region of integration of $f(x,y)$ be a rectangle given by

$$\{(x,y) : a \le x \le b \text{ and } c \le y \le d\},$$

then the double integral of f over E is given by

$$\iint_E f(x,y)dA = \iint_E f(x,y)dxdy = \int_c^d \int_a^b f(x,y)dxdy = \int_a^b \int_c^d f(x,y)dydx,$$

where the order of integration has been reversed. In each of the repeated integrals the inner integral is to be evaluated first.

If E is not a rectangle but it is quadratic in both x and y, then also the double integral of a function f over E can be evaluated in two repeated integrals of the types

$$I_1 = \int_c^d \int_{x_1}^{x_2} f(x,y)dxdy \quad \text{and} \quad I_2 = \int_a^b \int_{y_1}^{y_2} f(x,y)dydx,$$

which are equal, being equal to the double integral

$$I = \iint_E f(x,y)dA.$$

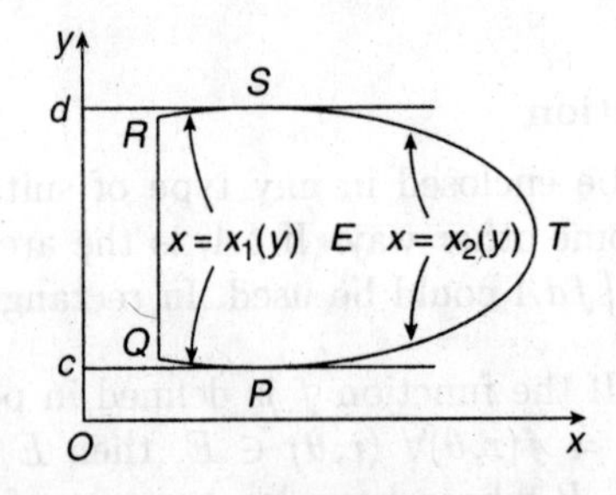

Fig. 5.9

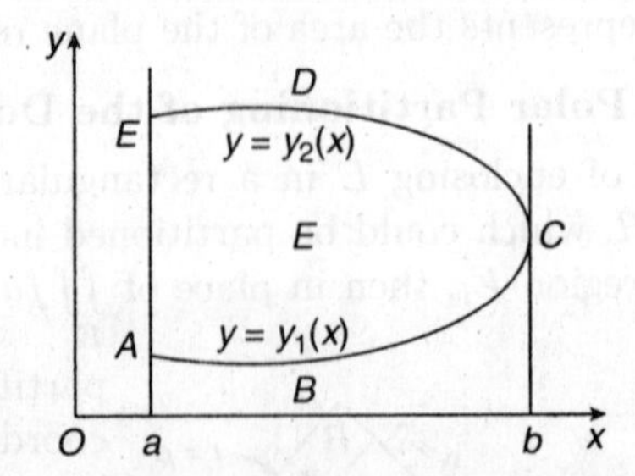

Fig. 5.10

For, if E is quadratic in x its boundary can be described as $x = x_1(y)$ representing the portion $PQRS$ of the boundary (Fig. 5.9), $x = x_2(y)$ representing the portion PTS, whereas y always lies between c and d. Here the functions x_1 and x_2 are so chosen that $x_1 \le x_2 \ \forall\, y \in [c,d]$. Hence

$$I = \int_c^d \int_{x_1}^{x_2} f(x,y)dxdy. \tag{1}$$

Similarly, if E is assumed to be quadratic in y, its boundary can be described as $y = y_1(x)$ representing the portion ABC of the boundary (Fig. 5.10), $y = y_2(x)$ representing the portion EDC of the boundary, the variable x always lying between a and b. So

$$I = \int_a^b \int_{y_1}^{y_2} f(x,y)dydx. \tag{2}$$

Thus to change the order of integral in (1), in which the limits c and d of the outer integral are fixed, first of all the fixed limits, viz., a and b of the variable x are to be determined so that the entire region E lies between the two straight lines $x = a$ and $x = b$. Then if E is quadratic in y $\forall$ $x \in [a, b]$, there will be at the most two values of y and hence the functions $y_1(x)$ and $y_2(x)$ $\forall$ $x \in [a, b]$ can be found out. If $y_1 \leq y_2$ for $\forall$ $x \in [a, b]$, y_1 and y_2 be the lower and upper limits of the inner integral in (2).

Examples 5.2. **1.** *Evaluate the integral* $I = \int\int [x^3 + y^3 - 3xy(x^2 + y^2)](x^2 + y)^{-3/2} dxdy$ *over the circular region given by* $x^2 + y^2 \leq 1$.

If E be the given circular region, then

$$E = \{(x, y) : x^2 + y^2 \leq 1\} = \{(x, y) : -1 \leq x \leq 1 \text{ and } -\sqrt{1 - x^2} \leq y \leq \sqrt{1 - x^2}\},$$

so that E is assumed quadratic in y. Then

$$I = \int_{-1}^{1} \int_{y_1}^{y_2} f(x, y) dy dx = \int_{-1}^{1} \phi(x) dx,$$

where $\phi(x) = x^3 \int_{y_1}^{y_2} \frac{dy}{(x^2 + y^2)^{3/2}}$, $y_1 = -\sqrt{1 - x^2}$ and $y_2 = \sqrt{1 - x^2}$,

since the other part of the integrand $f(x, y)$ viz., $(y^3 - 3x^3y - 3xy^3)/(x^2 + y^2)^{3/2}$ is an odd function of y and the interval of integration $[y_1, y_2]$ is symmetric with respect to the origin. Taking x as the parameter and $y = x \tan\theta$, where $-\pi/2 < \theta < \pi/2$, we have $dy = x \sec^2\theta$ so that

$$\int \frac{dy}{(x^2 + y^2)^{3/2}} = \int \frac{x \sec^2\theta d\theta}{x^3 \sec^3\theta} = \frac{1}{x} \int \cos\theta d\theta = \frac{\sin\theta}{x} = \frac{y}{x^2\sqrt{x^2 + y^2}}.$$

$$\therefore \quad \phi(x) = \left. \frac{xy}{\sqrt{(x^2 + y^2)}} \right|_{y=y_1}^{y_2} = 2x\sqrt{1 - x^2}.$$

Hence, $I = \int_{-1}^{1} 2x\sqrt{1 - x^2} dx = 0,$

since the integrand again is an odd function of x.

The same result will be obtained by taking

$$E = \{(x, y) : -\sqrt{1 - y^2} \leq x \leq \sqrt{1 - y^2} \text{ and } -1 \leq y \leq 1\}$$

when E is assumed quadratic in x.

2. *Evaluate the double integral of the function:* $f(x, y) = 1 + x + y$ *over a region E bounded by the lines*

$$y = -x, \ x = \sqrt{y} \ \text{ and } \ y = 2.$$

Let $y = -x$ represent the straight line OB, $x = \sqrt{y}$ represent the part OA of the parabola for which $x \geq 0$ and $y = 2$ is the straight line AB (Fig. 5.11). To determine A, we solve $x = \sqrt{y}$ and $y = 2$.

So, A is $(\sqrt{2}, 2)$.

To determine B, we solve $y = -x$ and $y = 2$.

So, B is $(-2, 2)$.

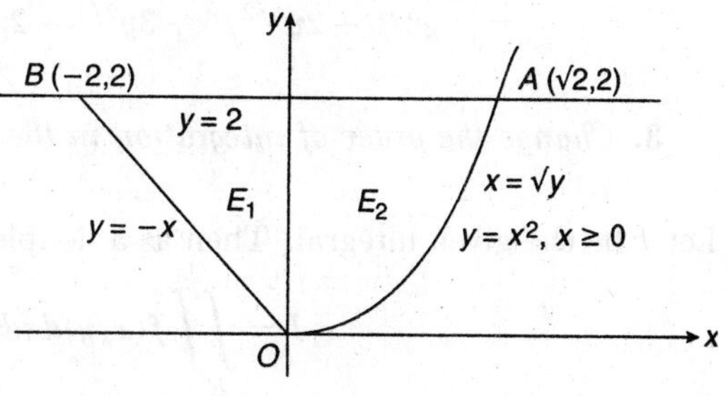

Fig. 5.11

So, $E = \{(x,y) : -2 \le x \le 0$ and $-x \le y \le 2$ or, $0 \le x \le \sqrt{2}$ and $x^2 \le y \le 2\}$
$= E_1 \cup E_2,$

where $E_1 = \{(x,y) : -2 \le x \le 0$ and $-x \le y \le 2\}$

and $E_2 = \{(x,y) : 0 \le x \le \sqrt{2}$ and $x^2 \le y \le 2\}$.

Here E has been divided into two subregions E_1 and E_2 each of which has been assumed quadratic in y.

So, if $I = \iint_E f(x,y)dxdy$, then $I = I_1 + I_2$,

where $I_1 = \iint_{E_1} f(x,y)dxdy$ and $I_2 = \iint_{E_2} f(x,y)dxdy$.

$$I_1 = \int_{-2}^{0}\int_{-x}^{2}(1+x+y)dydx = \int_{-2}^{0}[y(1+x)+y^2/2]_{-x}^{2}dx$$
$$= \int_{-2}^{0}[2+2(x+1)-x^2/2+x(1+x)]dx$$
$$= \int_{-2}^{0}(x^2/2+3x+4)dx = [x^3/6+3x^2/2+4x]_{-2}^{0} = \frac{10}{3}.$$
$$I_2 = \int_{0}^{\sqrt{2}}\int_{x^2}^{2}(1+x+y)dydx = \int_{0}^{\sqrt{2}}[y(1+x)+y^2/2]_{x^2}^{2}dx$$
$$= \int_{0}^{\sqrt{2}} 2(x+1)+2-x^2(1+x)-x^4/2]dx$$
$$= \int_{0}^{\sqrt{2}}(-x^4/2-x^3-x^2+2x+4)dx = 1+44\sqrt{2}/15$$

$\therefore$ $I = I_1 + I_2 = 13/3 + 44\sqrt{2}/15.$

Assuming $E = \{(x,y) : -y \le x \le \sqrt{y}$ and $0 \le y \le 2\}$,

$$I = \int_{0}^{2}\int_{-y}^{\sqrt{y}}(1+x+y)dxdy = \int_{0}^{2}(x+x^2/2+xy)_{x=-y}^{\sqrt{y}}dy$$
$$= \int_{0}^{2}[-y^2/2+y^{3/2}+3y/2+y^{1/2}]dy$$
$$= [-y^3/6+2y^{5/2}/5+3y^{2/4}+2y^{3/2}/3]_0^2 = 13/3+44\sqrt{2}/15.$$

3. *Change the order of integration in the integral* $\displaystyle\int_{-a}^{a}\int_{0}^{\sqrt{(a^2-y^2)}} f(x,y)dxdy.$

Let I be the given integral. Then as a double integral

$$I = \iint_E f(x,y)dxdy,$$

where $E = \{(x,y) : -a \le y \le a$ and $0 \le x \le \sqrt{a^2-y^2}\}$.

Now, $x = \sqrt{a^2 - y^2} \Rightarrow x^2 = a^2 - y^2$

$\Rightarrow \quad x^2 + y^2 = a^2$. So, $0 \le x \le \sqrt{a^2 - y^2}$ and $-a \le y \le a$ give a semicircle on the right side of y-axis (Fig. 5.12).

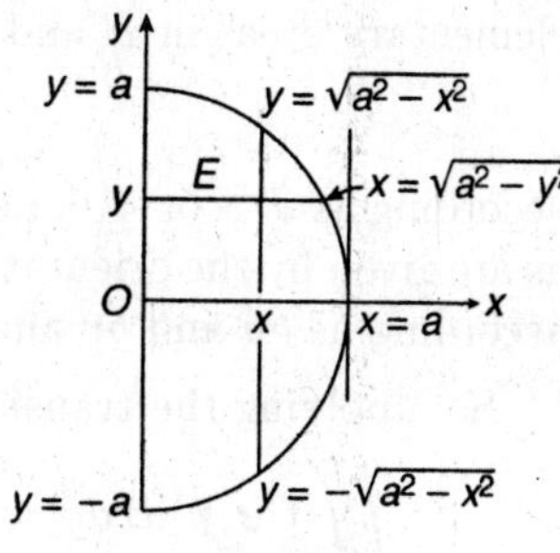

Fig. 5.12

The same region E lies between two fixed lines $x = 0$ and $x = a$ and for any value of $x \in [0, a]$ there are two values of y, viz.,

$$y = y_1 = \sqrt{a^2 - x^2} \quad \text{and} \quad y = y_2 = \sqrt{a^2 - x^2}.$$

$$\therefore \quad E = \{(x, y) : 0 \le x \le a \quad \text{and} \quad -\sqrt{a^2 - x^2} \le y \le \sqrt{a^2 - x^2}\}.$$

$$\therefore \quad I = \int_0^a \int_{-\sqrt{a^2-x^2}}^{\sqrt{a^2-x^2}} f(x, y) dy dx.$$

4. *Evaluate* $\iint r \sin\theta dr d\theta$ *over the cardioid:* $r = a(1 - \cos\theta)$ *above the initial line.*

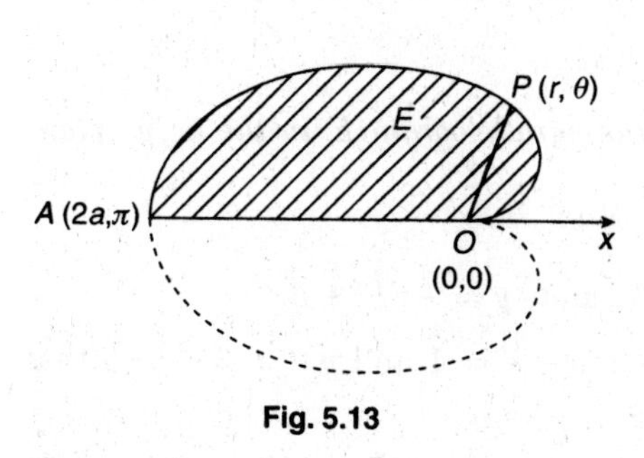

Fig. 5.13

If E is the upper half of the cardioid,

$$E = \{(r, \theta) : 0 \le \theta \le \pi, 0 \le r \le a(1 - \cos\theta)\}.$$

$\therefore$ the required integral is

$$\int_0^\pi \int_0^{a(1-\cos\theta)} r \sin\theta dr d\theta = \int_0^\pi \left[\frac{r^2}{2}\right]_0^{a(1-\cos\theta)} \sin\theta d\theta$$

$$= \frac{a^2}{2} \int_0^\pi (1 - \cos\theta)^2 \sin\theta d\theta = \frac{a^2}{6} [1 - \cos\theta)^3]_0^\pi$$

$$= \frac{8a^2}{6} = \frac{4a^3}{3}.$$

5.3.8 Change of Variables in a Double Integral

For the purpose of easy evaluation it is sometimes necessary to **apply** a change of variables. Given the integral

$$\iint_E f(x, y) dx dy, \tag{1}$$

we now study the effect of the transformation given by

$$x = x(u, v) \quad \text{and} \quad y = y(v, u). \tag{2}$$

Suppose, $T : E_1 \to E$ is the transformation given by (2), where E_1 is the domain and E is the range of T (Fig. 5.14).

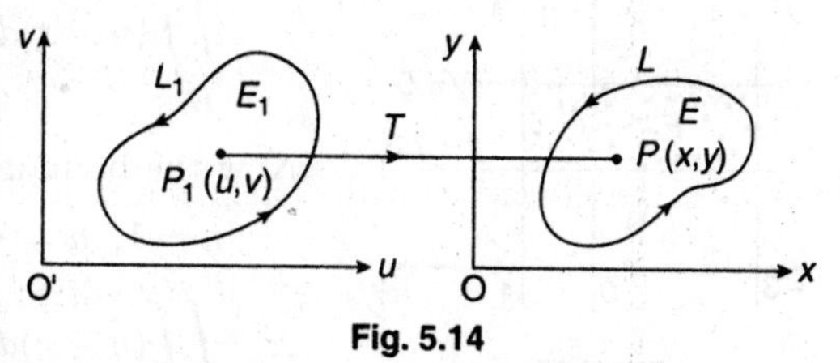

Fig. 5.14

Let $J(T) = \frac{\partial(x,y)}{\partial(u,v)}$, the Jacobian of the transformation T.

Let L_1 be the positively oriented piece-wise smooth boundary of E_1. If $J(T) \neq 0$ everywhere in E_1 (except for a finite number of points in E_1), then T induces an oriented boundary L of E. This induced orientation is positive or negative according as $J >$ or < 0. If $J > 0$, the boundary will be described in the direction shown in the figure. By such a transformation the elementary areas in E and E_1 are related by

$$\delta x \delta y = \pm J \delta u \delta v$$

according as $J >$ or < 0, i.e., according as the orientation of E and E_1 are the same or opposite as are given by the orientations of the boundary. It might be noted that J is positive or negative according as δu and δv and consequently $\delta x, \delta y$ have the same or opposite signs.

So, applying the transformation (2)

$$\iint_E f(x,y)dxdy = \iint_{E_1} f(u,v)|J|dudv,$$

where $F(u,v) = f(x(u,v), y(u,v))$.

For the transformation: $x = r\cos\theta,\ y = r\sin\theta,\ J = r > 0$, if $r > 0$ (see Ex. 1, § 4.17.2).

$$\therefore \quad \iint_E f(x,y)dxdy = \iint_{E_1} F(r,\theta)rdrd\theta.$$

Example. *Compute the integral* $\iint_E (y-x)dxdy$ *over the parallelogram* E *in the* (x,y) *plane bounded by the four straight lines*

$$y = x+1, \quad y = x-3, \quad y = -x/3 + 7/3 \quad \textit{and} \quad y = -x/3 + 5.$$

It will be lengthy to compute the integral directly. Since $y - x = 1$ and $y - x = -3$ so that $y - x =$ constant for the first two lines, we take

$$u = y - x. \tag{1}$$

Similarly, since for the last two lines $y + x/3 =$ constant, we put

$$v = y + x/3. \tag{2}$$

Since $\partial(u,v)/\partial(x,y) = \begin{vmatrix} -1 & 1 \\ 1/3 & 1 \end{vmatrix} = -4/3\ \forall\ (x,y) \in E,$

this Jacobian always maintains the same sign in E.

$$J = \partial(x,y)/\partial(u,v) = -3/4.$$

The equations (1) and (2) define a transformation T from (x,y) plane to the (u,v) plane. Let E_1 be the image of E (Fig. 5.15). Then the given integral:

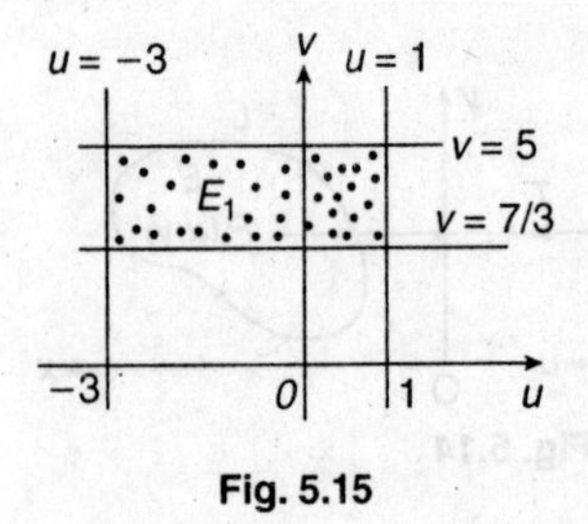

Fig. 5.15

$$\iint_E (y-x)dxdy = \iint_{E_1} u|J|dudv = \frac{3}{4}\iint_{E_1} ududv.$$

Now the boundary of E_1 consists of

$$u = 1,\ u = -3,\ v = 7/3,\ v = 5.$$

$$\therefore \quad \iint_E (y-x)dxdy = \frac{3}{4}\int_{-3}^{1} udu \int_{7/3}^{5} dv = \frac{3}{4}\left[\frac{u^2}{2}\right]_{-3}^{1} [v]_{7/3}^{5} = -8.$$

Exercises 5.1

1. Evaluate the double integral $\iint (x^2+y^2)dxdy$ over the square R, whose sides are $x+y=0$, $x+y=2$, $x-y=0$ and $x-y=2$.

2. Describe the region of integration and evaluate the following integrals:

(i) $\int_0^1\int_0^1 (x^2+y^2)dydx$, (ii) $\int_0^1\int_x^{2x} (x^2+y^2)dydx$, (iii) $\int_0^1\int_y^{y^2+1} x^2ydxdy$,

(iv) $\int_0^1 dx\int_0^x \sqrt{(x^2+y^2)}dy$.

3. Evaluate $\iint xydxdy$ over the region A bounded by the x-axis, ordinate $x=2a$ and the curve $x^2=4xy$, $a>0$.

4. Evaluate the following integrals:

(i) $\int_0^1\int_x^{\sqrt{x}} (x^2+y^2)dydx$, (ii) $\int_0^1\int_0^{\sqrt{(1+x^2)}} dydx/(1+x^2+y^2)$,

(iii) $\int_0^a\int_{x/a}^{\sqrt{(x/a)}} (x^2+y^2)dxdy$, (iv) $\int_0^3\int_1^{\sqrt{(4-y)}} (x+y)dxdy$, (v) $\int_0^{4a}\int_{x^2/4a}^{2\sqrt{ax}} dydx$.

5. Change the order of integration and then evaluate the following integrals:

(i) $\int_0^1\int_x^{\sqrt{(2-x^2)}} \frac{xdydx}{\sqrt{(x^2+y^2)}}$, (ii) $\int_0^{a/\sqrt{2}}\int_y^{\sqrt{(a^2-y^2)}} \ln(x^2+y^2)dxdy, a>0$,

(iii) $\int_0^a\int_{\sqrt{(ax)}}^a \frac{y^2dydx}{\sqrt{(y^4-a^2x^2)}}$, (iv) $\int_0^\infty\int_x^\infty \frac{e^{-y}}{y}dydx$, (v) $\int_0^\infty\int_0^x xe^{-x^2/y}dydx$.

6. Evaluate $\iint_E xydxdy$, where E is the first quadrant of the circle $x^2+y^2=a^2$.

7. Evaluate $\iint xy(x+y)dxdy$ over the region E, whose boundary consists of $x=y$ and $x^2=y$. **[WBUT 2001]**

8. Evaluate $\iint (x+y)^2dxdy$ over the region, whose boundary is the ellipse

$$x^2/a^2+y^2/6^2=1.$$

9. Evaluate $\iint \sqrt{\frac{1-x^2/a^2-y^2/b^2}{1+x^2/a^2+y^2/b^2}}dxdy$ over the positive quadrant of the ellipse

$$x^2/a^2+y^2/b^2=1.$$

10. Evaluate $\iint \sqrt{(a^2-x^2-y^2)}dxdy$ over the circular disc bounded by $x^2+y^2=ax$.

Answers

1. 8/3. **2.** **(i)** 2/3; **(ii)** 5/6; **(iii)** 67/120; **(iv)** $[\sqrt{2}+\ln(\sqrt{2}+1)]/6$. **3.** $a^4/3$.
4. **(i)** 3/35; **(ii)** $(\pi/4)\ln(\sqrt{2}+1)$; **(iii)** $a^3/28+a/20$; **(iv)** 241/60; **(v)** $16a^2/3$.
5. **(i)** $1-1/\sqrt{2}$; **(ii)** $(\pi a^2/4)(\ln a-1/2)$; **(iii)** $\pi a^2/6$; **(iv)** 1; **(v)** 1/2.
6. $a^4/8$. **7.** 3/56. **8.** $\pi ab(a^2+b^2)/4$. **9.** $\pi(\pi-2)ab/r$.
10. $a^2(3\pi-4)/9$.

5.4 Triple Integral

Let $f : R \to \mathbb{R}$, be a bounded function of three variables x, y and z, where the domain R of f is in the form of a rectangular parallelopiped, with its edges parallel to the coordinate axes, given by

$$R = \{(x, y, z) \in \mathbb{R}^3 : x_1 \le x \le x_2,\ y_1 \le y \le y_2 \quad \text{and} \quad z_1 \le z \le z_2\},$$

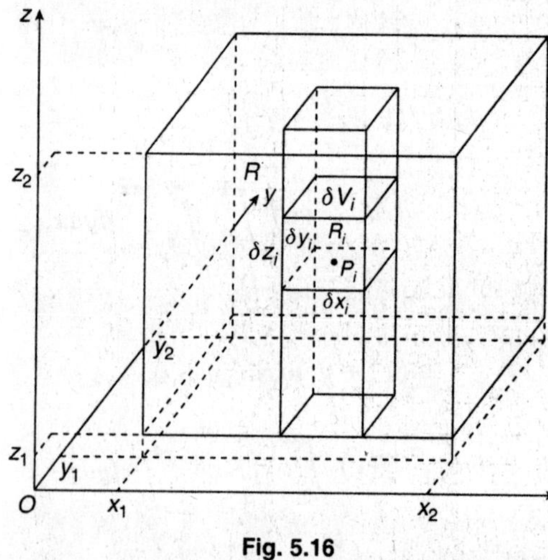

Fig. 5.16

where x_1, x_2, y_1, y_2, z_1 and z_2 are all constants (Fig. 5.16). This region R can be partitioned into a finite number n of parts by drawing planes parallel to the coordinate planes.

If R_i is one of such regions with edge lengths δx_i, δy_i and δz_i, then its volume

$$\delta V_i = \delta x_i \delta y_i \delta z_i.$$

Let P_i be any point chosen at random in R_i. We form the sum: $S_n \sum_{i=1}^{n} f(P_i)\delta v_i$.

If now we make finer partitions of R so that the greatest of the diagonals of R_i tends to zero, then n will tend to infinity. Then the limit: $\underset{n\to\infty}{\text{Lt}}\ S_n$, when it exists and is independent of the position of P_i in R_i for $i = 1, 2, \ldots, n$ then the limits of this sum is defined to be the triple integral of f over R and it is denoted by

$$\iiint_R f dV \quad \text{or} \quad \iiint_R f dx dy dz.$$

Thus $$\iiint_R f dx dy dz = \underset{n\to\infty}{\text{Lt}}\ S_n,$$

where $S_n = \sum f(P_i)\delta x_i \delta y_i \delta z_i$.

If $f(P) \ge 0\ \forall\ P \in R$, the triple integral represents the **mass** of the parallelopiped if $f(P)$ represents the density at P. If $f(P) = 1$, δV represents an elementary volume of the parallelopiped and hence the triple integral $\iiint_R dV$ or $\iiint_R dx dy dz$ represents the **volume V of the parallelopiped.**

The triple integral over an arbitrary closed region D in $\mathbb{R}^3$ can be defined in the same way as the double integral in § 5.3.4.

5.4.1 Evaluation of a Triple Integral. Change in the Order of Integration

If the domain of integration is the region R in the form of rectangular parallelopiped given by

$$R = \{(x,y,z) \in \mathbb{R}^3 : x_1 \le x \le x_2, y_1 \le y \le y_2 \text{ and } y_1 \le z \le z_2\},$$

then $$\iiint\limits_R f(x,y,z)dxdydz = \int_{x_1}^{x_2}\int_{y_1}^{y_2}\int_{z_1}^{z_2} f(x,y,z)dzdydx.$$

Since x_1, x_2, y_1, y_2 and z_1, z_2 are constants, the order of integration can be changed without affecting the value of the triple integral, provided the triple integral exists.

But if the domain of integration is D, not necessary rectangular, to evaluate $\iiint\limits_D f dxdydz$, D is enclosed in a rectangular region of the form R above and a function F is defined such that

$$\begin{aligned} F(P) &= f(P) \ \forall \ P(x,y,z) \in D \\ &= 0 \quad \forall \ P(x,y,z) \notin D \text{ but } \in R. \end{aligned}$$

Then $$\iiint\limits_D f dxdydz = \iiint\limits_R F dxdydz = \int_{x_1}^{x_2}\int_{y_1}^{y_2}\int_{z_1}^{z_2} F dzdydx.$$

Now, if D is quadratic with respect to the z-axis (Fig. 5.17) and its projection E in the xy-plane is also quadratic with respect to the y-axis, then the bounding surface of D may be expressed in the form:

$$x = a, \ x = b, \ a < b,$$

$$y = y_1(x), \ y = y_2(x) \ \forall \ x \in [a,b],$$

where $y_1(x) \le y_2(x) \ \forall x \in [a,b]$

and $z = z_1(x,y), \ z = z_2(x,y),$

where $z_1 \le z_2 \ \forall (x,y) \in E.$

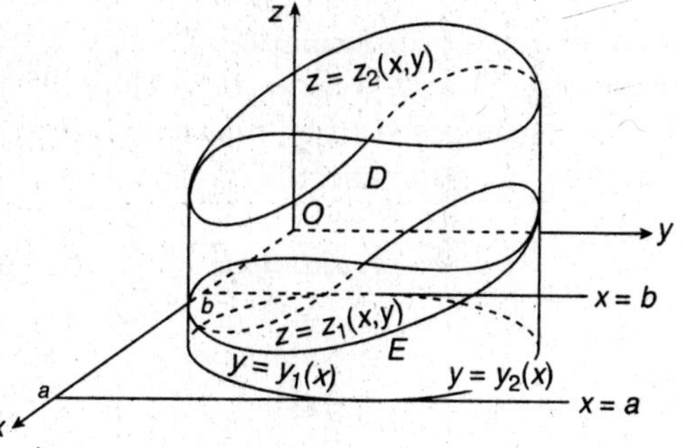

Fig. 5.17

Then the integral of f over D can be expressed as

$$\begin{aligned} \iiint\limits_D f dxdydz &= \iint\limits_E dxdy \int_{z_1}^{z_2} f(x,y,z)dz = \int_a^b dx \int_{y_1}^{y_2} dy \int_{z_1}^{z_2} f(x,y,z)dz \\ &= \int_a^b \int_{y_1}^{y_2}\int_{z_1}^{z_2} f(x,y,z)dzdydx = \int_a^b F(x)dx, \end{aligned}$$

where $F(x) = \int_{y_1}^{y_2} F_1(x,y)dy$,

where $F_1(x,y) = \int_{z_1}^{z_2} f(x,y,z)dz.$

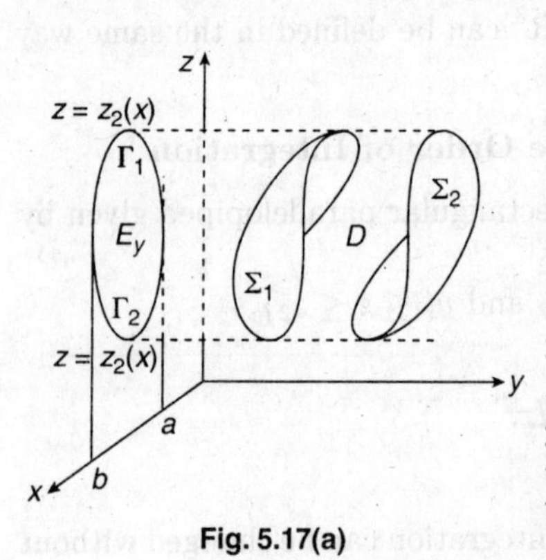

Fig. 5.17(a)

The order of integration in $\mathbb{R}^3$, as in the case of integration in $\mathbb{R}^2$, can be altered also, provided the domain D can be described accordingly. For example, if D is found to be quadratic in y [Fig. 5.17(a)], every straight line parallel to the y-axis will meet the bounding surface of D in not more than two points. This bounding surface can be split up into two surfaces $\sum_1$ and $\sum_2$ given by $y = y_1(x,z)$ and $y = y_2(x,z)$ respectively, where $y_1 \le y_2 \ \forall \ (x,z) \in E_y$, where E_y is the projection of D in the zx plane. So $\forall (x,y,z) \in D$, $y_1(x,z) \le y \le y_2(x,z)$. Next if E_y is quadratic in z as in the diagram, we get $z_1 = z_1(x)$ and $z = z_2(x)$ as two of its boundary lines Γ_1 and Γ_2 respectively. Then the entire region E_y will be between two straight lines $x = a$ and $x = b$. Then

$$D = \{(x,y,z) : y_1 \le y \le y_2 \ \forall (x,z) \in E_y, z_1 \le z \le z_2 \ \forall \ x \in [a,b], a \le x \le b\}$$

and $$\iiint_D f(x,y,z)dV = \int_a^b \int_{z_1}^{z_2} \int_{y_1}^{y_2} f(x,y,z)dydzdx.$$

5.4.2 Change of Variables in a Triple Integral

If the variables x, y, z of a triple integral $\iiint_D f(x,y,z)dxdydz$ are replaced by u, v and w given by

$$x = x(u,v,w), \ y = y(u,v,w) \text{ and } z = z(u,v,w) \ \forall \ (u,v,w) \in D', \tag{1}$$

then we get a transformation $T : (u,v,w) \to (x,y,z)$ so that $\forall \ (u,v,w) \in D'$ we get a unique point $(x,y,z) \in D$ given in (1). Then D' becomes the domain of T and let D be the range of T. If T is supposed to be bijective and x, y, z all possess continuous first order partial derives w.r. to u, v and w and the Jacobian $J(T)$ does not vanish anywhere in D', then we get

$$\iiint_D f(x,y,z)dxdydz = \iiint_D F(u,v,w)|J(T)|dudvdw,$$

where $F(u,v,w) = f(x(u,v,w), y(u,v,w), z(u,v,w))$ and $J(T) = \dfrac{\partial(x,y,z)}{\partial(u,v,w)}$.

For change of coordinates of Cartesian, spherical and cylindrical and the Jacobians of transformation, see § 4.17.3.

Examples 5.3. **1.** *Evaluate* $\iiint_D xyzdxdydz$, *where*

$$D = \{(x,y,z) : x \ge 0, \ y \ge 0, \ z \ge 0 \ \ \textit{and} \ \ x + y + z - 1 \le 0\}.$$

The domain has been shown in the adjacent diagram (Fig. 5.18). It is a tetrahedral region in the first octant whose bounding surfaces are the four triangles:

OBC, OCA, OAB and ABC, of which the first three faces lie in the three coordinate planes:

$$x = 0, \ y = 0 \ \text{ and } \ z = 0 \text{ respectively.}$$

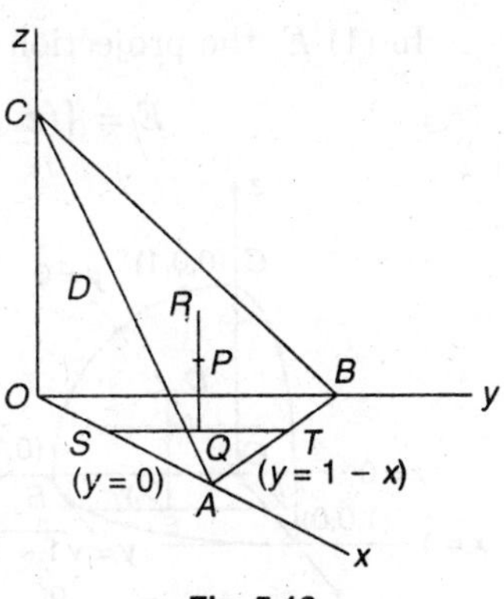

Fig. 5.18

The fourth bounding surface, the triangle ABC lies in the plane given by

$$x + y + z = 1.$$

These surfaces bounding D satisfy the conditions mentioned in the last article. It is quadratic in the z-axis. If through an interior point P of D a straight line is drawn parallel to the z-axis if meets the boundary of D in two points Q and R and not more. When P moves from Q to R, its z-coordinate varies from 0 (since Q lies in the plane: $z = 0$) to $1 - x - y$ (since R lies in the plane: $x + y + z = 1$). Thus the z-coordinate of P satisfies the relation

$$0 \le z \le 1 - x - y. \tag{1}$$

Next in the (x, y)-plane, $z = 0$. In this plane when Q moves parallel to the y-axis, its y-coordinate varies from 0 (which corresponds to S) to $1 - x$ (which corresponds to T). For T, $z = 0$ and $x + y + z = 1$. So, $x + y = 1$ or, $y = 1 - x$. Thus

$$0 \le y \le 1 - x. \tag{2}$$

Finally, when S moves along the x-axis from O to A, its x-coordinate varies form 0 to 1, i.e.,

$$0 \le x \le 1. \tag{3}$$

Since A lies on the x-axis, its y and z-coordinates are 0. But A lies in the plane $x+y+z = 1$. $\therefore\ x + 0 + 0 = 1$ or, $x = 1$.

So, $$\iiint_D xyz\,dx\,dy\,dz = \int_0^1 dx \int_0^{1-x} dy \int_0^{1-x-y} xyz\,dz$$

$$= \int_0^1 dx \int_0^{1-x} \left[xy\frac{z^2}{2}\right]_0^{1-x-y} dy = \frac{1}{2}\int_0^1 x\,dx \int_0^{1-x} y(1-x-y)^2 dy.$$

$$= \frac{1}{2}\int_0^1 x\,dx \int_0^{1-x} y\{(1-x)^2 - 2(1-x)y + y^2\}dy$$

$$= \frac{1}{2}\int_0^1 x\,dx \int_0^{1-x} \{(1-x)^2 y - 2(1-x)y^2 + y^3\}dy$$

$$= \frac{1}{2}\int_0^1 x\left[\frac{1}{2}(1-x)^2y^2 - \frac{2}{3}(1-x)y^3 + \frac{1}{4}y^4\right]_0^{1-x} dx$$

$$= \frac{1}{24}\int_0^1 x(1-x)^4 dx = \frac{1}{24}\left[-\frac{1}{5}x(1-x)^5 + \frac{1}{30}(1-x)^6\right]_0^1 = \frac{1}{720}.$$

2. *Express the triple integral* $\iiint_D f(x, y, z)\,dx\,dy\,dz$ *in the form of repeated integrals in two different ways, where D is the portion of the sphere* $\sum x^2 = 1$ *in the first octant and hence evaluate the integral when* $f(x, y, z) = (1 - x^2 - y^2 - z^2)^{-1/2}$.

The domain D can be described in different ways of which two are

(1) $D = \{(x, y, z) : 0 \le x \le 1,\ 0 \le y \le \sqrt{1 - x^2},\ 0 \le z \le \sqrt{1 - x^2 - y^2}\}$

and (2) $D = \{(x, y, z) : 0 \le y \le 1,\ 0 \le x \le \sqrt{1 - y^2},\ 0 \le z \le \sqrt{1 - x^2 - y^2}\}$.

In (1) E, the projection of D in the xy-plane has been described as

$$E = \{(x, y) \in xy\text{-plane: } 0 \le x \le 1 \text{ and } 0 \le y \le \sqrt{1-x^2}\}.$$

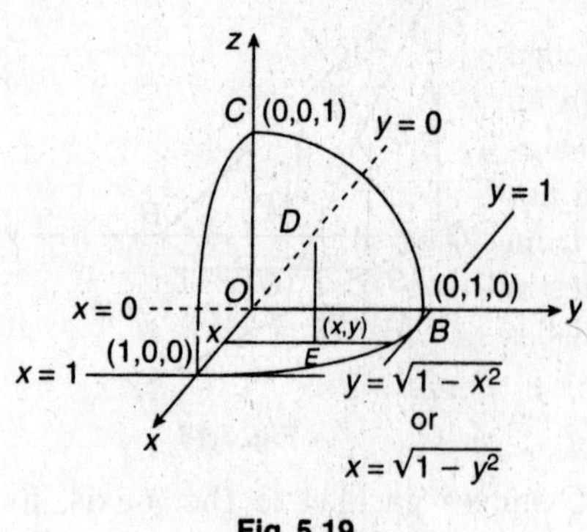

Fig. 5.19

So, E lies between the two straight lines $x = 0$ and $x = 1$ and the portion of the circle $x^2 + y^2 = 1$ lying in the first quadrant in the xy-plane, where both $x \ge 0$ and $y \ge 0$.

In (2) E is the same projection but it has been assumed to lie between the straight lines $y = 0$ and $y = 1$ and the same arc of the circle $x^2 + y^2 = 1$ given by $x = \sqrt{1-y^2}$.

Following the first description of D and taking $y_1 = \sqrt{1-x^2}$ and $z_1 = \sqrt{(1-x^2-y^2)}$ we get

$$I = \iiint_D f\,dxdydz = \int_0^1 \int_0^{y_1} \int_0^{z_1} \left(1 - \sum x^2\right)^{-1/2} dzdydx.$$

We know that $\displaystyle\int_0^a \frac{dz}{\sqrt{a^2-z^2}} = \sin^{-1}\frac{z}{a}\Big|_0^a = \frac{\pi}{2}$, where $a > 0$.

Taking $a = \sqrt{1-x^2-y^2}$, the value of the inner most integral w.r. to z is $\dfrac{\pi}{2}$.

$$\therefore \quad I = \int_0^1 \int_0^{y_1} \frac{\pi}{2} dydx = \frac{\pi}{2} \int_0^1 y|_0^{y_1} dx = \frac{\pi}{2} \int_0^1 \sqrt{(1-x^2)}dx$$

$$= \frac{\pi}{2}\left[\frac{x}{2}\sqrt{1-x^2} + \frac{1}{2}\sin^{-1} x\right]_0^1 = \frac{\pi^2}{8}.$$

Similarly, following the second description of D and taking $x_1 = \sqrt{(1-y^2)}$ and z_1 as before,

$$I = \int_0^1 \int_0^{x_1} \int_0^{z_1} \left(1 - \sum x^2\right)^{-1/2} dzdxdy,$$

giving the same value, viz., $\pi^2/8$.

3. *Find the volume of the region bounded by the sphere:* $\sum x^2 = a^2$ *and the cylinder:* $x^2 + y^2 = ay$, *where* a *is assumed positive.*

The plane $z = 0$ cuts the cylinder in a circle whose centre is $G(0, a/2, 0)$ and the radius is $a/2$. For, the equation

$$x^2 + y^2 = ay \Rightarrow x^2 + (y - a/2)^2 = (a/2)^2.$$

Let D be the region common to the sphere and the cylinder. Then in the figure, ABC is the region. In the Cartesian coordinates the region is

$$D = \left\{(x, y, z) : -\frac{a}{2} \le x \le \frac{a}{2},\ \frac{a}{2} - \sqrt{\frac{a^2}{4} - x^2} \le y \le \frac{a}{2} + \sqrt{\frac{a^2}{4} - x^2},\right.$$

$$\left. -\sqrt{a^2 - x^2 - y^2} \le z \le \sqrt{a^2 - x^2 - y^2}\right\}.$$

Since the limits of integration are complicated, determination of the volume, by using Cartesian coordinates will not be very convenient. Applying the transformation T from Cartesian to cylindrical polar coordinates given by

$$x = \rho\cos\phi,\ y = \rho\sin\phi,\ z = z, \quad \text{where} \quad \rho \ge 0,$$

the equations to the sphere and the cylinder become

$$\rho^2 + z^2 = a^2 \quad \text{and} \quad \rho^2 = a\rho\sin\phi$$

or, $\quad z = \pm\sqrt{a^2-\rho^2}$ and $\rho = a\sin\phi$ respectively.

If D' is the range of T, then

$$D' = \{(\rho,\phi,z) : 0 = \phi = \pi, 0 \le \rho \le a\sin\phi,$$
$$-\sqrt{a^2-\rho^2} \le z \le \sqrt{a^2-\rho^2}\}.$$

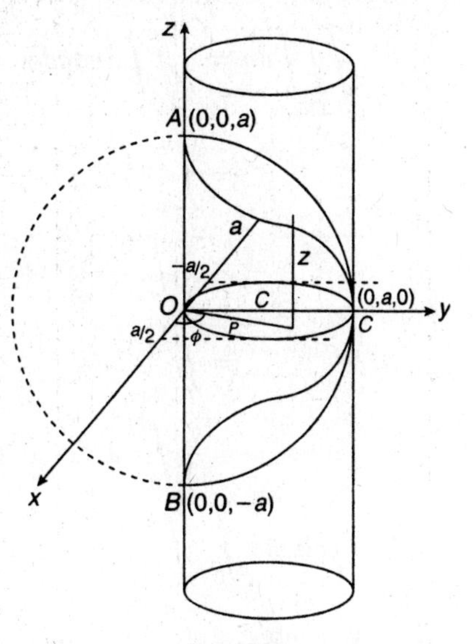

Fig. 5.20

Let $\quad \rho_1 = a\sin\phi, z_1 = -\sqrt{a^2-\rho^2}$ and $z_2 = \sqrt{a^2-\rho^2}$.

Since $J(T) = \begin{vmatrix} x_\rho & x_\phi & x_z \\ y_\rho & y_\phi & y_z \\ z_\rho & z_\phi & z_z \end{vmatrix} = \begin{vmatrix} \cos\phi & -\rho\sin\phi & 0 \\ \sin\phi & \rho\cos\phi & 0 \\ 0 & 0 & 1 \end{vmatrix} = \rho$

and $\quad \rho \ge 0, |J| = \rho$.

So, the required volume is given by

$$V = \iiint_D dxdydz = \iiint_{D'} \rho d\rho d\phi dz$$

$$= \int_0^{\pi} d\phi \int_0^{\rho_1} \rho d\rho \int_{z_1}^{z_2} dz$$

$$= 4\int_0^{\pi/2} d\phi \int_0^{\rho_1} \rho d\rho \int_0^{z_2} dz,$$

because of symmetry in ϕ and z

$$= 4\int_0^{\pi/2} d\phi \int_0^{\rho_1} \rho\sqrt{a^2-\rho^2}d\rho, \quad \because z_2 = \sqrt{a^2-\rho^2}$$

$$= -\frac{4}{3}\int_0^{\pi/2} [(a^2-\rho^2)^{3/2}]_0^{a\sin\phi} d\phi, \quad \because \rho_1 = a\sin\phi$$

$$= -\frac{4}{3}\int_0^{\pi/2} (a^3\cos^3\phi - a^3)d\phi = \frac{4a^3}{3}\int_0^{\pi/2}(1-\cos^3\phi)d\phi$$

$$= \frac{4a^3}{3}\left(\frac{\pi}{2}-\frac{2}{3}\right) = \frac{2a^3}{9}(3\pi-4).$$

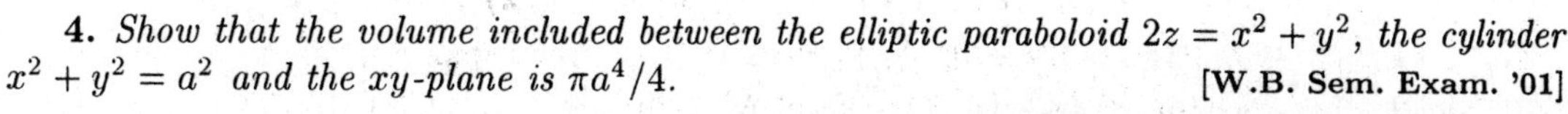

4. *Show that the volume included between the elliptic paraboloid $2z = x^2+y^2$, the cylinder $x^2+y^2=a^2$ and the xy-plane is $\pi a^4/4$.* **[W.B. Sem. Exam. '01]**

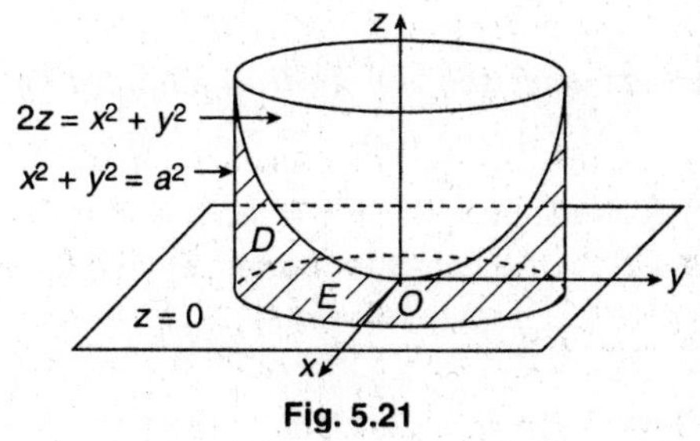

Fig. 5.21

If D is the region enclosed in the given surfaces, then it is quadratic in z and if E is its projection in the xy-plane, its boundary is the circle given by $x^2+y^2=a^2$ and $z=0$. Then the boundary of E is quadratic in y (also in x). The circle lies between the lines given by $x = \pm a$ and $z = 0$.

So, $D = \{(x,y,z), 0 \le z \le z_1, y_1 \le y \le y_2, -a \le x \le a\}$, where $z_1 = (x^2+y^2)/2, y_1 = -\sqrt{a^2-x^2}, y_2 = \sqrt{a^2-x^2}$.

Therefore the required volume is

$$\iiint_D dv = \iiint_D dxdydz = \int_{-a}^{a}\int_{y_1}^{y_2}\int_0^{z_1} dzdydx$$

$$= \int_{-a}^{a}\int_{y_1}^{y_2}\{(x^2+y^2)/2\}dydx, \quad \because\ z_1 = (x^2+y^2)/2$$

$$= \frac{1}{2}\int_{-a}^{a}[x^2y + y^3/3]_{y_1}^{y_2} = \frac{1}{2}\int_{-a}^{a}\{x^2(y_2-y_1) + (y_2^3 - y_1^3)/3\}dx$$

$$= \int_{-a}^{a}[x^2\sqrt{(a^2-x^2)} + (a^2-x^2)^{3/2}/3]dx$$

$$= \frac{2}{3}\int_0^a (2x^2+a^2)\sqrt{a^2-x^2}dx = \frac{2}{3}a^4\int_0^{\pi/2}(2\sin^2\theta + 1)\cos^2\theta d\theta = \frac{\pi a^4}{4}.$$

5. *Evaluate* $\int_0^a\int_0^x\int_0^y x^3y^2zdzdydx.$ **[W.B. Sem. Exam. '01]**

The given integral $= \int_0^a x^3dx\int_0^x y^2dy\int_0^y zdz$

$$= \frac{1}{2}\int_0^a x^3dx\int_0^x y^2[z^2]_0^y dy = \frac{1}{2}\int_0^a x^3dx\int_0^x y^4dy$$

$$= \frac{1}{10}\int_0^a x^3[y^5]_0^x dx = \frac{1}{10}\int_0^a x^8dx = a^9/90.$$

6. *Show that* $\iiint (1+x+y+z)^{-3}dxdydz = \frac{1}{2}\ln 2 - 5/16$, *where the region of integration is the interior of the tetrahedron bounded by* $x=0, y=0, z=0$ *and* $x+y+z=1$.

If I is the given integral

$$I = \int_0^1 dx\int_0^{1-x}dy\int_0^{1-x-y}(1+x+y+z)^{-3}dz.$$

The innermost integral $I_1 = -\frac{1}{2}(1+x+y+z)^{-2}\Big|_0^{1-x-y} = -\frac{1}{8} + \frac{1}{2}(1+x+y)^{-1/2}.$

So I_2, the next inner integral $= \left[-y/8 - \frac{1}{2}(1+x+y)^{-1}\right]_0^{1-x} = \frac{1}{2(1+x)} - \frac{3}{8} + \frac{x}{8}.$

$$\therefore\ I = \left[\frac{1}{2}\ln(1+x) - \frac{3}{8}x + \frac{1}{16}x^2\right]_0^1 = \frac{1}{2}\ln 2 - \frac{5}{16}.$$

7. *Evaluate* $\iiint (1-x-y-z)^{p-1}x^{q-1}y^{r-1}z^{s-1}dxdydz$ *over the tetrahedron bounded by the planes* $x=0,\ y=0,\ z=0$ *and* $x+y+z=1$.

It can be better evaluated by employing a special transformation $x+y+z=u,\ x+y=uv$ and $x=uvw$.

Solving for x, y and z in terms of u, v and w, we get

$$x = uvw,\ y = uv(1-w) \quad\text{and}\quad z = u(1-v).$$

Similarly, solving for u, v and w in terms of x, y and z, we get

$$u = x + y + z,\ v = (x+y)/(x+y+z) \quad\text{and}\quad w = x/(x+y).$$

Here $x \geq 0,\ y \geq 0,\ z \geq 0$ and $x + y + z \leq 1 \Rightarrow 0 \leq u, v, w \leq 1$.

So, if D be the given tetrahedral region, the transformed region D' is given by $0 \leq u \leq 1$, $0 \leq v \leq 1$ and $0 \leq w \leq 1$.

Also, $J = \partial(x,y,z)/\partial(u,v,w) = -u^2 v$.

So, the given integral

$$= \int_0^1 u^{q+r+s-1}(1-u)^{p-1}du \int_0^1 v^{q+r-1}(1-v)^{s-1}dv \int_0^1 w^{q-1}(1-w)^{r-1}dw$$

$$= B(p, q+r+s)B(q+r, s)B(q, r) = \Gamma(p)\Gamma(q)\Gamma(r)/\Gamma(p+q+r+s).$$

Exercises 5.2

1. Evaluate $\iiint (x+y+z+1)^4 dxdydz$ over the region defined by $x \geq 0,\ y \geq 0,\ z \geq 0$ and $x + y + z \leq 1$. **[W.B. Sem. Exam. '01]**

2. Evaluate $\iiint \sqrt{(1 - \sum x^2)/(1 + \sum x^2)}\, dxdydz$, where the region of integration is the positive octant of the sphere $\sum x^2 = 1$, where $\sum x^2 = x^2 + y^2 + z^2$.

3. Prove that the value of $\iiint \left(xyz/\sqrt{\sum x^2}\right) dxdydz$ taken through the positive octant of the ellipsoid $\sum x^2/a^2 = 1$ is

$$a^2b^2c^2(bc + ca + ab)/[15(b+c)(c+a)(a+b)].$$

4. Show that $\iiint z^2 dxdydz = 2a^5(15\pi - 16)/225$, the field of integration being bounded by the sphere $\sum x^2 = a^2$ and the cylinder $x^2 + y^2 = ax (a > 0)$.

5. Show that $\iiint (ax^2 + by^2 + cz^2)dxdydz = 4\pi R^5 \sum a/15$, when taken throughout $\sum x^2 \leq R^2$.

6. Show that the volume included between the elliptical paraboloid $2z = x^2/p^2 + y^2/q^2$, the cylinder $x^2 + y^2 = a^2$ and the xy-plane is $\pi a^4(1/p^2 + 1/q^2)/8$.

7. Show that the volume of the solid bounded by the cylinder $x^2 + y^2 = 2ax$ and the paraboloid $y^2 + z^2 = 4ax$ is $2a^3(3\pi + 8)/3$.

8. Find the volume of that part of the cylinder: $x^2 + y^2 = 2ax$, which is cut off by the surface $z^2 = 2ax$.

9. Prove that the volume enclosed by the two cylinders $x^2 + y^2 = a^2$ and $x^2 + z^2 = a^2$ is $16a^3/3$.

10. Show that the entire volume of the solid $\sum (x/a)^{2/3} = 1$ is $4\pi abc/35$.

Answers

1. $117/70$. **2.** $(\pi/8)[B(3/4, 1/2) - B(5/4, 1/2)]$. **8.** $128a^3/15$.

5.5 Physical Application of Multiple Integrals

The knowledge of multiple integrals can be used in the calculation of mass, centre of mass (CM) and moment of inertia (MI) of a body.

5.5.1 Calculation of Mass

(i) Plane Lamina

Let E be the region in the xy-plane occupied by a plane lamina.

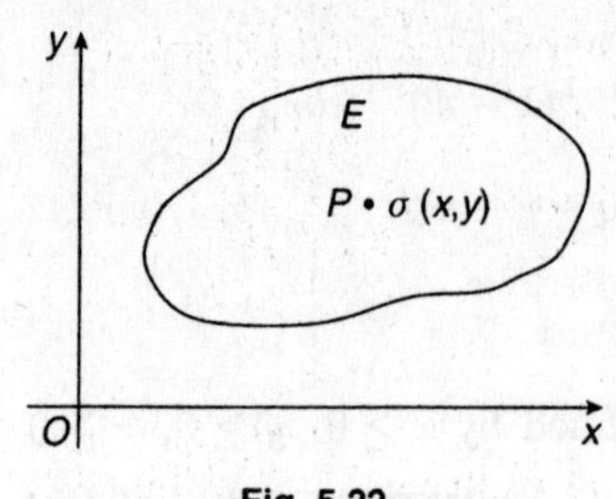

Fig. 5.22

If $\sigma(x, y)$ be the surface density of the lamina at the point $P(x, y)$, then the mass of the lamina is given by

$$M = \iint_E \sigma dxdy,$$

which is nothing but a double integral.

If σ is constant, then

$$M = \sigma \iint_E dxdy = \sigma A,$$

where A is the area of the lamina.

(ii) A Solid Body

If D is the space occupied by a solid body in the xyz-space and $\rho(x, y, z)$ is the density, i.e., the mass per unit volume at $P(x, y, z)$, then its mass is given by

$$M = \iiint_D \rho dxdydz,$$

a triple integral.

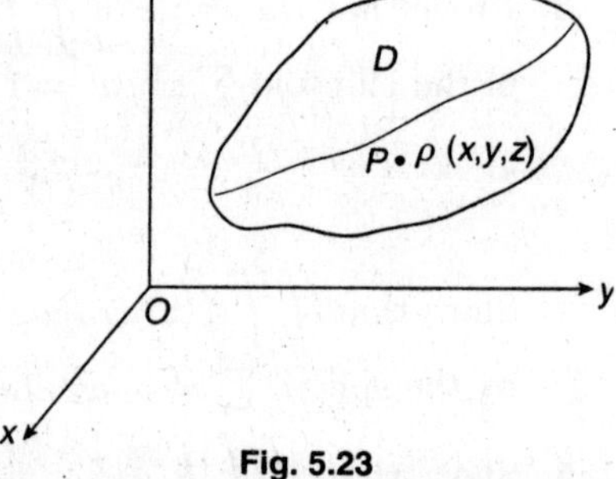

Fig. 5.23

Example 1. *Find the mass of the portion of the solid sphere given by $\sum x^2 \leq 9$ lying in the first octant where $x \geq 0$, $y \geq 0$ and $z \geq 0$, if the density at (x, y, z) is $2xyz$.*

If D be the region occupied by the octant of the sphere $\sum x^2 \leq 9$, then

$$D = \{(x, y, z) : 0 \leq x \leq 3,\ 0 \leq y \leq \sqrt{9 - x^2} \quad \text{and} \quad 0 \leq z \leq \sqrt{9 - x^2 - y^2}\} \quad \text{(Fig. 5.19)}.$$

The required mass:

$$M = \iiint_D 2xyzdxdydz = 2\int_0^3 xdx \int_0^{\sqrt{9-x^2}} ydy \int_0^{\sqrt{9-x^2-y^2}} zdz$$

$$= \int_0^3 xdx \int_0^{\sqrt{9-x^2}} y[z^2]_0^{\sqrt{9-x^2-y^2}} dy = \int_0^3 xdx \int_0^{\sqrt{9-x^2}} y(9 - x^2 - y^2)dy$$

$$= \int_0^3 x\left[\frac{9}{2}y^2 - \frac{x^2y^2}{2} - \frac{y^4}{4}\right]_0^{\sqrt{9-x^2}} dx = \int_0^3 x\left[\frac{9 - x^2}{2}(9 - x^2) - \frac{1}{4}(9 - x^2)^2\right] dx$$

$$= \frac{1}{4}\int_0^3 x(81 - 18x^2 + x^4)dx = \frac{1}{4}\left[\frac{81}{2}x^2 - \frac{9}{2}x^4 + \frac{x^6}{6}\right]_0^3$$
$$= \frac{1}{24}[x^2(243 - 27x^2 + x^4)]_0^3 = \frac{9}{24}[243 - 27 \times 9 + 81]$$
$$= \frac{3}{8} \times 81 = \frac{243}{8} = 30.375.$$

5.5.2 Centre of Mass (C.M.) or Centre of Gravity (C.G.) or Centroid

Referred to a rectangular Cartesian xy-frame if $G(\overline{x}, \overline{y})$ is the centroid of a distribution of particles of mass m_i situated at $P_i(x_i, y_i)$

$$\overline{x} = \sum m_i x_i \Big/ \sum m_i \quad \text{or,} \quad \left(\sum m_i x_i\right) \Big/ M$$

and $\overline{y} = \sum m_i y_i \Big/ \sum m_i$ or, $\left(\sum m_i y_i\right) \Big/ M$,

where $M = \sum m_i$, the total mass of all the particles.

For a distribution of particles in the 3-D Cartesian xyz-space the additional z-coordinate $\overline{z}$ of the centroid is given by

$$\overline{z} = \sum m_i z_i \Big/ \sum m_i = \left(\sum m_i z_i\right) \Big/ M.$$

For a continuous distribution of mass, the coordinates of the centroid are obtained by replacing $\sum$ by $\int$, x_i by x, y_i by y, z_i by z and m_i by dm so that

$$\overline{x} = \int x dm \Big/ \int dm \quad \text{or,} \quad I_1 \Big/ M,$$

$$\overline{y} = \int y dm \Big/ \int dm \quad \text{or,} \quad I_2 \Big/ M$$

and $\overline{z} = \int z dm \Big/ \int dm$ or, $I_3 \Big/ M$,

where the integrals are taken so as to cover the entire domain.

A continuous distribution of mass may be over

(i) a line Γ, **(ii)** a plane region or a surface $\sum$ or **(iii)** a portion D of the 3-D space.

If λ, σ and ρ be the constant mass per **(i)** unit length (linear density), **(ii)** unit area (surface density) and **(iii)** unit volume (space density), then dm is

(i) λdS, **(ii)** σdS and **(iii)** ρdV.

If Γ is a plane curve lying in the xy-plane, then

$$\overline{x} = \int_\Gamma x\lambda ds \Big/ \int_\Gamma \lambda ds = \int_\Gamma x ds \Big/ \int_\Gamma ds = \int_\Gamma x ds \Big/ l,$$

and a similar result for $\overline{y}$, where $ds = \sqrt{(1 + y'^2)}dx$ or, $\sqrt{(1 + x'^2)}dy$ and l is length of the entire arc Γ.

If E is the portion of a surface in the xy-plane, then

$$\overline{x} = \int_E x dx dy \Big/ S, \overline{y} = \int_E y dx dy \Big/ S, \quad \text{where } S = \int_E dx dy.$$

Similarly, if D is the region occupied by a solid body of volume V, then

$$\overline{x} = \int_D xdxdydz \Big/ V,\ \overline{y} = \int_D ydxdydz \Big/ V \ \text{ and } \overline{z} = \int_D zdxdydz \Big/ V,$$

where $V = \int_D dxdydz$.

For the continuous distribution of mass over any surface $\sum$ in the xyz-space

$$\overline{x} = \int_E xdS \Big/ S,\ \overline{y} = \int_E ydS \Big/ S \ \text{ and } \overline{z} = \int_E zdS \Big/ S,$$

where S is the entire surface area of $\sum$.

Examples 5.4. 1. *Find the centroid of the uniform arc of a circle of radius r subtending an angle of $2\alpha^c$ at its centre.*

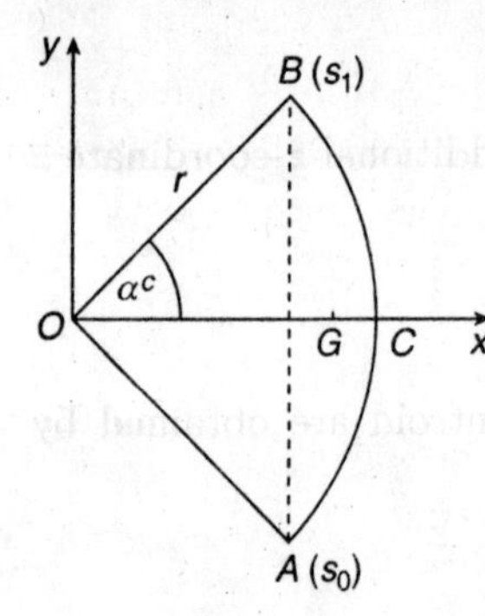

Fig. 5.24

Let the circle, of which the given arc is a part, be given by $x = r\cos\theta,\ y = r\sin\theta$, where $-\pi < -\alpha \le \theta \le \alpha < \pi$.

If $G(\overline{x}, \overline{y})$ is the centroid, by symmetry of the arc in the x-axis, $\overline{y} = 0$.

$$\overline{x} = I/l, \quad \text{where} \quad I = \int_{s_0}^{s} xds = \int_{-\alpha}^{\alpha} r\cos\theta r d\theta$$

$$= r^2[-\sin\theta]_{-\alpha}^{\alpha} = 2r^2 \sin\alpha \quad \text{and} \quad l = 2r\alpha.$$

$$\therefore \quad \overline{x} = r(\sin\alpha)/\alpha.$$

Thus the CG of the arc of a circle of radius r is at a distance $r(\sin\alpha)/\alpha$ units from the centre, on the line of symmetry of the arc, where $2\alpha^c$ is the angle subtended, at the centre, by the arc.

2. *Determine the centroid of the segment of the parabola $y^2 = 4ax$ cut off by its latus rectum.*

The given segment is the region E bounded by the parabola LAL' and its latus rectum LL'. Since E is symmetric in the x-axis $\overline{y} = 0$ and $\overline{x}$ is the same as that of the upper half ASL of the segment. If Δ is the area of ASL, then $\overline{x} = I/\Delta$, where

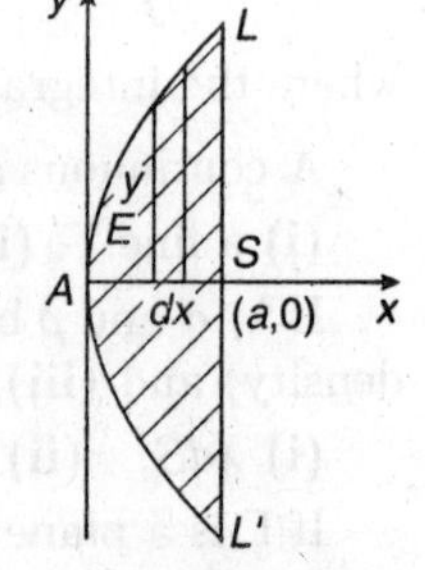

Fig. 5.25

$$I = \int_0^a x \cdot ydx = \int_0^a x \cdot 2\sqrt{ax}dx$$

$$= 2\sqrt{a}\int_0^a x^{3/2}dx = 4a^3/5$$

$$\text{and} \quad \Delta = \int_0^a ydx = 2\sqrt{a}\int_0^a x^{1/2}dx = 4a^2/3.$$

$$\therefore \quad \overline{x} = 3a/5.$$

3. *Show that the CG of a plane uniform lamina in the form of a circular sector with the central angle $2\alpha^c$ and radius R is at a distance $(2/3)R(\sin\alpha)/\alpha$, on the axis of symmetry from the vertex of the sector.*

Referred to the xy-frame let the equation to the sector OAB be $x^2 + y^2 = R^2$, where

$$r\cos\alpha \le x \le r \quad \text{and} \quad -r\sin\alpha \le y \le r\sin\alpha.$$

We divide the sector into elementary circular strips. Let PQ be such a strip of radius x and width dx. If σ is the constant surface density of the lamina, $\lambda(=\sigma dx)$ is the mass per unit length of the strip. By Example 1, the CG of this strip is at a distance of $x(\sin\alpha)/\alpha$ from O on the x-axis. So, the abscissa $\overline{x}$ of the CG of the entire lamina is

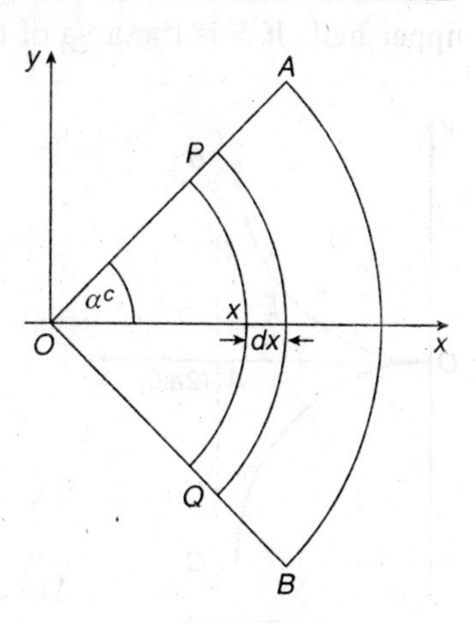

Fig. 5.26

$$\overline{x} = (1/M)\int_0^R x\frac{\sin\alpha}{\alpha}\sigma x 2\alpha dx = \frac{2}{M}\sigma\sin\alpha\frac{x^3}{3}\Bigg|_0^R$$

$$= (2\sigma\sin\alpha R^3)/3M,$$

where $M = (1/2)R^2\cdot 2\alpha\cdot\sigma = R^2\alpha\sigma$, the mass of the sector.

$$\therefore \qquad \overline{x} = \frac{2}{3}R\frac{\sin\alpha}{\alpha}.$$

From symmetry $\overline{y} = 0$.

4. *Find the centroid of the lamina in the form of the part of the astroid, given by $x^{2/3}+y^{2/3}=a^{2/3}$, lying in the first quadrant.* **[WBUT Sem. 2002]**

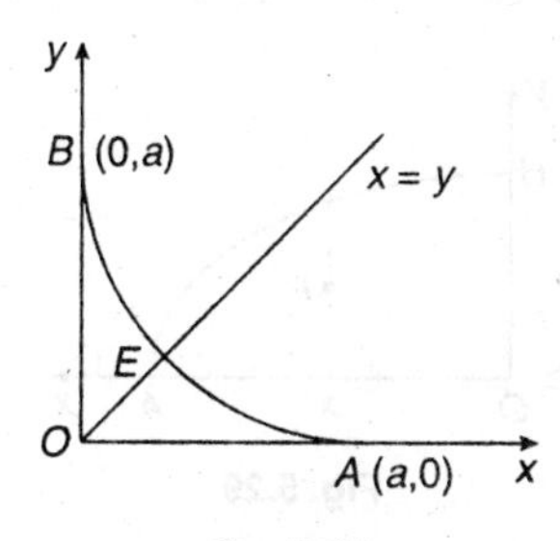

Fig. 5.27

If E is the region occupied by the lamina, then

$$E = \{(x,y)\in\mathbb{R}^2 : 0\le x\le a \text{ and } 0\le y\le (a^{2/3}-x^{2/3})^{3/2}\}.$$

Obviously, the region is symmetrical in the line $x = y$. So if $G(\overline{x},\overline{y})$ is the required CG, $\overline{x}=\overline{y}$.

The area S of the region is given by

$$S = \iint_E dxdy = \int_0^a dx\int_0^b dy, \quad \text{where} \quad b = (a^{2/3}-x^{2/3})^{3/2}$$

$$= \int_0^a (a^{2/3}-x^{2/3})^{3/2}dx.$$

Putting $x = a\cos^3\theta$, where $0\le\theta\le\pi/2$,

$dx = -3a\cos^2\theta\sin\theta d\theta$ and when $x = 0, \theta = \pi/2$ and when $x = a, \theta = 0$.

$$\therefore \qquad S = 3a\int_0^{\pi/2} a\sin^3\theta\cos^2\theta\sin\theta d\theta = 3a^2\int_0^{\pi/2}\sin^4\theta\cos^2\theta d\theta = 3a^2\cdot\frac{3.1.1}{6.4.2}\frac{\pi}{2} = \frac{3}{32}a^2\pi.$$

$$\text{Also,} \qquad I = \iint_E xdxdy = 3a^3\int_0^{\pi/2}\sin^5\theta\cos^4\theta d\theta = 8a^3/105.$$

$\therefore$ the centroid is $(256a/315\pi, 256/a/315\pi)$.

5. *Find the centroid of the area between the curve $y^2(2a-x) = x^3$ and its asymptote, where $a > 0$.*

Since y appears in even powers the curve is symmetrical in the x-axis (Fig. 5.28). If $x > 2a$, $x^3/(2a-x) < 0$. So, the corresponding value of y is imaginary. Similarly, when $x < 0$, y is again imaginary. Hence the entire curve lies between $x = 0$ and $x = 2a$. Also, $y^2 = x^3/(2a-x) \Rightarrow$ as $x\to 2a-0$, $y\to\infty$. So, $x = 2a$ is an asymptote of the curve. If $G(\overline{x},\overline{y})$ is the centroid of the given area $\overline{y} = 0$ and $\overline{x}$ will be the same as the centroid of the upper half. If S is the area of the upper half, then

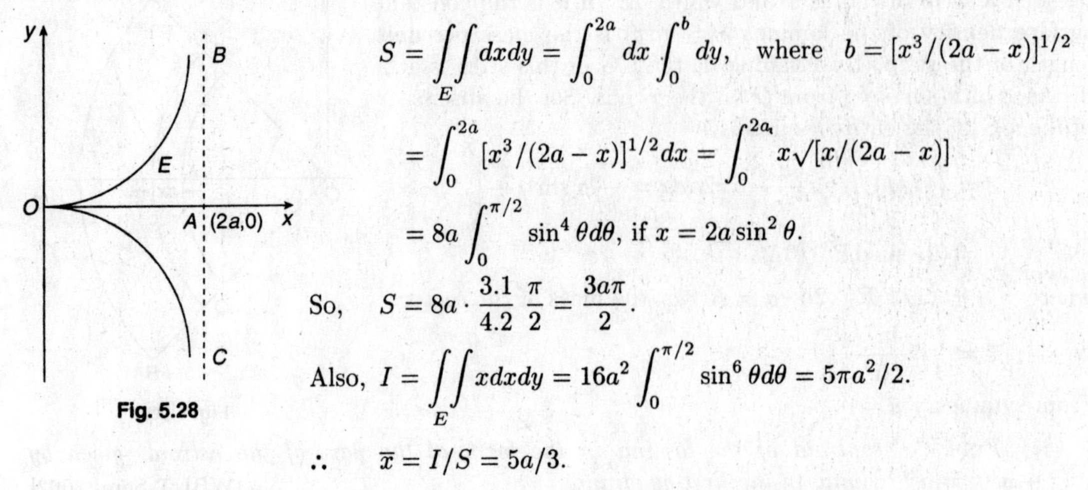

$$S = \iint_E dxdy = \int_0^{2a} dx \int_0^b dy, \quad \text{where} \quad b = [x^3/(2a-x)]^{1/2}$$

$$= \int_0^{2a} [x^3/(2a-x)]^{1/2} dx = \int_0^{2a} x\sqrt{[x/(2a-x)]}$$

$$= 8a \int_0^{\pi/2} \sin^4\theta d\theta, \text{ if } x = 2a\sin^2\theta.$$

So, $$S = 8a \cdot \frac{3.1}{4.2}\frac{\pi}{2} = \frac{3a\pi}{2}.$$

Also, $$I = \iint_E xdxdy = 16a^2 \int_0^{\pi/2} \sin^6\theta d\theta = 5\pi a^2/2.$$

Fig. 5.28

$$\therefore \quad \overline{x} = I/S = 5a/3.$$

6. *A plane lamina in the form of a quadrant of the ellipse $x^2/a^2 + y^2/b^2 = 1$ is lying in the first quadrant. The thickness of the lamina at a point varies as the product of the distances of the point from the axes. Show that the centroid of the lamina is $(8a/15, 8b/15)$.*

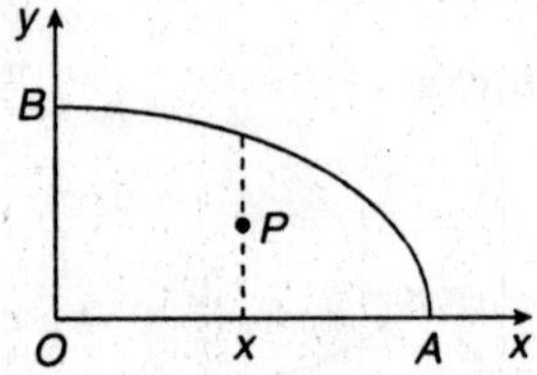

If σ is the mass per unit area of the lamina at $P(x,y), \sigma = kxy$, k being a constant (Fig. 5.29). So, the entire mass M of the lamina is given by

Fig. 5.29

$$M = \int_0^a dx \int_0^{y_1} kxydy, \; y_1(b/a)\sqrt{a^2-x^2}$$

$$= (kb^2/2a^2)\int_0^a x(a^2-x^2)dx$$

$$= (kb^2/2a^2)[a^2x^2/2 - x^4/4]_0^a = ka^2b^2/3.$$

$$I_1 = k\int_0^a dx \int_0^{y_1} x^2ydy = (kb^2/2a^2)\int_0^a x^2(a^2-x^2)dx$$

$$= (kb^2/2a^2)[a^2x^3/3 - x^5/5]_0^a = ka^3b^2/15.$$

$$I_2 = k\int_0^b y^2dy \int_0^{x_1} xdx, \; x_1 = (a/b)\sqrt{(b^2-y^2)}$$

$$= ka^22b^2 \int_0^b y^2(b^2-y^2)dy = ka^2b^3/15.$$

$$\therefore \quad \overline{x} = I_x/M = 8a/15 \quad \text{and} \quad \overline{y} = I_y/M = 8b/15.$$

7. *Find the centroid of the solid tetrahedron bounded by the coordinate planes and the plane $\sum x = 1$, the density at a point varying as its distance from the xy-plane.*

If ρ is the density at a point $\rho(x,y,z)$, then $\rho \propto z$ or, $\rho = kz$, k being a constant. Let $G(\overline{x}, \overline{y}, \overline{z})$ be the centroid of the tetrahedron $OABC$ (Fig. 5.30). Then $\overline{x} = I_1/M$, $\overline{y} = I_2/M$ and $\overline{z} = I_3/M$, where M is the mass and $I_1 = \iiint_R \rho x dxdydz$, $I_2 = \iiint_R \rho z dxdydz$ and

$I_3 = \iiint\limits_R \rho z\,dx\,dy\,dz$, R being the region bounded by the faces of the tetrahedron. Then

$$R = \{(x, y, z) \in \mathbb{R}^3 : 0 \le x \le 1,\ 0 \le y \le 1 - x \ \text{ and } \ 0 \le z \le 1 - x - y\}.$$

$$\therefore \quad M = \int_0^1 dx \int_0^{1-x} dy \int_0^{1-x-1} kz\,dx\,dy\,dz$$

$$= k\int_0^1 dx \int_0^{1-x} dy \int_0^{1-x-y} dz$$

$$= (k/2)\int_0^1 \int_0^{1-x} (1 - x - y)^2\,dy\,dx$$

$$= -(k/6)\int_0^1 (1 - x - y)^3\Big|_{y=0}^{1-x} dx$$

$$= (k/6)\int_0^1 (1 - x)^3 dx = -(k/24)(1 - x)^4\Big|_0^1 = k/24.$$

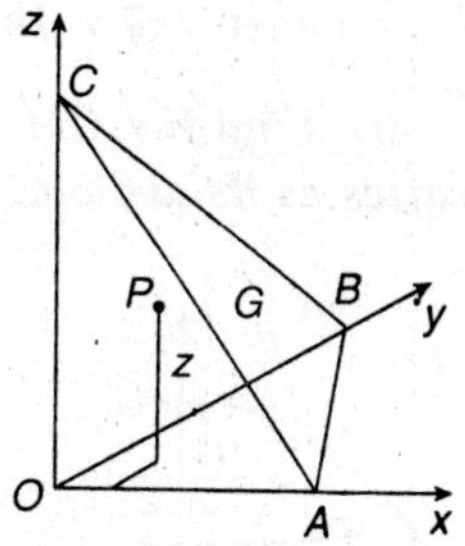

Fig. 5.30

$$I_1 = \int_0^1 dx \int_0^{1-x} dy \int_0^{1-x-y} kzx\,dx\,dy\,dz = (k/2)\int_0^1 x\,dx \int_0^{1-x} (1 - x - y)^2 dy$$

$$= -(k/6)\int_0^1 x(1 - x - y)^3\Big|_0^{1-x} dx = (k/6)\int_0^1 x(1 - x)^3 dx = (k/6)B(2, 4)$$

$$= (k/6)\Gamma(2)\Gamma(4)/\Gamma(6) = (k/6)\cdot 1!3!/5! = k/120.$$

$\therefore \quad \overline{x} = 1/5.$

8. *Find the CM of the portion of the solid ellipsoid $x^2/a^2 + y^2/b^2 + z^2/c^2 \le 1$ lying in the positive octant if the density at (x, y, z) varies as xyz.* **[ISM 2001]**

Let D be the positive octant of the interior of the ellipsoid. Then

$$D = \{(x, y, z) \in \mathbb{R}^3 : 0 \le x \le a, 0 \le y \le (b/a)\sqrt{(a^2 - x^2)}, 0 \le z \le c\sqrt{(1 - x^2/a^2 - y^2/b^2)}\}.$$

Let $\rho(= kxyz)$ be the density at $(x, y, z) \in D$.

Then $\overline{x} = I_1/M$, where M is the mass of the solid and

$$I_1 = \iiint\limits_D \rho x\,dx\,dy\,dz = k\iiint x^2yz\,dx\,dy\,dz = k\int_0^a x^2 dx \int_0^{y_1} y\,dy \int_0^{z_1} z\,dz,$$

where $y = 6\sqrt{(1 - x^2/a^2)}$, $z_1 = c\sqrt{(1 - x^2/a^2 - y^2/b^2)}$.

$$\therefore \quad I_1 = (kc^2/2)\int_0^a x^2 dx \int_0^{y_1} y(1 - x^2/a^2 - y^2/b^2)dy$$

$$= kc^2/2\int_0^a x^2[(1 - x^2/a^2)y^2/2 - y^4/4b^2]_0^{y_1} dx$$

$$= kc^2/2\int_0^a x^2[(1/2)(1 - x^2/a^2)b^2(1 - x^2/a^2) - (b^2/4)(1 - x^2/a^2)^2]dx$$

$$= (kb^2c^2/8)\int_0^a x^2(1 - x^2/a^2)dx = (ka^3b^2c^2/8)\int_0^1 t^2(1 - t^2)dt,\ x = at$$

$$= (ka^3b^2c^2/8)[t^3/3 - 2t^5/5 + t^7/7] = ka^3b^2c^2/105.$$

$$M = k \iiint_D xyzdxdydz = k \int_0^a xdx \int_0^{y_1} ydy \int_0^{z_1} zdz = ka^2b^2c^2/48.$$

$\therefore \quad \overline{x} = 48a/105 = 16a/35.$

By symmetry, $\overline{y} = 16b/35$ and $\overline{z} = 16c/35$.

9. *Find the CM of a hemispherical shell of radius R when the surface density at a point varies as its distance from the plane of its rim.*

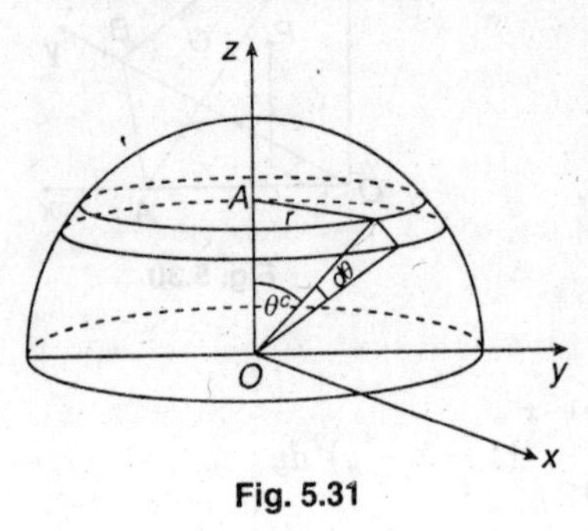

Fig. 5.31

Let us take a narrow circular strip of radius r, symmetrical in the axis z of symmetry of the hemisphere. If θ^c is the angle subtended by r at 0, the centre of the sphere, then $r = R\sin\theta$. If the height of the strip from the base is given by $z = R\cos\theta$, the width of the strip is $Rd\theta$. So, the area of the strip is

$$dA = 2\pi R\sin\theta Rd\theta = 2\pi R\sin\theta d\theta.$$

If σ is the surface density, $\sigma = kz = kR\cos\theta$.

$$\therefore \quad dm = \sigma dA = 2k\pi R^3 \sin\theta\cos\theta d\theta$$

and $\quad M = \int dm = \int_0^{\pi/2} 2k\pi R^3 \sin\theta\cos\theta d\theta = k\pi R^3.$

Also, $\quad I_3 = \int zdm = 2k\pi R^4 \int_0^{\pi/2} \sin\theta\cos^2\theta d\theta = (2/3)k\pi R^4.$

$\therefore \quad \overline{z} = I_z/M = 2R/3.$

Because of symmetry, $\overline{x} = \overline{y} = 0$.

10. *Show that the CM of a thin homogeneous right circular conical shell of altitude h is on the axis at a distance of h/3 units from the base.*

Let the centre of the base be chosen as the origin and the z-axis as the axis of the cone, with its semivertical angle α^c. Let PQR be a thin circular strip of width dl of the cone at a height z from the base. The radius of the strip is $(h-z)\tan\alpha$ and $dl = dz\sec\alpha$. So, the area of the strip is given by $dA = 2\pi(h-z)\sec\alpha\tan\alpha dz$.

$\therefore \quad \overline{z} = 2\pi \int_0^h (h-z)z\sec\alpha\tan\alpha dz/S$, where

$S = \pi rl$, the surface area of the cone.

$$\therefore \quad \overline{z} = (2/rl)[hz^2/2 - z^3/3]_0^h \sec\alpha\tan\alpha = \frac{\sec\alpha\tan\alpha}{3rl}h^3 = \frac{h}{3}.$$

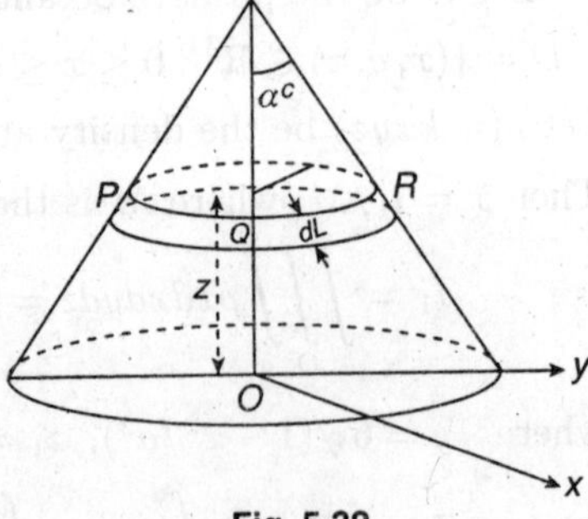

Fig. 5.32

Exercises 5.3

1. Find the centroid of the quadrant of a uniform circular wire of radius a.

2. Find the centroid of the region enclosed in the parabola $y^2 = 4ax$, the x-axis and the latus rectum in the first quadrant, a being positive.

3. Find the CG of the area bounded by the parabola $y^2 = 4ax$, $(a > 0)$, and the double ordinate $x = h$. **[W.B. Sem. Exam. '01, '03]**

4. Find the centroid of the area bounded by the coordinate axes and the curve $\sqrt{x}+\sqrt{y}=\sqrt{a}$.

5. Find the centroid of the region included between the parabolas $y^2 = 4ax$ and $x^2 = 4ay$, $(a > 0)$.

6. Find the centroid of a semicircular lamina of radius a, if its surface density varies as the square of the distance from the diameter. **[WBUT Sem. 2002]**

7. A plane lamina is bounded by the coordinate axes and the part of the curve $(x/a)^{2/3} + (y/b)^{2/3} = 1$ in the first quadrant. Find its CG, if the surface density at a point (x, y) varies as the product of its distances form the coordinate axes.

8. Find the CG of the quadrant of a uniform elliptic lamina, lying in the positive quadrant, if the equation to the ellipse is $x^2/a^2 + y^2/b^2 = 1$, $(a > b > 0)$.

9. Find the CG of the segment of a solid sphere of radius a cut of by a plane at a distance $c(< a)$ from the centre.

10. Find the centroid of the solid right circular cone of height h, whose density at a point varies as its distance from the base.

11. Find the x-coordinate of the centroid of the homogeneous solid bounded by the coordinate planes, the parabolic cylinder $z = 4 - x^2$ and the plane $y = 6$.

12. Find the centroid of a homogeneous solid hemisphere of radius R.

13. Show that the centroid of a thin homogeneous hemispherical shell of radius R is at a distance of $R/2$ units, from the base, on the axis of its symmetry.

14. Show that the centroid of a homogeneous solid right circular cone of height h is at a distance of $h/4$ units from the base.

Answers

1. $\overline{x} = \overline{y} = 2\sqrt{2}a/\pi$. **2.** $(3a/5, 3a/4)$. **3.** $(3h/5, 0)$. **4.** $(a/5, a/5)$. **5.** $(9a/5, 9a/5)$.

6. At a distance of $32a/(15\pi)$ from the diameter on the line of symmetry.

7. $(128a/429, 128b/429)$. **8.** $(4a/3\pi, 4b/3\pi)$. **9.** $(3(a+c)^2/(8a+4c), 0)$.

10. At a distance of $3h/5$ units from the vertex on the axis of the cone. **11.** $3/4$.

12. At a distance of $3R/8$ units from the base on the axis of symmetry.

5.5.3 Pappus Theorems on Volume and Surface Area of Revolution

1. *The volume of the solid obtained by revolving a plane region E about an axis, not intersecting E, is the product of the area of E and the length of the path traced by the centroid of E (Fig. 5.33).*

2. *The area of the surface obtained by revolving a plane arc Γ of a curve about an axis lying in the same plane but not intersecting Γ, is the product of the length of Γ and the length of the path traced by the centroid of Γ.*

For the proofs of the theorems see § 5.56, Appendix B.

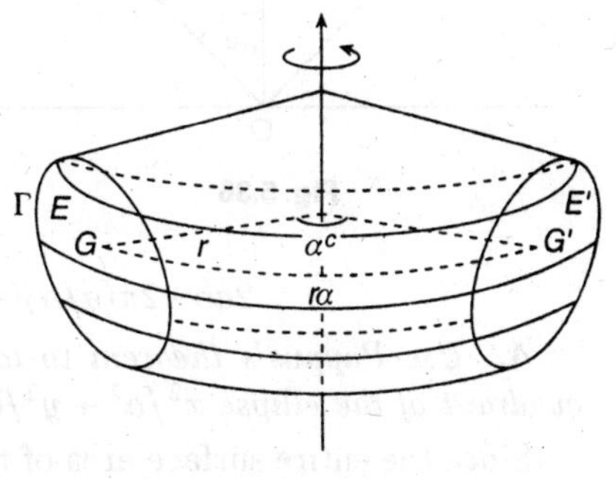

Fig. 5.33

Examples 5.5. 1. *The circular disc given by $x^2 + y^2 - 2ax - 2ay + a^2 \leq 0 (a > 0)$, revolves about the y-axis through an angle $\pi^c/2$. Find the volume of revolution, find also the area of the surface of revolution of the circumference of the disc.*

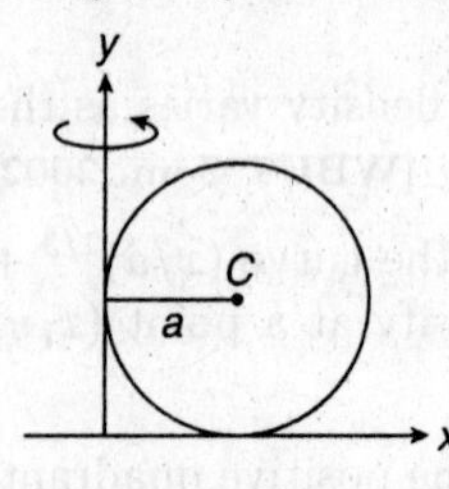

Fig. 5.34

Since the given inequation can be written as

$$(x - a)^2 + (y - a)^2 \leq a^2,$$

the centre is $C(a, a)$ and its radius is a. So, the centre of the disc, which is also the centroid of the disc is at a distance a from the y-axis, the axis of revolution.

Since the disc revolves through $\pi^c/2$, the length of the path of C is $\pi a/2$ units of length.

The area of the disc is πa^2. So by the Pappus theorem the required volume of revolution is $\pi^2 a^3/2$ units of volume.

Since the length of the circumference is $2\pi a$ and the centroid of the circumference is C, which is at a distance a from the y-axis the area of the surface of revolution of the circle: $x^2 + y^2 - 2ax - 2ay + a^2 = 0$ is $(2\pi a)(\pi a/2)$ or $\pi^2 a^2$.

2. *Use the theorem of Pappus to locate the CG of a thin uniform semicircular plate of radius a.*

Let AB be the diameter of the semicircular plate ABC. If it is rotated about AB, we get a complete sphere, a solid of revolution of volume $(4/3)\pi a^3$. The area of the plate is $\pi a^2/2$. Let $G(o, h)$ be the centroid of the plate. Then by the Pappus theorem

$$(4/3)\pi a^3 = (2\pi h)\pi a^2/2,$$

whence $h = 4a/3\pi$.

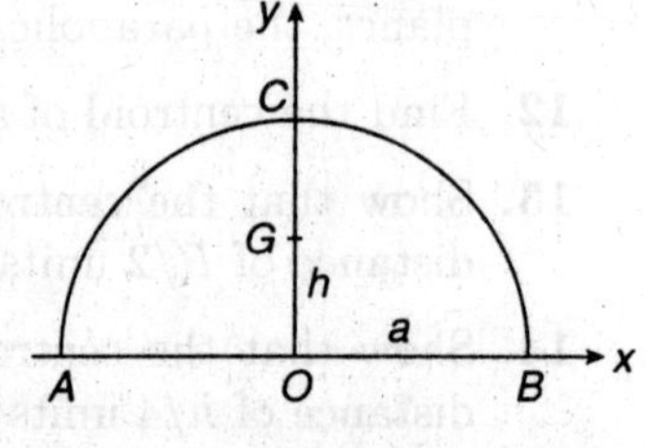

Fig. 5.35

3. *An arc of a circle of radius a revolves about its chord. If the length of the arc is $2a\alpha (0 < \alpha < \pi/2)$, show that the area of the surface generated for complete revolution is $4\pi a^2(\sin\alpha - \alpha\cos\alpha)$.*

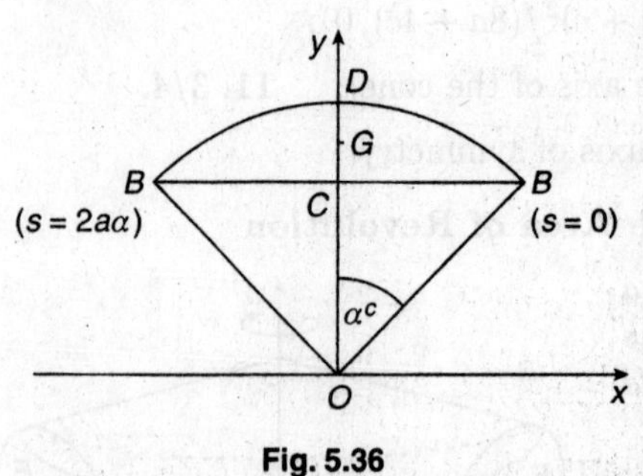

Fig. 5.36

Let ADB be the arc of a circle of radius a and centre O. Let C be the mid-point of the chord AB, the y-axis along OC and G the centroid of the arc. Then $m(\angle AOB) = 2\alpha^c$ (Fig. 5.36). We know that $OG = a(\sin\alpha)/\alpha$, Ex. 1, § 5.4.

Then $CG = a(\sin\alpha)/\alpha - CO = a(\sin\alpha)/\alpha - a\cos\alpha$

$= (a/\alpha)(\sin\alpha - \alpha\cos\alpha)$.

When the arc revolves once about AB, the length of the path that G describes is $2\pi CG$ or $2\pi(a/\alpha)(\sin\alpha - \alpha\cos\alpha)$. So, by the Pappus theorem the required surface area is

$$2a\alpha \cdot 2\pi(a/\alpha)(\sin\alpha - \alpha\cos\alpha) = 4\pi a^2(\sin\alpha - \alpha\cos\alpha).$$

4. *Use Pappus's theorem to locate the CG of a thin uniform lamina in the form of the quadrant of the ellipse $x^2/a^2 + y^2/b^2 = 1$ in the first quadrant.*

Since the entire surface area of the ellipse is πab, the surface area S of the quadrant is given by $S = \pi ab/4$ (Fig. 5.37).

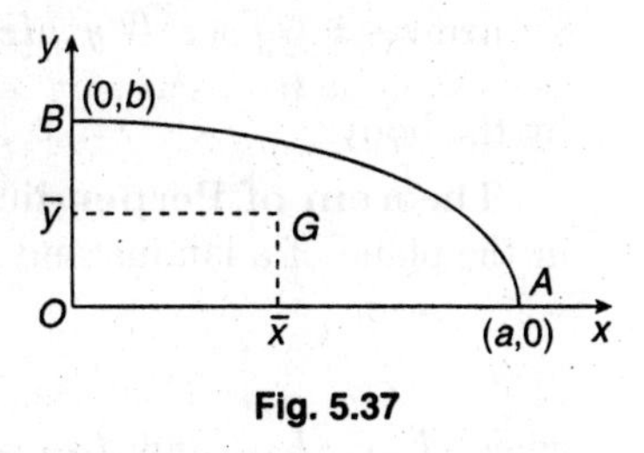

Fig. 5.37

Let $G(\overline{x}, \overline{y})$ be the centroid of the lamina. If the entire ellipse is revolved about the minor axis, the volume of the oblate spheroid is $4\pi a^2 b/3$. So, the volume V obtained by revolving the quadrant about the same minor axis is given by $V = (1/2).4\pi a^2 b/3 = 2\pi a^2 b/3$.

By Pappus's theorem, $V = 2\pi \overline{x} S$.

So, $\quad \overline{x} = V/(2\pi S) = 4a/(3\pi)$

Similarly, $\overline{y} = 4b/(3\pi)$.

5.5.4 Moment of Inertia

The **Moment of Inertia (MI)** of a particle of mass M situated at a distance r from a point O (or from α line Λ) is defined to be mr^2, the product of its mass and the square of its distance from the point (or from the line).

If there is a discrete system of particles, then the MI of the system is the sum of the moments of inertia of the particles. For a continuous system the MI of the body is given by

$$I = \int_D r^2 dm,$$

where D is the domain or the space occupied by the body.

If M is the mass of the body and $I = Mk^2$, then $k(\geq 0)$ is called the **radius of gyration**.

(a) MI of a plane lamina

Let the lamina occupy the space E in the xy-plane (Fig. 5.38) and $P(x, y)$ be any arbitrary point in E. Let δA be the area of an elementary region enclosing P. If $\sigma(x, y)$ be the surface density of the lamina at P, $\delta m = \sigma \delta A$.

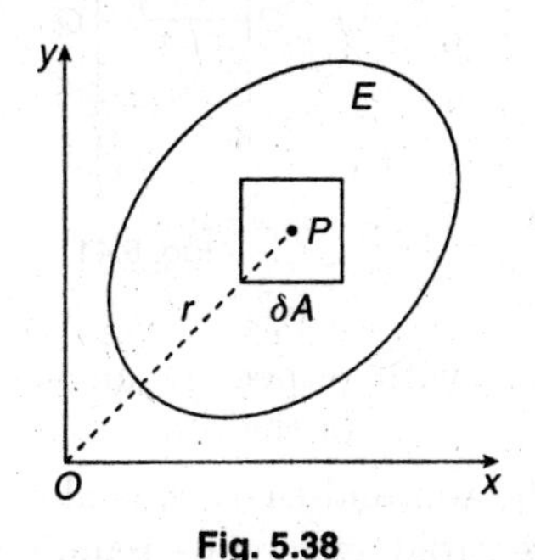

Fig. 5.38

Then the moment of inertia of the lamina about

(i) O is $I_o = \iint_E \sigma r^2 dA$ or, $\iint_E \sigma(x^2 + y^2) dxdy$,

(ii) the x-axis is $I_x = \iint_E \sigma y^2 dxdy$

and (iii) about the y-axis is $I_y = \iint_E \sigma x^2 dxdy$.

So, $I_o = I_x + I_y$.

(b) MI of a solid body about

(i) O, the origin is $I_o = \iiint_D \rho r^2 dxdydz$,

(ii) the x-, y- and z-axes are respectively

$$I_x = \iiint_D \rho(y^2 + z^2) dxdydz, \; I_y \iiint_D \rho(z^2 + x^2) dxdydz$$

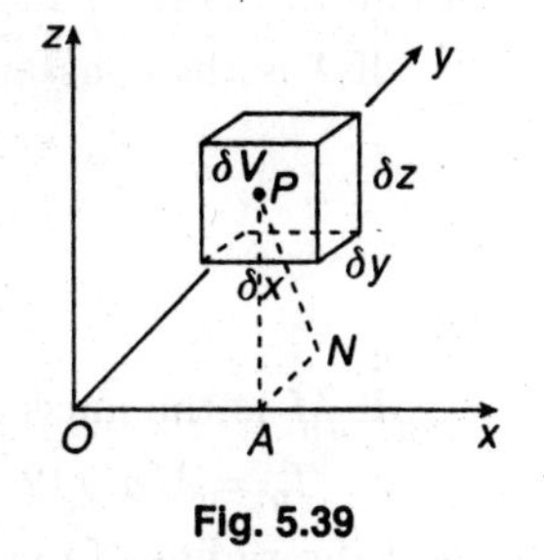

Fig. 5.39

and $I_z = \iiint\limits_D \rho(x^2+y^2)dxdydz$, ρ being the density at $P(x,y,z)$ and D, the space occupied by the body.

Theorem of Perpendicular Axis: If OA and OB are any two perpendicular lines lying in the plane of a lamina and OC is perpendicular, through O, is the plane of the lamina, then

$$I_{OC} = I_{OA} + I_{OB}, \quad \text{(Fig. 5.40)}$$

where I_{OA}, I_{OB} and I_{OC} are the MI's of the lamina about the axes OA, OB and OC respectively.

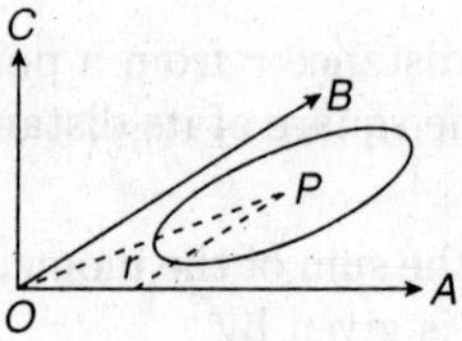

Fig. 5.40

If OA, OB and OC are the x-, y- z-axes respectively, we have already seen that

$$I_x = I_{OA} = \iint\limits_E \sigma y^2 dxdy, \; I_y = I_{OB} = \iint\limits_E \sigma x^2 dxdy.$$

$$\therefore \quad I_x + I_y = \iint\limits_E \sigma(x^2+y^2)dxdy = \iint\limits_E \sigma r^2 dxdy,$$

where $r = \sqrt{x^2+y^2}$, the distance of (x,y) form the z-axis.

Steiner's Theorem [The Theorem of Parallel Axes]: The MI I_o of a body about any axis Λ is equal to the sum of the MI I' of the body about a parallel axis Λ' through its CM and the product of the total mass and the square of the distance d between the parallel axes (Fig. 5.41):

$$I_o = I' + Md^2,$$

where I' = the MI about a parallel line Λ' through the centroid G of the body,

I_o = the MI about a given line Λ,

M = the mass of the body

and d = the distance between the parallel lines Λ and Λ'.

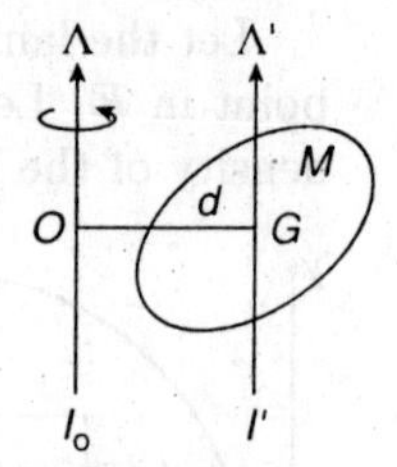

Fig. 5.41

5.5.5 A Few Simple Cases of M.I.

1. The moment of inertia of a thin homogeneous rod of length a with respect to an axis perpendicular to the rod passing through (a) the centroid, and (b) the end of the rod.

(a) Let us take the origin of the xy-frame at the centroid of the rod and the x-axis along the rod. Then the y-axis is perpendicular to the rod through the centroid and we are required to find I_y (Fig. 5.42).

If λ is the constant linear density of the rod

$$I_y = \int_{-a/2}^{a/2} \lambda x^2 dx = 2\lambda \int_0^{a/2} x^2 dx = 2\lambda \frac{x^3}{3}\Big|_0^{a/2} = \frac{\lambda a^3}{12}.$$

If M is the mass of the rod then obviously $M = \lambda a$.

$\therefore \quad I_y = Ma^2/12,$

and the radius of gyration $k = a/2\sqrt{3}$.

(b) Choosing the origin at the end O of the rod,

$$I_y = \int_0^a \lambda x^2 dx = \frac{\lambda}{3} x^3 \Big|_0^a = \frac{\lambda a^3}{3} = Ma^2/3 \quad \text{(Fig. 5.43)}.$$

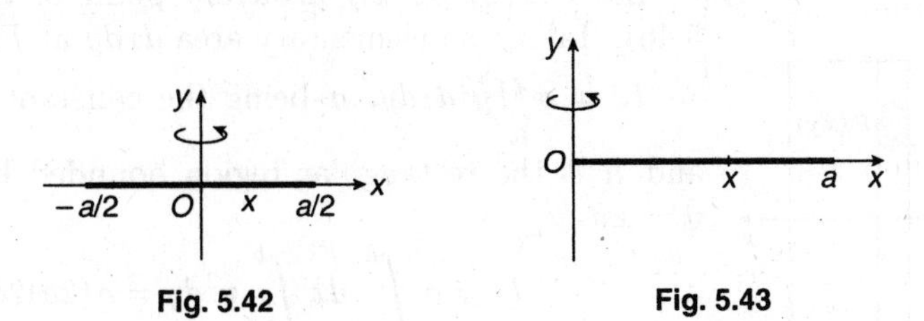

Fig. 5.42 Fig. 5.43

2. The MI of a homogeneous circular ring about the line through the centre and perpendicular to the plane of the circle.

Let the centre of the ring be chosen as the origin and x- and y-axes in the plane of the ring. So, we are required to find the MI of the ring about the z-axis,

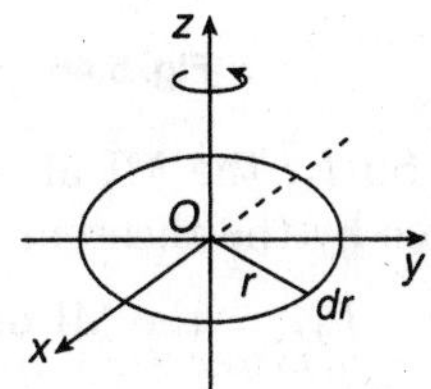

Fig. 5.44

$$I_z = \int_0^{2\pi r} \lambda r^2 ds = \lambda r^2 s \Big|_0^{2\pi r} = \lambda 2\pi r^3,$$

where λ is the linear density and r is the radius of the ring. Since M, the mass of the ring is $\lambda 2\pi r$, $I_z = Mr^2$.

Thus the MI of a circular ring about the line, through its centre, perpendicular to its plane is the same as that of a particle of the same mass as the ring placed at a point on its circumference.

3. The MI of a homogeneous circular disc about the axis, through the centre, perpendicular to the plane of the disc.

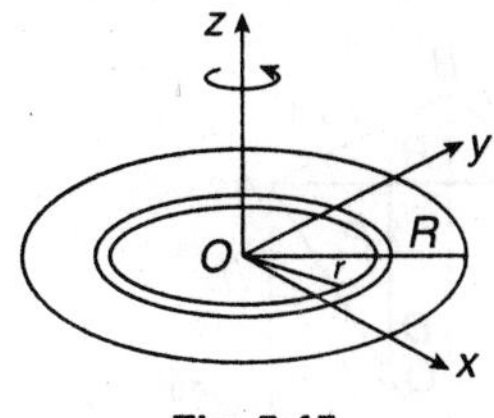

Fig. 5.45

Let us choose the origin at the centre O of the disc and z-axis perpendicular to the plane of the disc. Let R be the radius and M be the mass of the disc. If we consider a concentric ring of radius r and width dr, the mass of the ring is $\sigma 2\pi r dr$. The MI of this ring is the same as that of a particle of the same mass placed at a distance r. So, this MI is

$$\sigma \cdot r^2 2\pi r dr = 2\pi\sigma r^3 dr.$$

Hence the MI of the entire disc is

$$I = \int_0^R 2\pi\sigma r^3 dr = \pi\sigma R^4/2.$$

Since $M = \pi R^2 \sigma$, $I = MR^2/2$ and the radius of gyration is $k = R/\sqrt{2}$.

Cor. 1(a) The MI of a homogeneous, thin circular cylinder of mass M and radius r about its axis is Mr^2, since the cylinder may be regarded as composed of a large number of uniform circular rings.

(b) The MI of a homogeneous solid circular cylinder of mass M and radius R about its axis is $MR^2/2$, since the solid cylinder may be imagined to be composed of a large number of uniform circular discs.

4. The MI of a homogeneous rectangular lamina about a line of symmetry.

Let $ABCD$ be the lamina with $AB = 2a$ and $AD = 2b$, lying in the xy-plane. The centre of the rectangle is chosen as the origin at coordinates and the x- and y-axes parallel to AB and AD respectively.

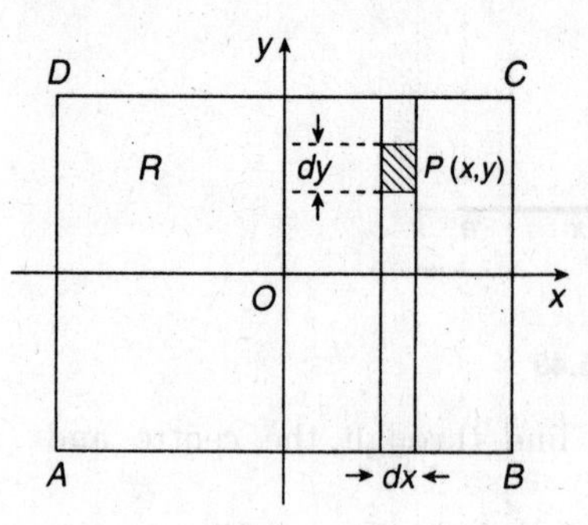

Fig. 5.46

Let $P(x, y)$ be any arbitrary point of the lamina (Fig. 5.46). Taking an elementary area $dxdy$ at P,

$I_x = \sigma \iint_R y^2 dxdy$, σ being the constant surface density and R is the rectangular region bounded by $x = \pm a$ and $y = \pm b$.

$$\therefore \quad I_x = \sigma \int_{-a}^{a} dx \int_{-b}^{b} y^2 dy = \sigma(2a)2b^3/3$$

$$= (4ab\sigma)b^2/3 = Mb^2/3,$$

where M is the mass of the lamina.

Similarly, $I_y = Ma^2/3$.

So I_z, the MI about the axis through O and perpendicular to both the x- and y-axes is given by the theorem of perpendicular axes as $M(a^2 + b^2)/3$. Also by Steiner's theorem

I_{AB} = the MI of the lamina about $AB = Mb^2/3 + Ma^2$

and I_{AD} = ········ ········ $AD = Ma^2/3 + Mb^2$.

Alternatively, dividing the rectangle into a large number of strips of the same width parallel to the y-axis and taking the sum of the moments of inertia of all of them about the x-axis we get, by 1(a) of § 5.5.5, $\sum m(2b)^2/12$ or, $\sum mb^2/3$, where m is the mass of each strip. So, the required MI: $I_x = Mb^2/3$.

Examples 5.6. 1. *Find the MI, about the z-axis, of a homogeneous tetrahedron bounded by the planes $x = 0$, $y = 0$, $z = x + y$ and $z = 1$.*

In the adjoining diagram (Fig. 5.47), $OABC$ is the tetrahedron. The equations of the faces OBC, OCA, OAB and ABC are respectively $x = 0$, $y = 0$, $z = x + y$ and $z = 1$. If D is the space bounded by these faces of the tetrahedron, its projection E in the xy-plane is $OA'B'$, where the equations of OB' and OA' in the xy-plane are $x = 0$ and $y = 0$. To find the equation of $A'B'$, we find that it is the projection of AB, whose equations are $x + y = z$ and $z = 1$. So in the xy-plane, the equation of $A'B'$ is $x + y = 1$. So,

$$E = \{(x, y) \in xy\text{-plane: } 0 \le x \le 1 \text{ and } 0 \le y \le 1 - x\}$$

and $D = \{(x, y, z) : 0 \le x \le 1, 0 \le y \le 1 - x \text{ and } x + y \le z \le 1\}$.

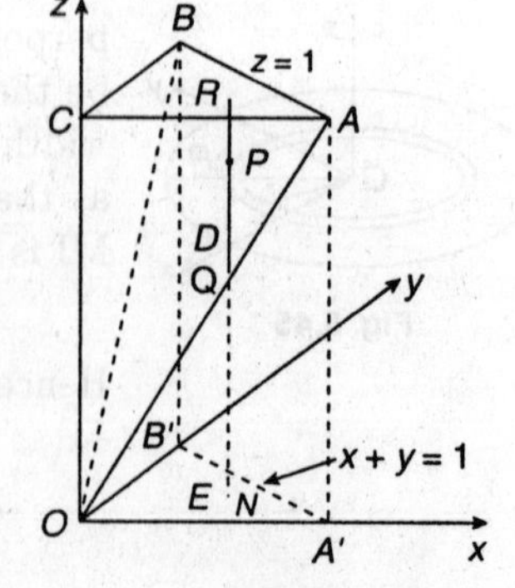

Fig. 5.47

If $P(x, y, z)$ is a point (Fig. 5.47) within the tetrahedron, for the fixed x and y, P will move along the segment QR and $PN(= z)$ will vary within $QN(= x + y)$ and $RN(= 1)$, where Q is the point where the line RPN through P parallel to the z-axis meets the face OAB. We are required to find I_z. If ρ is the uniform density,

$$I_z = \iiint_D \rho(x^2 + y^2)dxdydz = \rho \int_0^1 dx \int_0^{1-x} (x^2 + y^2)dy \int_{x+y}^{1} dz$$

$$= \rho \int_0^1 dx \int_0^{1-x} (x^2 + y^2)(1 - x - y)dy$$

$$= \rho \int_0^1 \left[x^2(1-x)y - \frac{1}{2}x^2y^2 + \frac{1}{3}(1-x)y^3 - \frac{1}{4}y^4 \right]_0^{1-x}$$

$$= \frac{1}{12}\rho \int_0^1 (1-x)^2(7x^2 - 2x + 1)dx$$

$$= \frac{1}{12}\rho \int_0^1 (1 - 4x + 12x^2 - 16x^3 + 7x^4)dx$$

$$= \frac{1}{12}\rho \cdot \left[x - 2x^2 + 4x^3 - 4x^4 + \frac{7}{5}x^5 \right]_0^1 = \rho/30.$$

2. ***Find the MI of a homogeneous circular plate of mass M about a tangent.***

Let the y-axis be the tangent to the circle of radius a at the origin. Let the plate occupy the circular disc given by $x^2 + y^2 \leq 2ax$. The required MI is given by

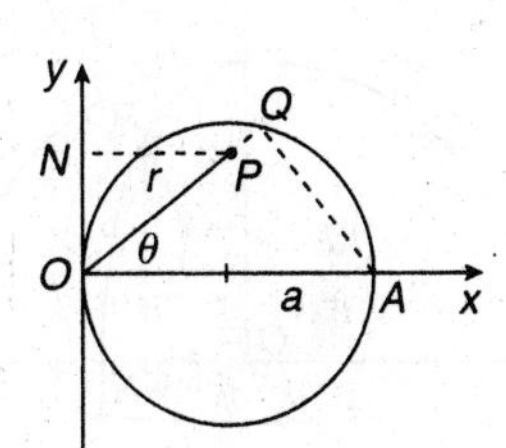

Fig. 5.48

$$I = \iint_E x^2 dm.$$

In polar coordinates $x = r\cos\theta$ and

$dm = \sigma r dr d\theta$, σ being the surface density,

where $-\pi/2 \leq \theta \leq \pi/2$ and $0 \leq r = OP = OQ = 2a\cos\theta$ (Fig 5.48).

$$\therefore \quad I = \int_{-\pi/2}^{\pi/2} d\theta \int_0^{2a\cos\theta} r^2 \cos^2\theta \sigma r dr = \sigma \int_{-\pi/2}^{\pi/2} \cos^2\theta \left[\frac{r^4}{4} \right]_0^{2a\cos\theta} d\theta$$

$$= \frac{\sigma}{4} \int_{-\pi/2}^{\pi/2} 16a^4 \cos^6\theta d\theta = 4a^4\sigma \cdot 2 \int_0^{\pi/2} \cos^6\theta d\theta$$

$$= 8\sigma a^4 \cdot \frac{5.3.1}{6.4.2}\frac{\pi}{2} = \frac{5}{4}\sigma\pi a^4.$$

Since $\quad M = \sigma\pi a^2$, $I = \dfrac{5}{4}Ma^2$.

3. ***Find the MI, about the initial line, of the lamina in the form of the cardioid*** $r = a(1 + \cos\theta)$.

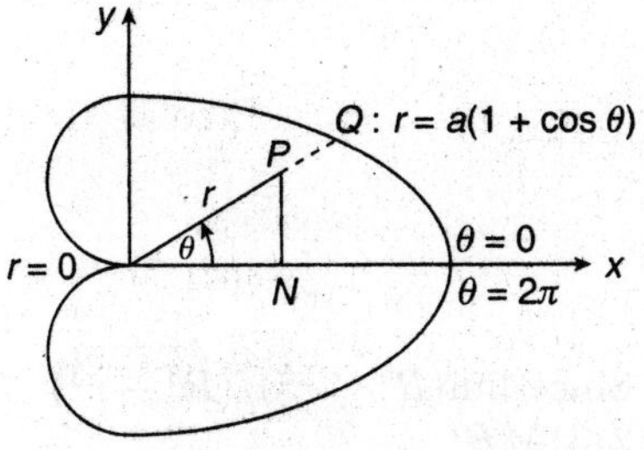

Fig. 5.49

$$I = \iint PN^2 dm \quad \text{(Fig. 5.49)}$$

$$= \iint r^2 \sin^2\theta \sigma r dr d\theta = \sigma \int_0^{2\pi} \sin^2\theta d\theta \int_0^{a(1+\cos\theta)} r^3 dr$$

$$= \sigma \frac{a^4}{4} \int_0^{2\pi} \sin^2\theta (1 + \cos\theta)^4 d\theta$$

$$= \sigma \frac{a^4}{2} \int_0^{\pi} \sin^2\theta (1 + \cos\theta)^4 d\theta$$

$$= \sigma \frac{a^4}{2} \int_0^{\pi/2} \sin^2 2\phi (2\cos^2 \phi)^4 \cdot 2d\phi, \quad \text{where } \phi = 2\phi$$

$$= 64\sigma a^4 \int_0^{\pi/2} \sin^2 \phi \cos^{10} \phi d\phi = 64\sigma a^4 \cdot \frac{1.9.7.5.3.1}{12.10.8.6.4.2} \frac{\pi}{2}$$

$$= 21\pi\sigma a^4/32.$$

4. *Find the MI of a homogeneous solid sphere of radius R and mass M about its diameter. Also obtain the corresponding radius of gyration.* **[WBUT Sem. 2003]**

If ρ is the uniform density, $dm = \rho dv$.

For the MI about the z-axis, the distance of a point $P(r, \theta, \phi)$ from it is $ON = r \sin\theta$.

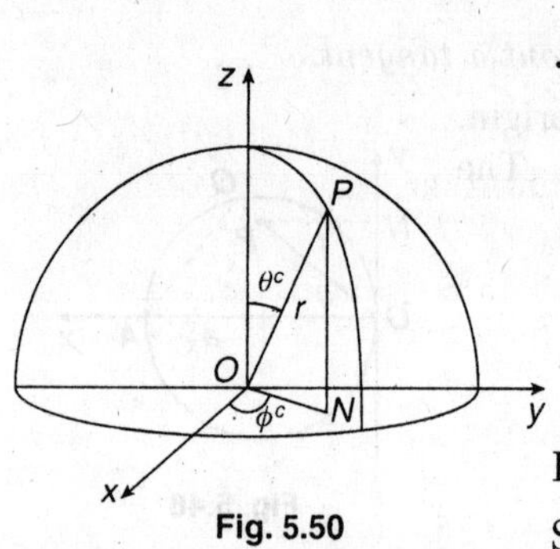

Fig. 5.50

$$\therefore \quad I_z = \int_v r^2 \sin^2 \theta dm = \rho \int_v r^2 \sin^2 \theta dv = \rho \int_v r^4 \sin^3 \theta dr d\theta d\phi$$

$$= \rho \int_0^R r^4 dr \int_0^{\pi} \sin^3 \theta d\theta \int_0^{2\pi} d\phi$$

$$= \rho (R^5/5) 2 \int_0^{\pi/2} \sin^3 \theta d\theta 2\pi$$

$$= 8\pi\rho R^5/15.$$

But $M = \rho v = \rho (4/3)\pi R^3$.

So, $I_z = (2/5)MR^2$

and the radius of gyration is $R\sqrt{2/5}$.

5. *Find the MI of a spherical shell about its diameter.*

Let r and R be the respective inner and outer radii of the shell and ρ its density. So, its mass $M = (4/3)\pi(R^3 - r^3)\rho$.

Let I_1 and I_2 be the respective moments of inertia of the solid inner sphere and the entire solid sphere of radius R; then

$$I_2 = I_1 + I_z,$$

where I_z is the required MI of the shell about the z-axis, which is also chosen as one of its diameters.

$\therefore \quad I_z = I_2 - I_1 = (2/5)(M_2 R^2 - M_1 r^2),$

where $M_2 = (4/3)\pi R^3 \rho$ and $M_1 = (4/3)\pi r^3 \rho$,

the masses of the whole and the inner spheres respectively.

So, $\quad I_z = (2/5)(4/3)\pi(R^5 - r^5)\rho = (8/15)\pi(R^5 - r^5)\rho.$

But $\quad (4/3)\pi(R^3 - r^3)\rho = M$

$\therefore \quad I_z = (2/5)M(R^5 - r^5)/(R^3 - r^3).$

Cor. When $r \to R$, the shell becomes a thin one of radius R. Since $\lim(R^5 - r^5)/(R^3 - r^3) = (5/3)R^2$, the MI of such a shell about one of its diameters is $(2/3)MR^2$.

6. *Find the MI of the positive octant of the homogeneous solid ellipsoid:* $x^2/a^2 + y^2/b^2 + z^2/c^2 \leq 1$ *about the x-axis.* **[JNTV, 1989]**

The required MI is given by

$$I_x = \iiint_D \rho(y^2 + z^2)dxdydz,$$

where D is the space occupied by the positive octant.

Referring to the figure (Fig. 5.51), if E is the projection of D in the xy-plane, then

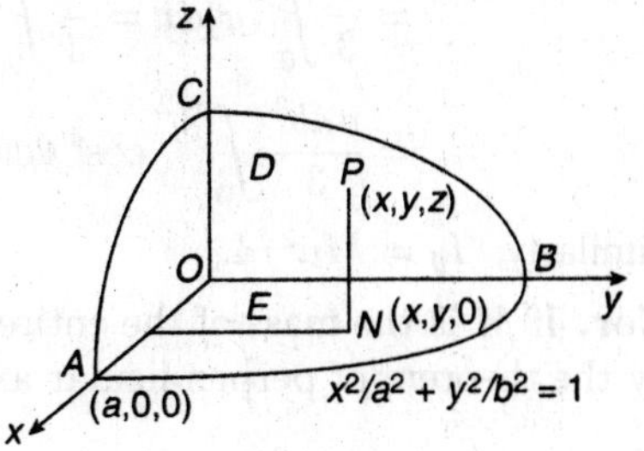

Fig. 5.51

$$E = \{(x, y) : 0 \le x \le a$$

and $\quad 0 \le y \le b\sqrt{1 - x^2/a^2}\ \forall\, x \in [0, a]\}$

and $\quad D = \{(x, y, z) : 0 \le x \le a,\ 0 \le y \le b\sqrt{1 - x^2/a^2} \forall x \in [0, a]$

and $\quad 0 \le z \le c\sqrt{1 - x^2/a^2 - y^2b^2}\ \forall (x, y) \in E\}.$

Taking $y_1 = b\sqrt{1 - x^2/a^2}$ and $z_1 = c\sqrt{1 - x^2/a^2 - y^2/b^2}$, we find that

$$y_1^2/b^2 = 1 - x^2/a^2 \text{ and } z_1^2/c^2 = y_1^2/b^2 - y^2/b^2.$$

$$\therefore \quad y_1 = (b/a)\sqrt{a^2 - x^2} \text{ and } z_1 = (c/b)\sqrt{y_1^2 - y^2}.$$

$$\therefore \quad I_x = \rho \int_0^a dx \int_0^{y_1} dy \int_0^{z_1} (y^2 + z^2)dz = \rho \int_0^a dx \int_0^{y_1} \left[y^2 z + \frac{1}{3}z^3 \right]_0^{z_1} dy$$

$$= \rho \int_0^a dx \int_0 \left(y^2 z_1 + \frac{1}{3} z_1^3 \right) dy = \rho \int_0^a dx \int_0^{y_1} \left[\frac{c}{b} y^2 (y_1^2 - y^2)^{1/2} + \frac{c^3}{3b^3} (y_1^2 - y^2)^{3/2} \right] dy$$

$$= \rho \int_0^a dx \int_0^{\pi/2} \left[y_1^2 \sin^2\theta \cdot \frac{c}{b} y_1 \cos\theta + \frac{c^3}{3b^3} y_1^3 \cos^3\theta \right] y_1 \cos\theta d\theta,\ y = y_1 \sin\theta$$

$$= \rho \int_0^a y_1^4 dx \int_0^{\pi/2} \left(\frac{c}{b} \sin^2\theta \cos^2\theta + \frac{c^3}{3b^3} \cos^4\theta \right) d\theta$$

$$= \rho \int_0^a y_1^4 \left(\frac{c}{b} \cdot \frac{1.1}{4/2} \frac{\pi}{2} + \frac{c^3}{3b^3} \frac{3.1}{4.2} \frac{\pi}{2} \right) dx$$

$$= \rho \frac{\pi bc(b^2 + c^2)}{16a^4} \int_0^a (a^4 - 2a^2x^2 + x^4)dx, \quad \because\ y_1 = (b/a)\sqrt{(a^2 - x^2)}$$

$$= \rho\pi abc(b^2 + c^2)/30.$$

7. ***Find the MI of a plane uniform lamina in the form of a quadrant of an ellipse about one of its plane edges.***

Let the equation to the ellipse be $(x/a)^2 + (y/b)^2 = 1, a > 0, b > 0$. The area of the ellipse is πab.

If σ be the surface density and M, the mass of the lamina

$$M = \pi ab\sigma/4 \quad \text{or,} \quad \sigma = 4M/(\pi ab).$$

$$\therefore \quad I_x = \iint_D y^2 dm = \iint_D y^2 \sigma dxdy = \sigma \int_0^a dx \int_0^{y_1} y^2 dy, y_1 = (b/a)\sqrt{a^2 - x^2}$$

$$= \frac{\sigma}{3}\int_0^a y_1^3 dx = \frac{\sigma}{3}\int_0^a \frac{b^3}{a^3}(a^2 - x^2)^{3/2} dx$$

$$= \frac{\sigma ab^3}{3}\int_0^{\pi/2} \cos^4\theta d\theta = \frac{\pi\sigma abb^2}{16} = Mb^2/4.$$

Similarly, $I_y = Ma^2/4$.

Cor. If M is the mass of the entire elliptic lamina, then also $I_x = Mb^2/4$ and $I_y = Ma^2/4$ and by the theorem of perpendicular axis,

$$I_z = M(a^2 + b^2)/4.$$

Exercises 5.4

1. Using double integration, find the moments of inertia of a uniform lamina, in the form of a triangle bounded by the coordinate axes and the line $x/a + y/b = 1$, $a > 0$, $b > 0$, about the coordinate axes if the surface density at a point is proportional to the product of its distances from the coordinate axes.

2. Find the MI of a circular disc of radius a about a diameter when

(i) the disc is uniform,

(ii) the surface density at a point is proportional to its distance from the centre.

3. A thin metal ring of diameter 6 cm and mass 100 g starts from rest and rolls down an inclined plane. Its linear velocity on reaching the foot of the plane is 50 cm/s. Calculate the MI of the ring.

[MI does not depend on the velocity.]

4. Show that the MI of a solid sphere of radius R and mass M about a tangent line is $(7/5)MR^2$, whereas that of a thin hemispherical shell is $(5/3)MR^2$.

5. Find the moments of inertia of an annular disc about (i) the axis passing through the centre and perpendicular to the plane of the disc, (ii) a diameter.

6. Find the MI of a uniform lamina in the form of a parallelogram $ABCD$ about the diagonal AC, where

$$AB = a,\ AD = b \text{ and } \angle CAB = \alpha.$$

7. Find the moments of inertia of a homogeneous solid right-circular cone about

(i) its axis,

(ii) an axis through the vertex and perpendicular to the axis of the cone.

8. Show that the radius of gyration of the area bounded by the parabola $y = 20x(1-x)$ and the line $y = 0$ about the z-axis is $\sqrt{0.3}$.

9. Find the radius of gyration of the area bounded by $x = 0$, $x = \pi$, $y = 0$ and $y = \sin x$ about the y-axis.

10. Find the MI about the axis, of the frustum of a cone of thickness h units and r and R units of length as the radii of the bases.

Answers

1. $(ka^2b^4/120, ka^4b^2/120)$. **2. (i)** $Ma^2/4$; **(ii)** $3Ma^2/10$, M being the mass of the disc.

3. 900 g cm^2. **5. (i)** $M(R^2+r^2)/2$;

(ii) $M(R^2+r^2)/4$, where M is the mass, r and R are the inner and outer radii of the annulus.

6. $M(a^2/b)\sin^2\alpha$. **7. (i)** $(3/10)Mr^2$; **(ii)** $(M/20)(3r^2+2h^2)$;

(iii) $(3M/20)(r^2+4h^2)$, where M is the mass, h is the height and r is the radius of the base of the cone.

9. $\sqrt{(\pi^{-2}/2-2)}$. **10.** $(\pi h/10)(R^5-r^5)/(R-r)$.

Appendix A

5.5.6 Proofs of Pappus Theorems

1. Volume of Revolution

Let us take the x-axis as the axis of revolution and let us first suppose that the boundary of the region E be quadratic in y and E lies between the two lines $x=a$ and $x=b$, $a<b$. Then every line parallel to the y-axis will meet the boundary of E in not more than two points. Let the ordinate through $M(x)$ meet this boundary in two points P_1 and P_2 of ordinates y_1 and y_2 respectively, where $a \le x \le b$ (Fig. 5.52).

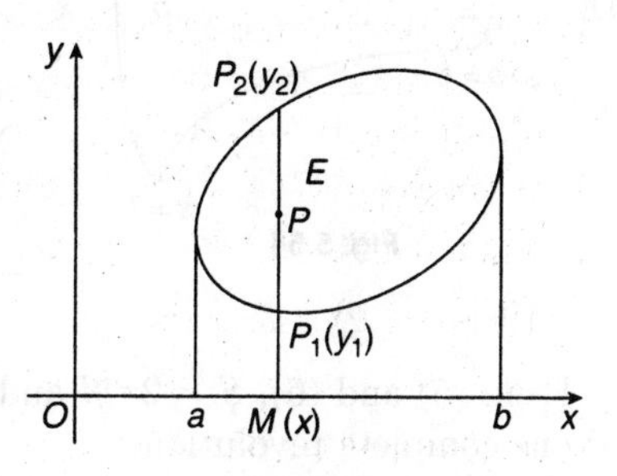

Fig. 5.52

Then the volume V, of the solid of a complete revolution about the x-axis is given by

$$V=\pi\int_a^b y_2^2dx-\pi\int_a^b y_1^2dx=\pi\int_a^b (y_2^2-y_1^2)dx. \tag{1}$$

If $\bar{y}$ is the ordinate of the centroid of E,

$$\bar{y}=(1/A)\int_a^b \frac{1}{2}(y_1+y_2)(y_2-y_1)dx,$$

since the ordinate of the mid-point P of P_1 and P_2 is $(1/2)(y_1+y_2)$.

$$\therefore\ \bar{y}=(1/2A)\int_a^b (y_2^2-y_1^2)dx,\ A \text{ being the area of } E. \tag{2}$$

From (1) and (2), $v=2\pi\bar{y}A$.

But the length of the path of the centroid of E for a complete revolution is obviously $2\pi\bar{y}$. Hence the theorem.

Cor. If the region revolves through an angle α^c about the x-axis, the elementary volume is

$$\frac{1}{2}y^2\alpha dx.$$

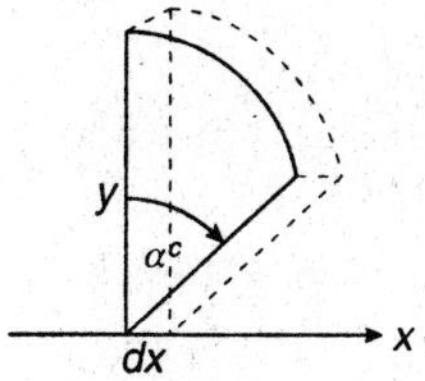

$$\therefore\quad V=(d/2)\int_a^b (y_2^2-y_1^2)dx \tag{3}$$

$$\text{and}\quad \bar{y}=(1/A)\int_a^b \frac{1}{2}(y_1+y_2)(y_2-y_1)dx$$

$$=(1/2A)\int_a^b (y_2^2-y_1^2)dx. \tag{4}$$

Fig. 5.53

From (3) and (4), $V = \overline{y}\alpha A$.

But $\overline{y}\alpha$ is the length of the path of the centroid. Hence the result.

Note: If the boundary of E is not quadratic, it should then be divided into parts, each of which is quadratic in y.

2. Surface Area of Revolution

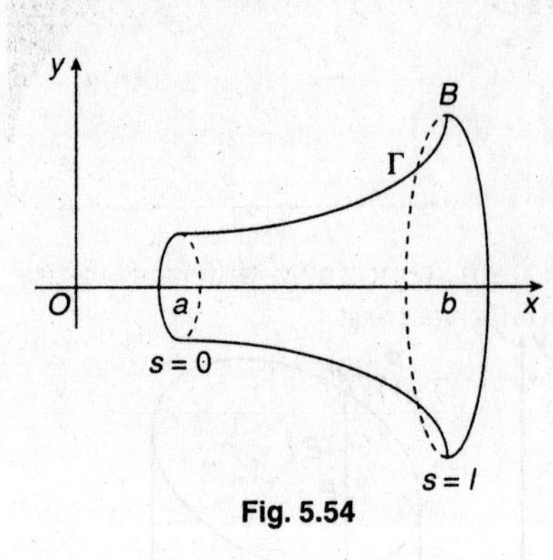

Fig. 5.54

Let Γ be a plane curve of length l with and points A and B lying in the xy-plane. Let it be revolved about the x-axis. If Γ makes one complete revolution, then the surface S of revolution is given by

$$S = \int_0^l 2\pi y dx, \tag{5}$$

assuming, initially, the entrie arc Γ to be above the x-axis. Now the centroid $\overline{y}$ of the curve is given by

$$\overline{y} = (1/l)\int_0^l yds. \tag{6}$$

From (5) and (6), $S = 2\pi\overline{y}l$ and we know that $2\pi\overline{y}$ is the length of the path of the centroid for one complete revolution.

The result for a revolution through an angle α^c can be proved similarly.

6 Vector Analysis

Vector analysis, i.e., the calculus part of vectors deals will limit, continuity, differentiability of vector functions and some results of integral calculus involving vectors.

DIFFERENTIAL CALCULUS OF VECTORS

6.1 Scalar and Vector Field

6.1.1 Scalar Function

An ordinary function

$$f : D \to \mathbb{R}$$

is said to be a scalar function if the value of f at every point of D is independent of any coordinate system and depends only upon the point, where D is a subset of $\mathbb{R}, \mathbb{R}^2, \mathbb{R}^3$, etc.

Examples 6.1. **1.** *The distance of a variable point P from another given point P_0 is a scalar function. For, if D is the one-dimensional Cartesian space represented by the x-axis and f is the function mapping D into $\mathbb{R}$ such that*

$$f(P) = f(x) = |x - x_0|,$$

where P and P_0 are the points on the x-axis represented by x and x_0 respectively, referred to some origin O.

Shifting the origin from O to H given by

$$x = x' + h,$$

Fig. 6.1

x_0 will be given by $x'_0 + h$ and

$$f(x) = |x - x_0| = |(x' + h) - (x'_0 + h)| = |x' - x'_0| = f(x').$$

Thus f, i.e., the distance of a point from a given point is a scalar function depending only on the position of P relative to that of P_0.

In $\mathbb{R}^2$, if P and P_0 are (x, y) and (x_0, y_0), $f(x, y)$ may be taken as $\sqrt{[(x - x_0)^2 + (y - y_0)^2]}$.

By shifting the origin or by applying a rotation to the xy-frame, it can be verified that $f(x, y)$ remains unchanged.

Note that the distance function will not be a scalar function if we consider the distance of a point P from the **origin**. Because, with the shifting of the origin the distance of P will not in general, remain the same, although the position of P relative to the origin remains the same.

2. *A physical example of a scalar function is the temperature of a body B at a particular instant of time. If f is the temperature of the body, then its domain is B.*

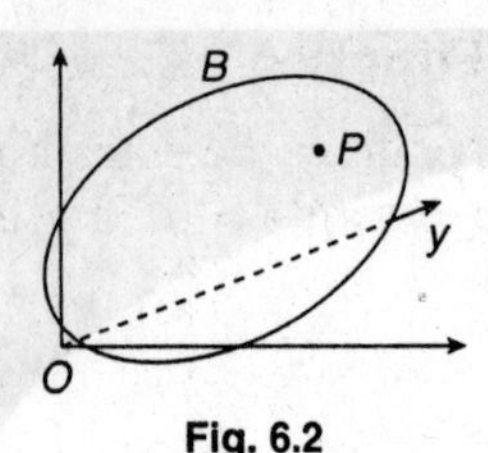

Fig. 6.2

Referred to a rectangular Cartesian coordinate system x, y, z, the value of f at a point $P(x, y, z)$ is $f(x, y, z)$. Referred to a cylindrical coordinate system (u, θ, z) with the same origin O (Fig. 6.2), $x = u\cos\theta$, $y = u\sin\theta$, and the value of f at P is $f(u\cos\theta, u\sin\theta, z) = \phi(u, \theta, z)$. But the temperature of the body at P will not definitely depend upon the coordinate system and

$$f(x, y, z) = \phi(u, \theta, z).$$

Because of this, the value of a scalar function f at a point P will be denoted by $f(P)$, without using any specific coordinates.

Referred to a rectangular Cartesian coordinate system x, y, z, a point R is expressed as (x, y, z). If $\overline{r}$ is the position vector (p.v.) of R, i.e., if $\overline{r} = \overrightarrow{OR}$, the value of f at R is sometimes expressed as $f(\overline{r})$. Thus

$$f(R) = f(\overline{r}).$$

Scalar Field: Given a scalar function f with D as its domain, we say that a **scalar field** is given in D. In other words, **a set of points D (which may be a line, a surface, or a part of the 3-D space) along with a scalar function f with D as its domain is a scalar field. Such a field may be denoted by (D, f).**

6.2 Vector Function

Let D be a set of points in $\mathbb{R}$, $\mathbb{R}^2$ or $\mathbb{R}^3$ and V, a set of vectors $\overline{u}, \overline{v}, \overline{w}$, etc. Then the function

$$\overline{f} : D \to V$$

is said to be a vector function with D as its domain, if $\forall\ P \in D$, $\overline{v}(= \overline{f}(P))$ is independent of any choice of coordinate system.

Vector Field: Given a vector function $\overline{f}$ with D as its domain, we say that a vector field has been given in D. In other words, **a set of points D along with a vector function $\overline{f}$ with D as its domain is a vector field, which may be denoted by $(D, \overline{f})$.**

Since the vector $\overline{f}$ assigned to each point P of the domain is independent of any coordinates, such a vector may be expressed as $\overline{f}(P)$ in place of $\overline{f}(x, y, z)$ or $\overline{f}(r, \theta, \phi)$. Moreover, if $\overline{r}$ is the position vector of P relative to some origin, $\overline{f}(P) = \overline{f}(\overline{r})$.

Examples 1. Gravity Field: At every point P of the universe there is a definite vector, the attractive force of earth on a unit mass placed at P. So, if D is the entire universe and $\overline{g}$ is the force of earth's attraction at a point, then $\overline{g}$ is a vector function with D as its domain. So, we have $(D, \overline{g})$ as the gravity field.

2. Velocity Field of a Rotating Body: Let a body B be rotating about an axis Oz (Fig. 6.3). Then every point P (except 0) of the body is moving with a definite linear velocity $\overline{v}$ at a particular instant of time. So, at a particular instant of time $\overline{v}$ is a vector function with B as its domain and we get a field $(B, \overline{v})$, called the **velocity field** of the rotating body.

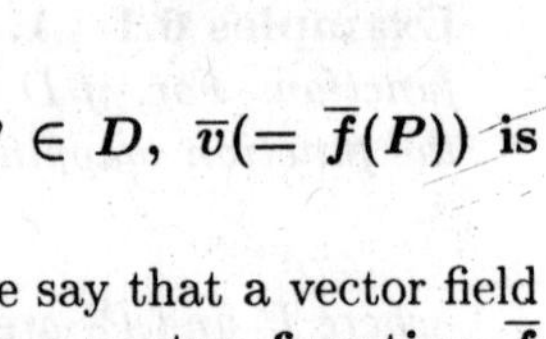

Fig. 6.3

Similarly, we may have other vector fields like field of force, **gravitational, magnetic** and **electric**.

6.3 Limit, Continuity and Differentiability of a Vector Function

Limit: Let $\overline{f}$ be a vector function of a single variable x. Let a be a limit point of the domain D of $\overline{f}$. Also, let $\overline{l}$ be a vector. Then the vector $\overline{l}$ is called the limit of $\overline{f}$ at $x = a$ if $\forall\, \epsilon > 0, \exists \delta > 0$:

$|\overline{f}(x) - \overline{l}| < \epsilon$ whenever $0 < |x - a| < \delta,\ x \in \text{Dom}\, f$ (Fig. 6.4).

Symbolically, $\underset{x \to a}{\text{Lt}}\ \overline{f}(x) = \overline{l}$.

Note that $\overline{f}(x) - \overline{l}$ is a vector and $|\overline{f}(x) - \overline{l}|$ or $|\overline{l} - \overline{f}(x)|$ is the length of that vector whereas $|x - a|$ is the numerical value of the real number $x - a$, which is the distance PA between P and A represented by x and a respectively.

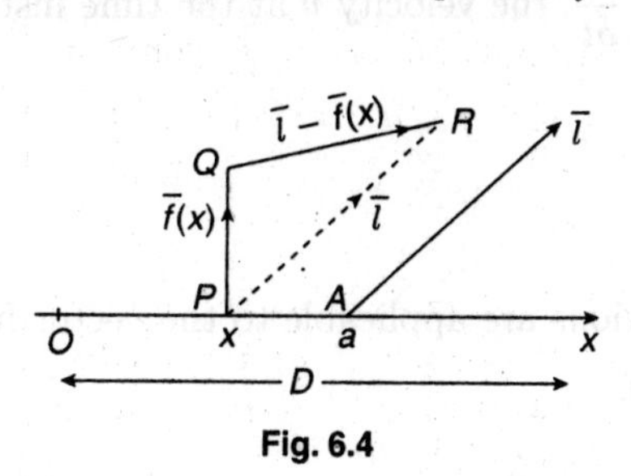

Fig. 6.4

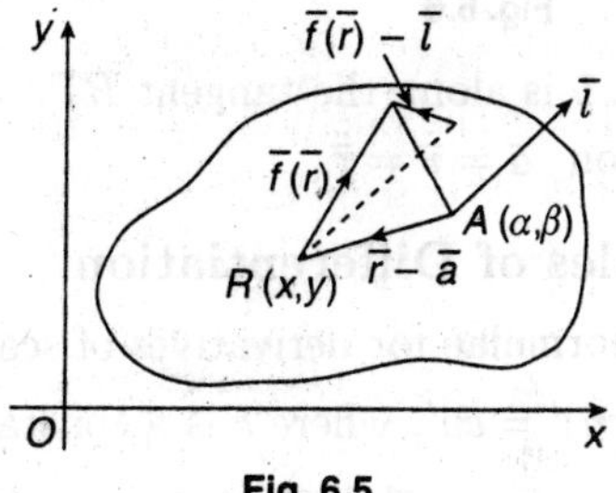

Fig. 6.5

If $\overline{f}$ is a function of two real variables x and y, and a point $A(\alpha, \beta)$ is a limit point of the domain D of $\overline{f}$ then a vector $\overline{l}$ is said to be the limit of $\overline{f}$ at A if $\forall\, \epsilon > 0, \exists \delta > 0$:

$$|\overline{f}(\overline{r}) - \overline{l}| < \epsilon \quad \text{whenever } 0 < |\overline{r} - \overline{a}| < \delta,\ \overline{r} \in \text{Dom}\, f,$$

where $\overline{r}$ and $\overline{a}$ are the p.v.'s of $R(x, y)$ and $A(\alpha, \beta)$ (Fig. 6.5).

Exactly similar is the definition of the limit of a vector function in the 3-D space.

Continuity: A vector function $\overline{f}$ is said to be continuous at a point $A(\overline{a}) \in \text{Dom}\, \overline{f}$, if $\forall\, \epsilon > 0, \exists \delta > 0$:

$|\overline{f}(\overline{r}) - \overline{f}(\overline{a})| < \epsilon$ whenever $|\overline{r} - \overline{a}| < \delta,\ \overline{r} \in \text{Dom}\, \overline{f}$. If A is a limit point of the domain, then, in this case

$$\underset{\overline{r} \to \overline{a}}{\text{Lt}}\ \overline{f}(\overline{r}) = \overline{f}(\overline{a}).$$

Differentiation: If $\overline{f}$ is a vector function of a single variable x and $a \in \text{Dom}\, \overline{f}$, then the limit

$$\underset{x \to a}{\text{Lt}}\ \frac{\overline{f}(x) - \overline{f}(a)}{x - a}, \tag{1}$$

when it exists, is defined to be the **derivative** or the **differential coefficient** of $\overline{f}$ with respect to x at a. It is denoted by symbols like

$$\overline{f}'(a),\ \left.\frac{d\overline{f}}{dx}\right|_{x=a} \quad \text{or} \quad \frac{d\overline{f}(a)}{dx} \quad \text{or } D\overline{f}(a), \text{ etc.}$$

Thus, $$\underset{x \to a}{\text{Lt}}\ \frac{\overline{f}(x) - \overline{f}(a)}{x - a} = \underset{\delta x \to 0}{\text{Lt}}\ \frac{\overline{f}(a + \delta x) - \overline{f}(a)}{\delta x} = \overline{f}'(a).$$

The derivative of a vector function $\overline{f}$ depending on the real variable t (time variable) is denoted by $\dot{\overline{f}}$ in place of $\overline{f}'$.

Velocity and Acceleration

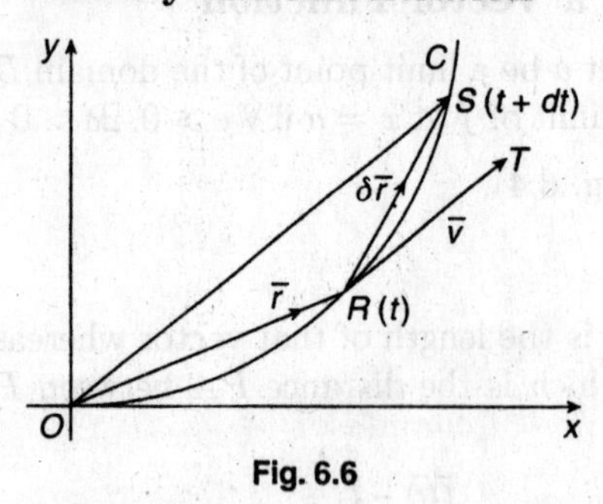

Fig. 6.6

If $\overline{r}$ is the position vector of a particle at R moving along a smooth curve C, its p.v. at a time instant t is given by

$$\overline{r} = \overline{r}(t).$$

If $\overline{r}+\delta\overline{r}(=\overline{r}(t+\delta t))$ is the p.v. of the same moving particle at a subsequent instant $t+\delta t$, then

$$\overrightarrow{RS} = \delta\overline{r} = \overline{r}(t+\delta t) - \overline{r}(t) \quad \text{(Fig. 6.6)}.$$

$$\therefore \quad \dot{\overline{r}} = \underset{\delta t \to 0}{\text{Lt}} \frac{\delta \overline{r}}{\delta t}, \text{ the velocity } \overline{v} \text{ at the time instant } t.$$

Obviously, $\overline{v}$ is along the tangent RT.

Acceleration $\overline{a} = \dot{\overline{v}} = \ddot{\overline{r}}$.

6.4 Rules of Differentiation

Ordinary formulae for derivatives of scalar functions are applicable to the vector functions:

(i) $(c\overline{u})' = c\overline{u}'$, where c is a constant scalar,

(ii) $(\overline{u} \pm \overline{v})' = \overline{u}' \pm \overline{v}'$,

$$\left(\sum \overline{u}_i\right)' = \sum \overline{u}_i',$$

(iii) $(\overline{u}\cdot\overline{v})' = \overline{u}'\cdot\overline{v} + \overline{u}\cdot\overline{v}'$,

(iv) $(\overline{u}\times\overline{v})' = \overline{u}'\times\overline{v} + \overline{u}\times\overline{v}'$,

(v) $[\overline{u}\,\overline{v}\,\overline{w}]' = [\overline{u}'\overline{v}\,\overline{w}] + [\overline{u}\,\overline{v}'\overline{w}] + [\overline{u}\,\overline{v}\,\overline{w}']$

and **(vi)** $((\overline{u}\times\overline{v})\times\overline{w})' = (\overline{u}'\times\overline{v})\times\overline{w} + (\overline{u}\times\overline{v}')\times\overline{w} + (\overline{u}\times\overline{v})\times\overline{w}'$.

6.4.1 Differential of a Vector Function

If $y = f(x)$ defines y as a scalar function of x, then $dy(=f'(x)dx)$ is called the **differential of** y, dx is called the **differential of** x and $f'(x)$ is the **derivative** or the **differential coefficient** of f or y with respect to x.

Similarly, if $z = f(x,y)$, $dz = f_x dx + f_y dy$ or if $u = f(x,y,z)$, $du = f_x dx + f_y dy + f_z dz$, where dx, dy, dz, du are all differentials of the variables x, y, z and u respectively.

Similarly, if $\overline{u} = \overline{f}(x)$ defines a vector function of the variable x

$$d\overline{u} = \overline{f}'(x)dx$$

and $d\overline{u}$ is called the **differential of the vector** $\overline{u}$.

If $\overline{u} = \overline{f}(x,y)$, $d\overline{u} = \overline{f}_x dx + \overline{f}_y dy$

and $\overline{u} = \overline{f}(x,y,z)$, $d\overline{u} = \overline{f}_x dx + \overline{f}_y dy + \overline{f}_z dz$,

where $d\overline{u}$ is the differential of the vector function $\overline{u}$.

If $\overline{r}$ is the p.v. of a point R and $\overline{r} = \overline{r}(t)$, then

$$d\overline{r} = \dot{\overline{r}}dt$$

implies that $d\overline{r}$ is parallel to $\dot{\overline{r}}$, i.e., the velocity of the point.

If $\overline{v}(x,y,z) = v_1(x,y,z)\hat{i} + v_2(x,y,z)\hat{j} + v_3(x,y,z)\hat{k}$

then $\partial\overline{v}/\partial x = \hat{i}\partial v_1/\partial x + \hat{j}\partial v_2/\partial x + \hat{k}\partial v_3/\partial x$,

$\partial\overline{v}/\partial y = \hat{i}\partial v_1/\partial y + \hat{j}\partial v_2/\partial y + \hat{k}\partial v_3/\partial y$,

$\partial\overline{v}/\partial z = \hat{i}\partial v_1/\partial z + \hat{j}\partial v_2/\partial z + \hat{k}\partial v_3/\partial z$,

$\partial^2\overline{v}/\partial y\partial z = \hat{i}\partial^2 v_1/\partial y\partial z + \hat{j}\partial^2 v_2/\partial y\partial z + \hat{k}\partial^2 v_3/\partial y\partial z$, etc.

6.5 Directional Derivative, Gradient of a Scalar Field

The directional derivative of a scalar function is the rate of change of its value with respect to distance in a particular direction.

Let P and Q be any two points in the domain $D(\subseteq \mathbb{R}^3)$ of a scalar function f. Then as a particle moves from P to Q, $f(Q) - f(P)$ is the change in the value of the function and

$$\frac{f(Q) - f(P)}{PQ} \tag{1}$$

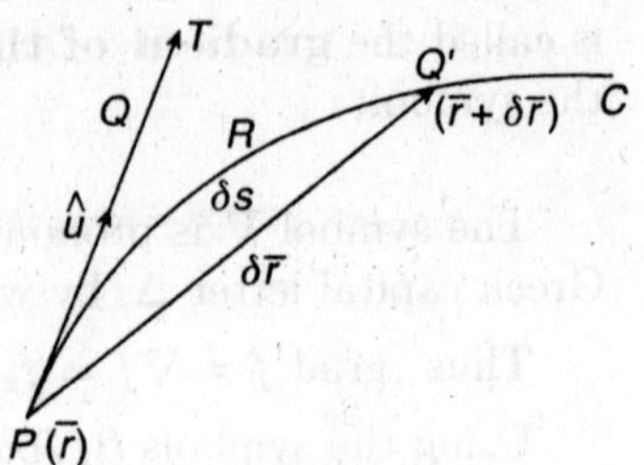

Fig. 6.7

is its average rate of change. Now, as the point Q moves towards P along the straight line joining Q and P, and ultimately coincides with P, the limit of the above ratio, when it exists is the directional derivative of f in the direction of $\overrightarrow{PQ}$. Now let us take any curve C through P having its tangent PT at P passing through Q and let a point Q' on C approach P along C such that the ultimate direction of $\overrightarrow{PQ'}$ is the same as that of $\overrightarrow{PQ}$. Then the limit of

$$[f(Q') - f(P)]/PQ' \tag{2}$$

as Q' approaches P will give the same directional derivative of f at P.

Let $\overline{r}$ and $\overline{r}+\delta\overline{r}$ be the p.v.'s of P and Q', then (2) can be expressed as $[f(\overline{r}+\delta\overline{r}) - f(\overline{r})]/|\delta\overline{r}|$.

If the equation to the curve C is $\overline{r} = r(s)$, where s is a parameter representing arc length, then

$$\overline{r} + \delta\overline{r} = \overline{r}(s+\delta s) \quad \text{and} \quad \delta\overline{r} = \overline{r}(s+\delta s) - r(s).$$

In the diagram (Fig. 6.7), $\delta\overline{r} = \overrightarrow{PQ'}$ and $\delta s =$ the length of the arc PRQ'. Now, as $Q' \to P, |\delta\overline{r}| = PQ' \approx \delta s$, so that

$$\underset{\delta s\to 0}{\text{Lt}}\ |\delta\overline{r}|/\delta s = 1.$$

The limiting vector $\overline{r}'(s)$ or $d\overline{r}/ds$ is, therefore, a unit vector $\hat{u}$ tangential to C at P and $\hat{u} = \overrightarrow{PQ}/PQ$. So the directional derivative of f with respect to distance in the direction $\overline{PQ}$ is

$$\underset{Q\to P}{\text{Lt}}\ \frac{f(Q) - f(P)}{PQ} = \underset{Q'\to P}{\text{Lt}}\ \frac{f(Q') - f(P)}{PQ'} = \underset{\delta\overline{r}\to\overline{0}}{\text{Lt}}\ \frac{f(\overline{r}+\delta\overline{r}) - f(\overline{r})}{|\delta\overline{r}|}$$

$$= \underset{\delta s\to 0}{\text{Lt}}\ \frac{f(\overline{r}+\delta\overline{r}) - f(\overline{r})}{(|\delta\overline{r}|/\delta s)\delta s} = \underset{\delta s\to 0}{\text{Lt}}\ \frac{f(\overline{r}(s+\delta s)) - f(\overline{r}(s))}{\delta s}$$

$$= \frac{df(\overline{r}(s))}{ds} = f'(\overline{r}(s)). \tag{3}$$

Taking $\bar{r} = \bar{r}(s) = \hat{\imath}x(s) + \hat{\jmath}y(s) + \hat{k}z(s)$,

$$f(\bar{r}(s)) = f(x(s), y(s), z(s)).$$

$$\therefore \quad f'(\bar{r}(s)) = f_x x'(s) + f_y y'(s) + f_z z'(s), \tag{4}$$

where $'$ represents derivative w.r.t. s. The R.H.S. of (4) can be formally expressed as the dot product of two vectors (?)

$$f_x\hat{\imath} + f_y\hat{\jmath} + f_z\hat{k} \quad \text{and} \quad x'\hat{\imath} + y'\hat{\jmath} + z'\hat{k}.$$

But $\hat{u} = r'(s) = x'\hat{\imath} + y'\hat{\jmath} + z'\hat{k}$.

And it can be proved that $f_x\hat{\imath} + f_y\hat{\jmath} + f_z\hat{k}$ is actually a vector function independent of any coordinate transformation. Such a vector function which is obtained from a scalar field (D, f) is called the **gradient of the scalar field** or **of the scalar function** f and it is denoted by the symbol:

$$\textbf{grad } f \quad \text{or} \quad \nabla f.$$

The symbol ∇ is pronounced as **del** or **nabla** or **atled** (which is obtained from delta, the Greek capital letter Δ, by writing its letters in the reversed order).

Thus grad $f = \nabla f = f_x\hat{\imath} + f_y\hat{\jmath} + f_z\hat{k}$.

Using the symbols $\partial f/\partial x, \partial f/\partial y$ and $\partial f/\partial z$

$$\begin{aligned}\nabla f &= \hat{\imath}\partial f/\partial x + \hat{\jmath}\partial f/\partial y + \hat{k}\partial f/\partial z \\ &= (\hat{\imath}\partial/\partial x + \hat{\jmath}\partial/\partial y + \hat{k}\partial/\partial z)f \\ &= (\hat{\imath}D_x + \hat{\jmath}D_y + \hat{k}D_z)f = (\textstyle\sum \hat{\imath}D_x)f.\end{aligned}$$

Thus we have the following identity of the two operators:

$$\nabla = \sum \hat{\imath}\partial/\partial x = \sum \hat{\imath}D_x.$$

Note that the vector operator ∇ behaves like scalar multiplication in gradient of f, i.e., ∇f. But this operation is non-commutative and $f\nabla$ is meaningless.

Using this, the directional derivative of f in the direction of $\hat{u}$ is

$$\boldsymbol{f'(\bar{r}(s)) = \hat{u} \cdot \nabla f}. \tag{5}$$

Taking $\hat{u} = \hat{\imath}$, i.e., $x'\hat{\imath} + y'\hat{\jmath} + z'\hat{k} = \hat{\imath}$

$$x' = 1,\ y' = 0,\ z' = 0.$$

$$\therefore \quad \hat{\imath}\nabla f = \hat{\imath}(f_x\hat{\imath} + f_y\hat{\jmath} + f_z\hat{k}) = f_x.$$

Thus the directional derivative of f in the direction of $\hat{\imath}$, i.e., in the positive direction of the x-axis is nothing but the 1st partial derivative of f with respect to x. Similarly, the 1st partial derivatives f_y and f_z are the directional derivatives of f in the positive directions of y- and z-axes respectively.

6.5.1 Interpretation of the Gradient of a Scalar Function

Given a scalar function f, the equation:

$$f(x, y, z) = c, \tag{1}$$

where c is a parameter, represents a family of surfaces, called **level surfaces**. For a real value of c, say $c = 1$, the equation (1) gives

$$f(x, y, z) = 1,$$

which represents a surface at every point of which the value of f is the same, viz., 1 (Fig. 6.8).

Physical examples of level surfaces are isotherms, isobars, etc. A family of surfaces is called isotherms or isobars according as the temperature or pressure over each of the surfaces is constant. Similarly, a family of surfaces over each of which the potential is maintained constant everywhere is called **equipotential family**.

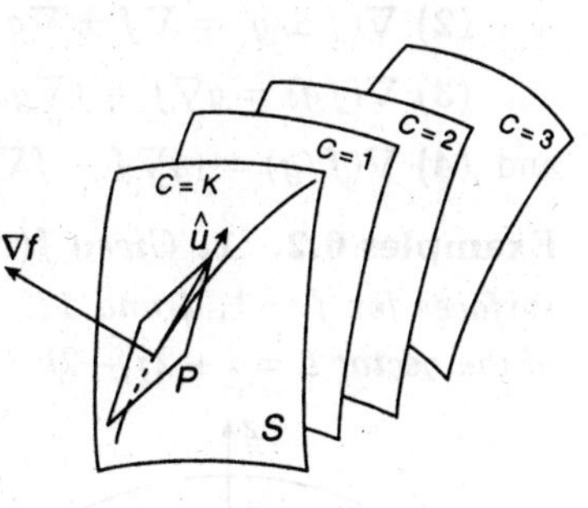

Fig. 6.8

Let P be any point on **one** of the level surfaces given by (1).

Let $\bar{r} = \bar{r}(s)$ be any curve, through P, lying over the surface S given by,

$$f(x, y, z) = k \quad \text{(Fig. 6.8)}.$$

Then $f(\bar{r}(s)) = k$, for all values of s. Differentiating with respect to s we have

$$f'(\bar{r}(s)) = 0,$$

which implies that

$$\hat{u} \cdot \nabla f = 0, \tag{2}$$

where $\hat{u}$ is the unit tangent vector at P. So, the gradient of f at P, i.e., ∇f is perpendicular to $\hat{u}$. If the surface is smooth at P, then all the tangents to the curves, through P, lying in the surface, will form a plane called the **tangent plane** at P. So the vector ∇f is **normal** to the level surface at P (Fig. 6.8).

We have, therefore, a geometrical interpretation of the direction of the vector ∇f, given by the following theorem:

Theorem 1. If a scalar function f with domain D is such that through a given point P in D there passes a unique smooth level surface S, then ∇f, if not $\bar{0}$, has the direction of the normal to S at P.

To get a physical interpretation of grad f we regard the derivative $f'(\bar{r}(s))$ as the rate of change of a scalar function f with respect to s, the arc length. But

$$f'(\bar{r}(s)) = \hat{u} \cdot \nabla f = |\hat{u}||\nabla f| \cos\theta = |\nabla f| \cos\theta,$$

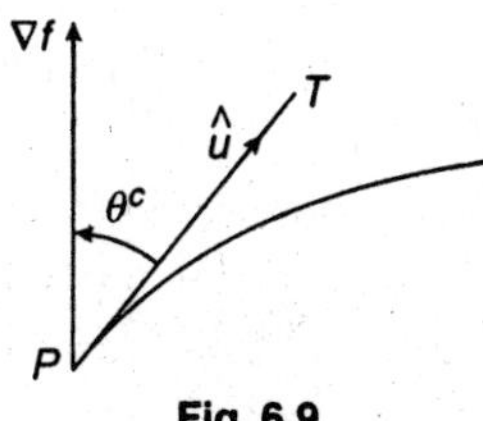

Fig. 6.9

where θ^c is the angle which the unit tangent vector $\hat{u}$, to a path through P, makes with ∇f (Fig. 6.9). So, $f'(\bar{r}(s))$ is maximum, being equal to $|\nabla f|$, when $\theta = 0$, i.e., dir $\hat{u} =$ dir ∇f. Thus ∇f is a vector whose magnitude is the maximum value of $f'(\bar{r}(s))$ and whose direction is given by that in which the rate of change of f is maximum. Thus we have the following theorem giving the physical interpretation of the gradient of a scalar function.

Theorem 2. The gradient of a scalar function at a point P, if not $\bar{0}$ is a vector whose magnitude is the maximum value of the rate of change of the function with respect to arc length, i.e., distance, and the direction of the vector is the direction in which the rate is maximum.

Properties of the gradient of a scalar function:

(1) $\nabla(cf) = c\nabla f$, c being a constant,

(2) $\nabla(f \pm g) = \nabla f \pm \nabla g$,

(3) $\nabla(fg) = g\nabla f + f\nabla g$

and **(4)** $\nabla(f/g) = (g\nabla f - f\nabla g)/g^2,\ g \neq 0$.

Examples 6.2. 1. *Given* $f(\overline{r}) = f(x,y,z) = x^2+y^2+z^2 = \sum x^2\ \forall\ (x,y,z) \in \mathbb{R}^3$, *draw level surfaces for* $f = 1, 4$ *and* 14. *Find the directional derivative of* f *at* $P(1,2,3)$ *in the direction of the vector* $\overline{a} = \hat{\imath} + 2\hat{\jmath} - 2\hat{k}$.

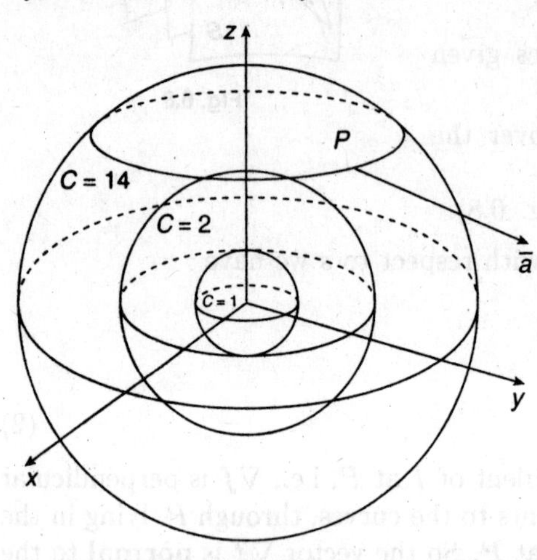

Fig. 6.10

The family of level surfaces is given by $f(x,y,z) = c$. For $c = 1$, 4 and 14 we get three concentric spherical surfaces

$$\sum x^2 = 1,\ \sum x^2 = 2^2 \ \text{ and } \ \sum x^2 = (\sqrt{14})^2$$

of radii 1, 2 and $\sqrt{14}$.

Since $1^2+2^2+3^2 = 14$, P lies on the level surface $\sum x^2 = 14$.

The directional derivative of f in the direction of a unit vector $\hat{u}$ is $\hat{u}\cdot\nabla f$. So the same in the direction of $\overline{a}$ is

$$\hat{a}\cdot\nabla f,$$

where $\hat{a}$ is a unit vector having the direction of $\overline{a}$ (Fig. 6.10).

So, $\hat{a} = (\hat{\imath}+2\hat{\jmath}-2\hat{k})/\sqrt{1^2+2^2+2^2} = \frac{1}{3}(\hat{\imath}+2\hat{\jmath}-2\hat{k})$.

Also, $f_x = 2x,\ f_y = 2y$ and $f_z = 2z$ and $f_x(P) = 2,\ f_y(P) = 4,\ f_z(P) = 6$.

$\therefore$ the required directional derivative is

$$\frac{1}{3}(\hat{\imath}+2\hat{\jmath}-2\hat{k})\cdot(2\hat{\imath}+4\hat{\jmath}+6\hat{k}) = \frac{2}{3}(1+4-6) = \frac{2}{3}.$$

2. *Find the directional derivative of*

$$f(x,y,z) = 2x^2+3y^2+z^2$$

at the point (2, 1, 3) in the direction of the vector $\hat{\imath} - 2\hat{k}$. **[W.B. Sem. Exam. 2001]**

Let $\overline{u} = \hat{\imath} - 2\hat{k}$. Then $\hat{u} = \overline{u}/|\overline{u}| = (\hat{\imath}-2\hat{k})/\sqrt{5}$. So, the directional derivative of f in the direction of $\overline{u}$ is

$$\nabla f\cdot\hat{u} = (f_x\hat{\imath}+f_y\hat{\jmath}+f_z\hat{k})\cdot(\hat{\imath}-2\hat{k})/\sqrt{5}$$
$$= (4x\hat{\imath}+6y\hat{\jmath}+2z\hat{k})\cdot(\hat{\imath}-2\hat{k})/\sqrt{5}$$
$$= 4(x-8)/\sqrt{5} = 4(2-3)/\sqrt{5} = -4/\sqrt{5}.$$

3. *In which direction from the point* $P(1,1,-1)$ *is the directional derivative of* $\phi(x,y,z) = x^2-2y+4z^2$ *maximum? What is the magnitude of this directional derivative?* **[W.B. Sem. Exam. 2003]**

The direction, in which the directional derivative of ϕ is maximum, is the direction of its gradient $\nabla\phi$. Now

$$\nabla\phi = \phi_x\hat{\imath}+\phi_y\hat{\jmath}+\phi_z\hat{k} = 2x\hat{\imath}-2\hat{\jmath}+8z\hat{k} = 2\hat{\imath}-2\hat{\jmath}-8\hat{k} \text{ at } P.$$

The corresponding unit vector is $\nabla\phi/|\nabla\phi| = (\hat{\imath} - \hat{\jmath} - 4\hat{k})/\sqrt{18}$. So, the direction cosines of $\nabla\phi$ are $1/3\sqrt{2}, -1/3\sqrt{2}, -4/3\sqrt{2}$. The magnitude of the directional derivative is

$$|\nabla\phi| = \sqrt{(2^2 + 2^2 + 8^2)} = 6\sqrt{2}.$$

4. *Find the constants a and b so that the surfaces $ax^2 + byz = (a+2)x$ and $4x^2y + z^3 = 4$ are mutually orthogonal at the point $P(1,-1,2)$.*

Let $f = ax^2 + byz - (a+2)x$ and $\phi = 4x^2y + z^3 - 4$. Then the equations to the surfaces are $f = 0$ and $\phi = 0$. For the orthogonality of the surfaces at P, first of all $f(P) = 0$ and $\phi(P) = 0$. So,

$$a + 2b - (a+2) = 0 \Rightarrow b = 1 \text{ and hence } f = ax^2 + yz - (a+2)x.$$

$$\therefore \quad \nabla f = \{2ax - (a+2)\}\hat{\imath} - z\hat{\jmath} - y\hat{k} = (a-2)\hat{\imath} - 2\hat{\jmath} + \hat{k} \text{ at } P.$$

Similarly, $\nabla\phi = 8xy\hat{\imath} + 4x^2\hat{\jmath} + 3z^2\hat{k} = -8\hat{\imath} + 4\hat{\jmath} + 12\hat{k}$ at P.

Since ∇f and $\nabla\phi$ are two vector orthogonal to the two surfaces $f = 0$ and $\phi = 0$ at P, these surfaces will intersect orthogonally at P if $\nabla f \perp \nabla\phi$, i.e., if $\nabla f \cdot \nabla\phi = 0$.

But $\quad \nabla f \cdot \nabla\phi = -8(a-2) - 8 + 12 = 0 \Rightarrow a = 5/2.$

Thus $\quad a = 5/2$ and $b = 1$.

5. *Find the angle between the tangent planes to the surfaces given by $f = 0$ and $g = 0$ at $P(1,0,1)$, where*

$$f = 2xe^y - z - 1 \quad \textit{and} \quad g = x^2 - xy - z.$$

The angle between the tangent planes to the two surfaces at a point P common to both of then is equal to the angle between their normals at P. The vectors ∇f and ∇g are normals to the two surfaces $f = 0$ and $g = 0$ respectively.

Now, $\nabla f = 2e^y\hat{\imath} + 2xe^y\hat{\jmath} - \hat{k} = 2\hat{\imath} + 2\hat{\jmath} - \hat{k}$ at $P(1,0,1)$

and $\quad \nabla g = (2x - y)\hat{\imath} - x\hat{\jmath} - \hat{k} = 2\hat{\imath} - \hat{\jmath} - \hat{k}$ at P.

The corresponding unit normal vectors are

$$\hat{u} = (2\hat{\imath} + 2\hat{\jmath} - \hat{k})/3 \quad \text{and} \quad \hat{v} = (2\hat{\imath} - \hat{\jmath} - \hat{k})/\sqrt{6}.$$

If θ^c is the angle between the tangent planes, then it is also the angle between $\hat{u}$ and $\hat{v}$. But

$$\hat{u} \cdot \hat{v} = \cos\theta \Rightarrow \cos\theta = 1/\sqrt{6} \quad \text{or,} \quad \theta = \cos^{-1} 1/\sqrt{6}.$$

Note: The angle between the tangent planes to two given surfaces at a point P common to both of them is also called **the angle between the surfaces** thereat. Also, the angle between the tangent planes is equal to the angle between the normals at the same point.

6. *Find the angle between the surfaces $z = x^2 + y^2$ and $x + y = 2$ at $P(1,1,2)$.*

Taking $f = x^2 + y^2 - z$ and $g = x + y - 2$, we find that P lies on both of them.

Now, $\quad \nabla f = 2x\hat{\imath} + 2y\hat{\jmath} - \hat{k} = 2\hat{\imath} + 2\hat{\jmath} - \hat{k}$ at P

and $\quad \nabla g = \hat{\imath} + \hat{\jmath}$ at P.

If $\hat{u} = \nabla f/|\nabla f|$ and $\hat{v} = \nabla g/|\nabla g|$, then

$$\hat{u} = (2\hat{\imath} + 2\hat{\jmath} - \hat{k})/3 \quad \text{and} \quad \hat{v} = (\hat{\imath} + \hat{\jmath})/\sqrt{2}$$

$$\cos\theta = \hat{u} \cdot \hat{v} = 4/3\sqrt{2} \quad \text{and} \quad \theta = \cos^{-1}(4/3\sqrt{2}).$$

Exercises 6.1

1. Find the directional derivative of $f(x, y, z) = xyz$ at (1,1,1) in the direction of $2\hat{i} - \hat{j} - 2\hat{k}$.
2. Find the directional derivative of $\phi(x, y, z) = 4e^{2x-y+z}$ at $P(-1, -1, 1)$ in the direction $\overrightarrow{PQ}$, where Q is $(3, -5, -6)$.
3. Find a unit normal to the surface $z = x^2 + y^2$ at $(-1, -2; 5)$.
4. If $\overline{r} = x\hat{i} + y\hat{j} + z\hat{k}$, find $\overline{r}.\ \nabla\phi$, where $\phi = x^3 + y^3 + z^3 - 3xyz$.
5. (i) If $\overline{r} = x\hat{i} + y\hat{j} + z\hat{k}$ and $r = |\overline{r}|$, prove that

$$\nabla r^n = nr^{n-2}\overline{r}.$$

(ii) Show that $\nabla r = -\overline{r}/r^3$.

6. Find $\nabla\phi$ if $\phi = \ln(x^2 + y^2 + z^2)$.
7. In what direction from $(3, 1, -2)$ is the directional derivative of ϕ maximum if $\phi = x^2y^2z^4$? Find also the magnitude of this maximum derivative.
8. Find the angle between the normals to the surface $xy = z^2$ at (4,1,2) and $(3, 3, -3)$.
9. Find the angle between the tangent planes to the surfaces $x \ln z = y^2 - 1$ and $x^2y = 2 - z$ at (1,1,1).
10. Find the angle between the surfaces $x^2 + y^2 + z^2 = 9$ and $z = x^2 + y^2 - 3$ at $(2, -1, 2)$.
11. For any vector $\overline{u}$ and for any differentiable scalar function ϕ, prove that (i) $(\overline{u} \cdot \nabla)\phi = \overline{u} \cdot (\nabla\phi)$, (ii) $(\overline{u} \cdot \nabla)\overline{r} = \overline{u}$, (iii) $\nabla(\overline{a} \cdot \overline{r}) = \overline{a}$, (iv) $\nabla \overline{r}^2 = 2\overline{r}$, where $\overline{r} = x\hat{i} + y\hat{j} + z\hat{k}$ and $\overline{a}$ is a constant vector.
12. If u and $\overline{v}$ are both differentiable functions of x, y and z, prove that $(\overline{v} \cdot \nabla u)\nabla u/(\nabla u)^2$ is a component of the vector $\overline{v}$ normal to the surface $u =$ constant.

Answers

1. $-1/3$. **2.** $41/9$. **3.** $(2\hat{i} + 4\hat{j} + \hat{k})/\sqrt{21}$. **4.** ϕ. **6.** $2(x\hat{i} + y\hat{j} + z\hat{k})/\sum x^2$.
7. The D.C.'s are $1/\sqrt{19}$, $3/\sqrt{19}$, $-3/\sqrt{19}$, $96/\sqrt{19}$. **8.** $\cos^{-1}(1/\sqrt{22})$.
9. $\cos^{-1}(1/\sqrt{30})$. **10.** $\cos^{-1}\{(8/63)\sqrt{21}\}$.

6.6 Divergence of a Vector Field

If $\overline{v}(= v_1\hat{i} + v_2\hat{j} + v_3\hat{k})$ is a vector function possessing first partial derivatives in a region D with respect to x, y and z then the scalar function $\partial v_1/\partial x + \partial v_2/\partial y + \partial v_3/\partial z$ is defined to be the divergence of the vector field $(D, \overline{v})$ and it is denoted by symbols like **div** $\overline{v}$ or $\boldsymbol{\nabla} \cdot \overline{v}$.

That the divergence of $\overline{v}$ can be expressed formally as $\nabla \cdot \overline{v}$ is obvious from the fact that

$$\nabla \cdot \overline{v} = (\hat{i}\partial/\partial x + \hat{j}\partial/\partial y + \hat{k}\partial/\partial z) \cdot (\hat{i}v_1 + \hat{j}v_2 + \hat{k}v_3) = \partial v_1/\partial x + \partial v_2/\partial y + \partial v_3/\partial z.$$

Thus the divergence of a vector function $\overline{v}$ can be regarded as the scalar product of vector operator ∇ and the vector function $\overline{v}$.

It can be shown that the divergence of a vector function is a scalar function, the value of which at every point of its domain is independent of any coordinate system.

If f is a scalar function possessing second order partial derivatives with respect to x, y and z, then since ∇f is a vector function, its divergence $\nabla \cdot \nabla f$ is again a scalar function denoted by $\nabla^2 f$.

Since $\nabla f = \hat{i} f_x + \hat{j} f_y + \hat{k} f_z$

$$\nabla^2 f = \nabla \cdot \nabla f = \partial f_x/\partial x + \partial f_y/\partial y + \partial f_z/\partial z = f_{xx} + f_{yy} + f_{zz}$$
$$= (\partial^2/\partial x^2 + \partial^2/\partial y^2 + \partial^2/\partial z^2) f.$$

So we have the identity of the two operators:

$$\nabla^2 = \partial^2/\partial x^2 + \partial^2/\partial y^2 + \partial^2/\partial z^2.$$

Each of these operators is known as the **Laplacian** and it is sometimes denoted by ∇.

Thus the Laplacian of f = div (grad f) = $\nabla \cdot \nabla f = \nabla^2 f = \nabla f = f_{x^2} + f_{y^2} + f_{z^2}$.

6.6.1 Physical Interpretation of Divergence

Suppose, some amount of fluid is injected at a point P. If the fluid is compressible, the amount of fluid leaving the point may not be equal to that entering the point. In that case, we say that there is divergence of the flow at P. Obviously, the density will fall in that case.

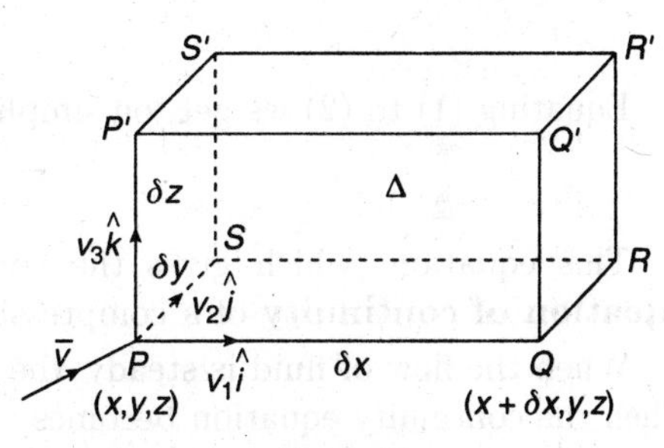

Fig. 6.11

To have a clear idea of divergence at a point $P(x, y, z)$ we imagine an elementary region Δ in the form of a rectangular parallelopiped $PQRSP'Q'R'S'$ with P at one of its corners. The point P could be taken anywhere else interior to Δ. But it is immaterial because ultimately, in the limit, the entire region Δ will shrink to the point P (Fig. 6.11). Let the lengths of the edges PQ, PS and PP' be δx, δy and δz respectively. Its volume is then given by

$$\delta V = \delta x \delta y \delta z.$$

Suppose, some compressible fluid like gas or vapour is flowing past the region Δ having no **sources** or **sinks** so that apart from the given fluid no extra fluid is produced in or disappears from Δ. Let the velocity of the fluid at P be

$$\overline{v} = v_1 \hat{i} + v_2 \hat{j} + v_3 \hat{k}.$$

If ρ is the density of fluid at P and $\overline{u} = \rho \overline{v}$, then $\overline{u}$ will be the mass of fluid entering the point P per unit area of cross-section perpendicular to $\overline{v}$ per unit time. If

$$\overline{u} = u_1 \hat{i} + u_2 \hat{j} + u_3 \hat{k},$$

then $u_1 = \rho v_1$, $u_2 = \rho u_2$ and $u_3 = \rho v_3$.

The amount of fluid entering Δ through the face $PSS'P'$ perpendicular to the x-axis in time δt is

$$u_1(x, y, z) \delta y \delta z \delta t.$$

The amount of fluid leaving Δ through the opposite face $QRR'Q'$ is

$$u_1(x + \delta x, y, z) \delta y \delta z \delta t.$$

So, the net amount of fluid leaving Δ in the direction of the x-axis in time δt is

$$[u_1(x+\delta x, y, z) - u_1(x, y, z)]\delta y \delta z \delta t \approx (\partial u_1/\partial x)\delta x \delta y \delta z \delta t = (\partial u_1/\partial x)\delta V \delta t.$$

Similarly, the net amounts of fluid leaving Δ in time δt in the directions of y- and z-axis are approximately

$$(\partial u_2/\partial y)\delta V \delta t \quad \text{and} \quad (\partial u_3/\partial z)\delta V \delta t$$

respectively. Taking into account all these three directions of flow the net loss of mass in Δ in time δt is approximately equal to

$$(\partial u_1/\partial x + \partial u_2/\partial y + \partial u_3/\partial z)\delta V \delta t \quad \text{or,} \quad \nabla \cdot \overline{u}\delta V \delta t \quad \text{or,} \quad \nabla \cdot \rho\overline{v}\delta V \delta t. \tag{1}$$

This loss of mass will cause a fall in the density. Since the density ρ is the mass per unit volume, $\partial\rho/\partial t$ is the time rate of increase of mass per unit volume. So, $-\partial\rho/\partial t$ is the loss of mass per unit time per unit volume. Hence the loss of mass in Δ of volume δV in time δt is

$$-(\partial\rho/\partial t)\delta V \delta t. \tag{2}$$

Equating (1) to (2) we get, on simplification, the equation

$$\nabla \cdot \rho\overline{v} + \partial\rho/\partial t = 0. \tag{3}$$

This equation, which gives the condition for conservation of mass, is the well-known **equation of continuity** of a compressible fluid flow in hydrodynamics.

When the flow of fluid is steady, the variables do not change with time, so that $\partial\rho/\partial t = 0$. Then the continuity equation becomes

$$\nabla \cdot (\rho\overline{v}) = 0. \tag{4}$$

If, further, the fluid is incompressible, ρ is constant and (4) reduces to

$$\nabla \cdot \overline{v} = 0. \tag{5}$$

This equation, therefore, gives the **condition for incompressibility** of fluid.

If the flux of a vector field $\overline{v}$ entering each point of the domain is the same as that leaving the point, i.e., if div $\overline{v} = 0$ at every point of the domain, then the field is called **solenoidal.**

6.7 Some Properties Involving Both Gradient and Divergence

If f and g are scalar functions and $\overline{v}$ is a vector function, then

(1) div (grad f) $= \nabla^2 f =$ Laplacian of f,

(2) div $(f\overline{v}) = f$ div $\overline{v} + \overline{v} \cdot \nabla f$,

(3) div $(f\nabla g) = \nabla f \cdot \nabla g + f\nabla^2 g$

and **(4)** div $(f\nabla g) -$ div $(g\nabla f) = f\nabla^2 g - g\nabla^2 f$.

(1) grad $f = \hat{i}f_x + \hat{j}f_y + \hat{k}f_z = \nabla f$.

$$\therefore \quad \text{div (grad } f) = \nabla \cdot \nabla f = \left(\hat{i}\frac{\partial}{\partial x} + \hat{j}\frac{\partial}{\partial y} + \hat{k}\frac{\partial}{\partial z}\right) \cdot (\hat{i}f_x + \hat{j}f_y + \hat{k}f_z)$$

$$= f_{xx} + f_{yy} + f_{zz} = \partial^2 f/\partial x^2 + \partial^2 f/\partial y^2 + \partial^2 f/\partial z^2$$

$$= (\partial^2/\partial x^2 + \partial^2/\partial y^2 + \partial^2/\partial z^2)f = \nabla^2 f.$$

(2) If $\overline{v} = v_1\hat{\imath} + v_2\hat{\jmath} + v_3\hat{k}, f\overline{v} = (fv_1)\hat{\imath} + f(v_2)\hat{\jmath} + (fv_3)\hat{k}$.

$\therefore \quad \text{div } (f\overline{v}) = \partial(fv_1)/\partial x + \partial(fv_2)/\partial y + \partial(fv_3)/\partial z.$

But $\partial(fv_1)/\partial x = f\partial v_1/\partial x + v_1\partial f/\partial x.$

$$\therefore \quad \text{div } (f\overline{v}) = f\sum \partial v_1/\partial x + \sum v_1 \partial f/\partial x$$
$$= f\nabla\cdot\overline{v} + \overline{v}\cdot\nabla f = f \text{ div } v + \overline{v}\cdot \text{ grad } f.$$

(3) $\text{div } (f\nabla g) = \nabla\cdot(f\overline{v}), \quad \text{where } \overline{v} = \nabla g$
$$= f\nabla\cdot\overline{v} + \overline{v}\cdot\nabla f \quad \text{by (2)}$$
$$= f\cdot\nabla\cdot\nabla g + \nabla g\cdot\nabla f = \nabla f\cdot\nabla g + f\nabla^2 g.$$

(4) Interchanging f and g in (3)
$$\text{div } (g\nabla f) = \nabla f\cdot\nabla g + g\nabla^2 f.$$

Subtracting the required result is obtained.

Examples 6.3. **1.** *Find the divergence of $x^2\hat{\imath} + y^2\hat{\jmath} + z^2\hat{k}$.*

If $\overline{v} = v_1\hat{\imath} + v_2\hat{\jmath} + v_3\hat{k} = x^2\hat{\imath} + y^2\hat{\jmath} + z^2\hat{k}$,

$$v_1 = x^2,\ v_2 = y^2, v_3 = z^2.$$

$\therefore \quad \partial v_1/\partial x = 2x, \partial v_2/\partial x = 2y,\ \partial v_3/\partial z = 2z,$

$\because \quad \text{div } \overline{v} = 2(x+y+z).$

2. *Find the directional derivative of div $\overline{v}$ at the point $P(4,4,2)$ in the direction of the outer normal of the sphere*

$$x^2 + y^2 + z^2 = 36, \quad \textit{where} \quad \overline{v} = x^3\hat{\imath} + y^3\hat{\jmath} + z^3\hat{k}.$$

Let $f =$ div $\overline{v}$. Then $f = 3(x^2+y^2+z^2)$ and $\nabla f = 6(x\hat{\imath} + y\hat{\jmath} + z\hat{k})$. The directional derivative in the direction of $\hat{n}$, a unit vector at P normal to the surface $\sum x^2 = 36$ is

$$\hat{n}\cdot\nabla f.$$

But if $g = \sum x^2 - 36$, the vector ∇g is normal to the surface $g = 0$. So, the unit normal vector as P in the direction of ∇g is

$$\hat{n} = \left.\frac{\nabla g}{|\nabla g|}\right|_P = \left.\frac{2(x\hat{\imath} + y\hat{\jmath} + z\hat{k})}{2\sqrt{\sum x^2}}\right|_P = \frac{4\hat{\imath} + 4\hat{\jmath} + 2\hat{k}}{6} = \frac{1}{3}(2\hat{\imath} + 2\hat{\jmath} + \hat{k}).$$

Also, ∇f at P is $6(4\hat{\imath} + 4\hat{\jmath} + 2\hat{k}) = 12(2\hat{\imath} + 2\hat{\jmath} + \hat{k})$.

$$\therefore \quad \hat{n}\cdot\nabla f = \frac{1}{3}(2\hat{\imath} + 2\hat{\jmath} + \hat{k})\cdot 12(2\hat{\imath} + 2\hat{\jmath} + \hat{k}) = 4(4+4+1) = 36.$$

or, $\quad \hat{n}\cdot\nabla\nabla\cdot\overline{v} = 36.$

Exercises 6.2

1. Find the divergence of the following vector functions:

(i) $x^2\hat{\imath} + y^2\hat{\jmath} - z\hat{k}$, **(ii)** $2yz(\hat{\imath} + \hat{\jmath} + \hat{k})$, **(iii)** $yz^2\hat{\imath} + zx^2\hat{\jmath} + xy^2\hat{k}$,

(iv) $(x\hat{\imath} + y\hat{\jmath} + z\hat{k})/\left(\sum x^2\right)^{3/2}$, **(v)** $e^x(\hat{\imath}\sin y + \hat{\jmath}\cos y)$.

2. Find the directional derivative of div $\overline{u}$ at (4, 4, 2) in the direction of the corresponding outer normal of the sphere $\sum x^2 = 36$, where

   (i) $\overline{u} = x^4\hat{i} + y^4\hat{j} + z^4\hat{k}$, (ii) $\overline{u} = zx\hat{i} + xy\hat{j} + yz\hat{k}$.

3. If the vector $\overline{u}$ and the scalar ϕ are both differentiable, prove that div $(\phi\overline{u}) = \phi$ div $\overline{u}$ – $\overline{u}$ grad ϕ.

4. Find div $\overline{v}$ if $\overline{v} =$ grad $(x^3 + y^3 + z^3 - 3xyz)$.

   [**Hint:** $\nabla\overline{v} = \nabla^2\phi = \phi_{x^2} + \phi_{y^2} + \phi_{z^2}$, where $\phi = \sum x^3 - 3\Pi x$.]

5. If $\overline{r} = x\hat{i} + y\hat{j} + z\hat{k}$ and $r = |\overline{r}|$, prove that

   (i) $\nabla \cdot (r^3\overline{r}) = 6r^3$, (ii) $\nabla[\nabla \cdot (\overline{r}/r)] = -2\overline{r}/r^3$, (iii) $\nabla^2 r^n = n(n+1)r^{n-2}$.

6. Find the value of the parameter a if $\nabla \cdot \overline{u} = 0$, where $\overline{u} = (ax^2 + z)y\hat{i} + x(y^2 - z^2)\hat{j} + 2xy(z - xy)\hat{k}$.

7. If $\overline{u}$ and $\overline{v}$ are two vectors joining the points $P(x_1, y_1, z_1)$ and $Q(x_2, y_2, z_2)$ respectively to a variable point $R(x, y, z)$, show that div $(\overline{u} \times \overline{v}) = 0$.

8. Show that $\nabla \cdot (f\nabla g - g\nabla f) = f\nabla^2 g = -g\nabla^2 f$, where f and g are scalar functions both possessing second order derivatives.

9. Show that the following vector functions are solenoidal:

   (i) $(2x^2 - 3yz)\hat{i} + (2y^2 - 3zx)\hat{j} - 4(x + y)z\hat{k}$, (ii) $(x + 3y)\hat{i} + (y - 3z)\hat{j} + (x - 2z)\hat{k}$.

Answers

1. **(i)** $2x + 2y - 1$; **(ii)** $\sum yz$; **(iii)** 0; **(iv)** 0; **(v)** 0.

2. **(i)** 272; **(ii)** 5/3. **4.** $6\sum x$. **6.** –2.

6.8 Curl of a Vector Field

We have already seen that the vector operator ∇ behaves like a scalar multiplication in gradient and like a scalar product in divergence. This suggests that the same operator may behave like a vector product with a vector function.

Let $\overline{F}(= X\hat{i} + Y\hat{j} + Z\hat{k})$ be a vector function possessing first order partial derivatives in a certain domain D. Then it can be proved that $(Z_y - Y_z)\hat{i} + (X_z - Z_x)\hat{j} + (Y_x - X_y)\hat{k}$ is actually a vector function forming a field in D. This vector function is called the **curl** or **rotation** of the vector function $\overline{F}$ and it is denoted by **curl $\overline{F}$** or **rot $\overline{F}$**. Formally, curl $\overline{F}$ can be expressed as $\nabla \times \overline{F}$, where

$$\nabla \times \overline{F} = \begin{vmatrix} \hat{i} & \hat{j} & \hat{k} \\ \partial/\partial x & \partial/\partial y & \partial/\partial z \\ X & Y & Z \end{vmatrix} \quad \text{or} \quad \begin{vmatrix} \hat{i} & \hat{j} & \hat{k} \\ D_x & D_y & D_z \\ X & Y & Z \end{vmatrix}. \tag{1}$$

It is to be noted that like $f\nabla$ and $\overline{F} \cdot \nabla$, $\overline{F} \times \nabla$ is also meaningless.

6.8.1 Physical Interpretation of Curl

We know that for a body rotating about an axis with the angular velocity $\overline{\omega}$

$$\overline{v} = \overline{\omega} \times \overline{r},$$

where $\overline{v}$ is the linear velocity of a point, in the body, whose p.v. relative to any point on the axis of rotation is $\overline{r}$. Let $\overline{r}$ be the p.v. of a point R of the body rotating about an axis through the origin O, with an angular velocity $\overline{\omega} = \theta\hat{i} + \phi\hat{j} + \psi\hat{k}$. If $\overline{v}$ be the linear velocity of the point, we know that

$$\overline{v} = \overline{\omega} \times \overline{r} = \begin{vmatrix} \hat{i} & \hat{j} & \hat{k} \\ \theta & \phi & \psi \\ x & y & z \end{vmatrix} = (z\phi - y\psi)\hat{i} + (x\psi - z\theta)\hat{j} + (y\theta - x\phi)\hat{k}.$$

$$\text{Now, } \operatorname{curl} \overline{v} = \nabla \times \overline{v} = \begin{vmatrix} \hat{i} & D_x & z\phi - y\psi \\ \hat{j} & D_y & x\psi - z\theta \\ \hat{k} & D_z & y\theta - x\phi \end{vmatrix} = 2(\theta\hat{i} + \phi\hat{j} + \psi\hat{k}) = 2\overline{\omega}.$$

Thus **for a body rotating about an axis with an angular velocity $\overline{\omega}$, the curl of the velocity field is a vector, which is $2\overline{\omega}$, i.e., twice the angular velocity of the body.**

The direction of curl $\overline{v}$ is the direction of $\overline{\omega}$, parallel to the axis of rotation such that the rotation is clockwise w.r.t. that direction. Since the curl of a vector field is closely associated with the rotation, it is sometimes called the **rotation of the vector field** and curl $\overline{v}$ is expressed as rot $\overline{v}$.

6.8.2 Properties of the Curl of a Vector

If f and g are scalar functions, whereas $\overline{u}$ and $\overline{v}$ are vector functions, then

(i) $\nabla \times (m\overline{v}) = m\nabla \times \overline{v}$, m being a constant,

(ii) $\nabla \times (f\overline{v}) = (\nabla f) \times \overline{v} + f\nabla \times \overline{v}$,

(iii) $\nabla \times (\overline{u} \pm \overline{v}) = \nabla \times \overline{u} \pm \nabla \times \overline{v}$,

(iv) $\nabla \cdot \nabla \times \overline{v} = 0$,

(v) $\nabla \cdot (\overline{u} \times \overline{v}) = \overline{u} \cdot \nabla \times \overline{u} + \overline{u} \cdot \nabla \times \overline{v}$

and **(vi)** $\nabla \times \nabla f = \overline{0}$.

$$\textbf{(i) } \nabla \times (m\overline{v}) = \begin{vmatrix} \hat{i} & \hat{j} & \hat{k} \\ D_x & D_y & D_z \\ mv_1 & mv_2 & mv_3 \end{vmatrix} = m \begin{vmatrix} \hat{i} & \hat{j} & \hat{k} \\ D_x & D_y & D_z \\ v_1 & v_2 & v_3 \end{vmatrix} = m\nabla \times \overline{v}.$$

$$\textbf{(ii) } \nabla \times (f\overline{v}) = \begin{vmatrix} \hat{i} & \hat{j} & \hat{k} \\ D_x & D_y & D_z \\ fv_1 & fv_2 & fv_3 \end{vmatrix} = \{D_y(fv_3) - D_z(fv_2)\}\hat{i} + \cdots$$

Now, $D_y(fv_3) - D_z(fv_2) = f_y v_3 + fD_y v_3 - f_z v_2 - fD_z v_2 = (f_y v_3 - f_z v_2) + f(D_y v_3 - D_z v_2)$.

$\therefore \quad \nabla \times (f\overline{v}) = [(f_y v_3 - f_z v_2)\hat{i} + \cdots] + f(D_y v_3 - D_z v_2)\hat{i}$.

$$\text{Now, } (\nabla f) \times \overline{v} = \begin{vmatrix} \hat{i} & \hat{j} & \hat{k} \\ f_x & f_y & f_z \\ v_1 & v_2 & v_3 \end{vmatrix} \text{ and } \nabla \times \overline{v} = \begin{vmatrix} \hat{i} & \hat{j} & \hat{k} \\ D_x & D_y & D_z \\ v_1 & v_2 & v_3 \end{vmatrix}.$$

$\therefore \quad \nabla \times (f\overline{v}) = (\nabla f) \times \overline{v} + f\nabla \times \overline{v}$.

$$\textbf{(iii) } \nabla \times (\overline{u} \pm \overline{v}) = \begin{vmatrix} \hat{i} & \hat{j} & \hat{k} \\ D_x & D_y & D_z \\ u_1 \pm v_1 & u_2 \pm v_2 & u_3 \pm v_3 \end{vmatrix} = \begin{vmatrix} \hat{i} & \hat{j} & \hat{k} \\ D_x & D_y & D_z \\ u_1 & u_2 & u_3 \end{vmatrix} \pm \begin{vmatrix} \hat{i} & \hat{j} & \hat{k} \\ D_x & D_y & D_z \\ v_1 & v_2 & v_3 \end{vmatrix}$$

$$= \nabla \times \overline{u} \pm \nabla \times \overline{v}.$$

(iv) $\nabla \cdot \nabla \times \overline{v} = (\hat{i}D_x+\hat{j}D_y+\hat{k}D_z)\cdot[\hat{i}(D_yv_3-D_zv_2)+\hat{j}(D_zv_1-D_xv_3)+\hat{k}(D_xv_2-D_yv_1)]$

$$= D_x(D_yv_3-D_zv_2)+D_y(D_zv_1-D_xv_3)+D_z(D_xv_2-D_yv_1)$$

$$= (D_yD_z-D_zD_y)v_1+(D_zD_x-D_xD_z)v_2+(D_xD_y-D_yD_x)v_3.$$

If we assume that $\overline{v}$ possesses continuous second order partial derivatives, then $\partial^2 v_1/\partial y\partial z = \partial^2 v_1/\partial z\partial y$, etc. Hence the result.

(v) $\nabla \cdot (\overline{u}\times\overline{v}) = D_x(u_2v_3-u_3v_2)+D_y(u_3v_1-u_1v_3)+D_z(u_1v_2-u_2v_1)$

$$= u_2D_xv_3+v_3D_xu_2-u_3D_xv_2-v_2D_xu_3+\cdots$$

$$= [v_1(D_yu_3-D_zu_2)+\cdots]+[u_1(D_yv_3-D_zv_2)+\cdots].$$

Now, $\overline{v}\cdot\nabla\times\overline{u} = v_1(D_yu_3-D_zu_2)+\cdots$

and $\overline{u}\cdot\nabla\times\overline{v} = u_1(D_yv_3-D_zv_2)+\cdots.$

$\therefore$ $\nabla\cdot(\overline{u}\times\overline{v}) = \overline{u}\cdot\nabla\times\overline{v}+\overline{v}\cdot\nabla\times\overline{u}.$

(vi) $\nabla\times\nabla f = \begin{vmatrix} \hat{i} & D_x & D_xf \\ \hat{j} & D_y & D_yf \\ \hat{k} & D_z & D_zf \end{vmatrix} = \hat{i}(D_yD_zf-D_zD_yf)+\cdots$

If f possesses continuous second order partial derivatives, then $f_{yz} = f_{zy}$ or, $D_yD_zf = D_zD_yf$. Hence the required result.

From the property (vi) above, we find that **if a vector field is the gradient of a scalar field, then the curl of such a vector field is always a null vector.**

Since the curl is always associated with the rotation of a vector field, if the curl of a vector field is seen to be a null vector at every point of the domain of the vector function, then such a vector field is called **irrotational**. So, the **gradient field is always irrotational.**

Conservative Field: A vector field $(D,\overline{v})$ is said to be **conservative** if v can be expressed as the gradient of a scalar function so that curl $\overline{v} = \overline{0}$ everywhere in D.

Example of a conservative field is the gravitational field.

Examples 6.4. 1. *If $\overline{r}$ is the p.v. of a point $R(x,y,z)$ relative to the origin O, show that rot $\overline{r} = \overline{0}$.*

$$\text{rot}\,\overline{r} = \nabla\times\overline{r} = \begin{vmatrix} \hat{i} & \hat{j} & \hat{k} \\ D_x & D_y & D_z \\ x & y & z \end{vmatrix} = (z_y-y_z)\hat{i}+(x_z-z_x)\hat{j}+(y_x-x_y)\hat{k}$$

$= \overline{0}$, since x, y and z are all independent.

2. *Find the curl of the following vectors:*

(i) $\overline{u} = xyz\hat{i}+3x^2y\hat{j}+(xz-y^2)z\hat{k}$,

(ii) $\overline{F} = (6xy+z^3)\hat{i}+(3x^2-z)\hat{j}+(3xz^2-y)\hat{k}$.

(i) $\nabla\times\overline{u} = \begin{vmatrix} \hat{i} & \hat{j} & \hat{k} \\ D_x & D_y & D_z \\ xyz & 3x^2y & (xz-y^2)z \end{vmatrix} = -2yz\hat{i}+(xy-z^2)\hat{j}+(6xy-xz)\hat{k}$

$$= -2yz\hat{i}+(xy-z^2)\hat{j}+x(6y-z)\hat{k}.$$

$$\text{(ii) } \nabla \times \overline{F} = \begin{vmatrix} \hat{i} & D_x & 6xy + z^3 \\ \hat{j} & D_y & 3x^2 - z \\ \hat{k} & D_z & 3xz^2 - y \end{vmatrix} = (-1+1)\hat{i} + (3z^2 - 3z^2)\hat{j} + (6x - 6x)\hat{k} = \overline{0}.$$

3. *Show that the following vector is irrotational:*

$$\overline{v} = (x^2 - yz)\hat{i} + (y^2 - zx)\hat{j} + (z^2 - xy)\hat{k}.$$

$$\text{rot } \overline{v} = \begin{vmatrix} \hat{i} & D_x & x^2 - yz \\ \hat{j} & D_y & y^2 - zx \\ \hat{k} & D_z & z^2 - xy \end{vmatrix} = (-x+x)\hat{i} + (-y+y)\hat{j} + (-z+z)\hat{k} = \overline{0}.$$

So, $\overline{v}$ is irrotational.

Exercises 6.3

1. Show that the curl of the gradient of a scalar function is always a null vector, provided the function possesses continuous second order partial derivatives. Verify by taking the function $f = x^2y + 2xy + z^2$. **[W.B. Sem. Exam. '03]**

2. Find the value of a if the vector function

$$(ax^2 + z)y\hat{i} + (y^2 - z^2)x\hat{j} + 2(z - xy)xy\hat{k}$$

is solenoidal. Hence find the curl of this solenoidal vector.

3. Find the rot of each of the following vectors:
   (i) $z^2\hat{i} + x^2\hat{j} + y^2\hat{k}$, **(ii)** $xz\hat{i} - yz\hat{j}$, **(iii)** $e^{xyz}(\hat{i} + \hat{j} + \hat{k})$, **(iv)** $\nabla(x^3 + y^3 + z^3 - 3xyz)$.

4. With respect to the right-handed Cartesian frame (x, y, z), find curl $(\overline{u} \times \overline{v})$, where $\overline{u} = y\hat{i} + z\hat{j} + x\hat{k}$ and $\overline{v} = xy\hat{i} + yz\hat{j} + zx\hat{k}$.

5. If $\overline{u}$ and $\overline{v}$ are the two vectors joining the fixed points $P(x_1, y_1, z_1)$ and $Q(x_2, y_2, z_2)$ respectively to a variable point $R(x, y, z)$, prove that

$$\text{curl } (\overline{u} \times \overline{v}) = 2(\overline{u} - \overline{v}).$$

6. If $\overline{r} = x\hat{i} + y\hat{j} + z\hat{k}$ and $\overline{a}$ is constant, prove that
   (i) curl $(\overline{a} \times \overline{r}) = 2\overline{a}$, **(ii)** rot $\{(\overline{a} \cdot \overline{r})\overline{r}\} = \overline{a} \times \overline{r}$, **(iii)** $\nabla \times (\overline{a} \cdot \overline{r})\overline{a} = \overline{0}$.

7. Show that
   (i) $\nabla \times (\overline{u} \times \overline{v}) = (\overline{v} \cdot \nabla)\overline{u} - (\overline{u} \cdot \nabla)\overline{v} + \overline{u}\nabla \cdot \overline{v} - \overline{v}\nabla \cdot \overline{u}$,
   (ii) $\nabla(\overline{u} \cdot \overline{v}) = \overline{u} \times (\nabla \times \overline{v}) + \overline{v} \times (\nabla \times \overline{u}) + (\overline{u} \cdot \nabla)\overline{v} + (\overline{v} \cdot \nabla)\overline{u}$, **[W.B. Sem. Exam. '03]**
   (iii) $\nabla \times (\nabla \times v) = \nabla(\nabla \cdot \overline{v}) - \nabla \overline{v}$.

Answers

2. -2.

3. **(i)** $2(y\hat{i} + z\hat{j} + x\hat{k})$; **(ii)** $y\hat{i} + x\hat{j}$; **(iii)** $e^{xyz}\{x(z-y)\hat{i} + y(x-z)\hat{j} + z(y-x)\hat{k}\}$; **(iv)** $\overline{z}$.

4. $(2yz - zx + xy)\hat{i} + (2zx - xy + yz)\hat{j} + (2xy - yz + zx)\hat{k}$.

INTEGRAL CALCULUS OF VECTORS

6.9 Oriented Curve. Arc Length

Let $x = x(t)$, $y = y(t)$ and $z = z(t)$ be the parametric equations of a certain curve. If the common domain of these functions be an interval $I_t = [t_0, t']$ on the t-axis so that for every point $T(t)$ on the t-axis we get a point $P(x(t), y(t), z(t))$ in xyz-space, then we get an arc Γ of the curve (Fig. 6.12). When t increases from t_0 to t', it might be, that a point P on Γ moves from one end point P_0 to another end point P' of Γ. This gives an orientation, i.e., a direction of description of the arc. If, as t **increases** from t_0 to t' a point moves from P' to P_0, then we get the same arc Γ with orientation reversed. If Γ is the curve with its orientation from P_0 to P', then the same curve with the reversed orientation will be denoted by $-\Gamma$, so that Γ and $-\Gamma$ are oppositely oriented arcs.

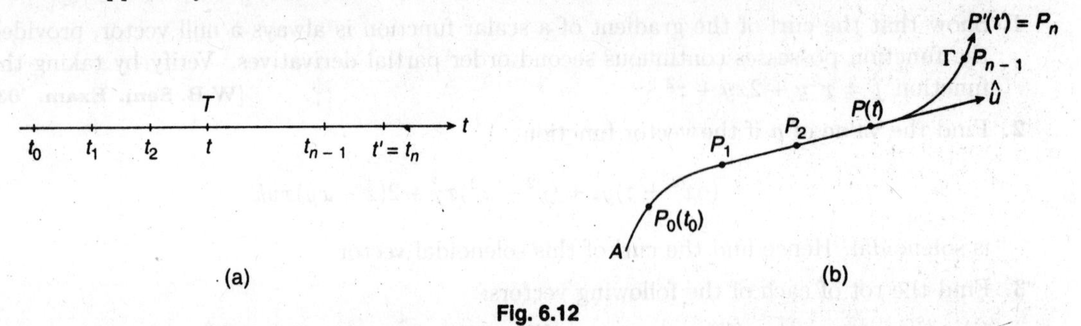

Fig. 6.12

The arc Γ is said to be **continuous** and **piecewise (or sectionally) smooth** if the three functions $x(t)$, $y(t)$ and $z(t)$ are all continuous and its derivatives $\dot{x}$, $\dot{y}$ and $\dot{z}$ are piecewise continuous and $\sum \dot{x}^2 \neq 0$ at any point in I_t and it is said to be **smooth** if the derivatives $\dot{x}$, $\dot{y}$ and $\dot{z}$ are continuous in I_t and $\sum \dot{x}^2 \neq 0$.

The vector equation of Γ is

$$\bar{r} = \bar{r}(t) = x(t)\hat{i} + y(t)\hat{j} + z(t)\hat{k}, \tag{1}$$

for all $t \in I_t$. Let s be the length of the arc AP, where A is some suitable point on the curve and P is on the arc Γ (Fig. 6.12). Then, by § 6.5, $d\bar{r}/ds = \hat{u}$, a unit vector tangential to Γ so that $|d\bar{r}/ds| = |\hat{u}| = 1$ and hence

$$\sum (dx/ds)^2 = 1.$$

To get the length of the arc Γ, we divide this arc into a finite number, n, of parts by means of a sequence of points $\{P_r\}_{r=0}^{n}$, where $P_n = P'$ and the arcs $\overset{\frown}{P_{r-1}P_r}$ maintaining the same orientation as that of Γ (Fig. 6.12). Let $S_n = \sum_{r=1}^{n} P_{r-1}P_r$. The sum of the lengths of the n **chords** $P_0P_1, P_1P_2, \ldots, P_{n-1}P_n$. For different values of n and for different ways of division of the arc into a finite number of parts, S_n will be different. The least upper bound (lub) of these sums when it exists, is said to be **rectifiable** and is defined to be **the length of the arc Γ**. If Γ is continuous and piecewise smooth, it can be shown that if all the chord lengths tend to zero, then

$$\operatorname*{Lt}_{n\to\infty} S_n = l,$$

the length of the arc Γ and it is denoted by the symbol

$$\int_{\Gamma} ds = \int_{P_0}^{P'} ds = \int_{s_0}^{s'} ds = s' - s_0,$$

where s_0 and s' are the lengths of the arcs $\widehat{AP_0}$ and $\widehat{AP'}$ respectively. Since

$$\dot{\bar{r}} = d\bar{r}/dt = (d\bar{r}/ds)(ds/dt) = \hat{u}\dot{s},$$

$$\dot{\bar{r}}^2 = \hat{u}^2\dot{s}^2 = \dot{s}^2 \quad \text{and} \quad |\dot{\bar{r}}| = \sqrt{\dot{\bar{r}}^2} = |\dot{s}| = \pm\dot{s} \tag{2}$$

according as $\dot{s} >$ or $< 0 \ \forall\, t \in (t_0, t')$. Then

$$l = \int_T ds = \pm \int_{t_0}^{t'} \frac{ds}{dt} dt = \pm \int_{t_0}^{t'} \dot{s} dt,$$

according as $\dot{s} >$ or < 0. So,

$$l = \int_{t_0}^{t'} |\dot{r}| dt.$$

If the arc Γ lies in the xy-plane, the equation (1) becomes

$$\bar{r} = \bar{r}(t) = x(t)\hat{\imath} + y(t)\hat{\jmath}.$$

So, $\dot{\bar{r}} = \dot{x}\hat{\imath} + \dot{y}\hat{\jmath},$

$$|\dot{\bar{r}}| = \sqrt{\dot{x}^2 + \dot{y}^2} = |\dot{s}| = \pm\dot{s} \tag{3}$$

according as $\dot{s} >$ or < 0

$$l = \int_{t_0}^{t'} |\dot{\bar{r}}| dt.$$

If $x = a\cos\theta$ and $y = a\sin\theta$, where $a > 0$, then

$$\bar{r} = a(\hat{\imath}\cos\theta + \hat{\jmath}\sin\theta),$$

which is the vector equation of the circle of radius a and centre at the origin. If $0 \le \theta \le \alpha \le 2\pi$, then the equation represents an arc $\widehat{PQ}$ of the circle with its orientation from P to Q which corresponds to θ increasing from 0 to α. So, its arc length

$$l = \int_P^Q ds = \int_0^{\alpha} \sqrt{\dot{x}^2 + \dot{y}^2} d\theta$$

$$= \int_0^{\alpha} \sqrt{(-a\sin\theta)^2 + (a\cos\theta)^2} d\theta = \int_0^{\alpha} a d\theta = a\alpha.$$

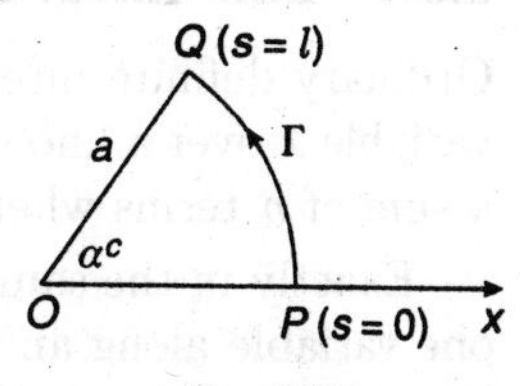

Fig. 6.13

If $x = a\cos(\alpha - \theta)$ and $y = a\sin(\alpha - \theta)$, where $0 \le \theta \le \alpha$, then again we get the same arc but, with its orientation reversed. In this case, as θ increases from 0 to α, the sense of description becomes from Q to P. So,

$$l = \int_{\Gamma} ds = -\int_0^{\alpha} \dot{s} d\theta, \text{ if } s \text{ is measured from } P \text{ to } Q,\ \dot{s} = ds/d\theta$$

$$= \int_0^{\alpha} |\dot{\bar{r}}| d\theta = a \int_0^{\alpha} |\hat{\imath}\sin(\alpha - \theta) - \hat{\jmath}\cos(\alpha - \theta)| d\theta$$

$$= a \int_0^{\alpha} d\theta = a\alpha.$$

Alternatively, since $s = a(\alpha - \theta), \dot{s} = -a$

and $\quad -\dot{s} = a \quad$ and $\quad l = \int_0^\alpha |\dot{s}| d\theta$

$$\int_0^\alpha -\dot{s} d\theta = a \int_0^\alpha d\theta = a\alpha.$$

If the equation to Γ is given in polar coordinates (r, θ),

$$\bar{r} = x\hat{i} + y\hat{j} = r\cos\theta\hat{i} + r\sin\theta\hat{j},$$

and $$\dot{\bar{r}} = \dot{r}(\hat{i}\cos\theta + \hat{j}\sin\theta) + r(-\hat{i}\sin\theta + \hat{j}\cos\theta)\dot{\theta},$$

where r and θ are both functions of a parameter t.

$$\therefore \qquad |\dot{\bar{r}}| = \sqrt{(\dot{r}^2 + r^2\dot{\theta}^2)} \tag{3}$$

and $$l = \pm \int_{t_0}^{t'} \sqrt{(\dot{\bar{r}}^2 + r^2\dot{\theta}^2)} dt,$$

according as $t_0 <$ or $> t'$.

If Γ is given in spherical polar coordinates (r, θ, ϕ),

$$x = r\sin\theta\cos\phi, \quad y = r\sin\theta\sin\phi \quad \text{and} \quad z = r\cos\theta,$$

$$\dot{x} = \dot{r}\sin\theta\cos\phi + r\cos\theta\cos\phi\dot{\theta} - r\sin\theta\sin\phi\dot{\phi},$$

$$\dot{y} = \dot{r}\sin\theta\sin\phi + r\cos\theta\sin\phi\dot{\theta} + r\sin\theta\cos\phi\dot{\phi}$$

and $$\dot{z} = \dot{r}\cos\theta - r\sin\theta\dot{\theta},$$

where x, y, z or r, θ and ϕ have been assumed to be functions of the parameter t in $I_t = [t_0, t']$.

Then $\dot{\bar{r}} = \dot{x}\hat{i} + \dot{y}\hat{j} + \dot{z}\hat{k}$

$$|\dot{\bar{r}}| = \sqrt{\sum \dot{x}^2} = \sqrt{(\dot{r}^2 + r^2\dot{\theta}^2 + r^2\sin^2\theta\dot{\phi}^2)} \tag{4}$$

So, $$l = \int_{t_0}^{t'} \sqrt{(\dot{r}^2 + r^2\dot{\theta}^2 + r^2\sin^2\theta\dot{\phi}^2)} dt.$$

6.9.1 Line Integral. Integration along a Line

Ordinary definite integral $\int_a^b f(x)dx$ can be regarded as the integral of a function of a single variable x over a line segment AB on the x-axis from $A(a)$ to $B(b)$. It is defined as the limit of a sum of n terms when n tends to infinity.

Exactly in the same way, we can have the idea of the integration of a function of more than one variable along an arc, called a line integral.

Let Γ be an arc of a curve, with P_0 and P' as its end points, lying in the xyz-space $\mathbb{R}^3$. Let us also assume that Γ is a continuous, piecewise smooth, rectifiable arc having its length s and having its orientation from P_0 to P' (Fig. 6.12). Let the arc be divided into n parts by a sequence of points $\{P_r\}_{r=0}^n$, where $P_n = P'$, such that the orientation of the elementary arc $P_{r-1}P_r$ is from P_{r-1} to P_r for each $r = 1, 2, \ldots, n$. Let the length of the arc

$$AP_r = s_r \text{ for } r = 0, 1, 2, \ldots, n$$

and $\delta s_r = s_r - s_{r-1} =$ the length of the arc $P_{r-1}P_r$. Then $\delta s_r > 0$ for all r. Let P_r' be any arbitrary point on the arc $P_{r-1}P_r$ and let $F(x, y, z)$ be a function whose domain includes the arc Γ so that the value of F for every point on Γ is known. Let us form the sum

$$S_n = \sum F(P'_r)\delta s_r. \tag{1}$$

For different ways of partitioning of Γ into a finite number of parts and for different position of P'_r on the rth elementary arc $P_{r-1}P_r$ we are expected to get different sums like S_n. Let us increase the number of parts, n, such that the greatest of the elementary arc lengths δs_r diminishes indefinitely. Then the limit of S_n as $n \to \infty$, if it exists and is independent of the choice of P'_r, is defined to be the **line integral of F along Γ from P_0 to P'** and it is denoted by

$$\int_\Gamma F ds = \int_{P_0}^{P'} F ds = \int_{s_0}^{s'} F ds, \tag{2}$$

where $F = F(x(s), y(s), z(s))$, so that $x = x(s)$, $y = y(s)$ and $z = z(s)$ are the parametric equations of the curve of which Γ is a part. It is often difficult to express x, y and z in terms of the parameter s. If instead of s, any other parameter t is taken so that the vector equation of Γ is

$$\bar{r} = \bar{r}(t) = x(t)\hat{i} + y(t)\hat{j} + z(t)\hat{k},$$

where $t_0 \le t \le t'$, then the line integral (2) becomes

$$\int_\Gamma F ds = \pm \int_{t_0}^{t'} \phi(t) dt, \tag{3}$$

according as $\dot{s} >$ or < 0, where

$$\phi(t) = F(x(t), y(t), z(t)\dot{s}.$$

Cor. 1. If $F(P) = 1 \ \forall \ P \in \Gamma$, then $\int_\Gamma F ds = \int_\Gamma ds = s$, the length of the arc Γ.

2. If $F(P) = \lambda(P)$ the linear density of Γ at $P, \lambda\delta s = \delta m$, the mass of an element of Γ at P, so the line integral

$$\int_\Gamma \lambda ds = \int_\Gamma dm = M,$$

the mass of the arc Γ.

In many practical applications, we have line integrals of the type

(i) $\int_\Gamma f dx$, **(ii)** $\int_\Gamma g dy$ or **(iii)** $\int_\Gamma h dx$.

The meaning of the symbol $\int_\Gamma f dx$ is given by

$$\int_\Gamma f dx = \pm \int_\Gamma f \frac{dx}{ds} ds = \pm \int_\Gamma F ds,$$

so that $F = F(x(s), y(s), z(s)) = f(x(s), y(s), z(s))dx/ds$, where the sign is + or − according as $dx/ds >$ or < 0.

Similarly, $\int_\Gamma g dy = \pm \int_\Gamma G ds$, where $G = g(x(s), y(s), z(s))dy/ds$ and

$\int_\Gamma h dz = \pm \int_\Gamma H ds$, where $H = h(x(s), y(s), z(s))dz/ds$.

6.9.2 Exact Differential Form

A differential form: $fdx + gdy + hdz$, where f, g and h are functions of x, y and z, is said to be exact, if it is the differential of some function u, i.e.,

$$du = fdx + gdy + hdz.$$

A condition for the differential form to be exact is that

$$\nabla \times (f\hat{i} + g\hat{j} + h\hat{k}) = \overline{0}$$

or, $h_y = g_z$, $f_z = h_x$ and $g_x = f_y$.

Examples 6.5. **1.** *$dx + dy + dz$ is exact. For, if $u = x + y + z$, then $du = dx + dy + dz$.*

2. *$xdx + ydy + zdz$ is exact. For, it is the differential of $u = \sum x^2/2$.*

3. *$yzdx + zxdy + xydz$ is exact, being the differential of $u = xyz$. Similarly, if f and g are functions of two variables x and y then $fdx + gdy$ is exact if there exists a function u such that $du = fdx + gdy$, or $f = u_x$ and $g = u_y$, or $f_y = g_x$.*

6.9.3 Vectorial Representation of a Line Integral

If $\overline{F} = f\hat{i} + g\hat{j} + h\hat{k}$ is a vector function of x, y and z, and its domain contains an arc Γ given by

$$\overline{r} = \overline{r}(t) = x(t)\hat{i} + y(t)\hat{j} + z(t)\hat{k}$$

for $t \in I_t = [t_0, t']$, Γ being continuous and piecewises smooth then since

$$d\overline{r} = \hat{i}dx + \hat{j}dy + \hat{k}dz = (\hat{i}\dot{x} + \hat{j}\dot{y} + \hat{k}\dot{z})dt,$$

$$\overline{F} \cdot d\overline{r} = fdx + gdy + hdz$$

and
$$\int_\Gamma \overline{F} \cdot d\overline{r} = \int_\Gamma fdx + gdy + hdz$$
$$= \int_\Gamma fdx + \int_\Gamma gdy + \int_\Gamma hdz. \qquad (4)$$

If Γ is a plane curve lying in the xy-plane, we have

$$\int_\Gamma \overline{F} \cdot d\overline{r} = \int_\Gamma fdx + gdy = \int_\Gamma fdx + \int_\Gamma gdy.$$

If Γ lies in the yz-plane, then

$$\int_\Gamma \overline{F} \cdot d\overline{r} = \int_\Gamma gdy + hdz = \int_\Gamma gdy + \int_\Gamma hdz.$$

6.9.4 Physical Interpretation of a Line Integral

If $\overline{F}$ is a force function and a particle is constrained to move along a given curve Γ from P_0 to P', then (4) represents the amount of work done by $\overline{F}$. Thus, if W is the work done by $\overline{F}$ in displacing a particle from P_0 to P' along Γ, then

$$W = \int_\Gamma \overline{F} \cdot d\overline{r}.$$

Conservative Field: A field of force $\overline{F}$ is said to be conservative in D, if for every pair of points P and Q in it, the work done by $\overline{F}$ in moving a particle from P to Q is independent of the path lying in D and connecting P and Q.

Theorem 1. The line integral along a path from P to Q is independent of the path, if $\overline{F}$ is the gradient of a scalar function.

Proof. Let Γ be any curve in the domain of $\overline{F}$ connecting P and Q. It is assumed that Γ is continuous and sectionally smooth. Let $\overline{F} = \nabla\phi$, ϕ being continuous, possessing continuous partial derivatives w.r.t. x, y and z. Then

$$\int_{\Gamma} \overline{F} \cdot d\overline{r} = \int_{\Gamma} \nabla\phi \cdot d\overline{r} = \int_{P}^{Q} d\phi = \phi(Q) - \phi(P), \tag{5}$$

which depends only on the values of ϕ at the end points P and Q. In particular, if $P = Q$, the path Γ is a closed path and we have

$$\oint_{\Gamma} \overline{F} \cdot d\overline{r} = 0.$$

Thus **for any conservative force field $(\overline{F}, D)$, the total amount of work done by $\overline{F}$ along every closed contour in D is always zero.**

Examples 6.6. 1. *Show that the integral* $\int_C (ydx + xdy)$ *is independent of the path C joining $P(0,1)$ and $Q(1,2)$.*

If $\overline{F} = y\hat{i} + x\hat{j}$ and $\overline{r} = x\hat{i} + y\hat{j}$, the p.v. of any point on C, then the p.v.'s of P and Q are

$\overline{r}_0 = 0\hat{i} + 1\hat{j} = \hat{j}$ and $\overline{r}_1 = 1\hat{i} + 2\hat{j}$ respectively.

Now, $\overline{F} \cdot d\overline{r} = (y\hat{i} + x\hat{j}) \cdot (\hat{i}dx + \hat{j}dy) = ydx + xdy$

$$= d(xy) = d\phi = \nabla\phi \cdot d\overline{r}, \phi = xy.$$

Then $I = \int_C (ydx + xdy) = \int_C F \cdot d\overline{r} = \int_P^Q d\phi = \phi(Q) - \phi(P)$

$$\phi(Q) - \phi(P) = (xy)_{(1,2)} - (xy)_{(0,1)} = 1.2 - 0 = 2,$$

which is true for any continuous piecewise smooth curve C. For example, if C is the segment PQ of a straight line then

$$PQ = \{(x,y) : \frac{x-0}{0-1} = \frac{y-1}{1-2}, \text{ where } 0 \le x \le 1 \text{ and } 1 \le y \le 2\} \text{ or}$$

$$PQ = \{(x,y) : x = -t \text{ and } y = -t+1, \ \forall\, t \in [-1,0]\}.$$

Here P and Q correspond respectively to $t = 0$ and $t = -1$. As a point moves from P to Q, t decreases from 0 to -1.

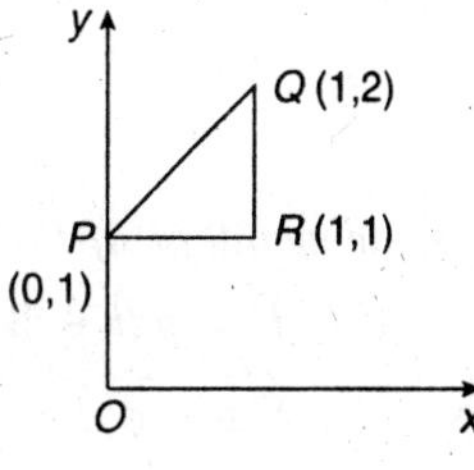

Fig. 6.14

So, $I = \int_C ydx + xdy = \int_0^{-1} \{(-t+1)(-dt) + (-t)(-dt)\}$

$$= \int_0^{-1} (2t-1)dt = [t^2 - t]_0^{-1} = 2 - 0 = 2.$$

Again, if $C = C_1 \cup C_2$, where $C_1 = PR = \{(x,y) : 0 \le x \le 1, y = 1\}$ and $C_2 = RQ = \{(x,y) : x = 1, 1 \le y \le 2\}$,

then $I = \int_{C_1} (1 \cdot dx + x \cdot 0) + \int_{C_2} (y \cdot 0 + 1 \cdot dy)$

$$= \int_0^1 dx + \int_1^2 dy = 1 + 1 = 2.$$

2. *Show that the line integral* $\int_P^Q x^2ydx + xyzdy + y^3dz$ *is not independent of the path of integration, where* P *and* Q *are respectively* $(0,0,0)$ *and* $(1,1,1)$.

The equations $y = x$ and $z = x$ for $0 \leq x \leq 1$ represent the line segment PQ. As x increases from 0 to 1, a point moves from P to Q so that P and Q correspond respectively to $x = 0$ and $x = 1$ respectively. Therefore, the given integral is

$$I = \int_0^1 x^2 \cdot xdx + x \cdot x \cdot xdx + x^3dx = 3\int_0^1 x^3dx = \frac{3}{4}. \qquad \text{(i)}$$

The equations $x = -t$, $y = t^2$ and $z = -t^3$, for $-1 \leq t \leq 0$, also represent a path connecting P and Q, which correspond respectively to $t = 0$ and $t = -1$. But as t increases from -1 to 0, a point moves from Q to P. So, for the orientation from P to Q, t must decrease from 0 to -1. Now,

$$I = I_1 + I_2 + I_3, \quad \text{where } I_1 = \int_P^Q x^2ydx,\ I_2 = \int_P^Q xyzdy \quad \text{and } I_3 = \int_P^Q y^3dz.$$

Now, $I_1 = \int_P^Q x^2ydx = \int_{-1}^0 (-t)^2 \cdot t^2 \cdot (-\dot{x})dt$, since $\dot{x} = -1 < 0$ for $-1 \leq t \leq 0$.

$$\therefore \quad I_1 = \int_{-1}^0 t^4dt = \frac{1}{5}.$$

$$I_2 = \int_{-1}^0 (-t)t^2 \cdot (-t^3)(-\dot{y})dt, \text{ since } \dot{y} = 2t < 0 \text{ for } -1 \leq t \leq 0.$$

$$\therefore \quad I_2 = -\int_{-1}^0 t^6 \cdot (2t)dt = -2\int_{-1}^0 t^7dt = -\frac{1}{4}t^8\Big|_{-1}^0 = \frac{1}{4}.$$

$$I_3 = \int_{-1}^0 t^6(-\dot{z})dt = -3\int_{-1}^0 t^8dt = \frac{1}{3}.$$

$$\therefore \quad I = I_1 + I_2 + I_3 = \frac{1}{5} + \frac{1}{4} + \frac{1}{3} = \frac{47}{60}. \qquad \text{(ii)}$$

From (i) and (ii) we find that the given integral is not independent of the path of integration. Moreover, it can be seen that $\nabla \times \overline{F} \neq \overline{0}$ so that $\overline{F}$ cannot be expressed as the gradient of any scalar function. Hence the integral cannot be independent of the path of integration.

3. *Show that the line integral*

$$\int_C yzdx + zxdy + xydz$$

is independent of the path of integration joining $A(0,0,0)$ *and* $B(1,1,1)$.

If $\overline{F} = yz\hat{i} + zx\hat{j} + xy\hat{k}$, then $\overline{F} = \nabla\phi$, where $\phi = xyz$. For,

$$\nabla\phi = \phi_x\hat{i} + \phi_y\hat{j} + \phi_z\hat{k} = yz\hat{i} + zx\hat{j} + xy\hat{k} = \overline{F}.$$

So, the given integral is independent of the path and its value is

$$\int_P^Q \nabla\phi \cdot d\overline{r} = \int_P^Q d\phi = \phi\Big|_P^Q = \phi(Q) - \phi(P)$$

$$= xyz|_Q - xyz|_P = 1.$$

4. *Find the work done by the force $\overline{F} = x\hat{i} - z\hat{j} + 2y\hat{k}$ in moving a particle from $A(0,0,2)$ to $B(1,2,2)$ along the arc of the parabola given by*

$$y = 2x^2 \quad and \quad z = 2 \ for \ 0 \le x \le 1.$$

Here both y and z may be regarded as functions of x:

$$y = y(x) = 2x^2 \quad \text{and} \quad z = z(x) = 2, \ \forall \ x \in [0,1].$$

The work done by $\overline{F}$ is given by

$$W = \int_A^B \overline{F} \cdot d\overline{r} = \int_A^B xdx - zdy + 2ydz$$

$$= \int_0^1 xdx - 2 \cdot \frac{dy}{dx}dx + 2 \cdot 2x^2 \cdot \frac{dz}{dx}dx,$$

since $dy/dx = 4x > 0$ for $0 < x < 1$ and $dz/dx = 0$.

$$\therefore \quad W = \int_0^1 (x - 2 \cdot 4x)dx = -7\int_0^1 xdx = -\frac{7}{2}x^2\Big|_0^1 = -7/2.$$

5. *Compute the line integral $\int ydx - xdy$ over an arc of the ellipse $x = a\cos t$, $y = b\sin t$, $0 \le t \le \pi$.*

When $t = 0$, $x = a$ and $y = 0$ and when $t = \pi$, $x = -a$, $y = 0$. So, if A and B are respectively $(a, 0)$ and $(-a, 0)$, the required arc Γ is the part of the ellipse lying above the x-axis and as t increases from 0 to π, a point moves along Γ from A to B.

Here $dx = -a\sin tdt$ and $dy = \cos tdt$.

$$\therefore \quad \int_\Gamma ydx - xdy = \int_0^\pi [b\sin t(-a\sin t) - a\cos tb\cos t]dt = -ab\int_0^\pi dt = -\pi ab.$$

6. *Compute the line integral $I = \int_\Lambda x^3dx + 3y^2zdy - x^2ydz$, where Λ is the oriented straight line segment joining $P(3,2,1)$ and $Q(0,0,0)$, the orientation being from P to Q.*

The scalar equations of the line through P and Q are $x/3 = y/2 = z/1 = t$ say. Then $x = 3t, y = 2t$ and $z = t$, which are the parametric equations to the line segment if $0 \le t \le 1$. As t increases from 0 to 1, a point moves from Q to P. So the oriented line segment is $-\Lambda$.

$$\therefore \quad I = -\int_{-\Lambda} fdx + gdy + hdz, \quad \text{where} \quad x = 3t, \ y = 2t, \ z = t \quad \text{and} \quad 0 \le t \le 1$$

$$= -\int_0^1 (27t^3 \cdot 3 + 3 \cdot t \cdot 4t^2 \cdot 2 - 9t^2 \cdot 2t \cdot 1)dt, \quad \because \ dx = 3, dy = 2, dz = 1$$

$$= -\int_0^1 (81t^3 + 24t^3 - 18t^3)dt = -\int_0^1 87t^3dt = -87/4.$$

Exercises 6.4

1. Compute the following integrals:

(i) $\int y^2dx + 2xydy$ along the entire circumference of the circle $x = a\cos t$, $y = a\sin t$, $0 \le t \le 2\pi$.

(ii) $\int xdx/(x^2 + y^2) - ydy/(x^2 + y^2)$ along the entire circumference of the circle $x = a\sin t,\ y = a\cos t,\ 0 \le t \le 2\pi$.

(iii) $\int (ydx + xdy)/(x^2 + y^2)$ along the oriented straight line segment given by $y = x$ and x increases from 1 to 2.

2. Show that the integrands in the following integrals are all exact and evaluate the integrals:

(i) $\int_{(0,0,0)}^{(1,1,1)} xdx + ydy - zdz,$ (ii) $\int_{(0,-1,-1)}^{(\pi,3,2)} \cos xdx + zdy + ydz,$

(iii) $\int_{(0,2,1)}^{(2,0,1)} ze^x dx + 2yzdy + (e^x + y^2)dz$

and (iv) $\int yzdx + xzdy + xydz$ along the arc of the circular helix: $x = a\cos t,\ y = a\sin t,\ z = kt$ as t varies from 0 to 2π.

Answers

1. (i) 0; (ii) 0; (iii) ln 2. **2.** (i) 1/2; (ii) 5; (iii) $e^2 - 5$.

6.10 Green's Theorem or Gauss's Theorem in a Plane

If two functions f and g of the two real variables x and y are continuous, possessing continuous first order partial derivatives in a domain E in the xy-plane bounded by a sectionally smooth positively oriented simple closed curve C, then

$$\oint_C (fdx + gdy) = \iint_E (g_x - f_y)dxdy.$$

The result will be true if we can prove that

$$\oint_C fdx = -\iint_E f_y dxdy \quad \text{and} \quad \oint_C gdy = \iint_E g_x dxdy.$$

Let us assume that the bounding curve C is quadratic in x. Then we shall have

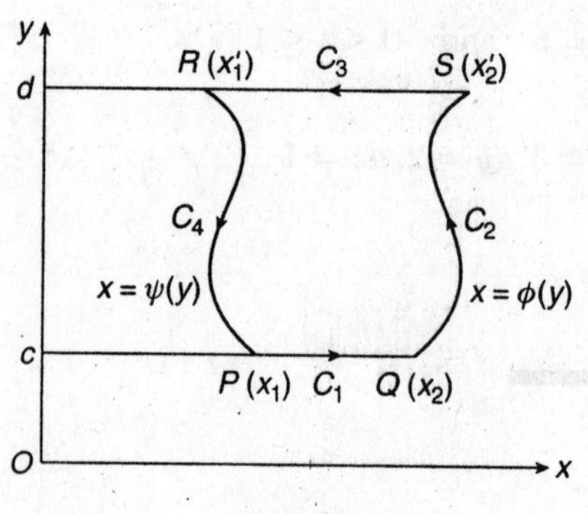

Fig. 6.15

$$C = C_1 \cup C_2 \cup C_3 \cup C_4,$$

where $C_1 : x_1 \le x \le x_2$ and $y = c$,

$C_2 : x = \phi(y)$ and $c \le y \le d$,

$C_3 : x_1' \le x \le x_2'$ and $y = d$

and $C_4 : x = \psi(y)$ and $c \le y \le d$.

The orientations of C_1, C_2, C_3 and C_4 are given by

(i) x increasing from x_1 to x_2,

(ii) y increasing from c to d,

(iii) x **decreasing** from x_2' to x_1'

and (iv) y decreasing from d to c (Fig. 6.15).

Let us prove first of all that $\oint_C g dy = \iint_E g_x dx dy.$ (1)

$$\iint_E g_x dx dy = \int_c^d dy \int_{\psi(y)}^{\phi(y)} g_x dx = \int_c^d g(x,y)\Big|_{\psi(y)}^{\phi(y)} dy$$

$$= \int_c^d [g(\phi(y), y) - g(\psi(y), y)] dy. \quad (2)$$

And $\quad \oint_C g dy = \left(\int_{C_1} + \int_{C_2} + \int_{C_3} + \int_{C_4}\right) g dy = I_1 + I_2 + I_3 + I_4,$

where $\quad I_1 = \int_{C_1} g dy = 0, \quad \because\ y = c \Rightarrow dy = 0,$

$$I_2 = \int_{C_2} g dy = \int_c^d g(\phi(y), y) dy,$$

$$I_3 = \int_{C_3} g dy = 0, \quad \because\ y = d \text{ on } C_3 \text{ so that } dy = 0$$

and $\quad I_4 = \int_{C_4} g dy = -\int_c^d g(\psi(y), y) dy,$ since y decreases from d to c for the orientation of C_4 from R to P.

$\therefore \quad I = I_2 + I_4 = \int_c^d \{g(\phi(y), y) - g(\psi(y), y)\} dy =$ RHS of (2).

$\therefore \quad \oint_C g dy = \iint_E g_x dx dy.$

If C is not quadratic in x then E can be divided into a finite number of subdomains E_1, E_2, E_3, etc., each of which will be quadratic in x. For example, if E is as shown in the adjacent diagram (Fig. 6.16), then C is not quadratic in x. For a line drawn parallel to the x-axis may cut the boundary in four points giving four values of x. Here E can be divided into two parts E_1 and E_2 by a line C_5 so that the boundary of each part is quadratic in x.

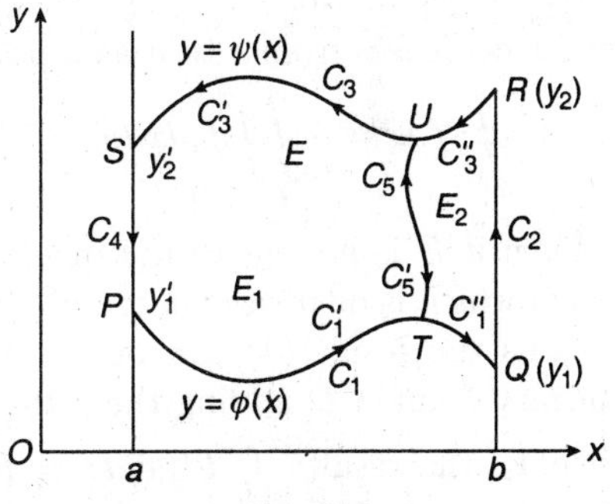

Fig. 6.16

If L_1 and L_2 are the two positively oriented boundaries of E_1 and E_2 respectively, then $L_1 = C_1' \cup C_5 \cup C_3' \cup C_4$, where C_1' is the arc $\widehat{PT}$ of C_1 and C_5 is the arc $\widehat{TU}$ and C_3' is the arc $\widehat{US}$ of C_3, and, $L_2 = C_1'' \cup C_2 \cup C_3'' \cup C_5'$, where C_1'' is the arc $\widehat{TQ}$, C_3'' is the arc $\widehat{RU}$ and $C_5' = -C_5$, the arc $\widehat{UT}$.

So, $\quad \int_C = \int_{L_1} + \int_{L_2} = \left(\int_{C_1'} + \int_{C_5} + \int_{C_3'} + \int_{C_4}\right) + \left(\int_{C_1''} + \int_{C_2} + \int_{C_3''} + \int_{C_5'}\right)$

$$= \left(\int_{C'} + \int_{C_1''}\right) + \int_{C_2} + \left(\int_{C_3'} + \int_{C_3''}\right) + \int_{C_4} + \left(\int_{C_5} + \int_{C_5'}\right) \quad (3)$$

$$= \int_{C_1} + \int_{C_2} + \int_{C_3} + \int_{C_4}, \quad \because\ \int_{C_5} + \int_{C_5'} = \int_{C_5'} - \int_{C_5}$$

so that there is no contribution from the last term in (3). This shows that the integrals along additional boundary lines have no contribution in the value of the given integral and as before

$$\iint_E g_x dxdy = \left(\int_{C_1} + \int_{C_2} + \int_{C_3} + \int_{C_4}\right) gdy = \oint_C gdy. \tag{4}$$

Next we shall see that $\displaystyle\iint_E f_y dxdy = -\oint_C fdx.$

For that, it is better to take the boundary of E quadratic in y, so that

$C = C_1 \cup C_2 \cup C_3 \cup C_4$ (Fig. 6.16),

where $C_1 = PQ : a \le x \le b$ and $y = \phi(x)$,

$C_2 = QR : x = b$ and $y_1 \le y \le y_2$,

$C_3 = RS : a \le x \le b$ and $y = \psi(x)$

and $C_4 = SP : x = a$ and $y_1' \le y \le y_2'$.

Then $$\iint_E f_y dxdy = \int_a^b dx \int_{\phi(x)}^{\psi(x)} f_y dy = \int_a^b [f(x,y)]_{\phi(x)}^{\psi(x)} dx$$

$$= \int_a^b [f(x,\psi(x)) - f(x,\phi(x))]dx.$$

Also, $$\oint_C fdx = \left(\int_{C_1} + \int_{C_2} + \int_{C_3} + \int_{C_4}\right) fdx$$

$$= \int_{C_1} fdx + \int_{C_3} fdx, \text{ since } dx = 0 \text{ on } C_2 \text{ and } C_4$$

$$= \int_a^b f(x,\phi(x))dx - \int_a^b f(x,\psi(x))dx,$$

since x decreases from b to a as a point moves along C_3 from R to S.

$$\therefore \quad \oint_C fdx = -\iint_E f_y dxdy.$$

Even if E is not quadratic in y, as before E can be divided into a finite number of parts each of which is quadratic in y and by taking the sums of both the line integrals and the double integrals the result (4) can be proved. Since both the results (2) and (4) are true for any arbitrary domain D adding them the required result is obtained.

Note 1. the result: $\int_a^b F'(x)dx = F(x)\big|_a^b = F(b) - F(a)$, in the theory of ordinary definite integration, gives a relation between the derivative of a function and the integral of the derivative. Integration cancels differentiation giving a result involving the boundary values of the function. Similarly, the result:

$$\iint_E f_y dxdy = -\oint_C fdx \quad \text{or,} \quad \iint_E g_x dxdy = \oint_C gdy$$

in Green's theorem gives a relation between the partial derivative of a function and the double integral of the derivative of the function over a region. Here also integration cancels differentiation giving the line integral of the function over the closed boundary of the region.

Note 2. Taking $\overline{F} = g\hat{i} - f\hat{j}, \nabla = \hat{i}D_x + \hat{j}D_y, \hat{\nu} = \hat{i}\frac{dy}{ds} - \hat{j}\frac{dx}{ds}$, the result of Green's theorem can be expressed in vector form as

$$\oint_C \overline{F} \cdot \hat{\nu} ds = \iint_E \nabla \cdot \overline{F} dA.$$

Since $|\hat{\nu}| = 1$ and $\hat{\nu} \cdot d\overline{r} = 0$, $\hat{\nu}$ is unit vector normal to $d\overline{r}$.

6.10.1 Area of a Plane Region as a Line Integral

Let A be the area of the plane region E bounded by a positively oriented continuous piecewise smooth curve C. Then putting $g(x, y) = x$ in (1), § 6.10

$$A = \iint_E dxdy = \oint_C xdy. \tag{1}$$

Similarly, putting $f(x, y) = y$ in (4), § 6.10

$$A = -\oint_C ydx. \tag{2}$$

From (1) and (2) we therefore have

$$\mathbf{A = \frac{1}{2}\oint_C xdy - ydx.}$$

Also, if $\nabla \cdot \overline{F} = 1$, i.e., $g_x - f_y = 1$, $\oint_C \overline{F} \cdot \hat{\nu} ds = \iint_E dA = A.$

Examples 6.7. **1.** *Verify Green's theorem for the following integral:*

$$\int x(x+y)dx + (x^2 + y^2)dy$$

along the entire positively oriented perimeter of the square whose vertices are given by $x = \pm 1$ and $y = \pm 1$.

Let A, B, C and D be the vertices with coordinates $(1, 1)$, $(-1, 1)$, $(-1, -1)$ and $(1, -1)$. Then referred to a right-handed frame the perimeter Γ described in the direction from A to B, B to C, C to D and from D back to A is positively oriented. Then we are required to prove that

$$\oint_\Gamma fdx + gdy = \iint_E (g_x - f_y)dxdy,$$

where E is the interior of the given square and

$$f = x(x+y) \quad \text{and} \quad g = x^2 + y^2.$$

Now, $$\oint_\Gamma fdx + gdy = \left(\int_A^B + \int_B^C + \int_C^D + \int_D^A\right) fdx + gdy$$

$$= \int_1^{-1} x(x+1)dx + \int_1^{-1} \{(-1)^2 + y^2\}dy \int_{-1}^1 x(x-1)dx + \int_{-1}^1 (1^2 + y^2)dy,$$

since y is constant over the paths AB and CD so that $dy = 0$ and similarly, x is constant over BC and DA so that $dx = 0$. Also, $y = 1$ over AB and $y = -1$ over CD. Similarly, $x = 1$ over

DA whereas $x=-1$ over BC. So,

$$\oint_\Gamma f dx + g dy = \left[\frac{x^3}{3}+\frac{x^2}{2}\right]_1^{-1} + \left[y+\frac{y^3}{3}\right]_1^{-1} + \left[\frac{x^3}{3}-\frac{x^2}{2}\right]_{-1}^{1} + \left[y+\frac{y^3}{3}\right]_{-1}^{1}$$

$$= -\frac{2}{3} + \left(-2-\frac{2}{3}\right) + \frac{2}{3}\left(2+\frac{2}{3}\right) = 0.$$

$$\iint_E (g_x - f_y) dx dy = \int_{-1}^{1} dx \int_{-1}^{1} (g_x - f_y) dy = \int_{-1}^{1} dx \int_{-1}^{1} (2x - x) dy$$

$$= 2\int_{-1}^{1} x dx = 2 \times 0 = 0.$$

Thus, $\oint_\Gamma f dx + g dy = \iint_E (g_x - f_y) dx dy$

and Green's theorem is verified.

2. *Verify Green's theorem for the line integral*

$$I = \oint_C (x+y) y dx + x^2 dy,$$

where C is the closed curve given by $y = x$ and $y = x^2$, $0 \le x \le 1$ described positively.

[W.B. Sem. Exam. '01]

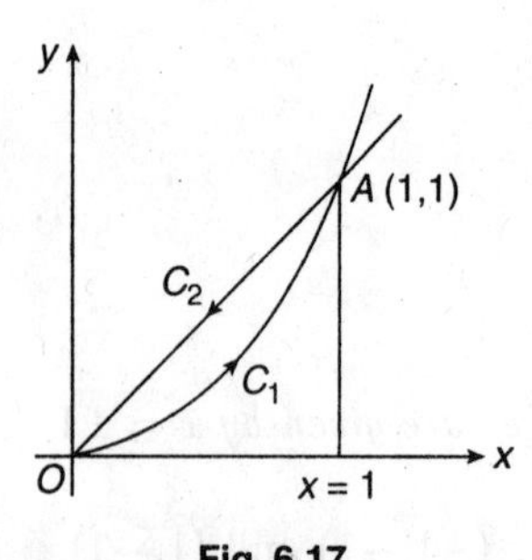

Fig. 6.17

Let C_1 be the arc given by $y = x^2$, $0 \le x \le 1$, x increasing from 0 to 1 and C_2 be the line segment $\overrightarrow{AO}$ given by $y = x$, $0 \le x \le 1$, where x decreases from 1 to 0. Then the entire positively oriented boundary

$$C = C_1 \cup C_2.$$

$\therefore$ the given integral $I = \left(\int_{C_1} + \int_{C_2}\right) f dx + g dy,$

where $f = (x+y)y$ and $g = x^2$. Thus, $I = I_1 + I_2$,

where $I_1 = \int_{C_1} f dx + g dy = \int_0^1 (x + x^2) x^2 dx + x^2 \cdot 2x dx$

$$= \int_0^1 (3x^3 + x^4) dx = \frac{3}{4} + \frac{1}{5} = \frac{19}{20}$$

and $I_2 = \int_{C_2} f dx + g dy = \int_1^0 (x+x) x dx + x^2 dx = \int_1^0 3x^2 dx = -1.$

$\therefore \quad I = I_1 + I_2 = \frac{19}{20} - 1 = -\frac{1}{20}.$

Also, $\iint_E (g_x - f_y) dx dy = \int\int_E \{2x - (x + 2y)\} dx dy$

$$= \int\int_E (x - 2y) dx dy = \int_0^1 \int_{x^2}^{x} (x - 2y) dy dx$$

$$= \int_0^1 (xy - y^2)_{x^2}^{x} dx = \int_0^1 (x^4 - x^3) dx = \frac{1}{5} - \frac{1}{4} = -\frac{1}{20}.$$

3. *Verify Green's theorem by evaluating, in two ways, the following integral:*

$$\oint_C [(x^3 + y^2)dx + (x^2 + y^3)dy]$$

where C *is the positively oriented pentagon whose vertices are successively* $O(0,0)$, $P(1,0)$, $Q(2,1)$, $R(1,2)$ *and* $S(0,1)$ *(Fig. 6.18).*

Here C consists of the directed line segments $\overrightarrow{OP}$, $\overrightarrow{PQ}$, $\overrightarrow{QR}$, $\overrightarrow{RS}$ and $\overrightarrow{SO}$ which are given by

(i) $y = 0$, x is increasing from 0 to 1,

(ii) $y = x - 1$, x is increasing from 1 to 2

or, $x = y + 1$, y is increasing from 0 to 1,

(iii) $y = -x + 3$ and x is decreasing from 2 to 1

or, $x = -y + 3$ and y is increasing from 1 to 2,

(iv) $y = x + 1$ and x is decreasing from 1 to 0

or, $x = y - 1$ and y is decreasing from 2 to 1,

and **(v)** $x = 0$ and y is decreasing from 1 to 0.

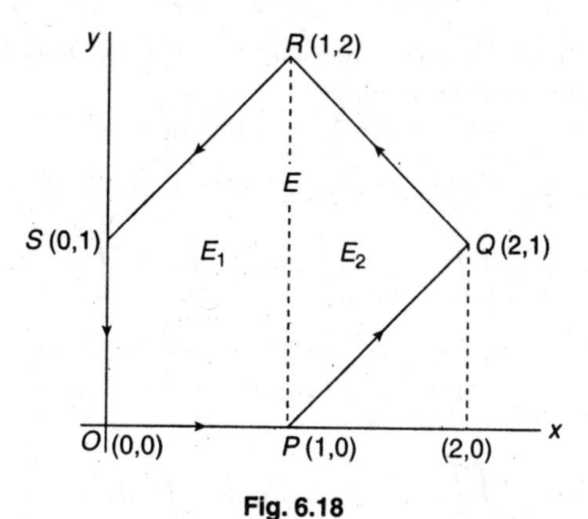

Fig. 6.18

Let $f = x^3 + y^2$ and $g = x^2 + y^3$.

Then $\oint_C f dx = \left\{ \int_{OP} + \int_{PQ} + \int_{QR} + \int_{RS} + \int_{SO} \right\} f dx.$

But $dx = 0$ along SO. So,

$$\oint_C f dx = \int_0^1 x^3 dx + \int_1^2 \{x^3 + (x-1)^2\}dx + \int_0^1 \{x^3 + (-x+3)^2\}dx + \int_1^0 \{x^3 + (x+1)^2\}dx$$

$$= \left\{ \int_0^1 + \int_1^2 + \int_2^1 + \int_1^0 \right\} x^3 dx + \int_1^2 \{(x-1)^2 - (x-3)^2\}dx \int_0^1 (x+1)^2 dx$$

$$= \int_1^2 (4x - 8)dx - \int_0^1 (x+1)^2 dx = [2x^2 - 8x]_1^2 - \frac{1}{3}[(x+1)^3]_0^1 = -2 - \frac{7}{3} = -\frac{13}{3}.$$

Next, $\oint_C g dy = \left\{ \int_{OP} + \int_{PQ} + \int_{QR} + \int_{RS} + \int_{SO} \right\} g dy.$

But along OP, $y = 0$ and $dy = 0$. So,

$$\oint_C g dy = \int_0^1 \{(y+1)^2 + y^3\}dy + \int_1^2 \{(-y+3)^2 + y^3\}dy + \int_2^1 \{(y-1)^2 + y^3\}dy$$

$$+ \int_1^0 (0^2 + y^3)dy$$

$$= \left\{ \int_0^1 + \int_1^2 + \int_2^1 + \int_1^0 \right\} y^3 dy + \int_0^1 (y+1)^2 dy + \int_1^2 [(y-3)^2 - (y-1)^2]dy$$

$$= \frac{1}{3}[(y+1)^3]_0^1 + \int_1^2 (-4y + 8)dy = \frac{7}{3} + 2 = \frac{13}{3}.$$

$$\therefore \quad \oint_C (f dx + g dy) = -\frac{13}{3} + \frac{13}{3} = 0.$$

Now, if E is the region within the pentagon $OPQRS$,

$$\iint_E (g_y - f_x) dx dy = \left\{ \iint_{E_1} + \iint_{E_2} \right\} (g_y - f_x) dx dy,$$

where E_1 is the quadrilateral $OPRS$ and E, is the triangle PQR. Their boundaries C_1 and C_2 can be described as

$$x = 0,\ x = 1,\ y = 0 \quad \text{and} \quad y = x + 1$$

and $\quad x = 1,\ x = 2,\ y = x - 1 \quad \text{and} \quad y = -x + 3.$

$$\therefore \quad \iint_{E_1} (g_y - f_x) dx dy = \int_0^1 dx \int_0^{x+1} 3(y^2 - x^2) = 3 \int_0^1 \left[\frac{1}{3} y^3 - x^2 y \right]_0^{x+1} dx$$

$$= 3 \int_0^1 \left[\frac{1}{3}(x+1)^3 - x^2(x+1) \right] dx = \left[\frac{1}{4}(x+1)^4 - \frac{3}{4}x^4 - x^3 \right]_0^1 = 2.$$

And $$\iint_{E_2} (g_y - f_x) dx dy = 3 \int_1^2 dx \int_{x-1}^{-x+3} (y^2 - x^2) dy = 3 \int_1^2 \left[\frac{y^2}{3} - x^2 y \right]_{x-1}^{-x+3} dx$$

$$= 3 \int_1^2 \left[-\frac{1}{3}\{(x-3)^3 + (x-1)^3\} - x^2(-2x+4) \right] dx = -2.$$

$$\therefore \quad \iint_E (g_y - f_x) dx dy = 0 = \oint_C (f dx + g dy).$$

Exercises 6.5

1. Apply Green's theorem to evaluate

$$\oint_\Gamma (2x^2 - y^2) dx + (x^2 + y^2) dy,$$

where Γ is the positively oriented closed boundary consisting of $x^2 + y^2 = a^2$, where $y \geq 0$ and $y = 0$, where $-a \leq x \leq a$.

2. Verify Green's theorem for

$$\oint_C (3x - 8y^2) dx + 2(2 - 3x) y dy,$$

where C is the positively oriented closed curve consisting of $x = 0,\ y = 0$ and $x + y = 1$, where $0 \leq x \leq 1$ and $0 \leq y \leq 1$.

3. Find the area of the ellipse $x^2/a^2 + y^2/b^2 = 1$ using line integrals. [**Hint:** Use $x = a\cos\theta,\ y = b\sin\phi$ in $\frac{1}{2}\oint_C x dy - y dx$.]

4. Apply Green's theorem to evaluate

$$\oint_C (y - \sin x) dx + \cos x dy,$$

where C is the positively oriented perimeter of the triangle with sides $x = \pi/2$, $y = 0$ and $y = 2x/\pi$.

5. Using Green's theorem evaluate the line integral $\oint_C f dx + g dy$, C being a positively oriented closed contour, where

(i) $f = \cos x \sin y - xy$ and $g = \sin x \cos y$, $C : x^2 + y^2 = 1$;

(ii) $f = x^{-1}e^y$ and $g = e^y \ln x + 2x$, and C is the boundary of the region enclosed in $y = x^4 + 1$ and $y = 2$.

Answers

1. $4a^3/3$. **3.** πab. **4.** $-(\pi^2 + 8)/(4\pi)$. **5. (i)** 0; **(ii)** 16/5.

6.11 Surface Integral. Integration over a Surface in $\mathbb{R}^3$

Concept of a Surface: A surface is defined to be the locus of a point having two degrees of freedom.

In xyz-space a surface is given by

$$\sum = \{(x, y, z) \in \mathbb{R}^3 : F(x, y, z) = 0 \;\forall\; (x, y, z) \in D \subseteq \mathbb{R}^3\}.$$

The equation:

$$F(x, y, z) = 0 \tag{1}$$

is called **the equation to the surface** $\sum$ and $\sum$ is called the **locus of the equation (1)**. Since the value or values of one of the variables x, y and z, determined by (1), depend on the values of two other variables, which can take different values, independently of each other, the equation (1) represents the locus of a point having two degrees of freedom. A point $P(x', y', z')$ lies on $\sum$ iff $F(x', y', z') = 0$. Set theoretically, $P \in \sum$ iff $F(P) = 0$. So, $P \notin \sum$ iff $F(P) \neq 0$.

It may be possible to isolate z in (1) and express it as a function of x and y in some domain E in the xy-plane (Fig. 6.19):

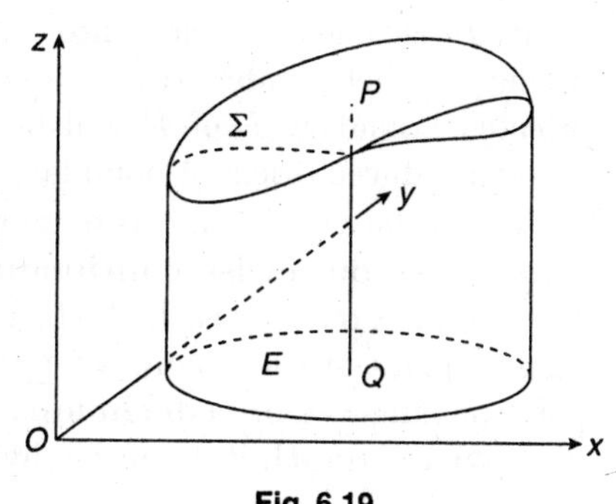

Fig. 6.19

$$z = h(x, y) \;\forall\; (x, y) \in E. \tag{2}$$

Such a surface is said to be **linear in z**. Every line through a point Q in E, parallel to the z-axis meets the surface in only one point P. The surface $\sum$ given by (1) is said to be **quadratic in z** if the equation can be expressed as

$$z^2 + az + b = 0, \tag{3}$$

where a and b are functions of x and y. In this case, a line parallel to the z-axis will meet the surface in two points at the most (Fig. 6.20).

If z_1 and z_2 are the roots of the equation (3) in z, then the surface $\sum = \sum_1 \cup \sum_2$, where $\sum_1$ and $\sum_2$ are the two surfaces given by

$$z = z_1(x, y) \quad \text{and} \quad z = z_2(x, y),$$

each of which is linear in z.

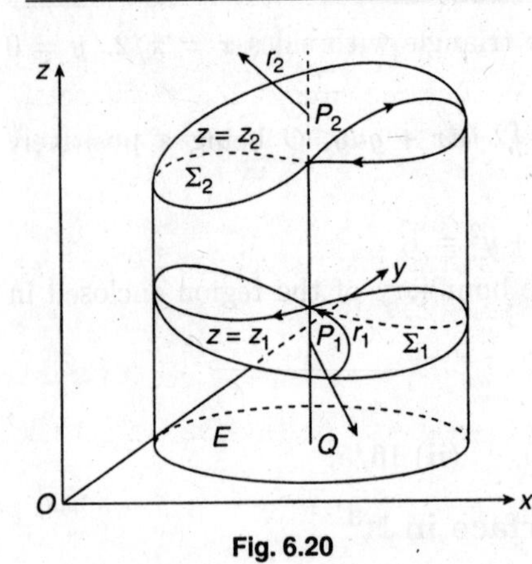

Fig. 6.20

All conicoids like sphere, spheroids, ellipsoids, paraboloids, hyperboloids, right-circular cones and cylinders are quadratic in some of the variables.

Another useful representation of a surface in $\mathbb{R}^3$ is the **parametric representation.** If

$$\sum = \{(x, y, z) \in \mathbb{R}^3 : x = x(u, v), y = y(u, v) \text{ and } z = z(u, v) \; \forall \; (u, v) \in E \subseteq \mathbb{R}^2\}$$

then $\sum$ is a surface. The equations

$$x = x(u, v), \; y = y(u, v) \; \text{ and } z = z(u, v), \tag{4}$$

each of which defines a function of u and v, called **parameters**, are the parametric equations of the surface $\sum$. Here again, $\sum$ is the locus of a point with two degrees of freedom, since the coordinates of each point on $\sum$ depend on two independent parameters u and v. Since for every point $P(u, v) \in E$ we get a unique image $P'(x, y, z) \in \sum$ (Fig. 6.21), the equations (4) define a transformation T mapping E into $\sum$:

$$T : E \to \sum.$$

If T is one-one and onto and E is bounded, then under this mapping, the oriented boundary L of E will be mapped onto an induced oriented boundary Γ of $\sum$.

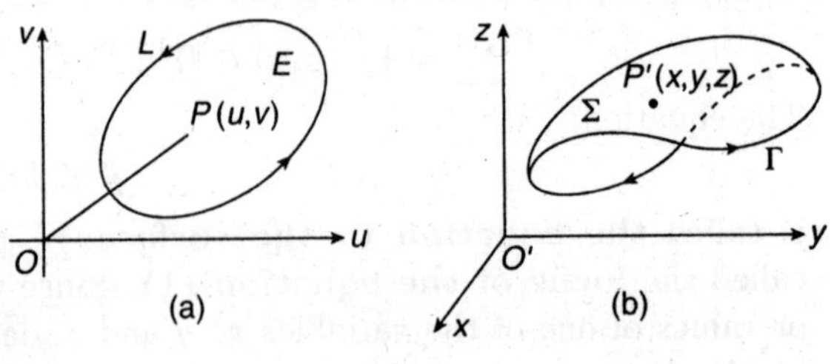

Fig. 6.21

The surface $\sum$ is said to be **continuous** if all the associated functions are continuous. It is said to be **piecewise continuous** or **sectionally continuous**, if $\sum$ can be divided into a finite number of parts each of which is continuous. It is said to be **smooth,** if there is a unique tangent plane at every point of $\sum$ and the **direction cosines (DC's) of the normal** at the point of tangency are **continuous functions of the associated variables.** The surface is **piecewise smooth** if it can be divided into a finite number of parts each of which is smooth.

6.11.1 Orientable Surface

As we can assign a certain orientation or sense of description or a direction to a curve, similarly we may assign a certain orientation to a surface. In vector algebra we have seen that the vector product $\overline{a} \times \overline{b}$ is associated with the area A of a parallelogram of which the vectors $\overline{a}$ and $\overline{b}$ are two adjacent sides. If $\hat{\nu} = (\overline{a} \times \overline{b})/|\overline{a} \times \overline{b}|$ then $\hat{\nu}$ is a unit vector normal to the plane of the parallelogram and

$$\overline{a} \times \overline{b} = A\hat{\nu}.$$

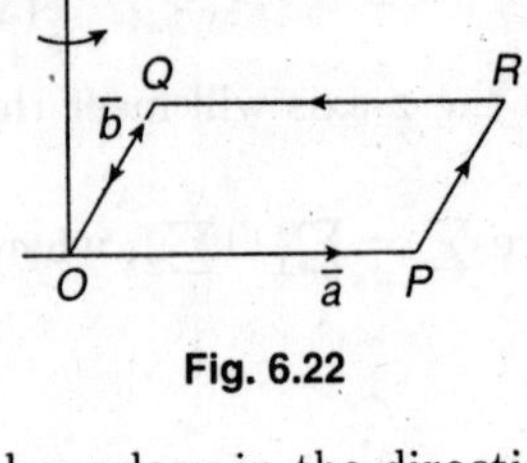

Fig. 6.22

Thus $A\hat{\nu}$ is an oriented area, i.e., an area with a definite direction. The direction of this area can also be given by the direction of description of the boundary of the parallelogram. This direction is the same as the direction in which $\overline{a}$ is to be rotated through the shortest path to bring it into coincidence with the direction of $\overline{b}$.

Note that the oriented area always lies to the left of a person with his or her head pointing towards $\hat{\nu}$ when he or she is walking along the boundary in the direction of its description.

A smooth surface $\sum$ is said to be **orientable**, if for every point $P \in \sum$ and for every closed curve through P, the direction of the unit normal at P remains the same when the normal, starting from P, moves continuously along each such closed path and comes back to the initial position.

All orientable surfaces are bilateral, having two distinct sides. If such a surface $\sum$ is represented by an equation of the type $F(x, y, z) = 0$, then two distinct points, P and Q not lying on $\sum$, will lie on the same side or on opposite sides of $\sum$ according as

$$F(P)F(Q) > \text{ or } < 0.$$

So, if $F(P) > 0$, the side of $\sum$ in which P lies may be considered positive and the other side negative, although it is purely relative. For, if

$$F = x - y + z - 1$$

then $F = 0$ represents a plane. Since $F(0) = -1$, the origin O may be considered to lie on the negative side of the plane. If now $G = -F = 1 - x + y - z$, then $G = 0$ also represents the same plane. But $G(0) = 1 > 0$. Thus O is found to lie on the positive side. Hence the assignment of positivity or negativity to a side of a bilateral surface is purely relative.

The two sides of a simple closed oriented surface are called **inside** and **outside**. For example, the centre of a spherical surface is supposed to lie inside the surface and the side of the surface in which the centre lies is therefore inside and the other side is outside.

If $F = x^2 + y^2 + z^2 - 1$, then $F = 0$ represents a spherical surface with its centre at the origin O and radius 1. Since $F(0) = -1$, the negative side of the surface is inside and consequently the positive side is outside.

It is erroneous to think that every surface in space is orientable or **bilateral**. For example, a **Möbius band** (Fig. 6.23) is **unilateral**, having only one side. And for such a surface there is nothing like positive and negative side or inside and outside. It can be formed by taking a

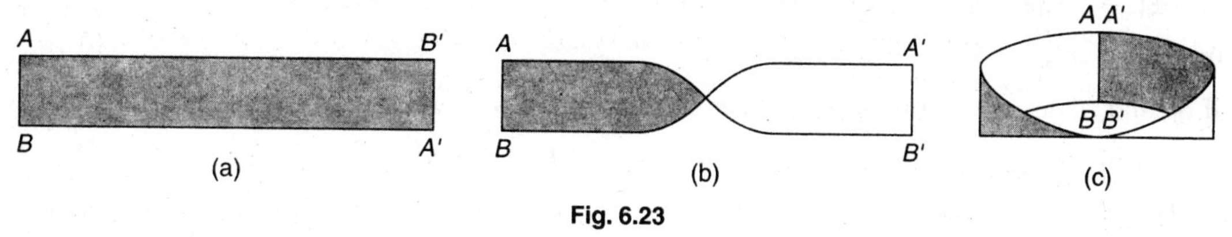

Fig. 6.23

long rectangular strip $ABA'B'$ of paper and pasting its two smaller edges AB and $A'B'$ together after giving it half a twist so that A and B coincide respectively with A' and B'. that it is unilateral can be verified by drawing a central line. Similarly, **Klein's bottle** is another example of a unilateral surface without any edge (Fig. 6.24).

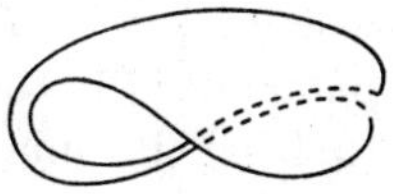

Fig. 6.24

It has neither any inside nor outside.

6.11.2 Tangent Plane and Normal. Line Element and First Fundamental Form

If a surface $\sum$ is represented by

$$F(x, y, z) = 0, \tag{1}$$

then we know that the grad F or ∇F, given by

$$\nabla F = F_x \hat{\imath} + F_y \hat{\jmath} + F_g \hat{k},$$

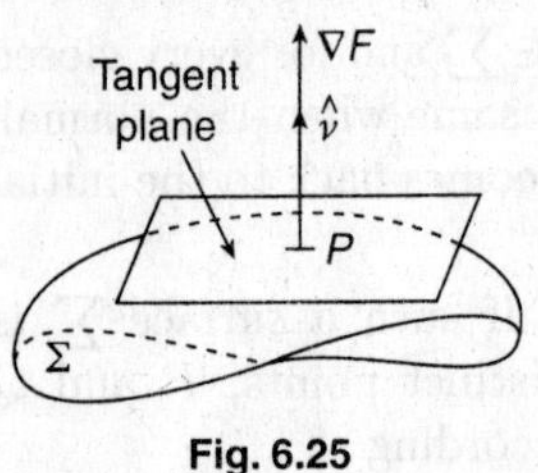

Fig. 6.25

calculated at a point $P \in \sum$, has the direction of the normal to $\sum$ at P (Fig. 6.25). If $\hat{\nu}$ is a unit vector normal to $\sum$ at P, then

$$\hat{\nu} = \pm \nabla F/|\nabla F| = \pm \sum F_x \hat{i}/\sqrt{\sum F_x^2}. \tag{2}$$

So, the D.C.'s of the normal at P are

$$\pm \frac{1}{\sqrt{\sum F_x^2}} (F_x, F_y, F_z).$$

The tangent plane at P is therefore given by

$$\sum (X - x) F_x = 0, \tag{3}$$

where (X, Y, Z) are the current coordinates and (x, y, z) are the coordinates of P.

If $\sum$ is given by

$$z = z(x, y), \tag{4}$$

then we may take $F(x, y, z) = z - z(x, y)$, where the first z is independent but $z(x, y)$ is a function of x and y.

$\therefore \quad F_x = 0 - z_x, \; F_y = 0 - z_y, \; F_z = 1.$

So, the D.C.'s of the normal to $\sum$ at P are $\pm(-z_x - z_y, 1)/\sqrt{(z_x^2 + z_y^2 + 1)}$. If the normal is so chosen that it makes a positive acute angle γ^c with $\hat{k}$ or the positive direction of the z-axis, then $\cos \gamma$ is positive and

$$\cos \gamma = 1/\sqrt{(z_x^2 + z_y^2 + 1)}. \tag{5}$$

Similarly, if $\sum$ is given by

$y = y(x, z) \quad$ or, $\quad x = x(y, z)$,

the D.C's of the normal are

$\pm(-y_x, 1, -y_z)/\sqrt{(y_x^2 + 1 + y_z^2)}$

or, $\pm(1, -x_y, -x_z)/\sqrt{(1 + x_y^2 + x_z^2)}$.

If α^c and β^c are the acute angles made by the normal with $\hat{i}$ and $\hat{j}$ respectively, then

$$\cos \alpha = 1/\sqrt{(1 + x_y^2 + x_z^2)} \quad \text{and} \quad \cos \beta = 1/\sqrt{(y_x^2 + 1 + y_z^2)}. \tag{6}$$

When $\sum$ is given by

$$\bar{r} = \bar{r}(u, v) = x(u, v)\hat{i} + y(u, v)\hat{j} + z(u, v)\hat{k}, \tag{7}$$

where $(u, v) \in E$, a region in the uv-plane, if u and v are both continuous functions of a parameter t in their common domain $I_t = [\alpha, \beta]$, then as $T(t)$ moves along the t-axis from $A(\alpha)$ to $B(\alpha)$, its image $S(u(t), v(t))$ traces an arc L of a curve in E (Fig. 6.26). If u and v are functions of t, x, y and z become all functions of t.

The image R of S traces a space curve Γ on $\sum$. Since

$$\bar{r} = \bar{r}(u(t), v(t)), \; \dot{\bar{r}} = \bar{r}_u \dot{u} + \bar{r}_v \dot{v}. \tag{8}$$

Since $\dot{\bar{r}}$ is tangential to Γ and it is a linear combination of $\bar{r}_u$ and $\bar{r}_v$, the three vectors are coplanar at $R(\bar{r}(t))$.

When v is constant and u is a variable the vector $\bar{r}$ in (7) becomes a function of a single variable describing a certain curve on $\sum$, called a u-curve. So $\bar{r}_u$ is tangential to this u-curve.

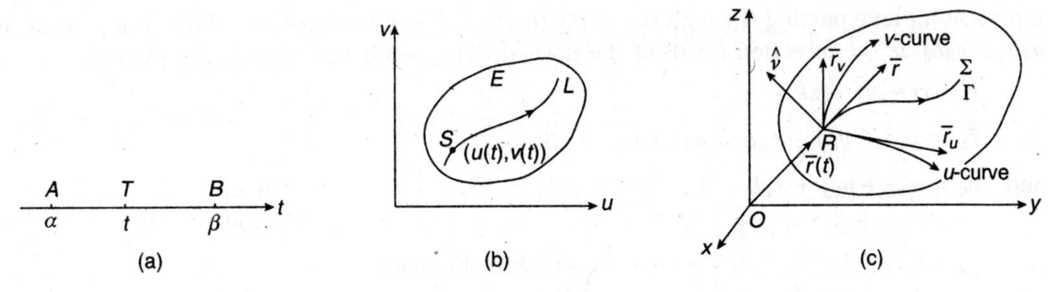

Fig. 6.26

Similarly, when u is constant and v is a variable we get a curve on $\sum$, called a v-curve and $\bar{r}_v$ is tangential to this curve. The plane determined by $\dot{\bar{r}}$, $\bar{r}_u$ and $\bar{r}_v$ at R is **the tangent plane** touching $\sum$ at R. If $\hat{\nu}$ is a unit normal to this tangent plane, then obviously

$$\hat{\nu} = \pm(\bar{r}_u \times \bar{r}_v)/|\bar{r}_u \times \bar{r}_v| \tag{9}$$

and the vector equation to the tangent plane at R is

$$(\bar{r} - \dot{\bar{r}}) \cdot \bar{r}_u \times \bar{r}_v = 0. \tag{10}$$

Expressed in terms of the partial derivatives of x, y and z,

$$\bar{r}_u \times \bar{r}_v = \begin{vmatrix} \hat{i} & \hat{j} & \hat{k} \\ x_u & y_u & z_u \\ x_v & y_v & z_v \end{vmatrix} = J_1\hat{i} + J_2\hat{j} + J_3\hat{k}, \tag{11}$$

where J_1, J_2 and J_3 are the Jacobians given by

$$J_1 = \partial(y,z)/\partial(u,v), \quad J_2 = \partial(z,x)/\partial(u,v) \quad \text{and} \quad J_3 = \partial(x,y)/\partial(u,v). \tag{12}$$

$$\therefore \qquad \hat{\nu} = \pm\sum J_1\hat{i}/\sqrt{\sum J_1^2}.$$

A smooth closed surface is said to be **positively oriented** if the direction of the unit normal vector $\hat{\nu}$ is outward at every point of the surface.

If s is the length of an arc Γ on $\sum$ then we have from (8), the elementary arc length ds, given by

$$ds^2 = d\bar{r}^2 = (\bar{r}_u du + \bar{r}_v dv)^2 = Edu^2 + 2Fdudv + Gdv^2, \tag{13}$$

where $E = \bar{r}_u^2 = \sum x_u^2$, $F = \bar{r}_u \cdot \bar{r}_v = \sum x_u x_v$

and $G = \bar{r}_v^2 = \sum x_v^2$.

The form on the RHS of (13) giving the elementary arc length is called the **First Fundamental Form** of a surface.

Since $(\bar{r}_u \times \bar{r}_v)^2 = \bar{r}_u^2\bar{r}_v^2 - (\bar{r}_u \cdot \bar{r}_v)^2 = EG - F^2$, so by (11)

$$\boldsymbol{EG - F^2 = (\bar{r}_u \times \bar{r}_v)^2 = \sum J_1^2}. \tag{14}$$

6.11.3 Some Surfaces Represented Parametrically and Their Normals

(a) If the variables x, y and z are all linear functions of two parameters u and v, then these functions will represent a plane. Thus the equations:

$$x = a_1 + a_2u + a_3v, \quad y = b_1 + b_2u + b_3v \quad \text{and} \quad z = c_1 + c_2u + c_3v$$

represent a plane passing through the point (a_1, b_1, c_1) and having $b_2c_3 - b_3c_2,\ c_2a_3 - c_3a_2$ and $a_2b_3 - a_3b_2$ as the direction ratios of the normal. The vector equation of the plane is

$$\bar{r} = x\hat{i} + y\hat{j} + z\hat{k}.$$

$$\because \quad \bar{r}_u = x_u\hat{i} + y_u\hat{j} + z_u\hat{k} = a_2\hat{i} + b_2\hat{j} + c_2\hat{k}$$

and $\bar{r}_v = x_v\hat{i} + y_v\hat{j} + z_v\hat{k} = a_3\hat{i} + b_3\hat{j} + c_3\hat{k}$,

$$\therefore \quad \bar{r}_u \times \bar{r}_v = \begin{vmatrix} \hat{i} & \hat{j} & \hat{k} \\ a_2 & b_2 & c_2 \\ a_3 & b_3 & c_3 \end{vmatrix}, \text{ which is normal to the plane.}$$

(b) $x = R\sin u\cos v,\ y = R\sin u\sin v$ and $z = R\cos u$, where $0 \le u \le \pi$ and $0 \le v \le 2\pi$ and R is a positive constant, represent **a spherical surface** with centre at the origin (pole) O and radius R (Fig. 6.27).

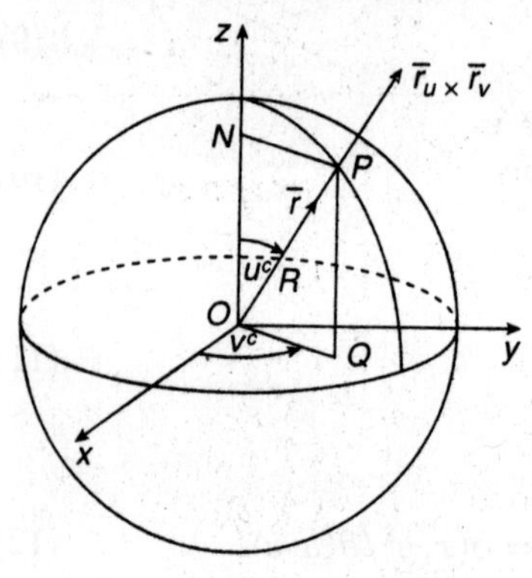

Fig. 6.27

Here, $$\bar{r}_u \times \bar{r}_v = \begin{vmatrix} \hat{i} & \hat{j} & \hat{k} \\ R\cos u\cos v & R\cos u\sin v & -R\sin u \\ -R\sin u\sin v & R\sin u\cos v & 0 \end{vmatrix}$$

$$= R^2\sin u(\hat{i}\sin u\cos v + \hat{j}\sin u\sin v + \hat{k}\cos u)$$

$$= R\sin u(x\hat{i} + y\hat{j} + z\hat{k})$$

$$= R(\sin u)\bar{r}.$$

So, the normal at $P(\bar{r})$ has the same direction as $\bar{r}$, which is radial to the sphere.

(c) $x = a\cos u,\ y = b\sin u$ and $z = v$, where a and b are constants and $0 \le u \le 2\pi$ and $v \in \mathbb{R}$, represent **an elliptic cylinder** with z-axis as its axis. If $a = b$, it becomes a right-circular cylinder.

$$\bar{r}_u \times \bar{r}_v = \begin{vmatrix} \hat{i} & \hat{j} & \hat{k} \\ -a\sin u & b\cos u & 0 \\ 0 & 0 & 1 \end{vmatrix} = b\cos u\hat{i} + a\sin u\hat{j}.$$

So, the normal to the surface is always parallel to the xy-plane (Fig. 6.28).

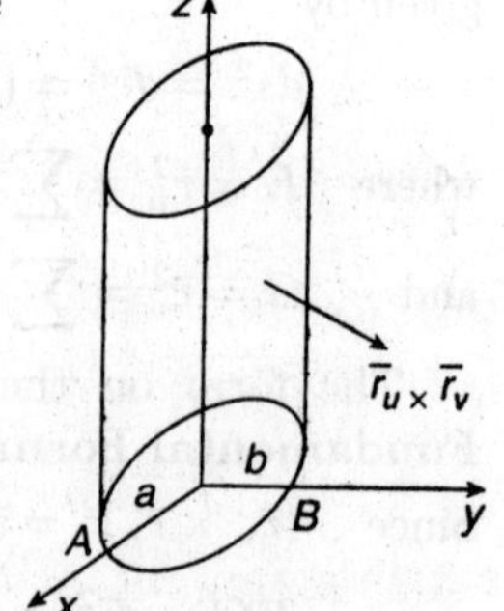

Fig. 6.28

(d) $x = (2 + v\sin u/2)\cos u$,

$y = (2 + v\sin u/2)\sin u$

and $z = v\cos u/2$, where $0 \le u \le 2\pi$ and $-1 \le v \le 1$

represent a non-orientable unilateral surface, a Möbius band.

When $u = 0,\ v = -1,\ P(x, y, z) = (2, 0, -1)$

and when $u = 2\pi,\ v = 1,\ Q(x, y, z) = (2, 0, -1)$.

So, a continuous curve drawn on the Möbius strip from P to Q, is a simple closed curve. But if

$$\bar{r}_u \times \bar{r}_v \text{ at } (0, -1) \quad \text{and} \quad \bar{r}_u \times \bar{r}_v \text{ at } (2\pi, 1)$$

are calculated, it will be found that one is the additive inverse of the other. In other words,

$$\bar{r}_u \times \bar{r}_v|_{(0,-1)} = -\bar{r}_u \times \bar{r}_v|_{(2\pi,1)}.$$

Although P and Q are the same point on the strip, the directions of the unit normal, when carried continuously along a closed curve from P to Q, will be found to be opposite to each other.

(e) The tetrahedral surface $\sum$ having its faces lying on the planes

$$S_1 : x = 0,\ S_2 : y = 0,\ S_3 : z = 0$$

and $S_4 : x/a + y/b + z/c = 1,\ a, b, c > 0$

is a continuous, piecewise smooth surface. In the adjoining diagram (Fig. 6.29), these triangular faces are

(i) $OCB : x = 0$,

(ii) $OAC : y = 0$,

(iii) $OBA : z = 0$

and **(iv)** $ABC : x/a + y/b + z/c = 1$.

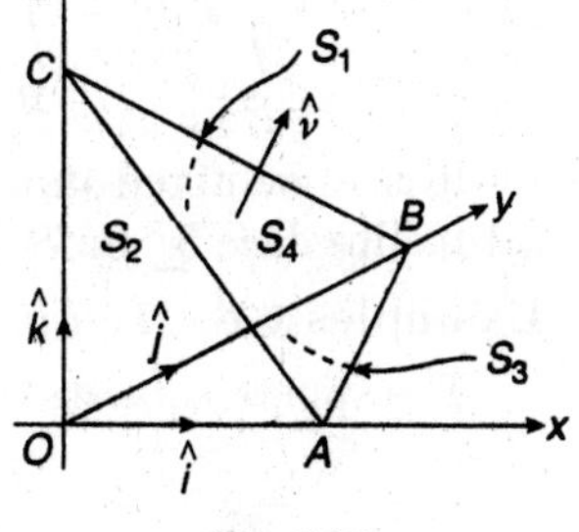

Fig. 6.29

If the boundaries of these faces are described in the order in which the letters appear, then the unit normals on these surfaces are

(i) $-\hat{\imath}$ on S_1,

(ii) $-\hat{\jmath}$ on S_2,

(iii) $-\hat{k}$ on S_3

and **(iv)** $(\hat{\imath}/a + \hat{\jmath}/b + \hat{k}/c)/\sqrt{\sum 1/a^2}$ on S_4.

Since these normals are all outward, the surface $\sum$ is positively oriented.

6.11.4 Area of an Oriented Surface

Let an oriented surface $\sum$ be given by $\bar{r} = \bar{r}(u, v)$:

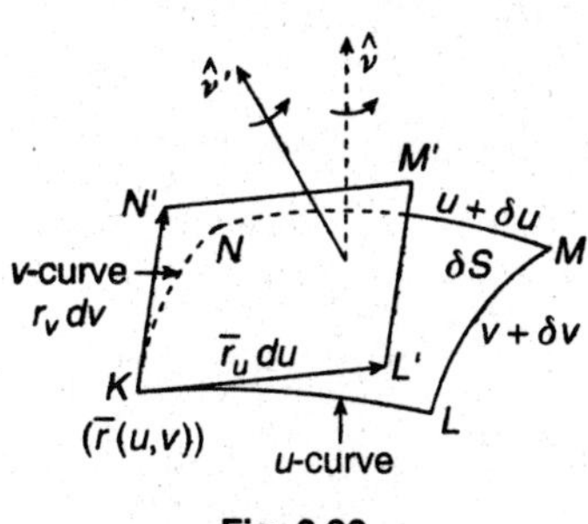

Fig. 6.30

To compute an element of area of the surface $\sum$ we consider a small portion of are δS on it bounded by two consecutive u-curves and two consecutive v-curves. Let the curves meet in K, L, M and N, whose p.v.'s are $\bar{r}(u, v)$, $\bar{r}(u + \delta u, v)$, $\bar{r}(u + \delta u, v + \delta v)$ and $\bar{r}(u, v + \delta v)$ respectively, where δu and δv **are** taken both positive (Fig. 6.30).

Since $\bar{r}_u$ is tangential to the u-curve at K if for small δu, $\overrightarrow{KL'} = \bar{r}_u \delta u$, $KL' \approx KL$. Similarly, if $\overrightarrow{KN'} = \bar{r}_v \delta v$, $KN' \approx KN$.

Now $(\bar{r}_u \delta u) \times (\bar{r}_v \delta v)$ is a vector representing the area of the parallelogram $KL'M'N'$ whose magnitude is approximately equal to δS.

Thus $\Delta S \approx |\bar{r}_u \delta u \times \bar{r}_v \delta v| = |\bar{r}_u \times \bar{r}_v| \delta u \delta v$

But $|\bar{r}_u \times \bar{r}_v| = \sqrt{\{\bar{r}_u^2 \bar{r}_v^2 - (\bar{r}_u \cdot \bar{r}_v)^2\}}$

$$= \sqrt{(EG - F^2)} = \sqrt{\sum J_i^2}.$$

$$\therefore \quad \delta S \approx \sqrt{(EG - F^2)} \delta u \delta v = \sqrt{\sum J_i^2 \delta u \delta v}. \tag{1}$$

If R is the domain of the function $\bar{r} = \bar{r}(u, v)$ and S is its image, then the area Δ of S is given by

$$\Delta = \iint_S dS = \iint_R \sqrt{(EG - F^2)}dudv = \iint_R \sqrt{\sum J_i^2}dudv. \tag{2}$$

If the surface S is given by $z = z(x, y) \ \forall \ (x, y) \in R$, then

since $\quad EG - F^2 = z_x^2 + z_y^2 + 1$,

$$\therefore \quad \Delta = \iint_S dS = \iint_R \sqrt{(z_x^2 + z_y^2 + 1)}dxdy. \tag{3}$$

It is to be noted that in this case R is the orthogonal projection of S in the xy-plane (Fig. 6.19). [Replace $\sum$ by S and E by R in the figure.]

Examples 6.8. 1. *Find the area of a sphere of radius R.*

The sphere consists of two hemispheres.

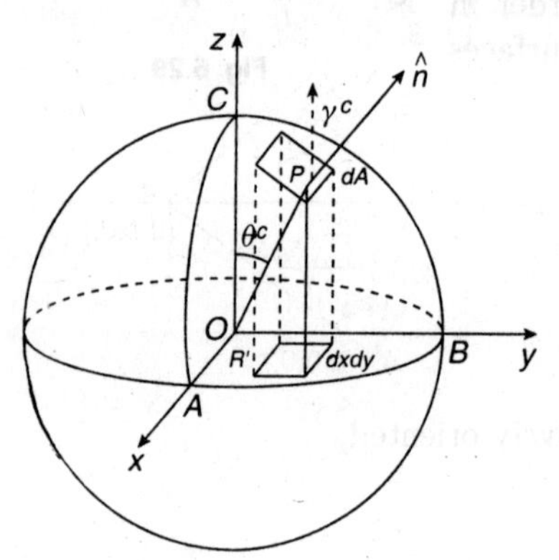

Fig. 6.31

The equation to a hemisphere of radius R may be taken $z = \sqrt{R^2 - x^2 - y^2}$, so that the origin O is at the centre of the circular base and the curved surface bulges in the positive direction of the z-axis.

Obviously, if Δ is the area of the hemisphere then

$$\Delta = 4\iint_S dS,$$

where S is the part of the hemisphere lying in the first octant, viz., ABC (Fig. 6.31). If γ^c is the positive acute angle made by the outward normal at P then by (3), § 6.11.4

$$\Delta = 4\iint_{R'} \sec\gamma dxdy, = 4\iint_{R'} \sqrt{1 + z_x^2 + z_y^2 dxdy},$$

where R' is the orthogonal projection of the surface ABC in the xy-plane. Now,

$$z_x = -x/z \quad \text{and} \quad z_y = -y/z$$

give $\quad \sqrt{(1 + z_x^2 + z_y^2)} = \sqrt{\left(1 + \frac{x^2}{z^2} + \frac{y^2}{z^2}\right)} = \frac{R}{z}, \quad \because \ z^2 = R^2 - x^2 - y^2.$

$$\therefore \quad \Delta = 4\int_0^R dx \int_0^{y_1} \frac{R}{\sqrt{y_1^2 - y^2}}dy, \quad \text{where } y_1 = \sqrt{R^2 - x^2}$$

$$= 4R\int_0^R \sin^{-1}\frac{y}{y_1}\Big|_0^{y_1} dx = 4R\frac{\pi}{2}\int_0^R dx = 2\pi R^2$$

Alternatively, using spherical polar coordinates, the position vector $\bar{r}$ of P (Fig. 6.31) is given by

$$\bar{r} = R(\sin\theta\cos\phi\hat{i} + \sin\theta\sin\phi\hat{j} + \cos\theta\hat{k}), \quad \text{where } 0 \le \theta \le \pi/2 \ \text{ and } \ 0 \le \phi \le 2\pi.$$

$$\therefore \quad d\bar{r} = R[(\cos\theta\cos\phi d\theta - \sin\theta\sin\phi d\phi)\hat{i} + (\cos\theta\sin\phi d\theta + \sin\theta\cos\phi d\phi)\hat{j} - \sin\theta d\theta\hat{k}]$$

and $\quad d\bar{r}^2 = Ed\theta^2 + 2Fd\theta d\phi + Gd\phi^2,$

where $\quad E = R^2[(\cos\theta\cos\phi)^2 + (\cos\theta\sin\phi)^2 + \sin^2\theta] = R^2,$

$F = 0$ and $G = R^2[\sin\theta\sin\phi)^2 + \sin\theta\cos\phi)^2] = R^2\sin^2\theta.$

$\therefore \quad EG - F^2 = R^4\sin^2\theta$ and $\sqrt{(EG - F^2)} = R^2\sin\theta.$

So by (2), § 6.11.4

$$\Delta = 4\int\int_R R^2\sin\theta d\theta d\phi = 4\int_0^{\pi/2} d\phi \int_0^{\pi/2} R^2\sin\theta d\phi = 4(\pi/2)R^2[-\cos\theta]_0^{\pi/2} = 2\pi R^2.$$

So, the area of the entire sphere is $4\pi R^2$.

2. *Using surface integrals in* $\mathbb{R}^3$,, *find the area of the tetrahedron whose faces are given by* $x = 0,\ y = 0,\ z = 0$ *and* $x/a + y/b + z/c = 1, a, b, c$ *being positive.*

The four faces OBC, OCA, OAB and ABC of the tetrahedron $OABC$ are

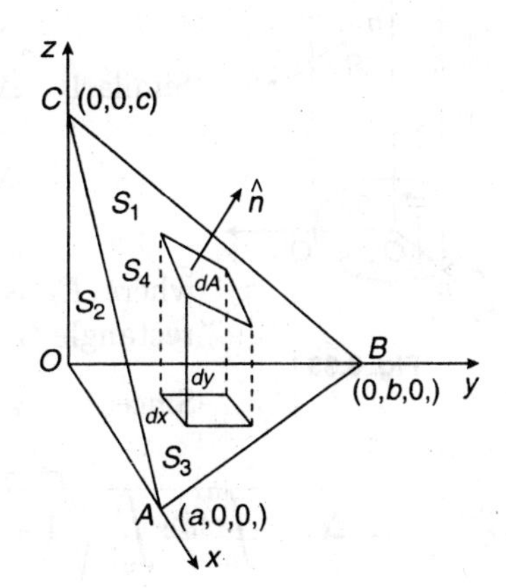

Fig. 6.32

$$S_1 : x = 0 \text{ and } 0 \le y/b + z/c \le 1,$$

$$S_2 : y = 0 \text{ and } 0 \le z/c + x/a \le 1,$$

$$S_3 : z = 0 \text{ and } 0 \le x/a + y/b \le 1$$

and $\quad S_4 : 0 \le x/a + y/b + z/c \le 1.$

Let $\quad S = S_1 \cup S_2 \cup S_3 \cup S_4.$

Then the required area is

$$\Delta = \iint_S dS = \sum_{i=1}^{4} \iint_{S_i} dS = \sum_{i=1}^{4} \Delta_i$$

Now, $$\Delta_1 = \iint_{S_1} dS = \iint_{S_1} dydz = \int_0^b dy \int_0^{c(1-y/b)} dz$$

$$= c\int_0^b (1 - y/b)dy = c\left[y - \frac{1}{2}y^2/b\right]_0^b = \frac{bc}{2}.$$

Similarly, $\Delta_2 = \dfrac{ca}{2}$ and $\Delta_3 = \dfrac{ab}{2}$,

$$\Delta_4 = \iint_{S_4} dS = \iint_{S_3}\int \sec\gamma dxdy = \iint_{S_3} \sqrt{(1 + z_x^2 + z_y^2)}dxdy,$$

where S_3 is the orthogonal projection of S_4 in the xy-plane and γ^c is the angle made by the outward normal $\hat{n}$ to the plane ABC. Since z is given by

$$z = c(1 - x/a - y/b),\ z_x = -c/a,\ z_y = -c/b$$

and $$\sqrt{(1 + z_x^2 + z_y^2)} = \sqrt{(1 + c^2/a^2 + c^2/b^2)} = c\sqrt{\sum 1/a^2}.$$

$$\therefore \quad \Delta_4 = c\sqrt{\sum 1/a^2}\int_0^a dx \int_0^{b(1-x/a)} dy = bc\sqrt{\sum 1/a^2}\int_0^a (1 - x/a)dx$$

$$= \frac{1}{2}abc\sqrt{\sum 1/a^2}.$$

$$\therefore \quad \Delta = \frac{1}{2}\left(\sum bc + abc\sqrt{\sum 1/a^2}\right).$$

3. *Find the area of the cylindrical surface given by $z = 0,\ z = h$ and $x^2 + y^2 = r^2, h, r > 0$, using surface integrals.*

If S is the entire surface of the cylinder, its area is four times the area lying in the first octant. The surfaces $S_1 : OP'Q'$, $S_2 : GPQ$ and $S_3 : GPP'Q'Q$ lie in the first octant (Fig. 6.33). If Δ_1, Δ_2 and Δ_3 are the areas of these surfaces respectively,

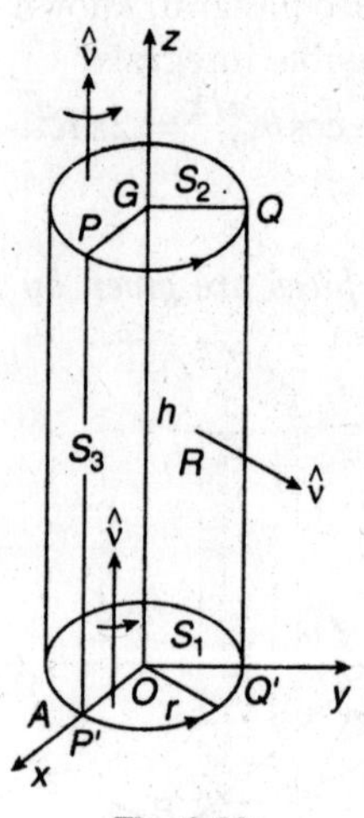

Fig. 6.33

$$\Delta_1 = \iint_{S_1} \sec\gamma dxdy = \iint_{S_1} dxdy, \quad \because \gamma = 0$$

$$= \int_0^r dx \int_0^{\sqrt{r^2-x^2}} dy, \text{ since for the arc } P'Q', y = \sqrt{r^2 - x^2}$$

$$= \int_0^r \sqrt{r^2 - x^2}dx = \left[\frac{1}{2}x\sqrt{r^2 - x^2} + \frac{r^2}{2}\sin^{-1}\frac{x}{r}\right]_0^r = \frac{\pi r^2}{4}.$$

Similarly, $\Delta_2 = \dfrac{\pi r^2}{4}$.

$$\Delta_3 = \iint_{S_3} \sec\gamma dydz = \iint_{E} \sqrt{1 + x_y^2 + x_z^2}dydz,$$

where E_3 is the orthogonal projection of S_3 in the yz-plane. It is the rectangle $GOQ'Q$.

Since $\quad x = \sqrt{r^2 - y^2},\ x_y = \dfrac{-y}{\sqrt{r^2 - y^2}} \quad$ and $\quad x_z = 0$.

$$\therefore \quad \Delta_3 = \int_0^h dz \int_0^r \sqrt{1 + \frac{y^2}{r^2 - y^2}}dy = rh\int_0^r \frac{dy}{\sqrt{r^2 - y^2}} = rh\sin^{-1}\frac{y}{r}\Big|_0^r = \frac{1}{2}\pi rh.$$

$$\therefore \quad \Delta_1 + \Delta_2 + \Delta_3 = \frac{\pi r^2}{2} + \frac{\pi rh}{2} = \frac{\pi r}{2}(r + h)$$

$$\therefore \quad \Delta = \text{the area of } S = 4\frac{\pi r}{2}(r + h) = 2\pi r(r + h),$$

which is what is to be expected from ordinary mensuration.

6.11.5 Surface Integral

Let $f(x, y, z)$ be a bounded function whose domain contains a piecewise smooth oriented surface S of a finite area. Suppose, S is divided into n pairwise disjoint elementary surfaces $S_1, S_2, \ldots, S_n$ each having the same orientation as S, of areas $\delta S_1, \delta S_2, \ldots, \delta S_n$. Let P_i be any arbitrary point of S_i for $i = 1, 2, \ldots, n$. Let us from the sum:

$$\sum_{i=1}^{n} f(P_i)\delta S_i \tag{1}$$

of n terms. If now the greatest of the diameters of S_i tends to zero, n must tend to infinity and if the limit of (1) exists, is finite and independent of the nature of partition of S into n parts and also of the choice of the point P_i in S_i, then this limit is defined to be the surface integral of f over the oriented surface S and it is denoted by the symbol:

$$\iint_S f(x, y, z)dS \tag{2}$$

or simply by

$$\iint_S f dS.$$

If the orientation of the surface is reversed, S becomes $-S$ and $\iint_{-S} f dS = -\iint_S dS$. For $f(P) = 1\ \forall\ P \in S$, the integral (2) will give the area of the surface S. Such a surface integral, known as the **surface integral of the first type**, is evaluated with the help of double integrals.

Let S be linear in z. Then it can be represented in the form:

$$z = z(x, y)\ \forall\ (x, y) \in R''' \quad \text{(Fig. 6.34)}. \tag{3}$$

Then R''' is the orthogonal projection of S on the xy-plane. The orientation of S is supposed to be induced in R''', so that the unit normal to R''' is $\hat{k}$ if $\hat{\nu}$ makes an acute angle with $\hat{k}$.

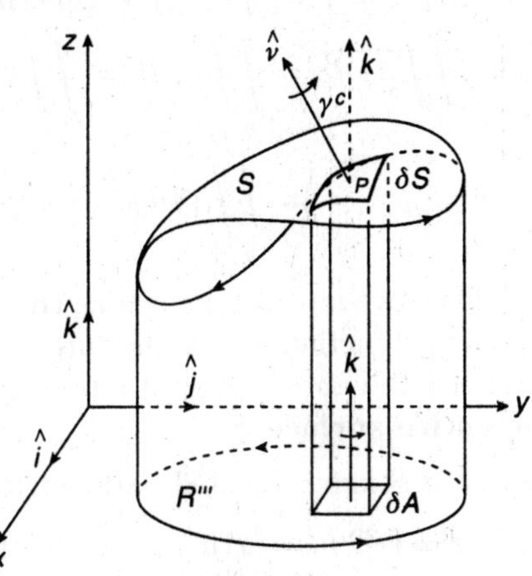

Fig. 6.34

The orthogonal projection of an elementary area δS at P of S on the same plane is an elementary area δA, where

$$\delta A = \delta S \cos\gamma = \delta S \hat{\nu} \cdot \hat{k}, \tag{4}$$

where γ^c is the angle made by the normal vector $\hat{\nu}$ with $\hat{k}$. If $h(x, y, z)$ is an integrable function whose domain includes S and $H(x, y) = h(x, y, z(x, y))$ then H is also integrable in R''' and we shall have

$$\iint_S h dS = \iint_{R'''} H dA \sec\gamma$$

$$= \iint_{R'''} H\sqrt{(1 + z_x^2 + z_y^2)}dxdy, \tag{5}$$

which is just a double integral over R''' in the xy-plane.

Also, $\iint_{R'''} H dxdy = \iint_{R'''} H dA = \iint_S H \cos\gamma dS = \iint_S h dxdy.$

Thus, if S is linear in z and $h(x, y, z)$ is integrable in a domain containing S, then the surface integral

$$\iint_S h dxdy \equiv \iint_S h\cos\gamma dS = \iint_{R'''} H dxdy, \tag{6}$$

where R''' is the orthogonal projection of S in the xy-plane.

If S is not linear in z, it can be divided into a finite number of parts each of which is linear in z. Then the integral of h over each such part can be expressed as a double integral in the xy-plane. Keeping the orientation of each part the same as that of S and adding the corresponding double integrals the integral over S is obtained.

Similarly, we get

$$\iint_S f dydz \equiv \iint_S f\cos\alpha dS = \iint_{R'} F dydz$$

and $$\iint_S g\,dz\,dx = \iint_S g\cos\beta\,dS = \iint_{R''} G\,dz\,dx,$$

where $F = f(x,(y,z),y,z)$, $G = g(x,y(z,x),z)$ and R' and R'' are respectively the orthogonal projections of S in the yz- and zx-planes.

Now, if $\overline{u} = f\hat{i} + g\hat{j} + h\hat{k}$ is a vector function whose domain includes the surface S, supposed to be piecewise smooth oriented and $\hat{\nu}$ is the unit normal vector at a point $P(x,y,z)$ on S and

$$u_\nu = \overline{u}\cdot\hat{\nu} = (f\hat{i} + g\hat{j} + h\hat{k})\cdot(l\hat{i} + m\hat{j} + n\hat{k}) = fl + gm + hn,$$

then $$\iint_S \overline{u}\cdot d\overline{S} = \iint_S u_\nu\,dS = \iint_S (f\cos\alpha + g\cos\beta + h\cos\gamma)\,dS$$

$$= \iint_S \{f\,dy\,dz + g\,dz\,dx + h\,dx\,dy\}. \tag{7}$$

The above integral, known as **the surface integral of the second type**, is an integral of the vector function $\overline{u}$ over the surface S. If $\overline{u}$ represents the velocity of a fluid flowing past the surface S the above integral represents physically **the total flux of the velocity $\overline{u}$ across the entire surface S.**

Finally, if S is represented parametrically, by

$$\overline{r} = \overline{r}(u,v) = \hat{i}x(u,v) + \hat{j}y(u,v) + \hat{k}z(u,v), \tag{8}$$

then since $\delta S \approx \sqrt{[EG - F^2]}\,\delta u\,\delta v$,

by (1), § 6.11.4 we have

$$\iint_S f\,dS = \iint_R X\sqrt{(EG - F^2)}\,du\,dv, \tag{9}$$

where S is the image of R under the transformation given by (8) and

$$X = X(u,v) = f(x(u,v), y(u,v), z(u,v)).$$

Also, $$\iint_S u_\nu\,dS = \iint_R (XJ_1 + YJ_2 + ZJ_3)\,du\,dv, \tag{10}$$

where J_1, J_2, J_3 are given by (12), § 6.11.2.

Examples 6.9. 1. *Evaluate* $\iint_S xyz\,dS$, *where S is the portion of the plane: $\sum x = 1$ lying in the first octant.*

Let R be the orthogonal projection of S in the xy-plane. Then the equation to S is

$$z = z(x,y) = 1 - x - y.$$

Since $z_x = -1,\ z_y = -1,\ \sqrt{1 + z_x^2 + z_y^2} = \sqrt{3}$,

$\therefore$ the given integral $= \iint_R xyz\sqrt{3}\,dx\,dy$,

where $z = -x - y$.

S meets the xy-plane in the line: $x + y = 1,\ z = 0$ and it meets the x-axis in the point $x = 1,\ y = z = 0$.

So, the region $R = \{(x, y) \in xy\text{-plane: } 0 \le x \le 1 \text{ and } 0 \le y \le 1 - x\}$.

$$\therefore \quad I = \iint_S xyz\,dS = \sqrt{3}\int_0^1 x\,dx \int_0^{1-x} y(1-x-y)\,dy = \sqrt{3}\int_0^1 xF(x)\,dx,$$

$$\text{where} \quad F(x) = \int_0^{1-x} y(1-x-y)\,dy = \left[\frac{1}{2}(1-x)y^2 - \frac{1}{3}y^3\right]_{y=0}^{y=1-x}$$

$$= \frac{1}{6}[y^2\{3(1-x) - 2y\}]_0^{1-x} = \frac{1}{6}(1-x)^2\{3(1-x) - 2(1-x)\}$$

$$= \frac{1}{6}(1-x)^3.$$

$$\therefore \quad I = \frac{\sqrt{3}}{6}\int_0^1 x(1 - 3x + 3x^2 - x^3)\,dx = \frac{\sqrt{3}}{120}.$$

2. *Evaluate* $\iint_S p \sum x^2 dS$, *where* S *is the entire surface of the ellipsoid* $x^2/a^2 + y^2/b^2 + z^2/c^2 = 1$, a, b, c *being* > 0 *and* p *is the length of the perpendicular drawn from the origin to the tangent plane of* S *at* (x, y, z).

Let $\quad f(x, y, z) = p\sum x^2 \cdot \forall\ (x, y, z) \in D$, which includes S.

Let us represent the surface S parametrically by

$$x = x(u, v) = a\sin u\cos v, \quad y = y(u, v) = b\sin u\sin v \quad \text{and} \quad z = c\cos u,$$

where $0 \le u \le \pi$ and $0 \le v \le 2\pi$.

The equation to the tangent plane at $P(x', y', z') \in S$ is $xx'/a^2 + yy'/b^2 + zz'/c^2 = 1$. So, the length p of the perpendicular drawn from the origin to this tangent plane is given by

$$p = 1/\sqrt{\textstyle\sum x'^2/a^4}. \tag{1}$$

If $\quad x' = a\sin u\cos v,\ y' = b\sin u\sin v,\ z' = c\cos u$ then,

$$p = 1/\sqrt{(1/a^2)\sin^2 u\cos^2 v + (1/b^2)\sin^2 u\sin^2 v + (1/c^2)\cos^2 u]}. \tag{2}$$

Now, $\quad EG - F^2 = \sum J_1^2$, by (14) § 6.11.2,

where $\quad J_1 = \partial(y, z)/\partial(u, v) = bc\sin^2 u\cos v,$

$\quad J_2 = \partial(z, x)/\partial(u, v) = ca\sin^2 u\sin v$

and $\quad J_3 = \partial(x, y)/\partial(u, v) = ab\sin u\cos v.$

$$\therefore \quad EG - F^2 = a^2b^2c^2[(1/a^2)\sin^2 u\cos^2 v = (1/b^2)\sin^2 u\sin^2 v + (1/c^2)\cos^2 u]$$

$$= a^2b^2c^2\sin^2 u/p^2 \text{ by (2)}$$

$$\therefore \quad \sqrt{(EG - F^2)} = abc\sin u/p. \tag{3}$$

Using (1), $f(x, y, z) = \sum x^2/\sqrt{\sum x^2/a^4}$,

which is symmetric in x, y and z. The surface S is also symmetric in x, y and z. So, the required surface integral is 8 times that over the portion of S lying in the 1st octant.

This portion is the image of R in the uv-plane, where

$$R = \{(u, v) : 0 \le u \le \pi/2, 0 \le v \le \pi/2\}.$$

So, the value of the required integral is

$$8 \int_0^{\pi/2} du \int_0^{\pi/2} p(a^2 \sin^2 u \cos^2 v + b^2 \sin^2 u \sin^2 v + c^2 \cos^2 u) \frac{abc \sin u}{p} dv$$

$$= 8abc \int_0^{\pi/2} du \int_0^{\pi/2} (a^2 \sin^3 u \cos^2 v + b^2 \sin^3 u \sin^2 v + c^2 \sin u \cos^2 u) dv$$

$$= 8abc \int_0^{\pi/2} \left(a^2 \cdot \frac{\pi}{2} \sin^3 u + b^2 \cdot \frac{\pi}{2} \sin^3 u + c^2 \frac{\pi}{2} \sin u \cos^2 u\right) du$$

$$= 4\pi abc \left[\frac{2}{3} a^2 + \frac{2}{3} b^2 + \frac{1}{3} c^2\right] = \frac{4}{3} \pi abc(2a^2 + 2b^2 + c^2).$$

3. *Evaluate* $\iint_S (xdydz + dzdx + xzy^2 dxdy)$, *where S is the part of the spherical surface: $x^2 + y^2 + z^2 = R^2$ lying in the first octant, with its orientation given by the outward drawn normal.*

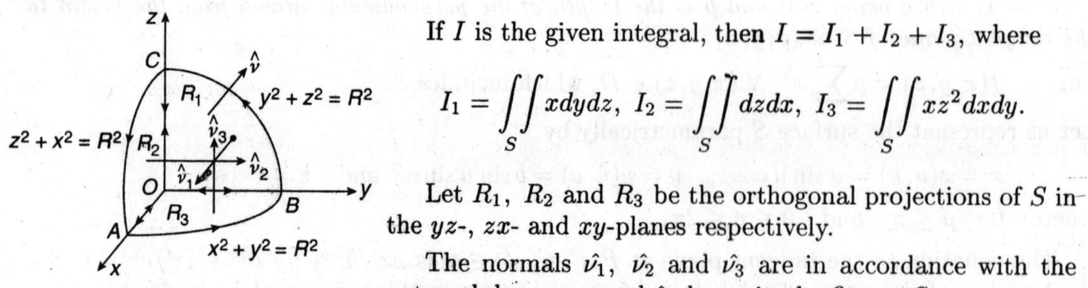

Fig. 6.35

If I is the given integral, then $I = I_1 + I_2 + I_3$, where

$$I_1 = \iint_S xdydz,\ I_2 = \iint_S dzdx,\ I_3 = \iint_S xz^2 dxdy.$$

Let R_1, R_2 and R_3 be the orthogonal projections of S in the yz-, zx- and xy-planes respectively.

The normals $\hat{\nu}_1$, $\hat{\nu}_2$ and $\hat{\nu}_3$ are in accordance with the outward drawn normal $\hat{\nu}$ shown in the figure. So,

$$I_1 = \iint_{R_1} xdydz = \int_0^R dy \int_0^{\sqrt{R^2-y^2}} \sqrt{R^2 - y^2 - z^2}\, dz$$

$$= \int_0^R \left[\frac{z}{2}\sqrt{R^2 - y^2 - z^2} + \frac{R^2 - y^2}{2} \sin^{-1} \frac{z}{\sqrt{R^2 - y^2}}\right]_0^{\sqrt{R^2-y^2}} dy$$

$$= \frac{1}{2} \int_0^R (R^2 - y^2) \frac{\pi}{2} dy = \frac{\pi R^3}{6},$$

$$I_2 = \iint_{R_2} dzdx = \text{Area of } OCA = \frac{\pi R^2}{4},$$

$$I_3 = \iint_{R_3} xz^2 dxdy = \iint_{R_3} x(R^2 - x^2 - y^2) dxdy$$

$$= \int_0^R xdx \int_0^{\sqrt{R^2-x^2}} (R^2 - x^2 - y^2)\, dy = \int_0^R x \left[R^2 y - x^2 y - \frac{2}{3} y^3\right]_0^{\sqrt{R^2-x^2}} dx$$

$$= \frac{2}{3}\int_0^R x(R^2 - x^2)^{3/2} dx = \frac{2}{15}R^5.$$

$$\therefore \quad I = \frac{\pi R^3}{6} + \frac{\pi R^2}{4} + \frac{2R^5}{15} = (\pi/60)R^2(15 + 10R + 4R^3).$$

4. *Evaluate* $\iint_S \overline{u} \cdot \hat{\nu} dS$, *where* $\overline{u} = 6z\hat{i} - 4\hat{j} + y\hat{k}$ *and* S *is the portion of the plane:* $x/6 + y/4 + z/2 = 1$ *lying in the first octant and* $\hat{\nu}$ *is the unit normal to* S *with its direction from the origin to* S.

If $\quad F(x, y, z) = x/6 + y/4 + z/2 - 1,$

then since

$$F_x = 1/6,\ F_y = 1/4,\ F_z = 1/2,$$

the D.R.'s of the normal to the plane S

are $\quad 1/6, 1/4, 1/2 \quad \text{or} \quad 2, 3, 6.$

$$\therefore \quad \hat{\nu} = \pm\frac{2\hat{i} + 3\hat{j} + 6\hat{k}}{\sqrt{(2^2 + 3^2 + 6^2)}} = \pm\frac{1}{7}(2\hat{i} + 3\hat{j} + 6\hat{k}).$$

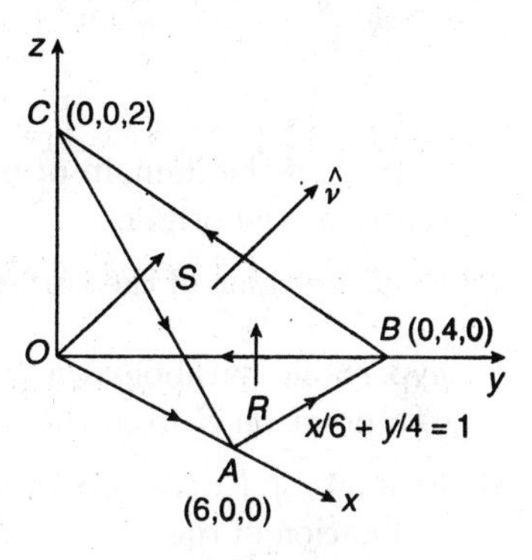

Fig. 6.36

But the plus sign will give the direction away from the origin. So, $\hat{\nu} = (1/7)(2\hat{i} + 3\hat{j} + 6\hat{k})$ and

$$u_\nu = \overline{u} \cdot \hat{\nu} = (1/7)(12z - 12 + 6y).$$

If R is the projection of S in the xy-plane, then in R

$$0 \le x \le 6 \quad \text{and} \quad 0 \le y \le 4(1 - x/6).$$

Since $\quad z = 2(1 - x/6 - y/4),\ z_x = -1/3, z_y = -1/2,$

$$\therefore \quad \sqrt{(1 + z_x^2 + z_y^2)} = \sqrt{(1 + 1/9 + 1/4)} = 7/6.$$

$$\therefore \quad \iint_S \overline{u} \cdot \hat{\nu} dS = \iint_R \frac{1}{7}\left[12.2\left(1 - \frac{x}{6} - \frac{y}{4}\right) - 12 + 6y\right] \cdot \frac{7}{6} dx dy$$

$$= \frac{1}{6}\int_0^6 dx \int_0^{4(1-x/6)} (12 - 4x) dy = \frac{2}{3}\int_0^6 (3 - x)4\left(1 - \frac{x}{6}\right) dx$$

$$= \frac{4}{9}\int_0^6 (18 - 9x + x^2) dx = 8.$$

Exercises 6.6

1. Represent the surface S in parametric form and then evaluate $\iint_S f dS$, where

(i) $f = xy,\ S = \{(x, y, z) : z = xy,\ 0 \le x \le 1,\ 0 \le y \le 1\},$

(ii) $f = 3x^3 \sin y,\ S = \{(x, y, z) : z = x^3,\ 0 \le x \le 1,\ 0 \le y \le \pi\},$

(iii) $f = xy,\ S = \{(x, y, z) : x^2 + y^2 = 4,\ -1 \le z \le 1\}$

and **(iv)** $f = x(z^2 + 12y - y^4),\ S = \{(x, y, z) : y^2 = z,\ 0 \le x \le 1,\ 0 \le y \le 1\}.$

2. Show that

$$\iint_S (lx + my + nz)dS = 3V$$

for every closed surface S, where l, m, n are the D.C.'s of the outward normal to the surface S at (x, y, z) and V is the volume of the space enclosed in S.

3. Find $\iint_S \sqrt{(x^2 + y^2)}dS$, where S is the lateral surface of the cone

$$x^2/a^2 + y^2/b^2 - z^2/c^2 = 0,\ 0 \le z \le b.$$

4. Show that the value of the integral

$$\iint xdydz + ydzdx + zdxdy$$

is $4\pi a^3$, if the domain of integration is the closed spherical surface of radius a having its centre at the origin.

5. Find the value of the surface integral $\iint_S dS/r$, where S is the portion of the surface of the hyperbolic paraboloid $z = xy$ cut off by the cylinder $x^2 + y^2 = a^2$ and r is the distance of a point on S from the z-axis.

6. Evaluate $\iint z \cos\gamma dS$ over the surface of the sphere $\sum x^2 = 1$, where γ^c is the angle of inclination of the outer normal at a point (x, y, z) of the sphere with the positive direction of the z-axis.

Answers

1. (i) $(3^{5/2} - 2^{7/2} + 1)/15$; **(ii)** $(10^{3/2} - 1)/9$; **(iii)** 0; **(iv)** $(5^{3/2} - 1)/2$.

3. $(2\pi a^2/3)\sqrt{(a^2 + b^2)}$. **5.** $\pi[a\sqrt{(a^2 + 1)} + \ln\{a + \sqrt{(a^2 + 1)}\}]$. **6.** $4\pi/3$.

6.12 Gauss's Divergence Theorem

Let a region $T(\subseteq \mathbb{R}^3)$ be bounded by a continuous, piecewise smooth closed surface S and let $\overline{F}$ be a continuous vector function, possessing continuous first partial derivatives in a domain containing T. Then the volume integral of the divergence of $\overline{F}$ over the region T is equal to the surface integral of the normal component of $\overline{F}$ over S:

$$\iiint_T \text{div}\, \overline{F} dV = \iint_S \overline{F} \cdot \hat{\nu} dS, \tag{1}$$

where $\hat{\nu}$ is the outward drawn unit normal vector at a point on S.

Proof. Let $\overline{F} = f\hat{i} + g\hat{j} + h\hat{k}$, where f, g, h are functions of x, y and z, supposed to be continuous possessing continuous first partial derivatives in a domain containing T. If $\hat{\nu} = l\hat{i} + m\hat{j} + n\hat{k}$ then $l(= \cos\alpha), m(= \cos\beta)$ and $n(= \cos\gamma)$ are the D.C.'s of the outward drawn normal and (1) becomes

$$\iiint_T \nabla \cdot \overline{F} dV \quad \text{or,} \quad \iiint_T (f_x + g_y + h_z)dxdydz = \iint_S (fl + gm + hn)dS$$

$$\text{or,} \quad \iint_S (fdydz + gdzdx + hdxdy). \tag{2}$$

(2) will be true if

(i) $\iiint\limits_T f_x dxdydz = \iint\limits_S f dydz,$

(ii) $\iiint\limits_T g_y dxdydz = \iint\limits_S g dzdx$

and (iii) $\iiint\limits_T h_z dxdydz = \iint\limits_S h dxdy.$

We shall prove (iii). The proofs of the two others will follow.

To prove (iii) we assume that the surface S enclosing T is quadratic in z. Then

$$\forall\ P(x,y,z) \in T,\ z_1 \le z \le z_2 \quad \text{(Fig. 6.37)},$$

where $z = z_1(x,y)$ and $z = z_2(x,y)$ represent the lower and the upper bounding surfaces S_1 and S_2 of T, where (x,y) belongs to the region R, in the xy-plane, which is the orthogonal projection of both S_1 and S_2 in the xy-plane. If S_1 and S_2 together do not enclose the region T, then there may exist a third boundary S_3, cylindrical with its generators parallel to the z-axis. Its projection in the xy-plane is just the boundary of R, having no area.

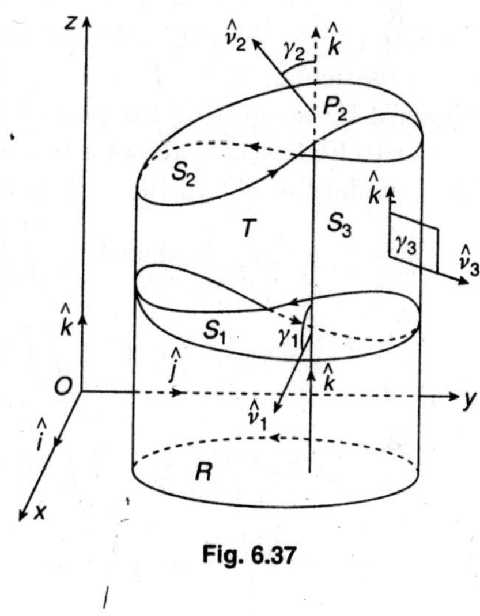

Fig. 6.37

Now $S = S_1 \cup S_2 \cup S_3$, where S_1, S_2 and S_3 are pairwise disjoint smooth surfaces except the common boundaries.

Also, $\iiint\limits_T h_z dV = \iiint\limits_T \frac{\partial h}{\partial z} dzdxdy$

$$= \iint\limits_R dxdy \int_{z_1}^{z_2} \frac{\partial h}{\partial z} dz = \iint\limits_R h(x,y,z)\Big|_{z_1}^{z_2} dxdy$$

$$= \iint\limits_R [h(x,y,z_2) - h(x,y,z_1)dA \qquad (3)$$

and $$\iint\limits_S h dxdy = \iint\limits_S hndS = \iint\limits_{S_1} hn_1 dS_1 + \iint\limits_{S_2} hn_2 dS_2 + \iint\limits_{S_3} hn_3 dS_3, \qquad (4)$$

where $\hat{\nu}_r = l_r\hat{i} + m_r\hat{j} + n_r\hat{k}$, for $r = 1,2,3$ so that $n_1 = \cos\gamma_1 < 0, n_2 = \cos\gamma_2 > 0$ and $n_3 = \cos\gamma_3 = 0$, since γ_1^c is obtuse, γ_2^c is acute and $\gamma_3^c = \frac{\pi^c}{2}$.

If δA is the projection of two elementary areas δS_1 and δS_2 of the surfaces S_1 and S_2 on the xy-plane

$$\delta A = -n_1 \delta S_1 = n_2 \delta S_2.$$

So, the R.H.S. of (4) becomes

$$-\iint_R h(x,y,z_1)dA + \iint_R h(x,y,z_2)dA,$$

which is the same as that of (3). Hence (iii) is proved under the assumption that the surface S bounding the region T is quadratic in z. If S is not quadratic in z, it can be divided into a finite number of parts each of which is quadratic in z. For each of these parts the result (iii) is true. Adding all these results, as in the case of a line integral along a closed contour, the result will be true for the entire surface S enclosing T.

Note. Gauss's divergence theorem connects a volume integral to a surface integral as Green's theorem (or Gauss's theorem in a plane), § 6.10, connects a double integral to a line integral. Since

$$\iiint_T h_x dV = \iint_S h dS \quad \text{or,} \quad \iiint_T (\partial h/\partial x) dx dy dz = \iint_S h dy dz,$$

here also we find that integration cancels differentiation reducing a volume integral of the partial derivative of a function to the surface integral of the function.

In particular, if div $\overline{F} = 1$, $\int_S \overline{F} \cdot \hat{\nu} dS = \int_T dV = V$, the volume of a region T can be obtained by integrating the normal component of $\overline{F}$ over the surface S enclosing T. This is similar to finding the area of a plane figure E by integrating the normal component of $\overline{F}$ over the boundary of E (§ 6.10.1). Also it might be noted that if

$f = x,\ g = h = 0$, then $V = \iint_S x dy dz$. Similarly, for

$f = 0,\ g = y,\ h = 0,\ V = \iint_S y dz dx$ and for

$f = 0 = g,\ h = z,\ V = \iint_S z dx dy.$

So, $V = (1/3)\iint_S x dy dz + y dz dx + z dx dy.$

6.12.1 Physical Interpretation of Divergence

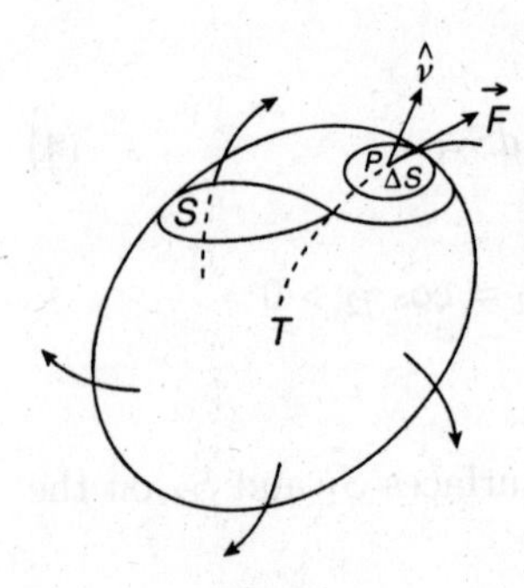

Fig. 6.38

A physical interpretation of Gauss's divergence theorem can be given by considering $\overline{F}$ as the velocity field of a steady flow of an incompressible fluid of constant density, say unity, through the region T bounded by the closed surface S. [A flow is steady if the state of motion is independent of time.]

If the amount of fluid flowing from inside to outside S through an element ΔS of surface of area ΔS also, per unit time is Δm then

$$\Delta m = \overline{F} \cdot \hat{\nu} \Delta S,$$

where $\hat{\nu}$ is the outward drawn unit normal to S at point P of ΔS. Accordingly, the total amount of fluid flowing across S from the inside to the outside per unit time is

$$\iint_S dm \quad \text{or,} \quad \iint_S \overline{F} \cdot \hat{\nu} dS.$$

By the divergence theorem, $\displaystyle\iiint_T \text{div } \overline{F} dV = \iint_S \overline{F} \cdot \hat{\nu} dS.$

So, the L.H.S. also represents the total amount of fluid leaving the region T in unit time. Since $\overline{F}$ is continuously differentiable, by the mean value theorem of integral calculus there is a point Q in T such that

$$\iiint_T \text{div } \overline{F} dV = V \text{ div } \overline{F}(Q),$$

where V is the volume of the region T. So

$$\text{div } \overline{F}(Q) = \frac{1}{V} \iiint_T \text{div } \overline{F} dV.$$

If R is any arbitrary point of T and the entire region T shrinks down on to R in such a way that Δ, the greatest of the distances of all the points of T from R tends to zero, then Q also must approach R and then

$$\text{div } \overline{F}(R) = \underset{Q \to R}{\text{Lt}} \text{ div } \overline{F}(Q) = \underset{\Delta \to 0}{\text{Lt}} \frac{1}{V} \iiint_T \text{div } \overline{F} dV = \underset{\Delta \to 0}{\text{Lt}} \frac{1}{V} \iint_S \overline{F} \cdot \hat{\nu} dS. \tag{1}$$

Since we are considering the steady flow of an incompressible fluid the total amount of fluid flowing outwards must be continuously supplied in T somehow. So, in the interior of T there must be sources producing fluid and from (1) we find that the divergence at a point R is the **source intensity** thereat. Hence, **physically the divergence of the velocity field $\overline{F}$ is the source intensity of the steady incompressible flow of fluid**. In the absence of any source or sink the divergence must therefore be zero.

Examples 6.10. 1. *Use Divergence Theorem to evaluate the surface integral* $\displaystyle\iint_S \overline{u} \cdot \hat{\nu} dS$, *where* $\overline{u} = x\hat{i} + y\hat{j} + z\hat{k}$ *and* S *is the cube given by* $S = \{(x, y, z) : 0 \le x, y, z \le a\}$ *and as usual* $\hat{\nu}$ *is the unit normal on* S *drawn outward.*

If T is the region enclosed in S then from the Divergence Theorem

$$\iint_S \overline{u} \cdot \hat{\nu} dS = \iiint_T (f_x + g_y + h_z) dV, \tag{i}$$

where $f = x$, $g = y$ and $h = z$. Since $f_x = g_y = h_z = 1$,

R.H.S. of (i) $= \displaystyle\iiint_T 3dV = 3V = 3a^3.$

$$\therefore \quad \iint_S \overline{u} \cdot \hat{\nu} dS = 3a^3.$$

2. *Use the Divergence Theorem to evaluate* $\iint_S \overline{F}\cdot d\overline{S}$, *where S is the closed cylindrical surface given by*

$$S = \{(x, y, z) : x^2 + y^2 = 4 \quad and \quad 0 \le z \le 3\}$$

and $\overline{F} = 4x\hat{i} - 2y^2\hat{j} + z^2\hat{k}$.

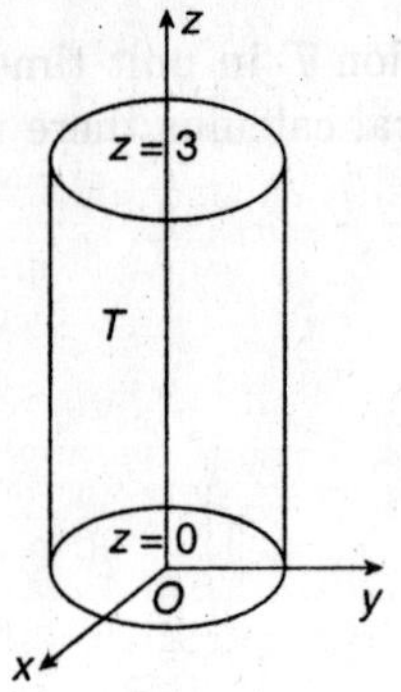

By the Divergence Theorem,

$$\iint_S \overline{F}\cdot d\overline{S} = \iiint_T \nabla\cdot\overline{F}dV,$$ where T is the region enclosed in the cylinder S (Fig. 6.39)

$$= \iiint_T (4 - 4y + 2z)dxdydz = \iint_R dxdy\int_0^3 (4 - 4y + 2z)dz$$

$$= \int_{-2}^{2} dx \int_{-\sqrt{4-x^2}}^{\sqrt{4-x^2}} [4z - 4yz + z^2]_0^3 dy$$

$$= \int_{-2}^{2} dx \int_{y_1}^{y_2} (12 - 12y + 9)dy$$

$$= \int_{-2}^{2} [21y - 6y^2]_{y_1}^{y_2} dx, \text{ where } y_1 = -\sqrt{4-x^2} \text{ and } y_2 = \sqrt{4-x^2}$$

$$= 3\int_{-2}^{2} 14\sqrt{4-x^2}dx = 42\times 2\int_0^2 \sqrt{4-x^2}dx$$

$$= 84\int_0^{\pi/2} 2\cos^2\theta d\theta = 84\pi.$$

Fig. 6.39

3. *Verify the Divergence Theorem for*

$$\overline{F} = (x^2 - yz)\hat{i} + (y^2 - zx)\hat{j} + (z^2 - xy)\hat{k}$$

taken over the rectangular parallelopiped given by

$$0 \le x \le a, \quad 0 \le y \le b \quad and \quad 0 \le z \le c.$$

Here, $\nabla\cdot\overline{F} = f_x + g_y + h_z = 2(x + y + z)$.

$$\therefore \quad \int_T \nabla\cdot\overline{F}dV = 2\iiint_T (x + y + z)dxdydz,$$

where T is the region enclosed in the parallelopiped.

$$\therefore \quad \int_T \nabla\cdot\overline{F}dV = 2\int_0^a dx\int_0^b dy\int_0^c (x + y + z)dz$$

$$= 2\int_0^a dx\int_0^b [(x + y)z + z^2/2]_0^c dy$$

$$= 2\int_0^a dx\int_0^b \{(x + y)c + c^2/2\}dy$$

$$= 2\int_0^a [cxy + c^2y/2 + cy^2/2]_0^b dx$$

$$= 2\int_0^a (cax + c^2b/2 + cb^2/2)dx = 2[cax^2/2 + bc(b+c)x/2]_0^a$$

$$= abc(a+b+c).$$

If S is the surface of the parallelopiped, then $S = \bigcup_{i=1}^6 S_i$, where S_1 $S_2, S_3, \ldots, S_6$ are the six faces of the parallelopiped. Then

$$\int_S = \sum_1^6 \int_{S_i}.$$

Let S_1 and S_2 be the faces given by $x = 0$ and $x = a$ respectively, where $0 \le y \le b$ and $0 \le z \le c$. Then

$$\int_{S_1} \overline{F}\cdot\hat{\nu}dS = \iint_{S_1} -(x^2 - yz)|_{x=0}dydz, \quad \because\ \hat{\nu} = -\hat{\imath} \text{ and } dS = dydz$$

$$\int_0^b dy \int_0^c yzdz = \int_0^b y[z^2/2]_0^c dy = (c^2/2)\int_0^b ydy = b^2c^2/4$$

and $$\int_{S_2} \overline{F}\cdot\hat{\nu}dS = \iint_{S_2} (x^2 - yz)|_{x=a}dydz, \quad \because\ \hat{\nu} = \hat{\imath}$$

$$= \int_0^b dy \int_0^c (a^2 - yz)dz = \int_0^b [a^2z - yz^2/2]_{z=0}^c dy$$

$$= \int_0^b (ca^2 - c^2y/2)dy = [ca^2y - c^2y^2/4]_0^b = a^2bc - b^2c^2/4.$$

$$\therefore \quad \int_{S_1} \overline{F}\cdot\hat{\nu}dS + \int_{S_2} \overline{F}\cdot\hat{\nu}dS = a^2bc.$$

If S_3 and S_4 are the faces given by $y = 0$ and $y = b$ then the sum of the integrals of $\overline{F}\cdot\hat{\nu}$ over them is b^2ca. Similarly, the contribution to the surface integral from the remaining faces is c^2ab. Adding all of them we get

$$\int_S \overline{F}\cdot\hat{\nu}dS = abc(a+b+c).$$

Thus the theorem is verified.

4. *Evaluate the surface integral* $\int_S \overline{F}\cdot d\overline{S}$, *where* $\overline{F} = x^3\hat{\imath} + y^3\hat{\jmath} + z^3\hat{k}$ *and* S *is the surface of the sphere* $\sum x^2 = a^2$, *using Gauss's divergence theorem.*

By the divergence theorem of Gauss,

$$\int_S \overline{F}\cdot d\overline{S} = \int_T \nabla\cdot\overline{F}dV.$$

But $\nabla\cdot\overline{F} = 3\sum x^2$. Transforming into spherical polar coordinates given by

$x = r\sin\theta\cos\phi$, $y = r\sin\theta\sin\phi$ and $z = r\cos\theta$, where

$0 \le r \le a$, $\quad 0 \le \theta \le \pi$ and $\quad 0 \le \phi \le 2\pi$,

$$dv = dxdydz = |J|drd\theta d\phi,$$

where $J = \partial(x,y,z)/\partial(r,\theta,\phi) = r^2\sin\theta$ and $\sum x^2 = r^2$.

$$\therefore \quad \int_T \nabla\cdot\overline{F}dV = 3\int_T r^2\cdot r^2\sin\theta drd\theta d\phi$$

$$= 3\int_0^a r^4dr\int_0^\pi \sin\theta d\theta\int_0^{2\pi} d\phi = 3[r^5/5]_0^a[-\cos\theta]_0^\pi 2\pi$$

$$= 12\pi a^5/5.$$

5. *Using the Divergence Theorem convert the surface integral*

$$\iint_S x^2(xdydz + ydzdx + zdxdy)$$

into a volume integral and then evaluate the integral, where S is the surface of the closed cylinder given by $x^2+y^2=a^2$, $z=0$ and $z=b$.

If T is the region bounded by S then

$$T = \{(x,y,z) : x^2+y^2\le a^2 \text{ and } 0\le z\le b\}.$$

By Gauss's divergence theorem, $\iint_S \overline{F}\cdot\hat{\nu}d'S = \iiint_T \nabla\cdot\overline{F}dV.$

Now, $\overline{F}\cdot\hat{\nu} = fl + gm + hn$, where $F = f\hat{i}+g\hat{j}+h\hat{k}, \hat{\nu} = l\hat{i}+m\hat{j}+n\hat{k}$

and $\iint_S \overline{F}\cdot\hat{\nu}dS = \iint_S (fl+gm+hn)dS = \iint_S fdydz + gdzdx + hdxdy).$

$\because \quad f = x^3,\ g = x^2y$ and $h = x^2z$,

$$\overline{F} = f\hat{i}+g\hat{j}+h\hat{k} = x^2(x\hat{i}+y\hat{j}+z\hat{k})$$

and $\nabla\cdot\overline{F} = f_x + g_y + h_z = 5x^2,$

$$\therefore \quad \iiint_T \nabla\cdot\overline{F}dV = 5\iiint_T x^2dV.$$

Transforming into cylindrical polar coordinates given by $x=\rho\cos\phi$, $y=\rho\sin\phi$, $z=z$, where $0\le\rho\le a$, $0\le\phi\le 2\pi, 0\le z\le b$,

$$dV = dxdydz = |J|d\rho d\phi dz,$$

where $J = \partial(x,y,z)/\partial(\rho,\phi,z) = \begin{vmatrix}\cos\phi & -\rho\sin\phi & 0\\ \sin\phi & \rho\cos\phi & 0\\ 0 & 0 & 1\end{vmatrix} = \rho.$

$$\therefore \quad dV = \rho d\rho d\phi dz$$

and $\int_T x^2dV = \int_T \rho^2\cos^2\phi\rho d\rho d\phi dz = \int_0^a \rho^3d\rho\int_0^{2\pi}\cos^2\phi d\phi\int_0^b dz = \pi a^4b/4.$

$$\therefore \quad \int_T \nabla\cdot\overline{F}dV = 5\pi a^4b/4.$$

6. *Verify Gauss's Divergence Theorem for $\overline{F} = y\hat{i} + x\hat{j} + z^2\hat{k}$ over the cylindrical region bounded by $x^2 + y^2 = 9, z = 0, z = 2$.* **[W.B. Sem. Exam. '03]**

$\because$ div $\overline{F} = 2z$, $\int_T \text{div}\,\overline{F}dV = \int_T 2zdxdydz$, where T is region enclosed in the given cylinder.

$$\therefore \quad \int_T \text{div}\,\overline{F}dV = \int_{-3}^{3} dx \int_{-\sqrt{9-x^2}}^{\sqrt{9-x^2}} dy \int_0^2 2zdz = 4 \cdot 2 \cdot 2 \int_0^3 \sqrt{9-x^2}dx$$

$$= 16 \int_0^{\pi/2} 3\cos\theta \cdot 3\cos\theta d\theta = 144 \int_0^{\pi/2} \cos^2\theta d\theta = 144 \cdot \frac{1}{2}\frac{\pi}{2} = 36\pi.$$

The corresponding surface integral is

$$\iint_S ydydz + xdzdx + z^2dxdy = I_1 + I_2 + I_3,$$

where $I_1 = \iint_S ydydz,\ I_2 = \iint_S xdzdx,\ I_3 = \iint z^2dxdy,$

S being the cylindrical surface.

To evaluate I_1, S is divided into two halves by the yz-plane, i.e., $x = 0$. The projection of each of them in the yz-plane in the rectangle $R_1 = \{(y,z) : -3 \le y \le 3 \text{ and } 0 \le z \le 2\}$. If I_1' and I_1'' are the integrals over the surfaces, then

$$I_1' = I_1'' = \int_0^2 dz \int_{-3}^{3} ydy = 0 \text{ and hence } I_1 = I_1' + I_1'' = 0.$$

Similarly, $I_2 = 0$.

To evaluate I_3 we project the surface S in the xy-plane, in which $z = 0$. This projection is the circular region

$$R_3 = \{(x,y) : x^2 + y^2 \le 9\}. \text{ Let } I_3 = I_3' + I_3'', \text{ where } I_3' = \int_{S_3'} z^2dxdy = \int_{R_3} 0^2dxdy,$$

if S_3' is the lower base of the cylinder and

$$I_3'' = \int_{S_3''} z^2dxdy = \int_{R_3} 2^2dxdy = 4\int_{R_3} dxdy = 4\pi 3^2 = 36\pi,$$

since $z = 2$ for the upper circular base S_3'' of the cylinder and area of R_3 is $\pi r^2 = \pi \cdot 3^2$.

7. *Using the integral over the surface of the sphere $\sum x^2 = a^2$ show that its volume is $(4/3)\pi a^3$.*

$$V = \frac{1}{3}\iint_S xdydz + ydzdx + zdxdy = \frac{1}{3} \cdot 3\iint_S xdydz.$$

Dividing the spherical surface S into two hemispheres S' and S'' by the yz-plane, the projection R of each of them on the same yz-plane is the circle $y^2 + z^2 \le a^2$.

$$I = \int_{S'} xdydz = \int_R \sqrt{(a^2 - y^2 - z^2)}dydz$$

$$= \int_{-a}^{a} dy \int_{-\sqrt{(a^2-y^2)}}^{\sqrt{(a^2-z^2)}} \sqrt{\{(a^2 - y^2) - z^2\}}dz$$

$$= 4\int_0^a dy \int_0^{\sqrt{(a^2-y^2)}} \sqrt{\{(a^2-y^2)-z^2\}}dz$$

$$= 4\int_0^a \left[\frac{z}{2}\sqrt{\{(a^2-y^2)-z^2\}} + \frac{a^2-y^2}{2}\sin^{-1}\frac{z}{\sqrt{(a^2-y^2)}}\right]_0^{\sqrt{(a^2-y^2)}} dy$$

$$= \pi\int_0^a (a^2-y^2)dy = (2\pi/3)a^3 = (2/3)\pi a^3.$$

Similarly, $\int_{S''} xdydz = (2/3)\pi a^3.$

$$\therefore \qquad V = \int_S xdydz = \int_{S'} xdydz + \int_{S''} xdydz = (4/3)\pi a^3.$$

Exercises 6.7

1. Using Gauss's Divergence Theorem show that

$$\iint zdydz + xdzdx + ydxdy$$

is zero when evaluated over a closed surface.

2. For any closed surface S prove that

$$\iint_S \left\{\sum x(y-z)\hat{i}\right\} \cdot d\overline{S} = 0.$$

3. Evaluate $\iint_S (lx+my+nz^2)dS$, where S is the closed surface bounded by the right-circular cone $x^2+y^2 = z^2$ and the plane $z = 1, l, m, n$ being the D.C.'s of the outward drawn normal to S at (x, y, z).

4. Apply the divergence theorem of Gauss to evaluate

$$\int_S (lx^2+my^2+nz^2)dS,$$

where S is the surface of the sphere : $(x-a)^2 + (x-b)^2 + (z-c)^2 = r^2$, l, m, n being the D.C.'s of the external normal to the surface at (x, y, z).

5. Show that for every closed surface S with volume V of the space enclosed in it

$$\int_S \overline{F}\cdot\hat{\nu}dS = V\sum a,$$

where $\overline{F} = ax\hat{i} + by\hat{j} + cz\hat{k}$.

6. Show that the integral

$$\iint xdydz + ydzdx + zdxdy$$

over the surface of a sphere equals there times its volume.

7. Prove that the integral

$$\iint x^2dydz + y^2dzdx + 2(xy-x-y)zdxdy$$

over the surface of the cube given by $0 \le x, y, z \le 1$ is 1/2.

8. Verify Gauss's theorem for the integral

$$\iint (2x - z)dydz + x^2ydzdx - z^2xdxdy$$

taken over the surface of the cube given by

$$x = 0,\ x = 1,\ y = 0,\ y = 1,\ z = 0 \text{ and } z = 1.$$

9. Verify the Divergence Theorem for the integral

$$\iint_S \overline{F} \cdot d\overline{S},$$

where $\overline{F} = 4zx\hat{\imath} - y^2\hat{\jmath} + yz\hat{k}$ and S is the surface of the cube given by $x = 0,\ x = 1,\ y = 0,\ y = 1,\ z = 0,\ z = 1$.

10. Use the Divergence Theorem of Gauss to evaluate $\int_S \overline{F} \cdot d\overline{S}$, where $\overline{F} = \sum x^2\hat{\imath}$ and S is $\sum x^2 = R^2$.

11. Show that $\iint x^2dydz + y^2dzdx + z^2dxdy = 0$, when carried over the surface of the ellipsoid $\sum x^2/a^2 = 1$.

12. Evaluate the integral

$$\iint z^2xdydz + (x^2y - z^3)dzdx + (2xy + y^2z)dxdy$$

carried over the entire surface of the hemispherical region T given by $T = \{(x, y, z) : -a \le x \le a, -\sqrt{(a^2 - x^2)} \le y \le \sqrt{(a^2 - x^2)}, 0 \le z \le \sqrt{(a^2 - x^2 - y^2)}\}$.

13. Verify Gauss's theorem for the surface integral

$$\iint (2xy + z)dydz + y^2dzdx - (x + 3y)dxdy$$

taken over the region bounded by $2x + 2y + z = 6,\ x = 0,\ y = 0,\ z = 0$.

14. Verify Gauss's divergence theorem for the function $\overline{F} = y\hat{\imath} + x\hat{\jmath} + z^2\hat{k}$ over the cylindrical region bounded by $x^2 + y^2 \le 9, 0 \le z \le 2$ and evaluate the integral.

Answers

3. $7\pi/6$. **4.** $(8\pi/3)r^3 \sum a$. **10.** 0. **12.** $2\pi a^5/5$. **14.** 36π.

6.13 Stokes Theorem

Let S be a piecewise smooth oriented surface bounded by a piecewise smooth oriented curve C such that the positive normal $\hat{\nu}$ of S is in conjunction with the sense of description of its boundary C by the right handed screw principle. Then if $\overline{F}$ is a vector function whose domain includes S and which is continuous possessing continuous first partial derivatives,

$$\iint_S \nabla \times \overline{F} \cdot \hat{\nu} dS = \oint_C \overline{F} \cdot d\overline{r}.$$

Proof. When written in the scalar form, Stokes formula becomes

$$\iint_S [(h_y - g_z)l + (f_z - h_x)m + (g_x - f_y)n]dS = \oint_C fdx + gdy + hdz, \tag{1}$$

if $\quad \overline{F} = f\hat{i} + g\hat{j} + h\hat{k}$ and $\hat{\nu} = l\hat{i} + m\hat{j} + n\hat{k}$. For

$$\nabla \times \overline{F} = (h_y - g_z)\hat{i} + (f_z - h_x)\hat{j} + (g_x - f_y)\hat{k}$$

and $\overline{F} \cdot d\overline{r} = (f\hat{i} + g\hat{j} + h\hat{k}) \cdot (\hat{i}dx + \hat{j}dy + \hat{k}dz) = fdx + gdy + hdz.$

So, the formula will be true if we can show that

$$\iint_S (f_z m - f_y n)dS = \oint_C fdx, \tag{2}$$

$$\iint_S (g_x n - g_z l)dS = \oint_C gdy \tag{3}$$

and $$\iint_S (h_y l - h_x m)dS = \oint_C hdz. \tag{4}$$

We shall prove the result (2) only. Two others will follow similarly. To prove (2) we assume S to be linear in z, so that its equation can be expressed in the form $z = z(x, y)$. Then the D.C.'s l, m and n of $\hat{\nu}$ at (x, y, z) are given by

$$l = -z_x/a, \quad m = -z_y/a \quad \text{and} \quad n = 1/a, \quad \text{where} \quad a = \sqrt{(z_x^2 + z_y^2 + 1)}.$$

$$\therefore \quad \oint_C fdx = \oint_{C'} f(x, y, z(x, y))dx = \oint_{C'} \phi(x, y)dx = -\iint_R \phi_y dxdy, \tag{5}$$

by Green's theorem in plane, § 6.10, where C' and R are projections of C and S respectively in the xy-plane along with the corresponding orientations.

But $\quad \phi(x, y) = f(x, y, z(x, y))$

$\Rightarrow \quad \phi_y(x, y) = f_y + f_z z_y = f_y - f_z m/n = (nf_y - mf_z)/n$

So from (5)

$$\oint_C fdx = \iint_R (f_z m - f_y n)(1/n)dxdy = \iint_S (f_z m - f_y n)dS.$$

Both the results (3) and (4) can be proved similarly. Hence (1) is true and the theorem is proved.

Note 1. If the surface S is closed, the boundary C disappears. So, both the surface and the line integrals in the theorem vanish:

$$\iint_S \nabla \times \overline{F} \cdot \hat{\nu} dS = 0 \quad \text{and} \quad \oint_C \overline{F} \cdot d\overline{r} = 0.$$

Since $\nabla \times \overline{F}$ and curl $\overline{F}$ are the same $\nabla \times \overline{F} \cdot \hat{\nu}$ is its component in the direction of $\hat{\nu}$, the normal to the surface. So, we find that the integral of the normal component of the curl of a continuous vector function possessing continuous partial derivatives, over a closed surface is always zero. Also, the line integral of the tangential component of such a function along a piecewise smooth closed curve is zero. For

$$\overline{F} \cdot d\overline{r} = \overline{F} \cdot (d\overline{r}/ds)ds = \overline{F}_t ds,$$

$\overline{F}_t$ being the component of $\overline{F}$ in the direction of the tangent to C.

Note 2. Green's theorem in plane is a particular case of Stokes theorem. It can be obtained from Stokes theorem by putting $z = 0$ and consequently, $dz = 0$.

Note 3. Although S has been assumed to be linear in a particular variable in the above proof of the theorem, the result is true for any piecewise smooth oriented surface. If S is not linear in a particular variable, it can be divided into a finite number of parts such that each part is linear in the same variable. Then the integral over S can be expressed us a sum of the integrals over all the parts.

6.13.1 Physical Interpretation of Stokes Theorem

As in the case of the physical interpretation of Gauss's divergence theorem here again we interpret $\overline{F}$ as the velocity field of a steady flow of an incompressible fluid and call

$$\oint_C \overline{F} \cdot d\overline{r} \quad \text{or} \quad \oint_C F_t ds,$$

as the **circulation** of the flow along the oriented curve C. Then Stokes theorem states that the circulation round a closed boundary C of a piecewise smooth oriented surface S is equal to the surface integral

$$\iint_S \nabla \times \overline{F} \cdot \hat{\nu} dS \quad \text{or} \quad \iint_S (\text{curl } \overline{F})_\nu dS$$

of the component $(\text{curl } \overline{F})_\nu$ of the curl in the direction of the positive normal $\hat{\nu}$ to S. Let S be divided into a large number of small surfaces all having the same orientation as that of S. Then, if $\sum$ be one of these small surfaces with its boundary Γ, by the Stokes theorem and the mean value theorem of integral calculus

$$\oint_\Gamma F_t ds = \iint_\Sigma (\text{curl } \overline{F})_\nu dS = A[\text{curl } \overline{F}(P)]_\nu,$$

where A is the area and P is some suitable point of $\sum$. So,

$$[\text{curl } \overline{F}(P)]_\nu = \frac{1}{A} \oint_\Gamma F_t ds.$$

Let Q be any point of $\sum$ and Δ be the greatest of all the distances of all the points of $\sum$ from Q then as $\Delta \to 0, A \to 0$, so that both $\sum$ and Γ shrink to the point Q we have

$$[\text{curl } \overline{F}(Q)]_\nu = \underset{\Delta \to 0}{\text{Lt}} \frac{1}{A} \oint_\Gamma F_t ds.$$

We find, therefore, that physically the normal component of the curl of velocity field at a point is the **specific circulation** or the **circulation density** of the flow in the surface at Q.

Another physical interpretation of the Stokes theorem can be given by taking $\overline{F}$ as mechanical or electrical field of force. Then the line integral of the Stokes theorem represents the work done by the field on a particle, subject to the force, when it is made to describe the boundary C. By the Stokes theorem the amount of this work is transformed into an integral over the surface S bounded by C, the integrand being the normal component of the curl of the field of force.

Since a field is irrotational when its curl vanishes, the irrotational nature of a field is a sufficient condition for the line integral of the tangential component of the vector round any closed curve to vanish.

Examples 6.11. 1. *Verify Stokes theorem for the vector field*

$$\overline{F} = (x^2 - y^2)\hat{\imath} + 2xy\hat{\jmath}$$

taking the surface S of the rectangle as the surface of integration, where

$$S = \{(x, y, z) : 0 \le x \le a \quad and \quad 0 \le y \le b \quad and \quad z = 0\}.$$

Taking $f = x^2 - y^2$, $g = 2xy$ and $h = 0$ and C as the positively oriented boundary of S

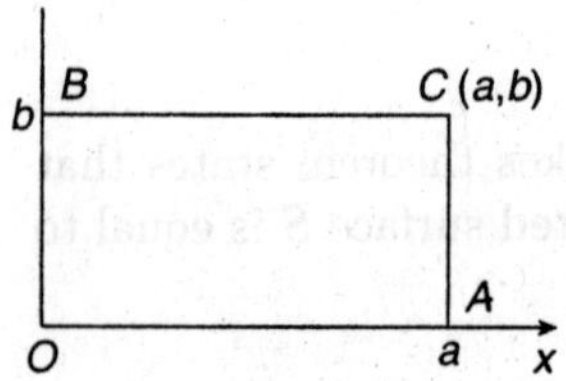

Fig. 6.40

$$\oint_C (f dx + g dy + h dz) = \oint_C (f dx + g dy)$$

$$= \left\{ \int_0^A + \int_A^C + \int_C^B + \int_B^0 \right\} (f dx + g dy)$$

$$= \int_0^a f(x, 0) dx + \int_0^b g(a, y) dy + \int_a^0 f(x, b) dx + \int_b^0 g(0, y) dy$$

$$= \int_0^a x^2 dx + \int_0^b 2ay dy - \int_0^a (x^2 - b^2) dx - \int_0^b 2 \cdot 0 \cdot y dy$$

$$= a^3/3 + ab^2 - (a^3/3 - ab^2) = 2ab^2. \tag{1}$$

Also, $\nabla \times \overline{F} = \begin{vmatrix} \hat{\imath} & \hat{\jmath} & \hat{k} \\ D_x & D_y & D_z \\ x^2 - y^2 & 2xy & 0 \end{vmatrix} = 4y\hat{k}.$

Since $\hat{\nu} = \hat{k}$ in this case, $\nabla \times \overline{F} \cdot \hat{\nu} = 4y$.

$$\therefore \quad \iint_S \nabla \times \overline{F} \cdot \hat{\nu} dS = \iint_S 4y dS = 4 \int_0^a dx \int_0^b y dy = 4ab^2/2 = 2ab^2. \tag{2}$$

Since R.H.S. of (1) = R.H.S. of (2), Stokes theorem is verified.

Moreover, since the function $\overline{F}$, the surface S and the boundary C satisfy all the conditions of Stokes theorem, this will be true in this particular case.

2. *Verify Stokes theorem for the vector field $\overline{F}$ given by $\overline{F} = (x^2 - y^2)\hat{\imath} + 2xy\hat{\jmath}$ over the surface of the rectangular parallelopiped whose bounding surface S consists of six distinct faces given by $x = 0, x = a$; $y = 0, y = b$; $z = 0, z = c$ with the face $z = 0$ removed.*

Considering S to be positively oriented

for the surface $S_1 : OBB'O' : x = 0, \hat{\nu}_1 = -\hat{\imath}$

for the surface $S_2 : OAA'O' \quad : y = 0, \hat{\nu}_2 = -\hat{j}$

.... $S_3 : ADD'A' \quad : x = a, \hat{\nu}_3 = \hat{i}$

.... $S_4 : DD'B'B \quad : y = b, \hat{\nu}_4 = \hat{j}$

.... $S_5 : O'A'D'B' : z = c, \hat{\nu}_5 = \hat{k}$

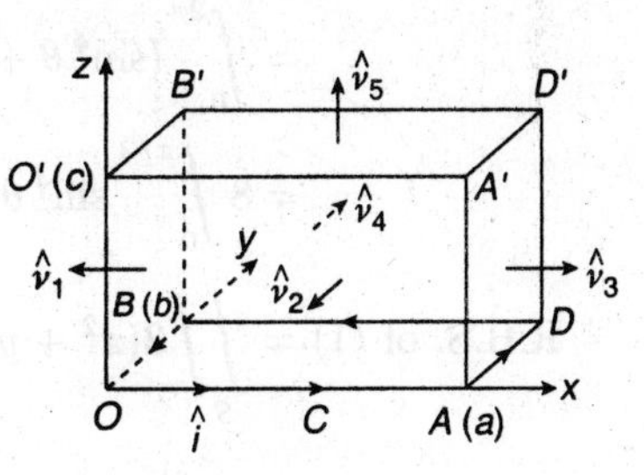

Fig. 6.41

The oriented boundary C of S, in conjunction with the corresponding normals consists of the directed line segments: $C_1 = \overrightarrow{OA}$, $C_2 = \overrightarrow{AD}$, $C_3 = \overrightarrow{DB}$ and $C_4 = \overrightarrow{BO}$.

Now, $\nabla \times \overline{F} = 4y\hat{k}$ and $\displaystyle\int_C (fdx + gdy + hdz) = \int_C (fdx + gdy), \quad \because h = 0$

$$= \sum_{r=1}^{4} \int_{C_r} (fdx + gdy) = \int_0^a f(x,0)dx + \int_0^b g(a,y)dy$$

$$+ \int_a^0 f(x,b)dx + \int_b^0 g(0,y)dy = 2ab^2, \quad (1)$$

already obtained in the solution of Ex. 1.

$$I = \iint_S \nabla \times \overline{F} \cdot \hat{\nu} dS = 4\iint_S y\hat{k} \cdot \hat{\nu} dS = 4\sum_{r=1}^{5} \iint_{S_r} y\hat{k} \cdot \hat{\nu} dS$$

$$= 4\Big[\iint_{S_1} y\hat{k} \cdot (-\hat{\imath})dS + \iint_{S_2} y\hat{k} \cdot \hat{\jmath} dS + \iint_{S_3} y\hat{k} \cdot \hat{\imath} dS + \iint_{S_4} y\hat{k} \cdot \hat{\jmath} dS + \iint_{S_5} y\hat{k} \cdot \hat{k} dS\Big]$$

$$= 4\iint_{S_5} ydS = 4\iint_R y\frac{1}{n}dxdy,$$

where R is the rectangle $OADB$, the orthogonal projection of S_5 in the xy-plane and $n = \hat{k} \cdot \hat{k} = 1$.

So, $\quad I = 4\displaystyle\int_0^a dx \int_0^b ydy = 2ab^2. \quad (2)$

Since (1) = (2), the Stokes theorem is verified.

3. *Verify Stokes theorem for the vector field defined by $\overline{F} = -y^3\hat{i} + x^3\hat{j}$ in the region: $x^2 + y^2 \le 1, z = 0$.*

It is actually a problem on Green's theorem or Gauss's theorem in a plane:

$$\oint_C \overline{F} \cdot dr = \iint_S \nabla \times \overline{F} \cdot \hat{k} dS$$

or, $$\oint_C (-y^3dx + x^3dy) = \iint_S \{D_x x^3 - D_y(-y^3)\}dxdy \quad (1)$$

Here C can be expressed parametrically by $x = \cos\theta,\ y = \sin\theta$, where $0 \le \theta \le 2\pi$.

$\therefore$ L.H.S. of (1) $= \displaystyle\int_0^{2\pi} [-\sin^3\theta(-\sin\theta) + \cos^3\theta\cos\theta]d\theta$

$$= \int_0^{2\pi} (\sin^4\theta + \cos^4\theta) d\theta = 4\int_0^{\pi/2} (\sin^4\theta + \cos^4\theta) d\theta$$

$$= 8\int_0^{\pi/2} \sin^4\theta = 8 \cdot \frac{3.1}{4.2}\frac{\pi}{2} = \frac{3\pi}{2}. \tag{2}$$

$$\text{R.H.S. of (1)} = \iint_S 3(x^2+y^2)dxdy = 3\int_{-1}^{1} dx \int_{-\sqrt{1-x^2}}^{\sqrt{1-x^2}} (x^2+y^2)dy$$

$$= 6\int_{-1}^{1} [x^2y + y^3/3]_0^{\sqrt{1-x^2}} dx = 6\int_{-1}^{1} \left[x^2 + \frac{1}{3}(1-x^2)\right]\sqrt{1-x^2}dx$$

$$= 2\int_{-1}^{1} (2x^2+1)\sqrt{1-x^2}dx = 4\left[2\int_0^1 x^2\sqrt{1-x^2}dx + \int_0^1 \sqrt{1-x^2}dx\right]$$

$$= 4\left[2\int_0^{\pi} \sin^2\theta\cos^2\theta d\theta + \int_0^{\pi/2} \cos^2\theta d\theta\right]$$

$$= 4\left[2 \cdot \frac{1}{4.2} \cdot \frac{\pi}{2} + \frac{1}{2} \cdot \frac{\pi}{2}\right] = \frac{3\pi}{2}. \tag{3}$$

That the Stokes theorem is applicable to the given function is verified from the equality of (2) and (3).

4. *Verify Stokes theorem for the vector field*

$$\overline{F} = (2x-y)\hat{i} - yz^2\hat{j} - y^2z\hat{k}$$

over the hemispherical surface of $\sum x^2 = 1$ *lying on the positive side of the xy-plane.*

The hemispherical surface S is given by

$$S = \{(x,y,z) : \sum x^2 = 1 \quad \text{and} \quad z \geq 0\}.$$

If $\hat{\nu}$ is a unit normal to S at $P(x,y,z)$ then

$$\hat{\nu} = \pm\nabla F/|\nabla F| = \pm(x\hat{i} + y\hat{j} + z\hat{k})/\sqrt{\sum x^2}$$

$$= \pm(x\hat{i} + y\hat{j} + z\hat{k}), \quad \because \sum x^2 = 1 \; \forall \; (x,y,z) \in S.$$

If $\hat{\nu}$ is so chosen that $\hat{\nu} \cdot \hat{k} \geq 0$, then since

$$\hat{\nu} \cdot \hat{k} = \pm z, \; \hat{\nu} = +(x\hat{i} + y\hat{j} + z\hat{k}), \text{ for } z \geq 0.$$

$$\text{Now, } \nabla \times \overline{F} = \begin{vmatrix} \hat{i} & \hat{j} & \hat{k} \\ D_x & D_y & D_z \\ 2x-y & -yz^2 & -y^2z \end{vmatrix} = \hat{k}.$$

$$\therefore \quad \nabla \times \overline{F} \cdot \hat{\nu} = \hat{k} \cdot \hat{\nu} = z.$$

$$\therefore \quad \iint_S \nabla \times \overline{F} \cdot \hat{\nu} dS = \iint_S z dS = \iint_E z\sqrt{1 + z_x^2 + z_y^2} dxdy$$

$$= \iint_E dxdy = \text{ Area of } E = \pi, \tag{1}$$

where E is the orthogonal projection of S on the xy-plane.

So, E is the circle: $\{(x, y, z) : x^2 + y^2 \leq 1, z = 0\}$.

Since E is a plane circular region with radius 1, its area $= \pi$.

Next, $\oint_C \overline{F} \cdot d\overline{r} = \oint_C [(2x - y)dx - yz^2dy - y^2zdz] = \oint_C (2x - y)dx$,

since C is the circular boundary of the hemisphere S, lying in the xy-plane,

$z = 0$ and $C = \{(x, y, z) : x^2 + y^2 = 1$ and $z = 0\}$.

$$\therefore \quad \oint_C \overline{F} \cdot d\overline{r} = \int_0^{2\pi} (2\cos\theta - \sin\theta)(-\sin\theta)d\theta, \quad \text{where } x = \cos\theta, y = \sin\theta$$

$$= \int_0^{2\pi} \sin^2\theta d\theta = \pi. \qquad (2)$$

Since from (1) and (2), $\iint_S \nabla \times \overline{F} \cdot \hat{\nu} dS = \oint_C \overline{F} \cdot d\overline{r}$.

Stokes theorem is true.

5. *If S is the surface of the sphere $\sum x^2 = R^2$, prove that* $\int_S$ *curl* $\overline{F} \cdot d\overline{S} = 0$.

Let S' be a part of S obtained by removing a small segment with circular boundary C (Fig. 6.42).

Then $S' \to S$ as C shrinks to zero.

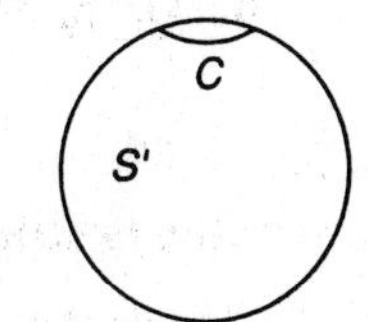

Fig. 6.42

Now, $\int_{S'} \text{curl } \overline{F} \cdot d\overline{S} = \oint_C \overline{F} \cdot d\overline{r}$.

But $\quad \underset{C \to 0}{\text{Lt}} \oint_C \overline{F} \cdot d\overline{r} = 0$. So $\underset{C \to 0}{\text{Lt}} \int_{S'} \text{curl } \overline{F} \cdot d\overline{S} = 0$

i.e., $\quad \int_S \text{curl } \overline{F} \cdot d\overline{S} = 0$.

Exercises 6.8

1. Apply Stokes theorem to evaluate $\oint_C \overline{F} \cdot d\overline{r}$, where F equals

(i) $yz\hat{i} + zx\hat{j} + xy\hat{k}$, C being the curve given by $x^2 + y^2 = 1$ and $y^2 = z$,

(ii) $y\hat{i} + z\hat{j} + x\hat{k}$, C being $\sum x^2 = a^2, x + z = a$, **[W.B. Sem. Exam. '01]**

(iii) $y\hat{i} + xz^3\hat{j} - y^3z\hat{k}$, C being $x^2 + y^2 = 4,\ z = 1.5$

and **(iv)** $(x + y)\hat{i} + (2x - z)\hat{j} + (y + z)\hat{k}$, C being the perimeter of the triangle whose vertices are $A(2, 0, 0)$, $B(0, 3, 0)$ and $C(0, 0, 6)$.

2. Using Stokes theorem transform the line integral

$$\oint_C ydx + zdy + xdz$$

into a surface integral over S with C as its boundary.

3. Evaluate the following line integrals directly and applying Stokes formula:

(i) $\oint_C (y+z)dx + (z+x)dy + (x+y)dz$, C being the circle $\sum x^2 = a^2$ and $\sum x = 0$,

(ii) $\oint_C x^2y^3dx + dy + zdz$, where C is the circle given by $x^2+y^2=R^2$ and $z=0$.

4. Verify Stokes theorem for $\overline{F} = (y-z+2)\hat{i} + (yz+4)\hat{j} - xz\hat{k}$ over the surface of the cube $x=y=z=0$, $x=y=z=2$ above the xy-plane.

5. Use Stokes theorem to find the surface integral

$$\iint_S (y-z)dydz + (z-x)dzdx + (x-y)dxdy,$$

where S is the surface given by $x^2+y^2-2ax+az=0, z \geq 0$.

Answers

1. **(i)** 0; **(ii)** $-\pi a^2/\sqrt{2}$; **(iii)** $19\pi/2$; **(iv)** 21. **2.** $-\iint_S (l+m+n)dS$.

3. **(i)** 0; **(ii)** $-\pi R^6/8$. **5.** πa^3.

7 Three-Dimensional Geometry

7.1 Introduction

In two-dimensional geometry or plane geometry we deal with geometric objects like points and lines lying in a plane; whereas in three-dimensional geometry we discuss, apart from points and lines, surfaces and solid figures.

7.2 Rectangular Cartesian Frame—Coordinate System

In plane coordinate geometry we require two mutually perpendicular straight lines, the x- and y-axes, for a rectangular Cartesian frame, which divide the entire plane into four parts called **quadrants**. In a 3-D space (i.e., three-dimensional space) it is not possible to divide the entire space by means of straight lines. Here we take three mutually perpendicular planes dividing the entire space into eight parts, called **octants**.

For convenience, let there be two vertical planes Π_1 and Π_2 intersecting mutually at right angles in a line, called the z-axis and the horizontal plane Π_3 intersecting Π_2 and Π_1 in lines called the x- and y-axes. As in the case of plane geometry, here also each of these lines, called **coordinate axes**, serves as a number line. The point O, the point of intersection of these three axes, is called the **origin**. This point O represents the real number 0 (zero). Choosing particular sides, with respect to O, of these axes as positive, positive real numbers are represented by the points on these sides of the axes. In Fig. 7.1 the positive sides of the axes are indicated by arrowheads.

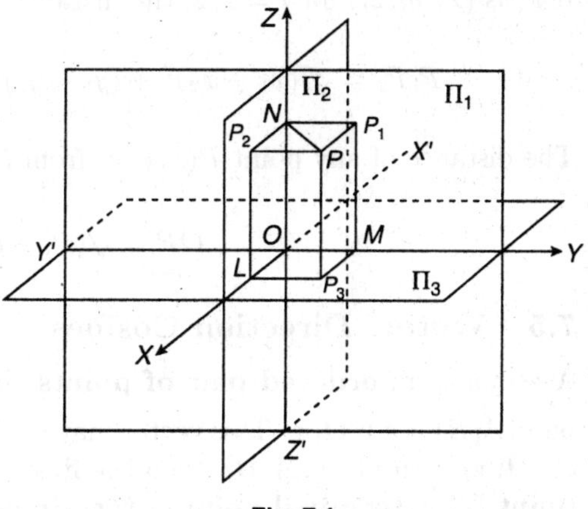

Fig. 7.1

Let us now take any point P in space and draw perpendiculars PP_1, PP_2 and PP_3 to the planes Π_1, Π_2 and Π_3. From P_1 perpendiculars P_1M and P_1N are drawn to the y- and z-axes respectively to meet them in M and N, which will represent two real numbers y and z. Similarly, from P_3, if P_3L is drawn perpendicular to the x-axis, then L also represents some real number x. Then these numbers x, y and z are called the **coordinates** of the point P. They are called the **abscissa, ordinate** and the **applicate** of P. The **ordered triple** (x, y, z) will be fixed for the point P. It can be easily verified that to each point in space there is a unique ordered triple and vice versa. So, as in the case of plane geometry, space geometry can be studied using real numbers (coordinates), equations and inequations.

The planes Π_1, Π_2 and Π_3 are called the **coordinate planes**. Since both the y- and z-axes lie in Π_1, it is called the yz-plane. Similarly, Π_2 and Π_3 are respectively called the zx- and xy-planes. Obviously, the coordinates of O are $(0,0,0)$ and the coordinates of a point on the x-, y- and z-axes are of the forms $(x,0,0)$, $(0,y,0)$ and $(0,0,z)$. Again, the x-coordinate of any point on the yz-plane is always zero, so that the coordinates of any point in this plane are of the form $(0,y,z)$. Similarly, the coordinates of points in the zx- and xy-planes are of the forms $(x,0,z)$ and $(x,y,0)$ respectively.

7.3 Division of a Line Segment

If $P_3(x_3,y_3,z_3)$ divides the line segment joining $P_1(x_1,y_1,z_1)$ and $P_2(x_2,y_2,z_2)$ in a ratio $m:n$, then

$$x_3 = (mx_2+nx_1)/(m+n), \quad y_3 = (my_2+ny_1)/(m+n), \quad z_3 = (mz_2+nz_1)/(m+n).$$

Also $(x_3-x_1)/(x_2-x_3) = (y_3-y_1)/(y_2-y_3) = (z_3-z_1)/(z_2-z_3) = m/n.$

The ratio $m/n >$ or < 0 according as P_3 divides P_1P_2 internally or externally. The conditions of collinearity of the three points are, therefore, given by

$$(x_1-x_3)/(x_2-x_3) = (y_1-y_3)/(y_2-y_3) = (z_1-z_3)/(z_2-z_3)$$

or, $\quad x_1-x_3 : y_1-y_3 : z_1-z_3 = x_2-x_3 : y_2-y_3 : z_2-z_3.$

7.4 Distance between Two Points

If P_i is (x_i,y_i,z_i) for $i=1,2$, the distance between P_1 and P_2:

$$P_1P_2 = \sqrt{[(x_1-x_2)^2+(y_1-y_2)^2+(z_1-z_2)^2]} \text{ or briefly } \sqrt{\sum(x_1-x_2)^2}.$$

The distance of any point $P(x,y,z)$ from O is, therefore,

$$OP = \sqrt{(x^2+y^2+z^2)} \quad \text{or,} \quad \sqrt{\sum x^2}.$$

7.5 Vector. Direction Cosines

A vector is an **ordered pair of points**. If A and B are any two points in space, the ordered pair (A,B) is a vector. This vector may be denoted by other symbols like $\overline{AB}, \overrightarrow{AB}, \overline{a}, \overline{b}, \overline{c}, \overline{u}, \overline{v}, \overline{w}$, etc. If we write $\overline{v} = (A,B)$, then the **first point** A is called the **initial point** and the **second point** B the **terminal point** or **terminus** of the vector $\overline{v}$. The magnitude of $\overline{v}$ is the distance between A and B, i.e., AB. The **magnitude** or **length** of $\overline{v}$ is denoted by $|\overline{v}|$ or simply by v. Thus $v = |\overline{v}| = AB$. The **direction** of $\overline{v}$ is given by the ordering of the points A and B. We say that the direction of $\overline{v}$ is from A, the initial point to B, the terminus.

The null vector $\overline{0}$ is a pair (A,A) of two coincident points. The sum of two vectors $\overline{u}$ and $\overline{v}$ is $\overline{w}$, i.e.,

$$\overline{u}+\overline{v} = \overline{w} = \overline{AC},$$

if $\overline{u} = \overline{AB}$, $\overline{v} = \overline{BC}$.

If $\overline{u} = \overline{AB}$, its additive inverse $-\overline{u}$ of $\overline{u}$ is $\overline{BA}$.

Note. An ordinary pair $\{A,B\}$ of two points A and B is called a **segment**, whereas an ordered pair (A,B) is called a **directed line segment** or a **vector**.

7.5.1 Scalar Multiplication

$\forall$ vector $\bar{a}$ and $\forall$ scalar $m \in \mathbb{R}$, we define the scalar multiple $m\bar{a}$ or $\bar{a}m$ of $\bar{a}$ by m to be a vector $\bar{b}$:

$$\bar{b} = m\bar{a} \text{ or } \bar{a}m$$

such that $\bar{b} = \bar{0}$ if either $m = 0$ or $\bar{a} = \bar{0}$, otherwise $\bar{b}$ is a proper vector with

$$|\bar{b}| = |m||\bar{a}|$$

and dir $\bar{b} = \pm$ dir $\bar{a}$, according as $m >$ or < 0.

(By dir $\bar{b} = \pm$ dir $\bar{a}$ we mean the direction of $\bar{b}$ is the same as, or opposite to, that of $\bar{a}$.)

7.5.2 Linear Combination and Linear Dependence of Vectors

For any finite number of vectors $\bar{u}_i$ for $i = 1, \ldots, n$, the expression $\sum_{i=1}^{n} x_i\bar{u}_i$, where $x_i \in \mathbb{R}$, is called a **linear combination** the vectors $\bar{u}_i$. Obviously, it is a vector $\bar{u}$ so that $\bar{u} = \sum x_i\bar{u}_i$. The vector $\bar{u}$ may be $\bar{0}$ when the scalars x_i are not all zero or, only when all the scalars are zero. In the former case the vectors $\bar{u}_i$ are said to be **linearly dependent** and in the latter case **linearly independent. If $\bar{u}_i$ are linearly dependent, then at least one of them can be expressed as a linear combination of the others. If they are linearly independent, none of them can be expressed as a linear combination of the remaining.**

In a 3-D space there are three linearly independent unit vectors $\hat{i}$, $\hat{j}$ and $\hat{k}$ whose directions are the positive directions of the x-, y- and z-axes of a rectangular Cartesian frame. These vectors are called the **base vectors.** Any vector in the 3-D space can be expressed as a linear combination of these base vectors. If $R(x, y, z)$ is any point, then the vector $\bar{r} = \overline{OR}$ is called the **position vector (p.v.)** of the point R relative to O, the origin. In the xyz-space by the position vector of a point we shall always mean its position referred to O. One can easily see that

$$\bar{r} = \overline{OR} = x\hat{i} + y\hat{j} + z\hat{k}, \text{ which is often expressed as } (x, y, z).$$

Thus $\bar{r} = x\hat{i} + y\hat{j} + z\hat{k} = (x, y, z)$, called the **current vector** and $\bar{r} = \bar{0}$ only when x, y, z are all zero.

If $R_i(x_i, y_i, z_i)$ then

$$\begin{aligned}\overline{R_1R_2} &= \overline{OR_2} - \overline{OR_1} = (x_2, y_2, z_2) - (x_1, y_1, z_1)\\ &= (x_2 - x_1, y_2 - y_1, z_2 - z_1)\end{aligned}$$

or $\quad (x_2 - x_1)\hat{i} + (y_2 - y_1)\hat{j} + (z_2 - z_1)\hat{k}$.

If $\bar{r}_i$, for $i = 1, 2, 3$ are the p.v.'s of $R_i(x_i, y_i, z_i)$ and R_3 divides R_1R_2 in the ratio $m : n$ then

$$\bar{r}_3 = (m\bar{r}_2 + n\bar{r}_1)/(m + n) = p\bar{r}_1 + q\bar{r}_2, \text{ say,}$$

where $p = n/(m + n)$ and $q = m/(m + n)$. Then $p + q = 1$.

So, the **condition of collinearly of three points R_1, R_2, R_3 in space is that if $\bar{r}_3 = p\bar{r}_1 + q\bar{r}_2$, then $p + q = 1$, the coefficient of $\bar{r}_3$.** The collinearity condition could also be expressed as follows: Three points R_1, R_2 and R_3 are collinear iff scalars x_1, x_2, x_3, **not all zero**, exist such that

$$\sum_{i=1}^{3} x_i\bar{r}_i = \bar{0}, \text{ where } \sum_{i=1}^{3} x_i = 0.$$

A necessary and sufficient condition for the coplanarity of four points R_i for $i = 1, 2, 3, 4$ is that

$$\sum_{i=1}^{4} x_i \overline{r}_i = \overline{0}, \text{ where } \sum_{i=1}^{4} x_i = 0 \text{ and } x_i \text{ are not all zero.}$$

7.5.3 Direction Cosines of a Vector and of a Line

Referred to a rectangular Cartesian frame $OXYZ$, with the associated base vectors $\hat{i}$, $\hat{j}$, $\hat{k}$, if $\overline{a}(= a_1\hat{i} + a_2\hat{j} + a_3\hat{k})$ is a non-null vector, then it makes definite angles α^c, β^c and γ^c with the positive directions of the x-, y- and z-axes or with $\hat{i}$, $\hat{j}$ and $\hat{k}$ respectively. Then these angles α^c, β^c and γ^c all lying between 0^c and π^c are called the **direction angles (D.A.'s)** of the vector $\overline{a}$. The cosines of those angles, i.e., $\cos\alpha$, $\cos\beta$ and $\cos\gamma$ are called the **direction cosines** of $\overline{a}$. They are usually denoted by the symbols l, m, n. If $\overline{r} = \overrightarrow{OR}$ and it has the same direction as that of $\overline{a}$, then its D.A.'s and D.C.'s will also be the same as that of $\overline{a}$. Any three real numbers proportional to the D.C.'s of a vector are called the **direction ratios** (D.R.'s) of the vector. It will be seen below that if $\overline{a} = a_1\hat{i} + a_2\hat{j} + a_3\hat{k} = (a_1, a_2, a_3)$, then a_1, a_2, a_3 are D.R.'s of the vector $\overline{a}$. Similarly, if $\overline{r} = x\hat{i} + y\hat{j} + z\hat{k} = (x, y, z)$, the three numbers x, y, z are D.R.'s of $\overline{r}$.

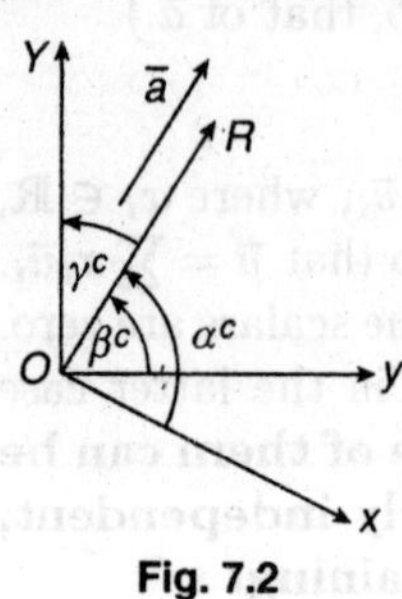

Fig. 7.2

Two different directions may be assigned to any line Λ. Any point P on Λ divides Λ into two rays or half-lines. If one of them is considered positive, then other is taken as negative. If Λ is also parallel to $\overline{a}$, then $\overline{a}$ is called a **direction vector** of Λ and direction ratios of Λ are the scalar components a_1, a_2, a_3 of $\overline{a}$ or any three numbers proportional to these numbers.

If $\alpha^c, \beta^c, \gamma^c$ are the angles made by $\overline{r}$ with the x-, y- and z-axes, we find that

$$\overline{r} \cdot \hat{i} = r\cos\alpha = rl, \quad \text{where } r = |\overline{r}| = \sqrt{\sum x^2},$$

$$\overline{r} \cdot \hat{j} = r\cos\beta = rm \quad \text{and} \quad \overline{r} \cdot \hat{k} = r\cos\gamma = rn.$$

But since $\overline{r} = x\hat{i} + y\hat{j} + z\hat{k}$, $\overline{r} \cdot \hat{i} = x$, $\overline{r} \cdot \hat{j} = y$ and $\overline{r} \cdot \hat{k} = z$.

So, $x = rl$, $y = rm$ and $z = rn$. Hence $x/l = y/m = z/n$.

Since x, y and z are proportional to the D.C.'s of $\overline{r}$, they are, therefore, the D.R.'s of $\overline{r}$. Thus the scalar components of a vector are the D.R.'s of that vector. Further

$$x^2 + y^2 + z^2 = r^2l^2 + r^2m^2 + r^2n^2$$

or, $$\sum x^2 = r^2 \sum l^2$$

But $r^2 = \sum x^2$. Hence we get $\sum \boldsymbol{l^2} = \mathbf{1}$.

Thus the sum of the squares of the D.C.'s of every vector is always 1.

Unit Vector: A vector of magnitude 1 is called a **unit vector**. Since for every non-null vector $\overline{u}$, $\left|\frac{1}{|\overline{u}|}\overline{u}\right| = \frac{|\overline{u}|}{|\overline{u}|} = 1$, $\frac{\overline{u}}{|\overline{u}|}$ is a unit vector. It will be denoted by $\hat{u}$. Thus

$$\hat{u} = \overline{u}/|\overline{u}|.$$

If $\hat{u} = l\hat{i} + m\hat{j} + n\hat{k}$, then the scalar components u_1, u_2, u_3 are the D.C.'s of the vector $\overline{u}$. For,

$$|\hat{u}| = 1 \Rightarrow \sqrt{\sum l^2} = 1 \text{ or, } \sum l^2 = 1.$$

Hence l, m, n are the D.C.'s of $\overline{u}$.

If $\quad \overline{u} = u_1\hat{i} + u_2\hat{j} + u_3\hat{k}$

$$\hat{u} = \frac{\overline{u}}{|\overline{u}|} = \frac{1}{\sqrt{\sum u_i^2}}(u_1\hat{i} + u_2\hat{j} + u_3\hat{k}).$$

$\therefore$
$$l = \frac{u_1}{\sqrt{\sum u_i^2}},\ m = \frac{u_2}{\sqrt{\sum u_i^2}},\ n = \frac{u_3}{\sqrt{\sum u_i^2}}$$

or,
$$\frac{u_1}{l} = \frac{u_2}{m} = \frac{u_3}{n}.$$

So, the scalar components u_1, u_2 and u_3 are the D.R.'s of $\overline{u}$.

Solved Problems

1. *Show that the centroid of the triangle with vertices $R_i(x_i, y_i, z_i)$ for $i = 1, 2, 3$, is $G(\overline{x}, \overline{y}, \overline{z})$, where*

$$\overline{x} = \sum_1^3 x_i/3,\quad \overline{y} = \sum_1^3 y_i/3,\quad \overline{z} = \sum_1^3 z_i/3.$$

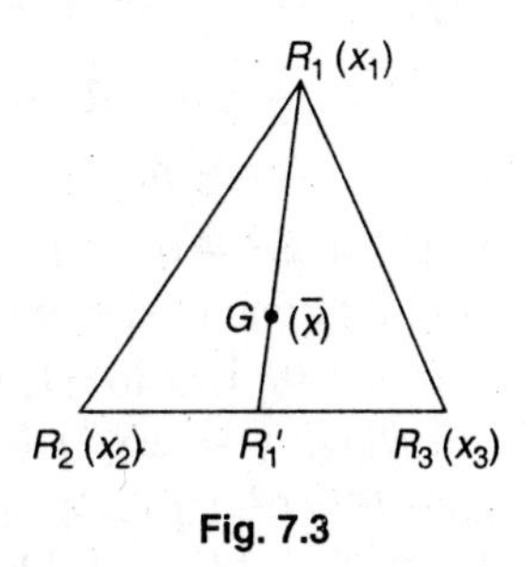

Fig. 7.3

Let $\quad R_1'(x_1', y_1', z_1')$ be the midpoint of R_2R_3.

Then $\ x_1' = (x_2 + x_3)/2, \ \ldots.$

Since G divides R_1R_1' in the ratio $2:1$,

so
$$\overline{x} = \frac{2x_i' + x_1}{2+1} = \frac{\sum x_i}{3}.$$

Similarly, for $\overline{y}$ and $\overline{z}$.

2. *Show that the centroid of the tetrahedron $R_1R_2R_3R_4$ is $G(\overline{x}, \overline{y}, \overline{z})$, where*

$$\overline{x} = \frac{1}{4}\sum_1^4 x_i,\quad \overline{y} = \frac{1}{4}\sum_1^4 y_i,\quad \overline{z} = \frac{1}{4}\sum_1^4 z_i$$

and the coordinates of R_i are (x_i, y_i, z_i) for $i = 1, 2, 3, 4$.

Fig. 7.4

Let $\quad R_1'$ be the centroid of the $\Delta R_2R_3R_4$.

Then $\ x_1' = (x_2 + x_3 + x_4)/3,\ y_1' = \ldots,\ z_1' = \ldots.$

Since the centroid G divides R_1R_1' in the ratio $3:1$,

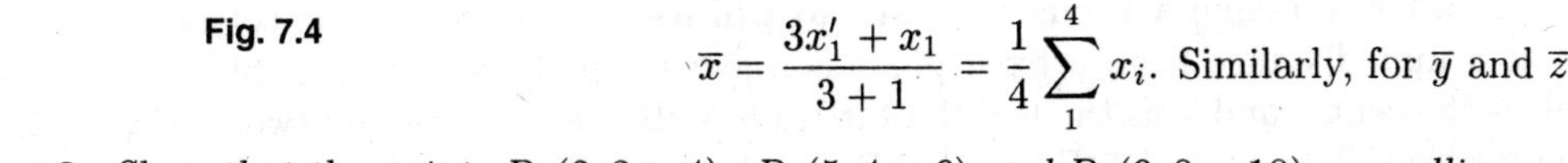

$$\overline{x} = \frac{3x_1' + x_1}{3+1} = \frac{1}{4}\sum_1^4 x_i.$$ Similarly, for $\overline{y}$ and $\overline{z}$.

3. *Show that the points $R_1(3, 2, -4)$, $R_2(5, 4, -6)$ and $R_3(9, 8, -10)$ are collinear, using the division formula of a segment.*

If $\overline{r}_i$ is the p.v. of R_i, then the three points R_1, R_2 and R_3 are collinear if the two vectors $\overline{R_1R_2}$ and $\overline{R_1R_3}$ are collinear. If R_1 divides R_2R_3 in the ratio $r:1$, then

$$\overline{r}_1 = (r\overline{r}_3 + 1\overline{r}_2)/(r+1).$$

So, there must exist a number r such that

$$(r+1)\overline{r}_1 = \overline{r}_2 + r\overline{r}_3,$$

or, $\quad r(\overline{r}_1 - \overline{r}_3) = \overline{r}_2 - \overline{r}_1,$

or, $r[(3,2-4)-(9,8,-10)]=(5,4,-6)-(3,2,-4)$,

or, $r(-6,-6,6)=(2,2,-2)$,

or, $-6r=2,\ -6r=2,\ 6r=-2$.

So, there is a unique value of r, viz., $-1/3$. Hence $R_1,\ R_2,\ R_3$ are collinear.

4. *The D.C.'s $l,\ m,\ n$ of two straight lines satisfy the equations*

$$(1)\ l+m+n=0 \quad \text{and} \quad (2)\ l^2+m^2-n^2=0.$$

Show that the acute angle between the lines is $\pi^c/3$.

Taking $\lambda=l/n$ and $\mu=m/n$, (1) and (2) reduce to

$$(3)\ \lambda+\mu+1=0 \quad \text{and} \quad (4)\ \lambda^2+\mu^2-1=0.$$

Eliminating μ between (3) and (4), we get $\lambda^2+\lambda=0$, so that either $\lambda=0$ or $\lambda=-1$.

When $\lambda=0,\ l=0, n\neq 0$. Then from (1) and (2)

$$m+n=0 \quad \text{and} \quad m^2-n^2=0.$$

$$\therefore \quad n=-m.$$

But $l^2+m^2+n^2=1$. So $0^2+m^2+n^2=1$ or, $m=\pm 1/\sqrt{2}$ and hence $n=\mp 1/\sqrt{2}$. So the D.C.'s of one of the lines are either

$$0,\ 1/\sqrt{2},\ -1/\sqrt{2} \quad \text{or} \quad 0, -1/\sqrt{2},\ 1/\sqrt{2}.$$

When $\lambda=-1,\ l/n=-1$ or, $n=-l$. So from (1) $m=0$. Again, since $\sum l^2=1$ we have $l^2+0^2+l^2=1$ or, $l=\pm 1/\sqrt{2}$. So the D.C.'s of the other line are $-1/\sqrt{2},\ 0,\ 1/\sqrt{2}$ or $1/\sqrt{2},\ 0,\ -1/\sqrt{2}$. Taking $0,\ 1/\sqrt{2},\ -1/\sqrt{2}$ and $1/\sqrt{2},\ 0,\ -1/\sqrt{2}$ as the D.C.'s of the lines and θ^c as the angle between them

$$\cos\theta=\sum l_1l_2=0\cdot 1/\sqrt{2}+(1/\sqrt{2})\cdot 0+(-1/\sqrt{2})(-1/\sqrt{2})=1/2$$

$$\therefore \quad \theta=\pi/3.$$

Changing the signs of one set of D.C.'s, $\sum l_1l_2=-\frac{1}{2}$. So, $\theta=\pi-\frac{\pi}{3}$. Hence the acute angle between the lines is $\pi^c/3$.

7.6 Locus

Although a locus is normally defined as the path of a moving point, by a path we mean a curve (including a straight line). But in a 3-D space we have surfaces and planes which appear as loci. So, it is better to define **a locus as a set of points**. For example, the set of all points which are at a unit distance from a fixed point is a locus. Obviously, it is a sphere with the fixed point as the centre and 1 as the length of its radius. If Λ and Γ are any two loci, they are sets of points also. Then since $\Lambda\cap\Gamma$ and $\Lambda\cup\Gamma$ are also sets of points, they are loci also. The complements Λ' and Γ' are also loci.

7.7 Plane

Referred to a rectangular Cartesian frame, a plane is the set of all those points (x,y,z) whose coordinates x,y,z satisfy a linear equation of the form $ax+by+cz+d=0$, where a,b,c and d are real numbers and a,b and c are not all zero. Set theoretically, a plane

$$\Pi=\{(x,y,z): F(x,y,z)=0,\ \text{where } F=ax+by+cz+d \text{ or } \sum ax+d,\ a,b,c \text{ being not all zero}\} \tag{1}$$

Thus every linear equation represents a plane. For example, $x = 0$ represents a plane. For, here $F = x = 1x + 0y + 0z + 0$. This plane is the yz-coordinate plane. Similarly, the equations of zx- and xy-coordinate planes are $y = 0$ and $z = 0$. If the plane Π is given by (1), then a point $R'(x', y', z')$ lies in the plane if $F(P)$, i.e., $F(x', y', z') = 0$. Symbolically, we say $R' \in \Pi$ if $F(R') = 0$. If $F(R') \neq 0$, $R' \notin \Pi$. Here F or $F(x, y, z)$ is a linear function of the three real variables x, y and z. The equation $\boldsymbol{F = 0}$ is called **the equation to Π** and the set Π is called the **locus of the equation**. There are different forms of the equation to a plane.

7.7.1 General Form

This form is given in the definition (1).

7.7.2 Intercept Form

If $F = x/a + y/b + z/c - 1$, then $F = 0$ is called the **intercept form**. The reason is that if Π meets the x-, y- and z-axes in A, B and C, then the intercepts made by Π on these axes are a, b and c respectively and the coordinates of A, B and C are $(a, 0, 0)$, $(0, b, 0)$ and $(0, 0, c)$ (Fig. 7.5).

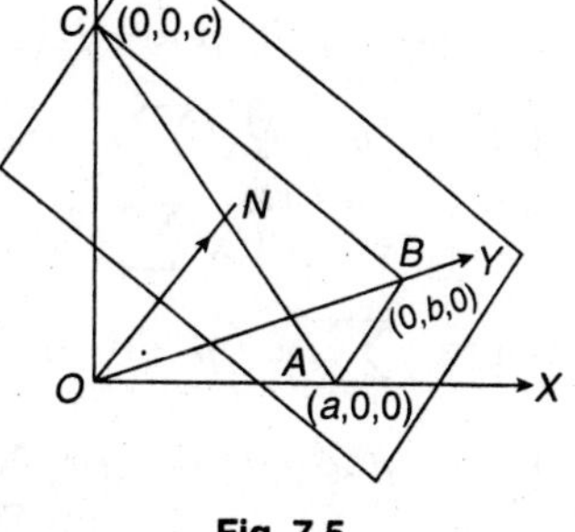

Fig. 7.5

7.7.3 Normal Form

If $F = lx + my + nz - p$, where l, m and n are the D.C.'s of the directed normal $\overrightarrow{ON}$ to Π and

$$p = |\overrightarrow{ON}| = ON > 0.$$

Then $F = 0$ is said to be the **normal form** of the equation to Π.

The corresponding vector form of the equation to Π is $\overline{r} \cdot \hat{\nu} = p$, where $\overline{r} = (x, y, z)$ and $\hat{\nu} = (l, m, n)$, the unit vector $\perp \Pi$.

7.7.4 Three-Point Form

If Π passes through three non-collinear points $R_i(x_i, y_i, z_i)$ then the function F is $\begin{vmatrix} x & y & z & 1 \\ x_1 & y_1 & z_1 & 1 \\ x_2 & y_2 & z_2 & 1 \\ x_3 & y_3 & z_3 & 1 \end{vmatrix}$, and $F = 0$ gives the equation to the plane. On expansion,

$F = \sum ax + d$, where a, b, c, d are the cofactors of x, y, z and 1 respectively. Using the properties of the determinant the equations to Π can also be written as

$$\begin{vmatrix} x - x_1 & y - y_1 & z - z_1 \\ x_2 - x_1 & y_2 - y_1 & z_2 - z_1 \\ x_3 - x_1 & y_3 - y_1 & z_3 - z_1 \end{vmatrix} = 0,$$

which, in vector form, is $(\overline{r}_1 - \overline{r}) \cdot (\overline{r}_1 - \overline{r}_2) \times (\overline{r}_1 - \overline{r}_3) = 0$ or

$$[\overline{r} - \overline{r}_1 \; \overline{r}_2 - \overline{r}_1 \; \overline{r}_3 - \overline{r}_1] = 0.$$

The plane passing through A, B, C (Fig. 7.5) is

$$\begin{vmatrix} x - a & y & z \\ 0 - a & b & 0 \\ 0 - a & 0 & c \end{vmatrix} = 0 \quad \text{or} \quad \sum x/a = 1.$$

7.7.5 Conversion of the General Form to the Normal Form

If $F = \sum ax + d = 0$ and $\sum lx - p = 0$, where $\sum l^2 = 1$ and $p > 0$ represent the same plane, then

$$a/l = b/m = c/n = -d/p. \tag{1}$$

$\therefore$ $l = -pa/d$, $m = -pb/d$ and $n = -pc/d$.

So, $\sum l^2 = 1 \Rightarrow \sum(p^2a^2/d^2) = 1$ or, $(p^2/d^2)\sum a^2 = 1$

$\therefore$ $p = |d|/\sqrt{\sum a^2}$.

Note that $d = a0 + b0 + c0 + d$ or $F(O)$ and $p = ON$. So, the distance of the plane from the origin is

$$|F(O)|/\sqrt{\sum a^2}.$$

Since $p > 0$, $p = \pm d/\sqrt{\sum a^2}$ according as $F(O)$ or d is +ve or −ve.

If $d > 0$, $p = d/\sqrt{\sum a^2}$. So from (1)

$$l = -a/\sqrt{\sum a^2},\ m = -b/\sqrt{\sum a^2} \text{ and } n = -c/\sqrt{\sum a^2}.$$

Hence $\sum lx - p = 0 \Rightarrow \sum -\frac{a}{\sqrt{\sum a^2}}x - \frac{d}{\sqrt{\sum a^2}} = 0$ or, $-F/\sqrt{\sum a^2} = 0$.

If $d < 0$, $p = -d/\sqrt{\sum a^2}$ and $\sum lx - p = 0 \Rightarrow F/\sqrt{\sum a^2} = 0$.

Note that $a/l = b/m = c/n \Rightarrow a, b, c$ are the D.R.'s of the normal to Π.

So, taking $\overline{\nu} = a\hat{i} + b\hat{j} + c\hat{k}$ or (a, b, c) and the **current vector** $\overline{r} = (x, y, z)$, the vector equation of the plane Π is $\overline{r} \cdot \overline{\nu} + d = 0$.

Example. *The normal form of the equation* $3x - 4y + 12z + 13 = 0$ *is* $-F/\sqrt{\sum a^2} = 0$, *i.e.,* $-(3x - 4y + 12z + 13)/\sqrt{3^2 + 4^2 + 12^2} = 0$

or $x(-3/13) + y(4/13) + z(-12/13) - 1 = 0$ *since* $c = 13 > 0$,

so that $(l, m, n) = (-3/13, 4/13, -12/13)$ *and* $p = 1 > 0$.

The normal form of $-3x + 4y - 7z - 16 = 0$ is $F/\sqrt{\sum a^2} = 0$, i.e.,

$$(-3x + 4y - 7z - 16)/\sqrt{3^2 + 4^2 + 7^2} = 0$$

or, $x \cdot (-3/8) + y(1/2) + z(-7/8) - 2 = 0$ since $c = -16 < 0$.

So, $(l, m, n) = (-3/8, 1/2, -7/8)$ and $p = 2 > 0$.

7.7.6 Position of a Point Relative to a Plane

Let P be any point (α, β, γ) and Π be any given plane whose equation is $F = 0$, where $F = \sum ax + d$. $P \in \Pi$ if $F(P) = 0$. $P \notin \Pi$ if $F(P) \neq 0$. Here $F(P) = \sum a\alpha + d$. Since $a, b, c, d, \alpha, \beta, \gamma$ are all real $F(P)$ is real. So if $F(P) \neq 0$, either $F(P) > 0$ or $F(P) < 0$. Accordingly, we say P lies on the positive or negative side of Π. So Π divides the entire 3-D space into two parts called the two **sides** of Π.

For all points on the **same side** of Π, F takes the **same sign** and for two points lying on **the opposite sides** of Π, F takes **opposite signs.**

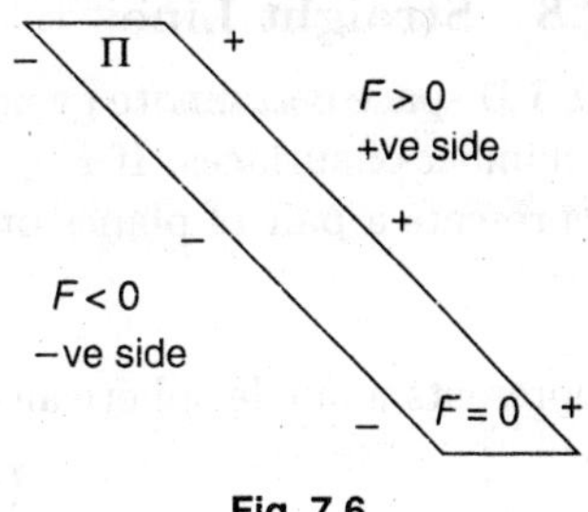

Fig. 7.6

For example, if $F = x$, $F = 0$ represents the yz-coordinate plane. If P and Q are $(1,2,3)$ and $(2,3,4)$, $F(P) = 1$ and $F(Q) = 2$. Since F takes the same sign (+ve) P and Q lie on the same positive side of Π.

If R is $(-1,0,0)$, $F(R) = -1 < 0$. So, P and R (or Q and R) lie on opposite sides of Π.

Let $R_1(x_1, y_1, z_1)$ and $R_2(x_2, y_2, z_2)$ be any two points such that $F(R_1) \neq 0$, $F(R_2) \neq 0$ if R_1R_2, or R_1R_2 produced, intersects Π in $R(\overline{x}, \overline{y}, \overline{z})$ then the coordinates of R and the ratio in which Π divides PQ can be found out.

If $\quad R_1R : R_2R = r : 1, \quad$ then $\overline{x} = (x_1 + rx_2)/(1+r)$, $\overline{y} = \ldots, \overline{z} = \ldots$.

Since $\quad R \in \Pi,\ F(R) = 0$, i.e., $\sum a\overline{x} + d = 0 \quad$ or, $\sum a\dfrac{x_1 + rx_2}{1+r} + d = 0$

or, $\quad \sum a(x_1 + rx_2) + (1+r)d = 0 \quad$ or, $\left(\sum ax_1 + d\right) + r\left(\sum ax_2 + d\right) = 0$

or, $\quad F(R_1) + rF(R_2) = 0$,

whence $r = -F(R_1)/F(R_2)$. After getting the value of the ratio $r, \overline{x}, \overline{y}, \overline{z}$ and hence R can be found out.

We know that R is an interior or an exterior point the segment according as $r >$ or < 0, i.e., according as $F(P)$ and $F(Q)$ are on opposite sides or on the same side of Π.

The Distance of a Given Point P from a Plane Π given by $F = 0$ is $|F(P)| \div \sqrt{\sum a^2}$.

The Image of a Given Point in a Plane

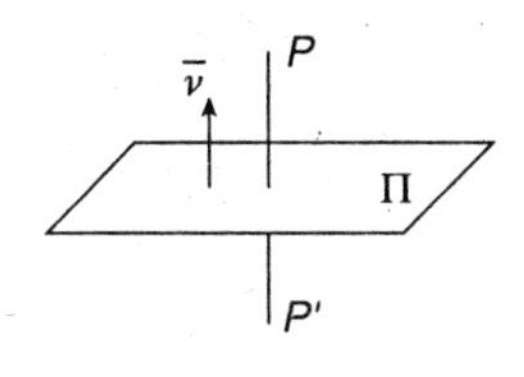

Fig. 7.7

Let $P'(\alpha', \beta', \gamma)$ be the image of $P(\alpha, \beta, \gamma)$ in the plane $\Pi : F = 0$, where $F = \sum ax + d$.

Then $\overline{\nu} = (a, b, c)$ is a normal to the plane. Since Π bisects P, P'

$$1 = -F(P)/F(P') \quad \text{or} \quad F(P') + F(P) = 0 \tag{1}$$

$$\because \quad \overline{\nu} || PP',\ (\alpha' - \alpha)/a = (\beta' - \beta)/b = (\gamma' - \gamma)/c. \tag{2}$$

By the property of ratio and proportion, each of these ratios

$$= [a(\alpha' - \alpha) + b(\beta' - \beta) + c(\gamma' - \gamma)] \div (a^2 + b^2 + c^2)$$

$$= \left[\left(\sum a\alpha' + d\right) - \left(\sum a\alpha + d\right)\right] \div \sum a^2$$

or, $\quad [F(P') - F(P)]/\sum a^2 = -2F(P)/\sum a^2$, using (1). $\qquad$ (3)

So, from (2) $\quad \alpha' = \alpha - 2aF(P)/\sum a^2$,

$\beta' = \beta - 2bF(P)/\sum a^2$

and $\quad \gamma' = \gamma - 2cF(P)/\sum a^2$.

Vectorially, $\quad \overline{p}' = \overline{p} - 2\overline{\nu}F(P)/|\overline{\nu}|^2$.

7.8 Straight Line

In 3-D space coordinate geometry a single equation in x, y and z always represents a surface or a number of surfaces. If $F_i = \sum a_i x + d_i$ for $i = 1, 2$, then the equation $F = 0$, where $F = F_1 F_2$ represents a pair of planes or plane surfaces. The equation

$$x^2 + y^2 + z^2 - 1 = 0$$

represents a single spherical surface. But the equation

$$(x^2 + y^2 + z - 1)(x - 2y + 3z - 2) = 0$$

represents two surfaces of which one is a sphere and the other is a plane.

A curve (including a straight line) is regarded as the intersection of two surfaces. If $F_1 = 0$ and $F_2 = 0$ represent the surfaces $\sum_1$ and $\sum_2$, then they intersect in a curve Γ. Set theoretically, $\Gamma = \sum_1 \cap \sum_2$.

$$\therefore \quad \Gamma = \{R(x, y, z) : R \in \textstyle\sum_1 \text{ and } \sum_2 \text{ both}\}$$
$$= \{R : F_1(R) = 0 \text{ and } F_2(R) = 0\}.$$

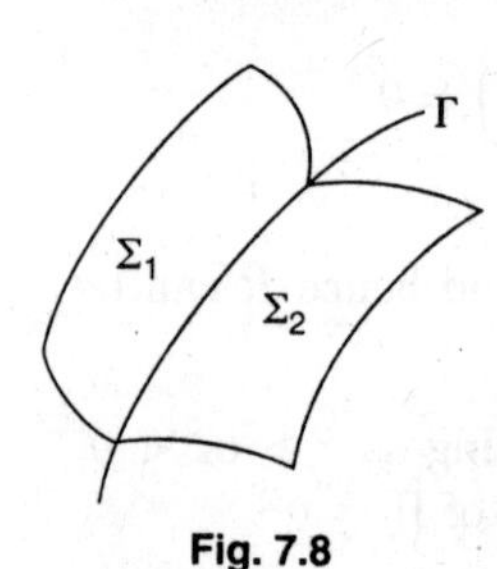

Fig. 7.8

So, Γ, the curve or line of intersection of two surfaces is represented by a pair of simultaneous equations $F_1 = 0$ and $F_2 = 0$.

If $F_i = \sum a_i x + d_i$ for $i = 1$ and 2, then $F_1 = 0$ and $F_2 = 0$ represent two planes Π_1 and Π_2 and if they intersect, i.e., if $\Pi_1 \cap \Pi_2 \neq \phi$, then their line of intersection Λ is a straight line. Its equations are $F_1 = 0$ and $F_2 = 0$.

So, $\sum a_1 x + d_1 = 0$ and $\sum a_2 x + d_2 = 0$ simultaneously represent the line Λ. These simultaneous equations are also written as $\boldsymbol{\sum a_1 x + d_1 = 0 = \sum a_2 x + d_2}$. The pairs of equations $y = 0 = z$, $z = 0 = x$, $x = 0 = y$ represent the x-, y- and z-axes respectively.

The equations of a straight line in the form mentioned are called the **asymmetric form** of the equations of a straight line.

The corresponding vector equations of Λ are

$$\boldsymbol{\overline{r} \cdot \overline{\nu}_i + d_i = 0},$$

where $\overline{\nu}_i = (a_i, b_i, c_i)$, $i = 1, 2$.

Example 7.8. 1. *Find a point on the straight line Λ given by*

$$x - 3y + 2z - 1 = 0 = 2x - 3y + z + 1.$$

There are two equations in three variables. We may assign any arbitrary value to anyone of the three variables and then we may easily solve the two equations thus obtained in two variables. For example, putting $z = 0$ we get

$$x - 3y - 1 = 0 \text{ and } 2x - 3y + 1 = 0,$$

which are solvable for x and y. $x = -2$, $y = -1$. Hence $(-2, -1, 0)$ is a point on Λ.

In fact, we may treat the two given equations in two variables x and y and solve them in terms of z:

$$x - 3y + (2z - 1) = 0$$
$$2x - 3y + (z + 1) = 0.$$

$$\therefore \qquad x = \frac{3(2z-1) - 3(z+1)}{-3+6} = \frac{3z-6}{3} = z - 2$$

and $$y = \frac{2(2z-1) - (z+1)}{3} = \frac{3z-3}{3} = z - 1.$$

Hence, with these values of x and y, $\forall\, z \in \mathbb{R}$, (x, y, z) is a point common to both the planes.

So, $\quad \forall z \in \mathbb{R}$, (x, y, z) is a point on Λ. Thus

$$\Lambda = \{(x, y, z), \text{ where } x = z - 2,\ y = z - 1 \,\forall\, z \in \mathbb{R}\}$$
$$= \{(z-2, z-1, z) \,\forall\, z \in \mathbb{R}\}.$$

Putting different values for z we get different points on Λ. $z = 0$ gives $x = -2, z = -1$ obtained above. $z = 3$ gives $x = 1, y = 2$, so that $(1, 2, 3)$ is a point on Λ.

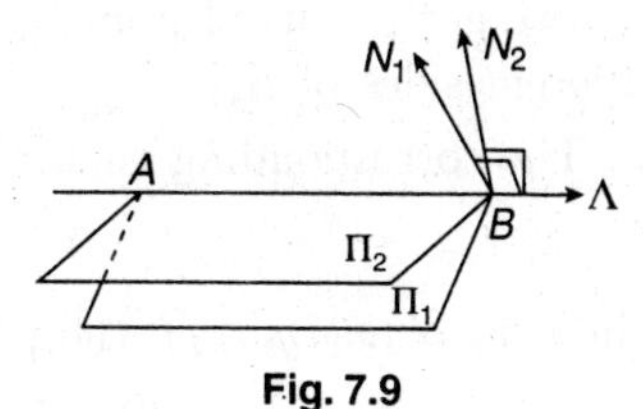

Fig. 7.9

The D.R.'s of Λ can be found out. Let l, m, n be the D.C.'s of Λ. The coefficients of x, y and z in the equation to a plane are the D.R.'s of the normal to the plane. So $1, -3, 2$ and $2, -3, 1$ are the D.R.'s of the normals to the given planes. Since Λ lies in both the planes, it is perpendicular to both these normals BN_1 and BN_2.

So, $\quad 1 \cdot l - 3m + 2n = -0$

and $\quad 2 \cdot l - 3m + 1n = 0,$

whence $\quad l/(-3 \cdot 1 + 3 \cdot 2) = m/(2 \cdot 2 - 1 \cdot 1) = n/(1(-3) - 2(-3))$

or, $\quad l/3 = m/3 = n/3$

or, $\quad l : m : n = 1 : 1 : 1.$

Hence $\quad 1, 1, 1$ are the D.R.'s of Λ.

If a_1, b_1, c_1 and a_2, b_2, c_2 are the coefficients of x, y and z in the equations to the two planes Π_1 and Π_2 and Λ is their line of intersection, then D.R.'s of Λ are obtained by the cross-multiplication from:

$$(a_1, b_1, c_1) \times (a_2, b_2, c_2) = \begin{vmatrix} \bar{i} & \bar{j} & \bar{k} \\ a_1 & b_1 & c_1 \\ a_2 & b_2 & c_2 \end{vmatrix} = \left\| \begin{matrix} a_1 & b_1 & c_1 \\ a_2 & b_2 & c_2 \end{matrix} \right\|$$
$$= (b_1c_2 - b_2c_1, c_1c_2 - c_2a_1, a_1b_2 - a_2b_1).$$

The D.R.'s are $b_1c_2 - b_2c_1$, $c_1a_2 - c_2a_1$ and $a_1b_2 - a_2b_1$.

7.8.1 Symmetric Form

If $R_1(x_1, y_1, z_1)$ is a point on the line Λ and the D.C.'s of the line Λ are l, m, n respectively, the equations to Λ can be easily obtained in a form called the **symmetric form**. If $R(x, y, z)$ is any arbitrary point on Λ then we have seen that the coordinate differences, viz., $x - x_1$, $y - y_1$ and $z - z_1$ are the D.R.'s of Λ. Since they are proportional to its D.C.'s, so we have

$$(x - x_1)/l = (y - y_1)/m = (z - z_1)/n, \tag{1}$$

which is the required symmetric form of the equations to Λ.

If another point $R_2(x_2, y_2, z_2)$ is known, then $x_2 - x_1$, $y_2 - y_1$ and $z_2 - z_1$ are also D.R.'s of Λ. So,

$$(x_2 - x_1)/l = (y_2 - y_1)/m = (z_2 - z_1)/n. \tag{2}$$

From (1) and (2) we therefore have

$$(x - x_1)/(x_2 - x_1) = (y - y_1)/(y_2 - y_1) = (z - z_1)/(z_2 - z_1)$$

as the symmetric form of the equations to Λ, when any two distinct points on it are given.

7.8.2 Parametric Form

Taking λ as the common value of the ratios in (1), §7.8.1, we have $x = x_1 + \lambda l$, $y = y_1 + \lambda m$ and $z = z_1 + \lambda n$ as the parametric equations of the same line Λ. Here λ is a parameter. For different real values of λ we get different points on Λ. For $\lambda = 0$, we get $x = x_1$, $y = y_1$, $z = z_1$, i.e., we get the fixed point R_1. For positive and negative values of λ we get points, on Λ, on different sides of R_1.

The corresponding vector equation of Λ is

$$\overline{r} = \overline{r}_1 + \lambda \overline{v},$$

where $\overline{r}_1 = (x_1, y_1, z_1)$, the p.v. of a fixed point R_1 on Λ

and $\quad \overline{v} = (l, m, n)$ a direction vector of Λ.

7.8.3 The Distance of a Point from a Line

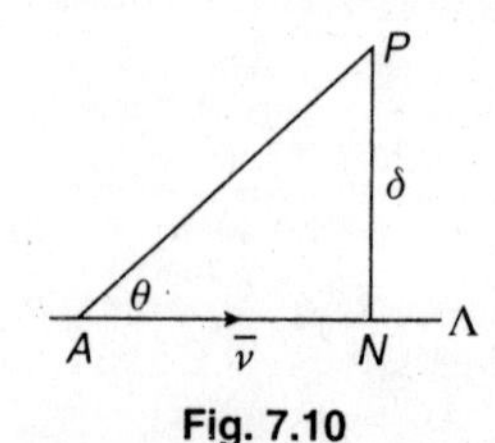

Fig. 7.10

Given any line $\Lambda : (x - a_1)/l = (y - a_2)/m = (z - a_3)/n$, the **distance δ of a point $P(x_1, y_1, z_1)$ from Λ** is given by

$$\delta = \textbf{mod} \begin{vmatrix} \hat{i} & \hat{j} & \hat{k} \\ x_1 - a_1 & y_1 - a_2 & z_1 - a_3 \\ l & m & n \end{vmatrix} \div \sqrt{\sum l^2}$$

$$= \sqrt{\sum \{n(y_1 - a_2) - m(z_1 - a_3)\}^2} \div \sqrt{\sum l^2}.$$

For, if A is (a_1, a_2, a_3), A lies on Λ. A direction vector of Λ is $\overline{v} = (l, m, n)$.

If $\overline{a} = (a_1, a_2, a_3)$ and $\overline{P} = (x_1, y_1, z_1)$, then

$$\delta = AP \sin\theta = |\overrightarrow{AP} \times \hat{v}| = \left|(\overline{P} - \overline{a}) \times \frac{\overline{v}}{|\overline{v}|}\right|$$

$$= |(\overline{P} - \overline{a}) \times \overline{v}| \div |\overline{v}| = \text{mod} \begin{vmatrix} \hat{i} & \hat{j} & \hat{k} \\ x_1 - a_1 & y_1 - a_2 & z_1 - a_3 \\ l & m & n \end{vmatrix} \div \sqrt{\sum l^2}.$$

Solved Problems

1. *A line passes through* $(1, 2, 3)$ *and* $(2, 3, 4)$. *Find its D.C.'s.*

The D.R.'s of the line are $2 - 1$, $3 - 2$, $4 - 3$ or $1, 1, 1$.

So, its D.C.'s are $\pm 1/\sqrt{1^2 + 1^2 + 1^2}$, $\pm 1/\sqrt{1^2 + 1^2 + 1^2}$, $\pm 1/\sqrt{1^2 + 1^2 + 1^2}$ or $\pm 1/\sqrt{3}$, $\pm 1/\sqrt{3}$, $\pm 1/\sqrt{3}$.

2. *Find D.C.'s of the normal to the plane*

$$3x - 12y + 4z + 7 = 0.$$

A direction vector of the normal to the plane is given by

$$\overline{v} = (3, -12, 4).$$

The corresponding unit vector $\hat{v} = \dfrac{\overline{v}}{|\overline{v}|} = \dfrac{(3, -12, 4)}{\sqrt{3^2 + 12^2 + 4^2}} = (3/13, -12/13, 4/13).$

The components of this unit vector, i.e., $3/13, -12/13$ and $4/13$ are the D.C.'s of the normal to the plane.

3. *Find the equation to the plane passing through the origin and containing the lines whose D.R.'s are* $1, -2, 0$ *and* $2, 3, -1$.

Any plane through the origin is $\sum ax = 0$. Then a vector normal to the plane is given by $\overline{\nu} = (a, b, c)$. Two direction vectors of the given lines are $\overline{\nu}_1 = (1, -2, 0)$ and $\overline{\nu}_2 = (2, 3, -1)$. $\overline{\nu}$ must be perpendicular to both $\overline{\nu}_1$ and $\overline{\nu}_2$. So, $\overline{\nu} \| \overline{\nu}_1 \times \overline{\nu}_2$.

$$\therefore \quad \overline{\nu} = \lambda \begin{vmatrix} \hat{i} & \hat{j} & \hat{k} \\ 1 & -2 & 0 \\ 2 & 3 & -1 \end{vmatrix} = \lambda(2, 1, 7).$$

$$\therefore \quad a : b : c = 2 : 1 : 7.$$

Hence the required plane is $2x + y + 7z = 0$.

4. *Write down the equation to the plane* $5x - 3y + 6z = 60$ *in the intercept form.*

The equation to a plane in the intercept form is

$$x/a + y/b + z/c = 1. \tag{1}$$

$\because$ (1) and the given equation represent the same plane,

$$\frac{5}{1/a} = \frac{-3}{1/b} = \frac{6}{1/c} = \frac{60}{1},$$

whence $a = 12, \ b = -20, \ c = 10$.

Hence the required equation is $x/12 + y/(-20) + z/10 = 1$.

5. *A plane meets the* x-, y- *and* z-*axes in* A, B *and* C *respectively. If the centroid of the triangle* ABC *is* $G(\overline{x}, \overline{y}, \overline{z})$, *show that the plane is* $x/\overline{x} + y/\overline{y} + z/\overline{z} = 3$.

If A, B and C are respectively $(a, 0, 0)$, $(0, b, 0)$ and $(0, 0, c)$, then $\overline{x} = \frac{1}{3}(a + 0 + 0) = a/3$, $\overline{y} = b/3$, $\overline{z} = c/3$. But a, b, c are the intercepts on the axes. So, the plane is $\sum x/a = 1$. But $a = 3\overline{x}$, $b = 3\overline{y}$ and $c = 3\overline{z}$. So, the plane is

$$\sum \frac{x}{3\overline{x}} = 1 \quad \text{or,} \quad \sum \frac{x}{\overline{x}} = 3.$$

6. *Find the direction angles of the normal drawn from the origin to the plane* $12x - 4y + 3z + 65 = 0$. *Find also the length of this normal.*

Since the constant term, viz., 65 is positive, the normal form of the equation to the plane is

$$(-12x + 4y - 3z - 65)/\sqrt{12^2 + 4^2 + 3^2} = 0$$

$$\text{or,} \quad x(-12/13) + y(4/13) + z(-3/13) - 5 = 0.$$

$$\therefore \quad \cos\alpha = -12/13, \ \cos\beta = 4/13 \quad \text{and} \quad \cos\gamma = -3/13$$

and $p = 5$.

The direction angles of the normal are $\pi - \cos^{-1}(12/13)$, $\cos^{-1}(4/13)$ and $\pi - \cos^{-1}(3/13)$ and the length of the normal is 5 units.

7. *Find the angle between the planes* $2x - y + z = 6$ *and* $x + y + 2z = 7$.

The unit normal vectors to the two planes are

$$\hat{\nu}_1 = (2, -1, 1)/\sqrt{2^2 + 1^2 + 1^2} \quad \text{and} \quad \hat{\nu}_2 = (1, 1, 2)/\sqrt{6}.$$

If θ^c is the angle between the planes, it is also the angle between their normal vectors. Since

$$\cos\theta = \hat{\nu}_1 \cdot \hat{\nu}_2 = (2.1 - 1.1 + 1.2)/(\sqrt{6})^2 = 3/6 = 1/2.$$

$$\therefore \quad \theta = \pi/3.$$

8. *Show that the distance of the point $P(2,3,-5)$ from the plane $x+2y-2z=9$ is 3.*

A normal vector of the plane is $\overline{\nu}(1,2,-2)$. The parametric form of the equations to the line through P perpendicular to the plane is

$$x = 2+\lambda 1, \quad y = 3+\lambda 2, \quad z = -5+\lambda(-2).$$

For the meeting point of this line and the plane,

$$2+\lambda+2(3+2\lambda)-2(-5-2\lambda) = 9 \text{ or, } \lambda = -1.$$

So Q, the point of intersection of the line and the plane, is $(2-1, 3-2, -5+2)$ or $(1,1,-3)$. The required distance is PQ and

$$PQ = \sqrt{(2-1)^2+(3-1)^2+(-5+3)^2} = \sqrt{1+4+4} = 3.$$

The **formula** for the distance of a point P from the plane $\Pi : F = 0$, where $F = \sum ax + d$ is $|F(P)|/\sqrt{\sum a^2}$. Using this formula we get immediately

$$PQ = |2+2\cdot 3-2(-5)-9| \div \sqrt{1+2^2+2^2} = 3.$$

9. *Show that the distance between two parallel planes $\Pi_i : \sum ax + d_i = 0$ for $i = 1,2$ is $|d_1 - d_2|/\sqrt{\sum a^2}$.*

If $P(\alpha,\beta,\gamma)$ be any point on Π_1, $\sum a\alpha + d_1 = 0$.

The distance of P from Π_2 is $|\sum a\alpha + d_2|/\sqrt{\sum a^2}$.

But $\sum a\alpha = -d_1$. So, the required distance is

$$|-d_1+d_2|/\sqrt{\textstyle\sum a^2} \quad \text{or,} \quad |d_1-d_2|/\sqrt{\textstyle\sum a^2},$$

the numerical value of the difference between the constant terms divided by the square root of the sum of the squares of the coefficients of x, y and z.

10. *Find the equations to the line through $P(1,2,3)$ and parallel to the line $x/3 = y/2 = z/1$.*

The D.R.'s of the required line are $3, 2, 1$. Hence the required equations of the line are

$$(x-1)/3 = (y-2)/2 = (z-3)/1.$$

11. *Find the equations to the line through the origin and intersecting the line given by $x-y+2z=2$ and $2x-y+3z=4$ at right angles.*

If l, m, n are the D.R.'s of the required line Λ, then (l,m,n) is also a point on the line. If this line intersects the given line Λ' at (l,m,n), then

(1) $l-m+2n-2=0$ and (2) $2l-m+3n-4=0$.

The D.R.'s of the given line are $\left\|\begin{matrix} 1 & -1 & 2 \\ 2 & -1 & 3 \end{matrix}\right\| = (-1,1,1)$.

$\Lambda \perp \Lambda'$ if (3) $l(-1)+m\cdot 1+n\cdot 1 = 0$.

Solving (1), and (2) for m and n interms of l

$m = 2-l$, $n = 2-l$.

Substituting in (3) $-l + 2 - l + 2 - l = 0$ or, $l = 4/3$

$\therefore \quad m = 2/3,\ n = 2/3.$

Hence the required equations are

$$\frac{x}{4/3} = \frac{y}{2/3} = \frac{z}{2/3} \quad \text{or,} \quad x/2 = y/1 = z/1.$$

12. *Find the condition that the planes* $\sum a_i x + d_i = 0$ *for* $i = 1, 2$ *are perpendicular to each other. Hence find the equation to the plane passing through* $(3, -2, 4)$ *and perpendicular to each of the planes* $7x - 3y + z - 5 = 0$ *and* $4x - y - z + 9 = 0$.

Normals to the two planes Π_1 and Π_2 are (a_1, b_1, c_1) and (a_2, b_2, c_2).

$\therefore \quad \Pi_1 \perp \Pi_2$ if $(a_1, b_1, c_1) \cdot (a_2, b_2, c_2) = 0$ or, $\sum a_1 a_2 = 0$.

Any plane Π through $P(3, -2, 4)$ is

$$a(x-3) + b(y+2) + c(z-4) = 0.$$

If $\Pi \perp \Pi_1$ and $\Pi \perp \Pi_2$, then

$$a \cdot 7 - b \cdot 3 + c \cdot 1 = 0$$

and $\quad a \cdot 4 - b \cdot 1 - c \cdot 1 = 0,$

whence $\quad a/4 = b/11 = c/5.$

So, the required equation to Π is

$$4(x-3) + 11(y+2) + 5(z-4) = 0$$

or, $\quad 4x + 11y + 5y - 10 = 0.$

13. *Is it possible for a straight line to make an angle of* $60°$ *with each of the coordinate axes? If it makes an angle of* $60°$ *with each of* y- *and* z-*axes, find the angle (or angles) which it may make with the* x-*axis.*

If the line makes an angle of $60°$ with each of the coordinate axes, then the D.C.'s of the line will be

$$l = \cos 60°,\ m = \cos 60° \text{ and } n = \cos 60°.$$

i.e., $\quad l = \frac{1}{2} = m = n$. But $\sum l^2$ must be 1,

i.e., $\quad (1/2)^2 + (1/2)^2 + (1/2)^2 = 1,$

i.e., $\quad \frac{3}{4} = 1$, which is absurd.

Hence it is impossible that a line makes such angles with the axes.

If α^c is the angle made by the line with the x-axis, then

$$l = \cos\alpha,\ m = 1/2 \text{ and } n = 1/2.$$

Since $l^2 + m^2 + n^2 = 1$, $\therefore \cos^2\alpha + \frac{1}{4} + \frac{1}{4} = 1$ or, $\cos^2\alpha = \frac{1}{2}$

$\cos\alpha = \pm 1/\sqrt{2}$. So, α is either $\pi/4$ or, $\pi - \pi/4$, i.e., $3\pi/4$.

Thus the angle which the line may make with the x-axis is either $45°$ or $135°$.

14. *Find the ratio in which the line-segment joining $R_1(x_1, y_1, z_1)$ and $R_2(x_2, y_2, z_2)$ is divided by the plane $\Pi : \sum ax + d = 0$ and find the coordinates of the points of intersection.*

Let $\quad F = \sum ax + d.$

Let $\quad \Pi$ divide R_1R_2 at $R(\overline{x}, \overline{y}, \overline{z})$ in the ratio $m : n$.

Then $\quad \overline{x} = (mx_2 + nx_1)/(m+n),\ \overline{y} = \ldots,\ \overline{z} = \ldots.$

Since R lies on Π, $F(R) = 0$, i.e.,

$$\sum a\overline{x} + d = 0 \quad \text{or,} \quad \sum a\frac{(mx_2 + nx_1)}{m+n} + d = 0$$

or, $$\sum a(mx_2 + nx_1) + (m+n)d = 0$$

or, $$m\left(\sum ax_2 + d\right) + n\left(\sum ax_1 + d\right) = 0.$$

In short, $mF(R_2) + nF(R_1) = 0$

$$\therefore \quad \frac{m}{n} = -\frac{F(R_1)}{F(R_2)}.$$

$$\therefore \quad \overline{x} = \frac{x_1F(R_2) - x_2F(R_1)}{F(R_2) - F(R_1)},\ \overline{y} = \ldots,\ \overline{z} = \ldots.$$

15. *Find the condition that the line*

$$\Lambda : (x - x_1)/l = (y - y_1)/m = (z - z_1)/n$$

entirely lies in the plane

$$\Pi : F = 0, \textit{ where } F = \sum ax + d.$$

The line Λ will lie entirely in the plane Π if only one of its points lies in Π and $\Lambda || \Pi$.

Now a point on Λ is $R_1(x_1, y_1, z_1)$. It lies in Π if

$$F(R_1) = 0, \text{ i.e., } \sum ax_1 + d = 0.$$

$\Lambda||\Pi$ if $\overline{v} \perp \overline{v}'$, where $\overline{v}$ is a normal to Π and $\overline{v}'$ is $||\Lambda$. Now (a, b, c) is normal to Π and if $\overline{v}' = (l, m, n)$ then the condition for $\Lambda||\Pi$ is $\overline{v} \cdot \overline{v}' = 0$ or $al + bm + cn = 0$ or briefly $\sum al = 0$.

So, the condition for Λ to lie in Π is that (1) $F(R_1) = 0$ and (2) $\sum al = 0$.

16. *Find the condition that the lines Λ_i given by*

$$(x - x_i)/l_i = (y - y_i)/m_i = (z - z_i)/n_i, \textit{ for } i = 1, 2$$

both lie in one and the same plane.

The condition that Λ lies in the plane $\Pi : \sum ax + d = 0$ is given by

$$(1) \sum ax_1 + d = 0 \quad \text{and} \quad (2) \sum al_1 = 0.$$

If Λ_2 also lies in the same plane, then

$$(3) \sum ax_2 + d = 0 \quad \text{and} \quad (4) \sum al_2 = 0.$$

From (1) and (3) we get

$$(5) \sum a(x_1 - x_2) = 0.$$

Eliminating a, b and c from (5), (2) and (4) we get

$$\begin{vmatrix} x_1 - x_2 & y_1 - y_2 & z_1 - z_2 \\ l_1 & m_1 & n_1 \\ l_2 & m_2 & n_2 \end{vmatrix} = 0,$$

which is therefore the required condition.

Taking $\overline{r}_1 = (x_1, y_1, z_1)$ and $\overline{r}_2 = (x_2, y_2, z_2)$

$$\overline{r}_1 - \overline{r}_2 = (x_1 - x_2, y_1 - y_2, z_1 - z_2).$$

So, vectorially, the above condition is

$$(\overline{\boldsymbol{r}}_1 - \overline{\boldsymbol{r}}_2) \cdot \overline{\boldsymbol{\nu}}_1 \times \overline{\boldsymbol{\nu}}_2 = \mathbf{0},$$

where $\overline{\nu}_1 = (l_1, m_1, n_1)$ and $\overline{\nu}_2 = (l_2, m_2, n_2)$.

17. *Show that the lines*

$$\Lambda_1 : (x-1)/2 = (y+1)/(-3) = (z+10)/8$$

and $\Lambda_2 : (x-4)/1 = (y+3)/(-4) = (z+1)/7$

intersect and find the plane through them.

If Λ_1 and Λ_2 intersect each other, then they must be coplanar. We find that

$$\begin{vmatrix} -1+4 & -1-3 & -10-1 \\ 2 & -3 & 8 \\ 1 & -4 & 7 \end{vmatrix} = 0.$$

Hence the lines lie in a plane. But since their D.R.'s are not proportional, they intersect. If the equation to the plane Π in which they lie is

$$(1)\ \sum ax + d = 0,$$

since Λ_1 lies Π, the point $(1, -1, -10)$ also lies in it. So,

$$(2)\ a \cdot 1 + b + (-1) + c(-10) + d = 0.$$

(1) − (2) gives $(3)\ a(x-1) + b(y+1) + c(z+10) = 0.$

But $(4)\ a \cdot 2 + b(-3) + c \cdot 8 = 0$

and $(5)\ a \cdot 1 + b(-4) + c \cdot 7 = 0.$

Eliminating a, b and c we get $\begin{vmatrix} x-1 & y+1 & z+10 \\ 2 & -3 & 8 \\ 1 & -4 & 7 \end{vmatrix} = 0$ as the required equation to Π.

18. *Find the distance of the point* $(1, -2, 3)$ *from the line through* $(2, -3, 5)$ *making equal angles with the positive directions of the coordinate axes.*

If P is $(1, -2, 3)$ and A is $(2, -3, 5)$, then A lies on the given line Λ. Since Λ makes equal angles with the positive directions of the coordinate axes, a direction vector of Λ is $\overline{\nu} = (1, 1, 1)$. So, the shortest distance

$$\delta = |\overline{AP} \times \overline{\nu}| \div |\overline{\nu}| = \begin{vmatrix} \hat{i} & \hat{j} & \hat{k} \\ -1 & 1 & -2 \\ 1 & 1 & 1 \end{vmatrix} \div \sqrt{(1^2 + 1^2 + 1^2)}$$

$$= \sqrt{\{(1+2)^2 + (-2+1)^2 + (-1-1)^2\}} \div \sqrt{3} = \sqrt{14/3}.$$

19. *If $\Lambda : (x-\alpha)/l = (y-\beta)/m = (z-\gamma)/n$ is an incident ray falling on the plane mirror $\Pi : F = 0$, where $F = \sum ax + d$, then find the reflected ray.*

The point $P(\alpha, \beta, \gamma)$ lies on Λ. If its image in Π is $P'(\alpha', \beta', \gamma')$, then we know that

$$\alpha' = \alpha - a \cdot 2F(P)/\sum a^2,\ \beta' = \beta - b \cdot 2F(P)/\sum a^2,\ \gamma' = \gamma - c \cdot 2F(P)/\sum a^2.$$

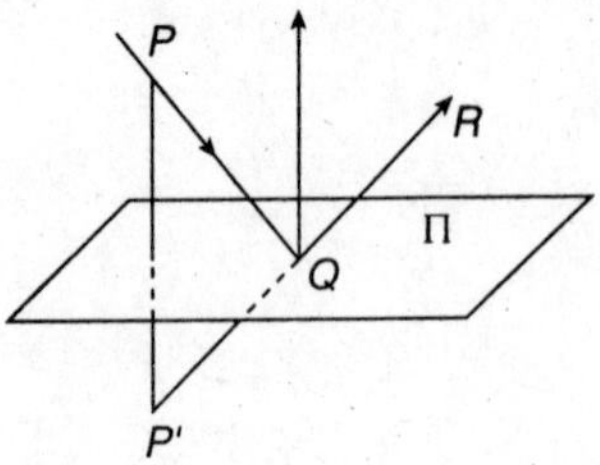

If $Q(x_0, y_0, z_0)$ is the point of incidence, then there is a number λ such that

$$x_0 = \alpha + \lambda l, \quad y_0 = \beta + \lambda m, \quad z_0 = \gamma + \lambda n.$$

Since Q lies on Π, $F(Q) = 0$ or, $\sum a x_0 + d = 0$

or, $\sum a(\alpha + \lambda l) + d = 0$ or, $\sum a\alpha + d + \lambda \sum al = 0$

or, $F(P) + \lambda \sum al = 0$ or, $\lambda = -F(P)/\sum al.$

Fig. 7.11

The reflected ray QR is the line joining P' and Q, i.e.,

$$(x - x_0)/(x_0 - \alpha') = (y - y_0)/(y_0 - \beta') = (z - z_0)/(z_0 - \gamma').$$

Exercises 7.1

1. If for a unit vector $\hat{\nu}$, $\hat{\nu} \cdot \hat{i} = 1/\sqrt{2}$, $\hat{\nu} \cdot \hat{j} = 1/2$, what is $\hat{\nu} \cdot \hat{k}$?
2. Find a direction vector of the line passing through $(3, 2, 5)$ and $(-1, 3, 2)$. Hence find the D.C.'s of the line.
3. Find the D.C.'s of the line perpendicular to the two lines whose direction vectors are $(1, -2, -2)$ and $(0, 2, 1)$.

   [**Hints.** If $\hat{\nu}(= (l, m, n))$ is a direction vector of the required line, $\hat{\nu} \| (1, -2, -2) \times (0, 2, 1)$.]
4. Prove that the angle between any two diagonals of a cube is $\cos^{-1}(1/3)$.
5. Show that the four points $A(3, 0, 0)$, $B(0, 3, 0)$, $C(0, 0, 3)$ and $D(1, 1, 1)$ are coplanar.

   [**Hints.** $\overline{AB} \cdot \overline{AC} \times \overline{AD} = 0$.]
6. Find the lengths of the intercepts made by the plane $20x + 15y - 12z - 60 = 0$ on the coordinate axes.
7. Find the ratio in which the segment joining $(1, 1, 1)$ and $(2, -3, 1)$ is divided by the plane $2x - 3y + 5z = 10$. Determine whether the plane divides the segment internally or externally.
8. Show that the equation to the plane through $(0, 1, 1)$, $(1, 0, 1)$ and $(1, 1, 0)$ is $\sum x - 2 = 0$.
9. Find the distance of the plane $6x - 2y + 3z + 7 = 0$ from the origin.
10. Find the image of the point $(1, 2, 3)$ in the plane $2x + 3y + 6z + 23 = 0$.
11. Show that the image of the origin in the plane $\sum x = 1$ is $(2/3, 2/3, 2/3)$.
12. A plane passes through $(1, 2, 3)$ and makes equal intercepts on the positive directions of the coordinate axes. Find the distance of the origin from the plane.
13. Find the equation to the plane through P and at right angles to OP, where P is $(2, 6, 3)$.
14. Find the equation to the plane through $(1, 0, -1)$ and bisecting the sides AB and AC of the triangle ABC, where A, B and C are $(1, 1, 1)$, $(1, 2, 3)$ and $(2, 3, 4)$ respectively.

15. Show that the planes $2x + y - z = 1$ and $x - y - 2z + 1 = 0$ include an angle of $60°$.

16. Find the distance of the point $(1, 2, -3)$ from the plane $2x - 3y + 6z - 6 = 0$.

17. Find the symmetrical form of the equations of the line given by $3y+4z-5 = 0 = 3z-4x+5$.

18. Find a point on the straight line

$$(x - 1)/6 = (y - 2)/3 = (z - 3)/2$$

at a distance of 7 units from the point $(1, 2, 3)$.

19. Show that the line: $(x+1)/1 = (y+3)/3 = (z-2)/(-2)$ meets the plane: $3x+4y+5z-5 = 0$ in $(1, 3, -2)$.

20. Find the distance of the point $(2, 3, -1)$ from the line $x/2 = (y - 2)/3 = (z + 1)/(-2)$.

21. A straight line through $P(1, 2, 3)$ meets the line $(x - 2)/1 = (y - 1)/2 = z/3$ in Q at right angles. Find Q and PQ.

22. Find the point where the line through $(5, -2, 3)$ and $(3, 0, 1)$ pierces the xy-plane.

23. Find the distance of the point $(1, -2, 3)$ from the plane measured parallel to the line: $x/2 = y/3 = z/6$.

24. Show that the straight line given by $x = -2 + \lambda$, $y = 3 - 4\lambda$ and $z = 5\lambda + 6$, where λ is a parameter, is parallel to the plane $x - y - z = 10$ but not lying in the plane.

25. Show that the line $(x + 1)/(-2) = (y + 2)/3 = (z + 5)/4$ lies in the plane $x + 2y - z = 0$.

26. A ray of light passing through the hole at $(1, 2, 3)$ falls on the plane mirror given by $3x - 2y - z - 1 = 0$ at $(0, -1, 1)$. Find the equation of the reflected ray.

Answers

1. $1/2$. **2.** $(4, -1, 3)$, $4/\sqrt{26}$, $-1/\sqrt{26}$, $3/\sqrt{26}$. **3.** $2/3$, $-1/3$, $2/3$. **6.** $3, 4, 5$.
7. $-3/4$, externally. **9.** 1. **10.** $(-3, -4, -9)$. **12.** $2\sqrt{3}$.
13. $2x + 6y + 3z = 49$. **14.** $x - 2y + z = 0$. **16.** 4 units.
17. $(x - 2)/9 = (y - 1/3)/(-16) = (z - 1)/12$. **18.** $(7, 5, 5)$ or $(-5, -1, 1)$.
20. $6/\sqrt{17}$. **21.** $(19/7, 17/7, 15/7)$, $3\sqrt{3/7}$. **22.** $(2, 1, 0)$. **23.** $7/5$.
26. $x/22 = (y + 1)/11 = (z - 1)/9$.

7.9 Elementary Ideas of Simple Quadric Surfaces

A linear equation in x, y and z represents a plane. Any surface represented by a second degree equation is called a **quadric surface**. The general second degree equation in x, y and z is

$$ax^2 + by^2 + cz^2 + 2fyz + 2gzx + 2hxy + 2ux + 2vy + 2wz + d = 0 \tag{1}$$

or briefly, $\phi + 2L + d = 0$, where $\phi = \sum ax^2 + 2\sum fyz$, a homogeneous quadratic function in x, y and z, and $L = \sum ux$, a homogeneous linear function in x, y, z.

The equation (1) represents, among others, surfaces like sphere, cone and cylinder.

7.9.1 Sphere

A sphere is defined to be the set of all the points in the 3-D space at a constant distance from a fixed point. The fixed point is called **the centre** and the fixed distance, **the radius** of the sphere. So, if $\Gamma = \{P : CP = R$, where C is a fixed point and P is a fixed number$\}$,

then Γ is a sphere. If C is (x_0, y_0, z_0) and P is (x, y, z), then

$$CP^2 = \sum(x - x_0)^2.$$

So, $\Gamma = \left\{(x, y, z) : \sum(x - x_0)^2 - R^2 = 0\right\}$.

Thus when the centre and radius of a sphere are given the equation to the sphere can be easily found out. The equation to Γ is

$$F = 0, \text{ where } F = \sum(x - x_0)^2 - R^2. \tag{1}$$

Here F is a second degree function of x, y and z. So, F may be written in the form (1) in § 7.9. If

$$F = \phi + 2L + d$$

then $\phi = x^2 + y^2 + z^2 = \sum x^2$,

$$L = -(x_0 x + y_0 y + z_0 z) = -\sum x_0 x$$

and $d = x_0^2 + y_0^2 + z_0^2 - R^2$.

From above, it is clear that the equation (1) represents a sphere if the coefficients of x^2, y^2 and z^2 are all equal, but not equal to zero, whereas the coefficients of the terms containing yz. zx and xy are all zero.

Thus the general equation to a sphere is

$$a \sum x^2 + 2L' + d' = 0,$$

where $a \neq 0, L' = \sum u'x$ and d' is a constant. Dividing by a the equation reduces to

$$\sum x^2 + 2Lx + d = 0, \tag{2}$$

where $L = \sum ux, \ u = u'/a, \ v = v'/a, \ w = w'/a, \ d = d'/a$.

Since $\sum x^2 + 2\sum ux + d = \sum(x^2 + 2ux + u^2) - (\sum u^2 - d) = \sum(x + u)^2 - R^2$,

where $R = \sqrt{\sum u^2 - d}$, the centre of the sphere given by (2) is $(-u, -v, -w)$ and its radius $R = \sqrt{\sum u^2 - d}$.

If the centre is chosen as the origin, the equation to the sphere becomes

$$\sum x^2 = R^2, \tag{3}$$

which is the **standard equation** of the sphere.

Example 7.9.1. **1.** *Find the centre and radius of the sphere*

$$2\sum x^2 - 2x + 4y + 2z + 1 = 0.$$

The equation may be written as $\sum x^2 - x + 2y + z + 1/2 = 0$.

So, the centre is $(1/2, -1, -1/2)$ and the radius $\sqrt{[(1/2)^2 + (-1)^2 + (-1/2)^2 - 1/2]} = 1$.

7.9.2 Sphere with Two Given Points as the Ends of a Diameter

If $R_i(x_i, y_i, z_i)$ for $i = 1, 2$ are the ends of a diameter, the equation of the sphere is

$$\sum (x - x_1)(x - x_2) = 0.$$

If $R(x, y, z)$ is any point on the surface of the sphere, one can always draw a great circle, i.e., a circle, on the surface of the sphere, with the centre, at the centre of the sphere, through R_1, R_2 and R. Then since R_1R_2 is the diameter of that circle, $\angle R_1RR_2$ is a right angle.

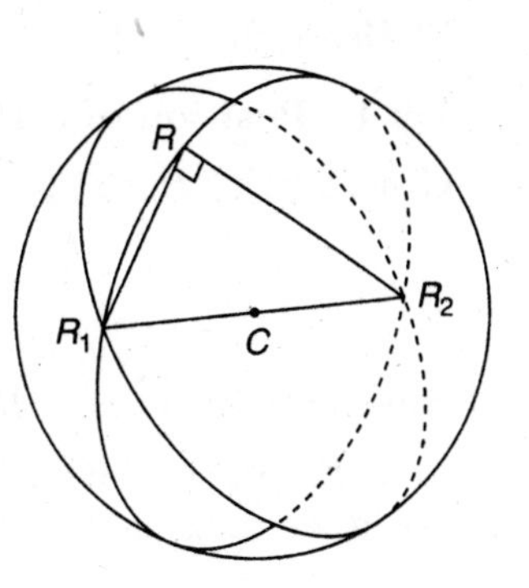

Fig. 7.12

Now the D.R.'s of R_1R and R_2R are $x - x_1$, $y - y_1$, $z - z_1$ and $x - x_2$, $y - y_2$, $z - z_2$. $R_1R \perp R_2R$ if

$$\sum (x - x_1)(x - x_2) = 0,$$

which is the required equation of the sphere.

Example 7.9.2. 1. *Find the sphere, the ends of a diameter of which are* $(3, 4, -1)$ *and* $(-1, 2, 3)$. *Find its centre and radius.*

The equation to the sphere is

$$(x - 3)(x + 1) + (y - 4)(y - 2) + (z + 1)(z - 3) = 0$$

or, $\sum x^2 - 2x - 6y - 2z + 2 = 0.$

The centre is $(1, 3, 1)$ and the radius is

$$\sqrt{1^2 + 3^2 + 1^2 - 2} = 3.$$

7.9.3 Sphere through Four Non-coplanar Points

If $R_i(x_i, y_i, z_i)$ for $i = 1, 2, 3, 4$, be four non-coplanar points, the sphere through them is

$$\begin{vmatrix} \sum x^2 & x & y & z & 1 \\ \sum x_1^2 & x_1 & y_1 & z_1 & 1 \\ \sum x_2^2 & x_2 & y_2 & z_2 & 1 \\ \sum x_3^2 & x_3 & y_3 & z_3 & 1 \\ \sum x_4^2 & x_4 & y_4 & z_4 & 1 \end{vmatrix} = 0.$$

If $\sum x^2 + 2\sum ux + d = 0$ is the sphere, since R_i lie on this $\sum x_i^2 + 2\sum ux_i + d = 0$ for $i = 1, 2, 3, 4$. Eliminating the four constants, viz., u, v, w and d from these five equations the required equation is obtained.

Example 7.9.3. 1. *The sphere through* O, $A(a, 0, 0)$, $B(0, b, 0)$ *and* $C(0, 0, c)$ *is* $\sum x^2 - \sum ax = 0$.

For, assuming

$$\sum x^2 + 2\sum ux + d = 0$$

as the equation to the sphere we find that since O lies on it,

$$\sum 0^2 + 2\sum u \cdot 0 + d = 0 \Rightarrow d = 0.$$

Since A lies on it

$$a^2 + 2ua = 0 \quad \text{whence} \quad u = -a/2.$$

Similarly, since B and C lie on it $v = -b/2$ and $w = -c/2$.

Hence the result.

7.9.4 Position of a Point Relative to a Sphere

Given a point $P(x_1, y_1, z_1)$ and a sphere

$$\Gamma : \sum x^2 - R^2 = 0, \tag{1}$$

P lies on Γ, if $\sum x_1^2 - R^2 = 0$ or, $F(P) = 0$, where $F = \sum x^2 - R^2$.

If P does not lie on Γ, then $F(P) \neq 0$. Then either $F(P) > 0$ or < 0. The centre O of Γ lies inside the sphere. But

$$F(0) = -R^2 < 0.$$

So, for any point P inside the sphere $F(P) < 0$ and if P is outside the sphere $F(P) = 0$.

Next the distance δ of P from the centre of the sphere is given by

$$\delta^2 = \sum x_1^2 = \sum x_1^2 - R^2 + R^2 = F(P) + R^2.$$

$$\therefore \quad F(P) = \delta^2 - R^2.$$

$\therefore$ *P* lies inside, on or outside Γ according as

$$\delta^2 - R^2 <, = \text{ or } > 0$$

or, $\boldsymbol{\delta^2 <, = \text{ or } > R^2}$.

7.9.5 Position of a Plane Relative to a Sphere

Let Π and Γ be a given plane and a given surface, their equations being

$$(1) \quad F = 0, \quad \text{where} \quad F = \sum ax + d$$

and $\quad (2) \quad G = 0, \quad \text{where} \quad G = \sum x^2 - R^2.$

Then $\Pi \cap \Gamma = \phi$ or, $\{P, \text{a point}\}$ or, a circle if $\Pi \cap \Gamma \neq \phi$. When $\Pi \cap \Gamma \neq \phi$, the equations to the circle are $F = 0$ and $G = 0$.

For any parameter λ, $G + \lambda F = 0$ represents a family of spheres, each one of which will pass through the circle of intersection of Π and Γ. For, the equation $G + \lambda F = 0$ is

$$\sum x^2 - R^2 + \lambda \left(\sum ax + d \right) = 0,$$

which contains no product terms and the coefficients of x^2, y^2 and z^2 are all equal. Since $\forall \lambda \in \mathbb{R}$, $G + \lambda F = 0$ is a sphere, we get a family of spheres when λ varies.

When $\Pi \cap \Gamma = \{P\}, \Pi$ intersects Γ in a point circle at P. So, in this case Π becomes a tangent plane of Γ. When $\Pi \cap \Gamma = \phi$, Π does not intersect Γ at all.

Examples 7.9.5. **1.** *Find the equation to the sphere whose centre is* $(0, 0, 4)$ *and which passes through the circle:*

$$\sum x^2 = 1 \quad \textit{and} \quad z = 0.$$

A sphere passing through the circle is given by

$$\sum x^2 - 1 + \lambda z = 0.$$

The centre of the sphere is $(0, 0, -\lambda/2)$. This coincides with the given centre, i.e., $(0, 0, -\lambda/2) = (0, 0, 4)$, when $\lambda = -8$.

Hence the required equation is $\sum x^2 - 8z - 1 = 0$.

2. *Show that the plane: Π touches the sphere Γ if $\boldsymbol{\delta = R}$, where δ is the distance of Π from the centre of Γ and R is the radius of Γ.*

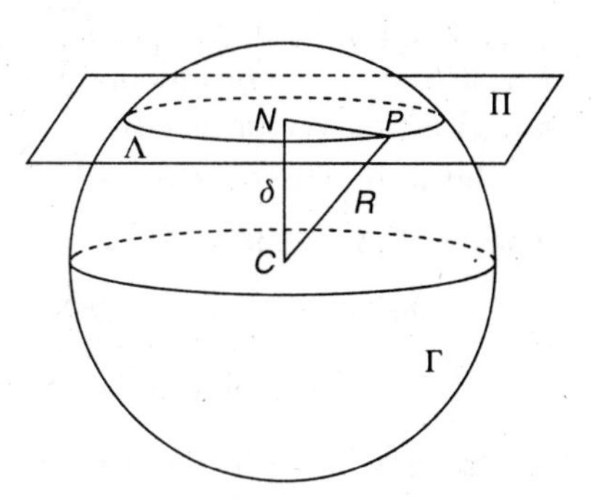

Fig. 7.13

Let C be the centre of Γ and CN be drawn perpendicular to Π. If P is any point on the curve Λ of intersection of Π and Γ, then $CP = R$. Since $PN^2 = CP^2 - CN^2$, $PN^2 = R^2 - \delta^2$, a constant. Since PN is constant, Λ is a circle, with N as its centre and PN as radius. When $PN = 0$, the circle becomes a point circle and Π becomes a tangent plane. The condition, therefore, for Π to be a tangent plane to Γ is

$$\boldsymbol{\delta = R}.$$

3. *Show that $2x - 6y + 3z - 49 = 0$ is a tangent plane to the sphere $\sum x^2 = 49$.*

The centre of the sphere is 0. So,

$$\delta = \frac{|F(0)|}{\sqrt{\sum a^2}} = \frac{|-49|}{\sqrt{2^2 + 6^2 + 3^2}} = \frac{49}{7} = 7$$

and $R = \sqrt{49} = 7$.

$\because \delta = R$, the plane touches the sphere.

The point of contact is the point where the normal to the plane through the centre of the sphere meets the sphere. The direction vector of a normal to Π is $(2, -6, 3)$. So, the normal to the plane through the centre is $x/2 = y/(-6) = z/3 = r$, say. So, any point on this normal is $(2r, -6r, 3r)$. It lies on the sphere if $(2r)^2 + (-6r)^2 + (3r)^2 = 49$ or, $r = \pm 1$. So, the point of contact is $(2, -6, 3)$ when $r = 1$. For $r = -1$ the coordinates are $(-2, 6, -3)$ which do not satisfy the equation to the plane.

7.9.6 Cone

A cone is defined to be the locus of a straight line always passing through a fixed point and intersecting a curve. The fixed point is called the **vertex** and the curve is called the **guiding curve.** Also, the variable line is called a **generator** of the cone.

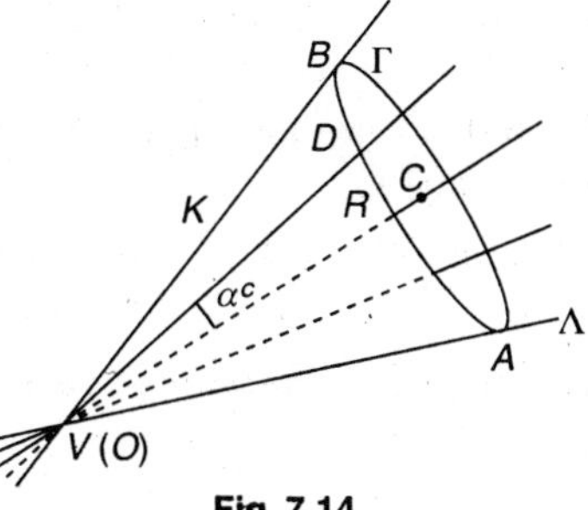

Fig. 7.14

If V is the vertex and Γ is the guiding curve, the cone K is given by

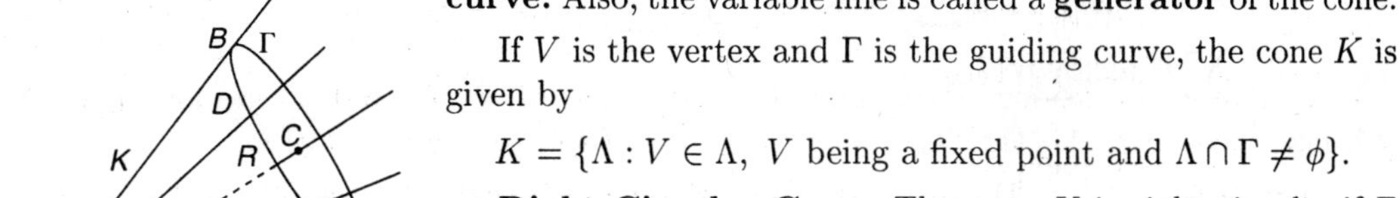

$$K = \{\Lambda : V \in \Lambda,\ V \text{ being a fixed point and } \Lambda \cap \Gamma \neq \phi\}.$$

Right Circular Cone: The cone K is right circular if Γ is a circle and V, called the **vertex**, lies on the perpendicular through the centre of the circle, called the **axis** of the cone (Fig. 7.14).

The lines VA, VB, VD are generators of the cone.

Every generator of a right circular cone makes a constant angle with the perpendicular VC, the axis of the cone. This angle is called the **semi-vertical angle.**

If V is the origin O of the frame $OXYZ$ and the equations to OC are $x/l = y/m = z/n$ and α^c is the semi-vertical angle of the right circular cone, then the equation of the cone is

$$\boldsymbol{\sum l^2 \left(\sum x^2\right) \cos^2 \alpha = \left(\sum lx\right)^2.}$$

For, if $R(x', y', z')$ be any point on the cone, the D.R.'s of VR are x', y', z' so that $\overline{r} = (x', y', z')$ is a direction vector of VR.

But a direction vector of VC, the axis of the cone is $\overline{v} = (l, m, n)$.

Now, $\quad \overline{r} \cdot \overline{v} = |\overline{r}||\overline{v}| \cos \alpha.$

$\therefore \quad lx' + my' + nz' = \sqrt{\sum x'^2} \sqrt{\sum l^2} \cos \alpha.$

Squaring and dropping the dashes we get

$$\left(\sum l^2\right)\left(\sum x^2\right) \cos^2 \alpha = \left(\sum lx\right)^2$$

as the equation to the cone.

Examples 7.9.6. 1. *Find the right circular cone whose vertex is the origin and which makes an angle of* 45° *with the* z*-axis.*

Here, $\quad l = 0,\ m = 0,\ n = 1,\ \alpha = 45°$. Hence the cone is

$$\sum x^2 \cos^2 45° = z^2$$

or, $\quad \sum x^2 = 2z^2 \quad$ or, $\quad \boldsymbol{x^2 + y^2 = z^2}$.

2. *Find the equation to the right circular cone if the coordinate axes are the generators.*

Since the coordinate axes are the generators and the axes intersect at the origin, the vertex of the cone is the origin also. Hence the equation of the cone is of the form

$$\sum l^2 \sum x^2 \cos^2 \alpha = \left(\sum lx\right)^2. \tag{1}$$

If $\overline{v} = (l, m, n)$, the direction vector of the axis, then since each coordinate axis makes the same angle α with $\overline{v}$, $\overline{v} \cdot \hat{i} = \overline{v} \cdot \hat{j} = \overline{v} \cdot \hat{k} = |\overline{v}| \cos \alpha$.

$\therefore \quad l = m = n = \sqrt{\sum l^2} \cos \alpha.$

$\therefore \quad l^2 + m^2 + n^2 = 3 \sum l^2 \cos^2 \alpha,$

i.e., $\quad \sum l^2 = 3 \sum l^2 \cos^2 \alpha,$

whence $\quad \cos \alpha = \pm 1/\sqrt{3}.$

From (1) $\ l^2 \sum x^2 = \left(l \sum x\right)^2$

or, $\quad \sum x^2 = \left(\sum x\right)^2$

or, $\quad \boldsymbol{\sum yz = 0}$, the required equation of the cone.

3. *Find the equation to the cone whose vertex is the origin and the guiding curve is* $\Gamma : \sum x^2 = 9$ *and* $\sum x = 1$.

Since the vertex is at the origin, a generator will be of the form:

$$x/l = y/m = z/n = r$$

or, $$x = rl, \quad y = rm, \quad z = rn. \tag{1}$$

Since the generator meets Γ, the values of x, y, z must satisfy both the equations of Γ. So,

$$\sum (rl)^2 = 9 \quad \text{or,} \quad r^2 \sum l^2 = 9 \tag{2}$$

and $$\sum rl = 1 \quad \text{or,} \quad r \sum l = 1. \tag{3}$$

Eliminating r between (2) and (3)

$$\sum l^2 = 9 \left(\sum l\right)^2. \tag{4}$$

Eliminating l, m, n between (1) and (4) we get

$$\sum (x/r)^2 = 9(\sum x/r)^2$$

or, $$\sum x^2 = 9 \left(\sum x\right)^2,$$

which is the required equation of the cone.

4. *If the vertex of a right circular cone is* $V(\alpha, \beta, \gamma)$ *and the axis is parallel to* $\overline{\nu} = (l, m, n)$ *and the semi-vertical angle is* α^c, *then show that the equation to the cone is*

$$\sum (x - \alpha)^2 \sum l^2 \cos^2 \alpha = \left\{\sum l(x - \alpha)\right\}^2.$$

If $\overline{r} = (x, y, z)$, the p.v. of any point on the cone

and $\overline{v} = (\alpha, \beta, \gamma)$, the p.v. of V, then

$$(\overline{r} - \overline{v}) \cdot \overline{\nu} = |\overline{r} - \overline{v}||\overline{\nu}| \cos \alpha$$

or, $$(\overline{r} - \overline{v})^2 \overline{\nu}^2 \cos^2 \alpha = \{(\overline{r} - \overline{v}) \cdot \overline{\nu}\}^2$$

or, $$\sum (x - \alpha)^2 \sum l^2 \cos^2 \alpha = \left\{\sum l(x - \alpha)\right\}^2.$$

7.9.7 Cylinder

A cylinder is the locus of a straight line which is always parallel to a fixed line and always meets a given curve.

The variable line is called the **generator** and the given curve is called the **guiding curve**.

A cylinder is called **right circular** if a plane perpendicular to its generator cuts it in a circle.

A right circular cylinder is also a surface generated by a straight line which is always parallel to, and at a constant distance from, a given line. Then the fixed line is called the **axis** and the fixed distance, the **radius** of the cylinder.

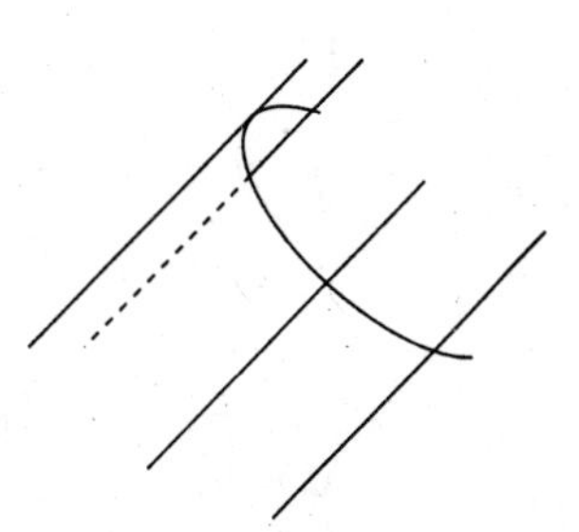

Fig. 7.15

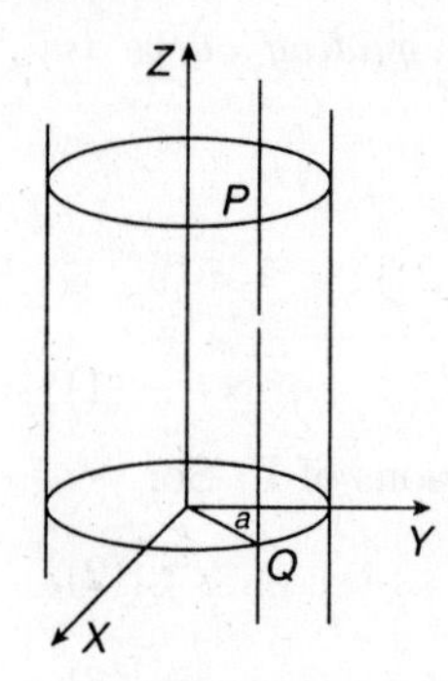

Fig. 7.16

Examples 7.9.7. 1. *Find the equation to the right circular cylinder whose axis is the z-axis and radius is a.*

Let $P(x', y', z')$ be any point on the cylinder and PQ is a generator.

Now the distance of P from the z-axis is $\sqrt{x'^2 + y'^2}$ and this distance is a. So,

$$\sqrt{x'^2 + y'^2} = a \quad \text{or,} \quad x'^2 + y'^2 = a^2.$$

Dropping the dashes the equation to the cylinder is $x^2 + y^2 = a^2$.

2. *Find the equation to the right circular cylinder whose cross-section is the circle through $A(1, 0, 0)$, $B(0, 1, 0)$, and $C(0, 0, 1)$.*

The plane in which the circle lies is $x/1 + y/1 + z/1 = 1$ or, $\sum x = 1$. The centre of the circle is the circum-centre or the centroid $(1/3, 1/3, 1/3)$ of the triangle ABC. The normal vector $\overline{\nu} = (1, 1, 1)$ to this plane must be parallel to the axis. So, the axis of the cylinder is

$$(x - 1/3)/1 = (y - 1/3)/1 = (z - 1/3)/1$$

or simply, $x/1 = y/1 = z/1$. (1)

Radius of this cylinder is the distance of $A(1, 0, 0)$ from the point $(1/3, 1/3, 1/3)$. So, the radius is $\sqrt{[(2/3)^2 + (1/3)^2 + (1/3)^2]} = \sqrt{\frac{2}{3}}$.

The distance of any point (x', y', z') from the axis (1) is $\sqrt{\dfrac{2}{3}}$.

$$\therefore \quad \text{mod} \begin{vmatrix} \hat{i} & \hat{j} & \hat{k} \\ x' - 0 & y' - 0 & z' - 0 \\ 1 & 1 & 1 \end{vmatrix} \div \sqrt{(1^2 + 1^2 + 1^2)} = \frac{\sqrt{2}}{3}$$

or, $(y' - z')^2 + (z' - x')^2 + (x' - y')^2 = 2.$

Dropping the dashes and simplifying we get $\sum x^2 - \sum yz = 1$ as the required equation.

3. *Show that equation to the right circular cylinder with its axis $(x - \alpha)/l = (y - \beta)/m = (z - \gamma)/n$ and radius a is $\sum \{n(y - \beta) - m(z - \gamma)\}^2 = a^2 \sum l^2$.*

Let $R(x, y, z)$ be any point on the cylinder and $\overline{r}$ is its p.v., $\overline{p} = (\alpha, \beta, \gamma)$, the p.v. of a point on the axis, RN is perpendicular to the axis and $\overline{\nu} = (l, m, n)$, a direction vector of the axis. Then

$$RN = |(\overline{r} - \overline{p}) \times \hat{\nu}|, \quad \text{where} \quad \hat{\nu} = \overline{\nu}/|\overline{\nu}|.$$

$$\therefore \quad \{(\overline{r} - \overline{p}) \times \hat{\nu}\}^2 = a^2$$

$$\text{or,} \quad \left\| \begin{matrix} x - \alpha & y - \beta & z - \gamma \\ l & m & n \end{matrix} \right\|^2 \div \sum l^2 = a^2$$

$$\text{or,} \quad \sum \{n(y - \beta) - m(z - \gamma)\}^2 = a^2 \sum l^2.$$

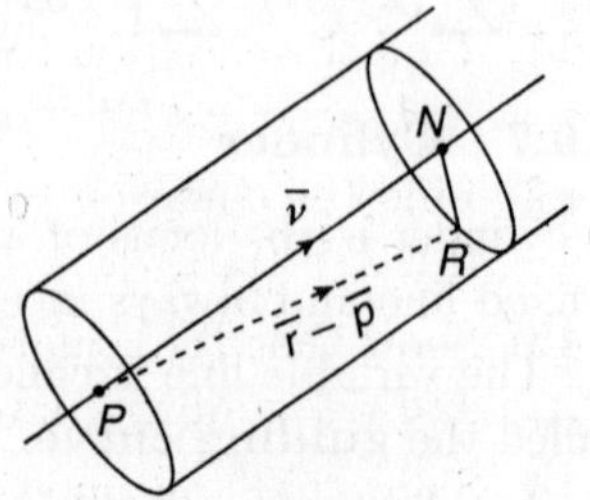

Fig. 7.17

4. *Find the equation to the right circular cylinder whose axis is $(x - 1)/2 = (y - 2)/(-3) = (z - 3)/6$ and radius 3.*

Here, $\overline{p} = (1, 2, 3)$, $\overline{\nu} = (2, -3, 6)$.

So, the required equation is

$$\left\| \begin{matrix} x-1 & y-2 & z-3 \\ 2 & -3 & 6 \end{matrix} \right\|^2 = 3^2(2^2+3^2+6^2)$$

or, $\{6(y-2)+3(z-3)\}^2+\{6(x-1)-2(z-3)\}^2+\{3(x-1)+2(y-2)\}^2 = 9\times 49$

or, $(6y+3z-21)^2+(6x-2z)^2+(3x+2y-7)^2 = 441.$

Exercises 7.2

1. Obtain the equation of the sphere with $A(2,-3,4)$ and $B(-5,6,-7)$ as the ends of one of its diameters.
2. Find the equation of the sphere whose centre is $(3,6,-4)$ and which touches the plane $2x-2y-z-10=0$.
3. Find the equation to the sphere passing through the origin and the points $(0,1,1)$, $(1,0,1)$ and $(1,1,0)$.
4. Show that the sphere passing through the point $(1,2,3)$ and the circle given by $\sum x^2 = 9$ and $2x+3y+4z=5$ is $3\sum x^2 - 2x - 3y - 4y - 22 = 0$.
5. Find the centre and radius of the sphere, if the circle $\sum x^2 + 10y - 4z = 8$, $\sum x = 3$ is one of its great circles.
6. Find the equation of the sphere which passes through $(1,1,4)$ and touches the coordinate planes.
7. Find the equation of the right circular cone whose vertex is the origin, axis is the x-axis and semi-vertical angle is $60°$.
8. Show that the semi-vertical angle of the right circular cone $4(x^2-y^2)-9z^2=0$ is $\tan^{-1}(3/2)$.
9. Find the equation to the right circular cone whose vertex is the origin, whose axis is $x/3=y/2=z/4$ and whose semi-vertical angle is $45°$.
10. Show that the right circular cone with its vertex at $(3,2,1)$ and semi-vertical angle $30°$ and axis parallel to $x/1=y/4=z/3$ is

$$2(x+4y+3z-14)^2 = 39\{(x-3)^2+(y-2)^2+(z-1)^2\}.$$

11. Find the equation to the right circular cylinder whose axis is the x-axis and radius equal to 1.
12. Find the equation to the right circular cylinder whose axis is the z-axis and which passes through $(3,4,5)$.
13. Show that the equation to the right circular cylinder whose axis is parallel to the y-axis and the guiding curve is the circle $z^2+x^2=1$, $y=5$, is $z^2+x^2=1$.
14. Show that the equation to the right circular cylinder, whose axis is $x/1=y/0=z/(-2)$ and radius 7, is

$$5\left(\sum x^2 - 49\right) = (x-2z)^2.$$

15. Find the equation to the right circular cylinder whose guiding curve is the circle through the points $(a,0,0)$, $(0,a,0)$ and $(0,0,a)$.

Answers

1. $\sum x^2 + 3(x - y + z) - 56 = 0.$ **2.** $\sum x^2 - 8x - 12y + 8z + 45 = 0.$ **3.** $\sum x^2 - \sum x = 0.$
5. $(2, -3, 4),\ 5.$ **6.** $\sum x^2 - 6\sum x + 18 = 0.$ **7.** $3x^2 = y^2 + z^2.$
9. $29\sum x^2 = 2(3x + 2y + 4z)^2.$ **11.** $y^2 + z^2 = 1.$ **12.** $x^2 + y^2 = 25.$
15. $\sum x^2 - \sum yz = a^2.$

Bibliography

1. Barnard, S. and Child, J. M., "Higher Algebra", Macmillan and Co. Ltd.
2. Courant, R., "Differential and Integral Calculus" vols. I and II, Blackie and Son Ltd.
3. Das, B. C. and Mukherjee, B. N., "Differential Calculus" and "Integral Calculus", U. N. Dhur and Sons Pvt. Ltd.
4. Edwards, Joseph, "The Differential Calculus", Macmillan and Co. Ltd.
5. Ghosh, R. K. and Maity, K. C., "An Introduction to Analysis, Differential Calculus Pt. I", "Integral Calculus", New Central Book Agency Pvt. Ltd. and "Differential Calculus Pt II", Books and Allied (P) Ltd.
6. Grewal, B. S., "Higher Engineering Mathematics", Khanna Publishers.
7. Hyslop, J. M., "Infinite Series", Olivers & Boyd Ltd.
8. James/James, "Mathematics Dictionary" CBS Publishers and Distributors.
9. Kreyszig, Erwin, "Advanced Engineering Mathematics", Wiley Eastern Ltd.
10. Malik, S. C. and Arora, S., "Mathematical Analysis", New Age International (P) Ltd. Publishers.
11. Mapa, S. K., "Introduction to Real Analysis", Sarat Book Distributors, Kolkata.
12. Murray, Daniel, A., "Introductory Course in Differential Equations", Orient Longman.
13. Myskis, A. D., "Advanced Mathematics for Engineers", MIR Publishers, Moscow.
14. Piskunov, N., "Differential and Integral Calculus" vols. I and II, MIR Publishers, Moscow.
15. Shanti Narayan
    - (i) "A Course of Mathematical Analysis"
    - (ii) "A Textbook of Matrices"
    - (iii) "Differential Calculus"
    - (iv) "Integral Calculus"

Index

D

E

H

I

J

K

L

T

U

V

W

Z

University Questions

(W.B.U. Tech)

2001

(Answer any *five* questions)

1. **(a)** If $y = 2\cos x(\sin x - \cos x)$, show that $(y_{10})_0 = 2^{10}$. [3]

(b) The region bounded by the parabola $y^2 = 4ax$, the lines $y = 0$ and $x = 2a$ revolves about the x-axis. Find the volume of the solid generated. [4]

(c) Calculate: $\int_{-1}^{1} |x| dx$. [3]

(d) Evaluate: $\int_0^{\pi/2} \int_0^{\pi} \sin(x + y) dx dy$ [4]

2. **(a)** If $y = \dfrac{x^2}{(x-1)(x-2)(x-3)}$, find y_n. [4]

(b) If $f(x) = \tan x$ and n is a positive integer, prove with the help of Leibnitz's theorem, that

$$f^n(0) - {}^nC_2 f^{n-2}(0) + {}^nC_4 f^{n-4}(0) - \cdots = \sin\left(\frac{n\pi}{2}\right) \quad [5]$$

(c) State and prove Cauchy's Mean Value Theorem. [5]

3. **(a)** State Taylor's theorem with Lagrange's form of remainder. Apply Maclaurin's theorem to the function $f(x) = (1 + x)^4$ to deduce that $(1 + x)^4 = 1 + 4x + 6x^2 + 4x^3 + x^4$. [2+5]

(b) If $I_n = \int_0^{\pi/2} \sin^{2n+1}\theta d\theta$, where n is a positive integer, show that $I_n = \dfrac{2n}{2n+1} I_{n-1}$.

Use this to evaluate $\int_0^{\pi/2} \sin^7 \theta d\theta$. [5+2]

4. **(a)** Test the convergency of the following series (any *two*) : [4+4]

(i) $\left(\frac{2^2}{1^2} - \frac{2}{1}\right)^{-1} + \left(\frac{3^3}{2^3} - \frac{3}{2}\right)^{-2} + \left(\frac{4^4}{3^4} - \frac{4}{3}\right)^{-3} + \cdots$

(ii) $\sum_{n=1}^{\infty} \frac{1}{\sqrt{n}} \sin \frac{1}{n}$

(iii) $1 + \frac{\sqrt{2} - 1}{1!} + \frac{(\sqrt{2} - 1)^2}{2!} + \frac{(\sqrt{2} - 1)^3}{3!} + \cdots$

(b) State Leibnitz's test for the convergence of an alternating series. Prove that the series

$$x - \frac{x^2}{2} + \frac{x^3}{3} - \cdots + (-1)^{n+1}\frac{x^n}{n} + \cdots$$

is absolutely convergent when $|x| < 1$ and conditionally convergent when $x = 1$. [6]

5. **(a)** Find : $\operatorname*{Lt}_{n\to\infty} \left\{\left(1 + \frac{1}{n}\right)\left(1 + \frac{2}{n}\right) \cdots \left(1 + \frac{n}{n}\right)\right\}^{\frac{1}{n}}$. [6]

(b) If $u = f(x^2 + 2yz, y^2 + 2zx)$, show that

$$(y^2 - zx)\frac{\partial u}{\partial x} + (x^2 - yz)\frac{\partial u}{\partial y} + (z^2 - xy)\frac{\partial u}{\partial z} = 0 \qquad [5]$$

(c) Verify Rolle's Theorem for $f(x) = x^2 - 5x + 6$ in $2 \leq x \leq 3$. [3]

6. (a) If $F(p, v, t) = 0$, show that

$$\left(\frac{dp}{dt}\right)_{v \text{ constant}} \times \left(\frac{dv}{dp}\right)_{t \text{ constant}} \times \left(\frac{dt}{dv}\right)_{p \text{ constant}} = -1 \qquad [5]$$

(b) If $U = \sin^{-1}\left[\frac{x^{\frac{1}{3}}+y^{\frac{1}{3}}}{x^{\frac{1}{2}}+y^{\frac{1}{2}}}\right]^{\frac{1}{2}}$, show that

$$x^2U_{xx} + 2xyU_{xy} + y^2U_{yy} = \frac{\tan U}{144}(13 + \tan^2 U) \qquad [5]$$

(c) If $U = \sqrt{xy}$, find the value of $\frac{\partial^2 U}{\partial x^2} + \frac{\partial^2 U}{dy^2}$. [4]

7. (a) Investigate the maxima and minima of the function

$$f(x, y) = x^3 + y^3 - 3x - 12y + 20 \qquad [5]$$

(b) Find a point in the plane $x + 2y + 3z = 13$ nearest to the point $(1, 1, 1)$ using the method of Lagrange's multipliers. [5]

(c) Evaluate $\iint xy(x + y)dx\,dy$ over the area bounded by $y = x^2$ and $y = x$. [4]

8. (a) Find div $\vec{F}$ and curl $\vec{F}$, where

$$\vec{F} = \text{grad } (x^3 + y^3 + z^3 - 3xyz) \qquad [8]$$

(b) Find the directional derivative of $f(x, y, z) = 2x^2 + 3y^2 + z^2$ at the point $(2, 1, 3)$ in the direction of the vector $\hat{\imath} - 2\hat{k}$. [6]

9. (a) State Stokes's Theorem. [2]

(b) Apply Stokes's Theorem to evaluate $\int_C (y\,dx + z\,dy + x\,dz)$, where C is the curve of intersection of $x^2 + y^2 + z^2 = a^2$ and $x + z = a$. [6]

(c) Verify Green's Theorem for $\int_C [(xy + y^2)dx + x^2dy]$, where C is bounded by $y = x$ and $y = x^2$. [6]

10. (a) State Fundamental Theorem of Integral Calculus. [2]

(b) Prove that $\int_0^1 dx \int_0^1 \frac{x-y}{(x+y)^3}dy \neq \int_0^1 dy \int_0^1 \frac{x-y}{(x+y)^3}dx$ [6]

(c) Find M.I. of a thin uniform straight rod of mass M and length $2a$ about its perpendicular bisector. [6]

Answers

1. (b) $8\pi a^3$ **(c)** 1 **(d)** 2

2. (a) $(-1)^n n! \left\{ \frac{1}{2} \cdot \frac{1}{(x-1)^{n+1}} - \frac{4}{(x-2)^{n+1}} + \frac{9}{2} \cdot \frac{1}{(x-3)^{n+1}} \right\}$

3. (b) $\frac{16}{35}$ **4. (a)** (i) convergent (ii) convergent (iii) convergent

5. (a) $\frac{4}{e}$ **6. (c)** $-\frac{1}{4} \frac{x^2+y^2}{x^{\frac{3}{2}} y^{\frac{3}{2}}}$

7. (a) minimum at (1.2), maximum at $(-1, -2)$ **(b)** $(\frac{3}{2}, 2, \frac{5}{2})$

(c) $\frac{3}{56}$ **8. (a)** $6(x + y + z)$ **8. (b)** $\frac{-4}{\sqrt{5}}$

2002

(Answer any *five* questions)

1. (a) If $x + y = 1$, prove that the nth derivative of $x^n y^n$ is

$$n! \left\{ y^n - ({}^nC_1)^2 y^{n-1} x + ({}^nC_2)^2 y^{n-2} x^2 - ({}^nC_3)^2 y^{n-3} x^3 + \cdots (-1)^n x^n \right\}. \quad [6]$$

(b) Evaluate $\lim_{n\to\infty} \left[\frac{1}{n} \left(\sin^2 \frac{\pi}{2n} + \sin^2 \frac{2\pi}{2n} + \cdots + \sin^2 \frac{n\pi}{2n} \right) \right]$. [4]

(c) State Rolle's Theorem. Examine whether the theorem is applicable on $f(x) = x^3 - 6x^2 + 11x - 6$ in $1 \le x \le 3$. [4]

2. (a) State and prove Lagrange's Mean Value Theorem. [5]

(b) Use M.V.T. to prove that $0 < \frac{1}{x} \log \frac{e^x - 1}{x} < 1$. [4]

(c) Expand a^x in a finite series with Lagrange's form of remainder. [5]

3. (a) Prove that if $u_n = \int_0^1 x^n \tan^{-1} x dx$, then

$$(n+1)u_n + (n-1)u_{n-2} = \frac{\pi}{2} - n. \quad [5]$$

(b) Evaluate $\iint \sqrt{4x^2 - y^2} dx\, dy$ over the triangle formed by the straight lines $y = 0$, $x = 1$ and $y = x$. [5]

(c) Find the length of the catenary $y = \frac{a}{2}\left(e^{\frac{x}{a}} + e^{\frac{-x}{a}}\right)$ between the points $(0, a)$ and (x, y). [4]

4. (a) State Cauchy's integral test for the convergence of infinite series.

Hence show that the series $\sum_{n=2}^{\infty} \frac{1}{n(\log n)^p}$, $(p > 0)$ converges if $p > 1$ and diverges if $p \le 1$. [6]

(b) Test the convergence of the following series (any *two*): [4+4]

(i) $\frac{1^2+2}{1^4}x + \frac{2^2+2}{2^4}x^2 + \frac{3^2+2}{3^4}x^3 + \cdots \infty,\ x > 0$

(ii) $\left(\frac{1}{3}\right)^2 + \left(\frac{1\cdot 2}{3\cdot 5}\right)^2 + \left(\frac{1\cdot 2\cdot 3}{3\cdot 5\cdot 7}\right)^2 + \cdots \infty$

(iii) $\sum_{n=1}^{\infty} \frac{\cos n\pi}{n^2+1}$.

5. (a) Find the maximum and minimum values of the function

$$f(x,y) = x^3 + y^3 - 3axy$$ [5]

(b) If v be a function of x and y, prove that

$$\frac{\partial^2 v}{\partial x^2} + \frac{\partial^2 v}{\partial y^2} = \frac{\partial^2 v}{\partial r^2} + \frac{1}{r}\frac{\partial^2 v}{\partial \theta^2}$$

where $x = r\cos\theta$, $y = r\sin\theta$. [5]

(c) Find the Taylor's series expansion for $f(x,y) = e^x \cos y$ about the origin up to second degree terms. [4]

6. (a) Evaluate $\int_0^a \int_0^{\sqrt{a^2-y^2}} (x^2+y^2)dy\,dx$ by changing to polar coordinates. [5]

(b) Find a point in the plane $x + 2y + 3z = 13$ nearest to the point $(1,1,1)$ using the method of Lagrange's multiplier. [4]

(c) Find the volume bounded by the cylinder $x^2 + y^2 = 4$ and the planes $y + z = 3$ and $z = 0$. [5]

7. (a) A particle moves on the curve $x = 2t^2, y = t^2 - 4t$ and $z = 3t - 5$, where t denotes time. Find the components of velocity and acceleration at time $t = 1$ in the direction $\vec{i} - 3\vec{j} + 2\vec{k}$. [5]

(b) Show that $\vec{A} = (6xy + z^3)\hat{\imath} + (3x^2 - z)\hat{\jmath} + (3xz^2 - y)\hat{k}$ is irrotational. Find the scalar function ϕ such that $\vec{A} = \vec{\nabla}\phi$. [5]

(c) If $A = (3x^2 + 6y)\hat{\imath} - 14yz\hat{\jmath} + 20xz^2\hat{k}$, evaluate $\int_C \vec{A}\cdot d\vec{r}$ from $(0,0,0)$ to $(1,1,1)$ along the path C given by $x = t$, $y = t^2$ and $z = t^3$. [4]

8. (a) State divergence theorem of Gauss. Verify divergence theorem for $\vec{F} = 4xz\hat{\imath} - y^2\hat{\jmath} + yz\hat{k}$ taken over the region bounded by $x = 0$, $x = 1$; $y = 0$, $y = 1$; $z = 0$, $z = 1$. [6]

(b) Verify Green's theorem for $\iint_C [(3x - 8y^2)dx + (4y - 6xy)dy]$, where C is boundary of the region bounded by $x = 0$, $y = 0$ and $x + y = 1$. [5]

(c) Use Stokes's theorem to prove div curl $F = 0$. [3]

9. (a) Find the volume of the largest rectangular parallelopiped that can be inscribed in the ellipsoid $\frac{x^2}{a^2} + \frac{y^2}{b^2} + \frac{z^2}{c^2} = 1$, using Lagrange's Multipliers. [5]

(b) Find the equation of the tangent plane and the normal line to the surface $2x^2 + y^2 + 2z = 3$ at the point $(2,1,-3)$. [5]

(c) Find the angle between the surfaces $xy^2z = 3x + z^2$ and $3x^2 - y^2 + 2z = 1$ at the point $(1,-2,1)$. [4]

10. **(a)** Find the centroid of the area of the astroid $x^{\frac{2}{3}}+y^{\frac{2}{3}}=a^{\frac{2}{3}}$ lying in the first quadrant. [6]

(b) Find the C.G. of a semicircular lamina of radius a if its density varies as the square of the distance from the diameter. [4]

(c) If s be the length of an arc of $3y^2=x(x-a)^2$ measured from the origin to the point (x,y), show that $3s^2=4x^2+4y^2$. [4]

Answers

1. **(b)** $\frac{1}{2}$ **2.** **(c)** $1+x\log a+\frac{x^2(\log a)^2}{2!}+\cdots+\frac{x^n}{n!}a^{\theta x}(\log a)^n, 0<\theta<1$

3. **(b)** $\frac{1}{3}\left(\frac{\sqrt{3}}{2}+\frac{\pi}{3}\right)$ **(c)** $a\sinh\frac{x}{a}$ **4.** **(b)** **(i)** convergent when $x\leq 1$

(ii) convergent **(iii)** convergent

5. **(a)** minimum at (a,a) when $a>0$, maximum at (a,a) when $a<0$

(c) $1+x+\frac{x^2-y^2}{2!}+\cdots$

6. **(a)** $\frac{\pi a^4}{8}$ **(b)** $\left(\frac{3}{2},2,\frac{5}{2}\right)$ **(c)** 12π

7. **(a)** $\frac{16}{\sqrt{14}},\frac{-2}{\sqrt{14}}$ **(b)** $3x^2y+z^3x-yz+c$ **(c)** 5

9. **(a)** $\frac{8abc}{3\sqrt{3}}$ **(b)** $4x+y+z=6$; $\frac{x-2}{4}=\frac{y-1}{1}=\frac{z+3}{1}$ **(c)** $\cos^{-1}\left(\frac{-\sqrt{3}}{7\sqrt{2}}\right)$

10. **(a)** $\left(\frac{256a}{315\pi},\frac{256a}{315\pi}\right)$ **(b)** $\left(\frac{32a}{15\pi},0\right)$

2003

(Answer any *five* questions)

1. **(a)** If $y=\tan^{-1}x$, then show that

(i) $(1-x^2)y_1=0$, **(ii)** $(1+x^2)y_{n+1}+2nxy_n+n(n-1)y_{n-1}=0$. [5]

(b) Examine the continuity and differentiability of the function

$$\begin{aligned} f(x) &= 1 && \text{when } x\leq 0 \\ &= 1+\sin x && \text{when } x>0, \text{ at } x=0 \end{aligned}$$ [4]

(c) Show that

$$\frac{d^n}{dx^n}\left(\frac{\log x}{x}\right)=(-1)^n\frac{n!}{x^{n+1}}\left(\log x-1-\frac{1}{2}-\frac{1}{3}-\cdots-\frac{1}{n}\right)$$ [5]

2. **(a)** State Rolle's theorem. Verify the theorem for the function $f(x)=|x|$ in $[-1,1]$ [5]

(b) State Cauchy's Mean Value Theorem. Apply this to the functions, $f(x)=e^x$ and $g(x)=e^{-x}$ in $[x,x+h]$ and obtain the value of θ. Interpret your result. [5]

(c) Use M.V.T. to prove the $\sin 46^\circ\sim\frac{1}{2}\sqrt{2}\left(1+\frac{\pi}{180}\right)$. Is this estimate high or low? [4]

3. **(a)** Prove that $\displaystyle\int_0^1\frac{x^6}{\sqrt{1-x^2}}dx=\frac{5\pi}{32}$ [4]

(b) If $u_n=\displaystyle\int_0^{\pi/4}\tan^n\theta\,d\theta$, then prove that $n(1_{n+1}+I_{n-1})=1$. [5]

(c) Find $D^n\{\sin(ax+b)\}$. Show that if $u = \sin ax + \cos ax$,

$$D^n u = a^n\{1+(-1)^n \sin 2ax\}^{\frac{1}{2}}, \text{ where } D = \frac{d}{dx}.$$ [5]

4. (a) Find out the length of the perimeter of the astroid $x^{\frac{2}{3}} + y^{\frac{2}{3}} = a^{\frac{2}{3}}$. [5]

(b) Find the C.G. of the homogeneous area bounded by the parabola $y^2 = 4ax$, the x-axis and the ordinate $x = h$. [5]

(c) Find out the moment of inertia of a sphere of radius a about one of its diameters. Also, obtain the corresponding radius of gyration. [4]

5. (a) State D'Alembert's ratio test.

Examine the convergence and divergence of the series

$$1 + \frac{x}{2} + \frac{x^2}{5} + \frac{x^3}{10} + \cdots \infty$$ [2+4]

(b) Test the convergence of the following series (any *two*): [5]

(i) $\frac{1}{1^p} - \frac{1}{2^p} + \frac{1}{3^p} - \frac{1}{4^p} + \cdots \infty$

(ii) $\sum_{n=1}^{\infty} \left(\sqrt[3]{n^3+1} - n\right)$

(iii) $\sum_{n=1}^{\infty} \frac{n!2^n}{n^n}$

(iv) $\left(\frac{2^2}{1^2} - \frac{2}{1}\right)^{-1} + \left(\frac{3^3}{2^3} - \frac{3}{2}\right)^{-2} + \left(\frac{4^4}{3^4} - \frac{4}{3}\right)^{-3} + \cdots \infty$

6. (a) If $u = \log(x^3 + y^3 + z^3 - 3xyz)$ show that [6]

(i) $\left(\frac{\partial}{\partial x} + \frac{\partial}{\partial y} + \frac{\partial}{\partial z}\right) u = \frac{3}{x+y+z}$

(ii) $\left(\frac{\partial^2}{\partial x^2} + \frac{\partial^2}{\partial y^2} + \frac{\partial^2}{\partial z^2}\right) u = -\frac{3}{(x+y+z)^2}$

(iii) $\left(\frac{\partial}{\partial x} + \frac{\partial}{\partial y} + \frac{\partial}{\partial y}\right)^2 u = -\frac{9}{(x+y+z)^2}$

(b) If $f(p, t, v) = 0$

$$\left(\frac{dp}{dt}\right)_{v=\text{const}} \times \left(\frac{dt}{dv}\right)_{p=\text{const}} \times \left(\frac{dv}{dp}\right)_{t=\text{constant}} = -1$$ [4]

(c) If $f(x, y) = xy\frac{x^2 - y^2}{x^2 + y^2}$ when $x \neq 0, y \neq 0$

$= 0$ when $x = 0, y = 0$

then show that $f_{xy}(0,0) \neq f_{yx}(0,0)$. [4]

7. (a) Show that the volume included between the elliptic paraboloid $2z = x^2 + y^2$, the cylinder $x^2 + y^2 = a^2$ and xy-plane is $\frac{\pi}{4}a^4$. [6]

(b) Evaluate $\int_0^a \int_0^x \int_0^y x^3y^2z\,dz\,dy\,dx$ [4]

(c) Evaluate $\iiint (x+y+z+1)^4 dx\,dy\,dz$ over the region defined by $x \geq 0,\ y \geq 0,\ z \geq 0,\ x+y+z \leq 1.$ [4]

8. (a) Prove that

$$\text{grad}\,(\vec{u}\cdot\vec{v}) = (\vec{v}\cdot\vec{\nabla})\vec{u} + (\vec{u}\cdot\vec{\nabla})\vec{v} + \vec{v}\times\text{curl}\,\vec{u} + \vec{u}\times\text{curl}\,\vec{v}$$ [5]

(b) Show that curl grad $f = 0$, where $f = x^2y + 2xy + z^2$. [4]

(c) In what direction from the point $(1,1,-1)$ is the directional derivative of $\phi(x,y,z) = x^2 - 2y^2 + 4z^2$ a maximum? Obtain the magnitude of that directional derivative. [4]

9. (a) Verify Stokes's theorem for $\vec{F} = (x^2+y^2)\hat{\imath} - 2xy\hat{\jmath}$ around the rectangle bounded by $x = \pm a,\ y = 0,\ y = b.$ [5]

(b) Verify Gauss's divergence theorem for $\vec{F} = y\hat{\imath} + x\hat{\jmath} + z^2\hat{k}$ over the cylindrical region bounded by $x^2+y^2 = 9,\ z = 0,\ z = 2.$ [6]

(c) Verify Green's theorem in the plane for $\oint_C (xy+y^2)dx + x^2dy$, where C is the closed curve of the region bounded by $y = x$ and $y = x^2$. [4]

Answers

1. (b) continuous but not differentiable at $x = 0$ 2. (b) $\theta = \frac{1}{2}$ (c) high

3. (c) $a^n \sin\left(\frac{n\pi}{2} + ax + b\right)$ 4. (a) $6a$ (b) $\left(\frac{3}{5}h, \frac{3}{4}\sqrt{ah}\right)$

(c) $\frac{2}{5}Ma^2, \frac{2}{5}a^2$, where M is the mass of the sphere

5. (a) convergent when $0 \leq x \leq 1$ and divergent when $x > 1$

(b) (i) convergent when $p > 0$ (ii) convergent (iii) convergent

(iv) convergent 7. (b) $\frac{a^9}{90}$ (c) $\frac{117}{70}$

8. (c) $2\hat{\imath} - 4\hat{\jmath} - 8\hat{k}, 2\sqrt{21}$

2004

1. Answer any *five* of the following :

(i) Prove that the function $f(x) = |x-1|, 0 < x < 2$ is continuous at $x = 1$ but not differentiable there.

(ii) If $z = \sin uv$, where $u = 3x^2$ and $v = \log x$, find $\frac{dz}{dx}$.

(iii) If the vector $\vec{A}$ and $\vec{B}$ be irrotational, then show that the vectors $\vec{A}\times\vec{B}$ is solenoidal.

(iv) Find the constant m such that the vectors $\vec{a} = 2\hat{\imath} - \hat{\jmath} + \hat{k},\ \vec{b} = \hat{\imath} + 2\hat{\jmath} - 3\hat{k},\ \vec{c} = 3\hat{\imath} + m\hat{\jmath} + 5\hat{k}$ are coplanar.

(v) Evaluate $\int_0^3 \int_1^{\sqrt{4-y}} (x+y)dx\,dy$.

(vi) Show that Rolle's theorem is not applicable to

$$f(x) = \tan x \text{ in } [0, \pi] \text{ although } f(0) = f(\pi).$$

(vii) Find the point where the straight line through the points $(5, -2, 3)$ and $(3, 0, 1)$ pierces the xy-plane.

2. **(a)** Use L'Hospital's rule to evaluate $\lim_{x\to 0} \frac{e^x + \sin x - 1}{\log(1+x)}$. [6]

(b) If $y = \cos(m \sin^{-1} x)$, then prove that

$$(1-x^2)y_{n+2} - (2n+1)xy_{n+1} + (m^2 - n^2)y_n = 0.$$

Find y_n for $x = 0$.

3. **(a)** State Lagrange's Mean Value Theorem. [2]

(b) Using Mean Value Theorem prove the inequalities:

$$1 + \frac{x}{2\sqrt{1+x}} < \sqrt{1+x} < 1 + \frac{x}{2}, \quad -1 < x < 0$$ [5]

(c) If $I_n = \int \frac{\sin n\theta}{\sin \theta} d\theta$, show that

$$(n-1)(I_n - I_{n-2}) = 2\sin(n-1)\theta$$ [5]

4. **(a)** Expand e^x in powers of x in infinite series. [4]

(b) Calculate the length of the perimeter of the ellipse $\frac{x^2}{16} + \frac{y^2}{25} = 1$ by integration. [4]

(c) Find the surface area of solid formed by revolving the cardioid $r = a(1 + \cos\theta)$ about the initial line. [4]

5. **(a)** State Cauchy's root test.

Discuss the convergence of the series $\sum_{n=1}^{\infty} \left(1 + \frac{1}{\sqrt{n}}\right)^{-n^{3/2}}$ [2+4]

(b) Test the convergence of *one* of the following series: [3]

(i) $\frac{\sqrt{1}}{a \cdot 1^{3/2} + b} + \frac{\sqrt{2}}{a \cdot 2^{3/2} + b} + \frac{\sqrt{3}}{a \cdot 3^{3/2} + b} + \cdots$, where $a > 0$

(ii) $1 + \frac{2^2}{3^2}x + \frac{2^2 \cdot 4^2}{3^2 \cdot 5^2}x^2 + \frac{2^2 \cdot 4^2 \cdot 6^2}{3^2 \cdot 5^2 \cdot 7^2}x^3 + \cdots (x \neq 1)$

(c) Show that the series $\sum_{n=1}^{\infty} \frac{\cos nx}{n^2}$ is absolutely convergent. [3]

6. **(a)** If $U = xf\left(\frac{y}{x}\right) + g\left(\frac{y}{x}\right)$, then show that

$$x\frac{\partial u}{\partial x} + y\frac{\partial u}{\partial y} = xf\left(\frac{y}{x}\right)$$

and $$x^2\frac{\partial^2 u}{\partial x^2} + 2xy\frac{\partial^2 u}{\partial x \partial y} + y^2\frac{\partial^2 u}{\partial y^2} = 0.$$ [5]

(b) If $f(u,v) = 3uv^2$, $g(u,v) = u^2 - v^2$, find the Jacobian $\dfrac{\partial(f,g)}{\partial(u,v)}$. [3]

(c) Find the extrema of the following function :

$$x^3 + 3xy^2 - 3y^2 - 3x^2 + 4$$ [4]

7. **(a)** Evaluate $\displaystyle\iint_R \frac{1}{\sqrt{x^2+y^2}}\,dx\,dy$, where $R = \{|x| \le 1,\ |y| \le 1\}$ [4]

(b) Evaluate $\displaystyle\iiint z^2 dx\,dy\,dz$ over the region defined by $z \ge 0, x^2 + y^2 + z^2 \le a^2$. [4]

(c) Find the centroid of a loop of the lemniscate $r^2 = a^2 \cos 2\theta$. [4]

8. **(a)** Show that $\vec{A} = (6xy + z^3)\hat{\imath} + (3x^2 - z)\hat{\jmath} + (3xz^2 - y)\hat{k}$ is irrotational. Find the scalar function ϕ such that $\vec{A} = \vec{\nabla}\phi$. [5]

(b) If $r = |\vec{r}|$, where $\vec{r} = x\hat{\imath} + y\hat{\jmath} + z\hat{k}$, prove that $\vec{\nabla}(r^n) = nr^{n-2}\vec{r}$. [3]

(c) In what direction from the point $(1, 2, 3)$, is the directional derivative of $f = x^2 - y^2 + 2z^2$ maximum? Also, find the value of this maximum directional derivative.

9. **(a)** Prove that $\text{div}(P\vec{Q}) = P \text{ div } \vec{Q} + \vec{Q} \text{ grad } P$. [4]

(b) Evaluate by Green's theorem

$$\oint_C \{(\cos x \sin y - xy)dx + \sin x \cos y\, dy\}$$ [3]

where C is the circle $x^2 + y^2 = 1$.

(c) Verify Stokes's theorem for $\vec{A} = 2y\hat{\imath} + 3x\hat{\jmath} - z^2\hat{k}$, where S is the upper half surface of the sphere $x^2 + y^2 + z^2 = 9$ and C is its boundary. [5]

Answers

1. **(ii)** $3x(2\ln x + 1)\cos uv$ **(iv)** $m = -4$ **(v)** $241/60$
(vii) $(2, 1, 0)$ **2.** **(a)** 2
(b) 0 or $\prod_0^{n-1}(4r^2 - m^2)$ according as n is odd or even
4. **(a)** $\sum_0^\infty x^n/n!$ **(b)** $20E(3/5, \pi/2)$, an elliptic integral of the second kind
(c) $32\pi a^2/5$ **5.** **(a)** con. **(b)** **(i)** div.
(ii) con. for $x < 1$ and div. for $x > 1$ **6.** **(b)** $-6v(2u^2 + v^2)$
(c) maximum value 4 at $(0, 0)$ and minimum value 0 at $(2, 0)$
7. **(a)** $8\ln(1 + \sqrt{2})$ **(b)** $2\pi a^5/15$ **(c)** $(a\pi\sqrt{2}/8, 0)$
8. **(a)** $\phi = 3x^2y + xz^3 - yz + c$ **(c)** in the direction of $\hat{\imath} - 2\hat{\jmath} + 6\hat{k}$, $2\sqrt{41}$
9. **(b)** 0

Engineering & Technology Examinations, December—2005

MATHEMATICS

SEMESTER—1

Time : 3 hours] [Full Marks : 70

The questions are of equal value.
The figures in the margin indicates full marks.
Candidates are required to give their answers in their own words as far as practicable.

Note: (i) Question No. **1** is compulsory.

(ii) Answer any *six* full questions from the remaining.

1. Answer any *five* of the following questions: $5 \times 2 = 10$

(i) Show that the sequence $\{U_n\}_{n \in N}$, where $U_n = 2(-1)^n$ does not converge.

(ii) Use L'Hospital's rule to evaluate $\lim\limits_{x \to 0} \dfrac{\sin x}{x}$.

(iii) If $u = \log(\tan x + \tan y)$, prove that $\sin 2x \dfrac{\partial u}{\partial x} + \sin 2y \dfrac{\partial u}{\partial y} = 2$.

(iv) Show that Lagrange's Mean Value Theorem is not applicable to the function

$$f(x) = \begin{cases} x \sin \dfrac{1}{x}, & \text{when } x \neq 0 \\ 0 & \text{when } x = 0 \end{cases}$$

in $[-1, 1]$.

(v) Evaluate the line integral $\int_C (x^2 dx + xy\, dy)$, where C is the line segment joining $(1, 0)$ and $(0, 1)$.

(vi) If α, β, γ are the angles which a line makes with the coordinate axes, prove that $\sin^2 \alpha + \sin^2 \beta + \sin^2 \gamma = 2$.

(vii) If $|\vec{\alpha}| = 3$ and $|\vec{\beta}| = 4$, then find the values of the scalar c for which the vectors $\vec{\alpha} + c\vec{\beta}$ and $\vec{\alpha} - c\vec{\beta}$ will be perpendicular to one another.

(viii) Find the unit vector normal to the surface $x^2 + y - z = 1$ at the point $(1, 0, 0)$.

2. (a) Test the convergence of any *two* of the following series : $2 \times 3 = 6$

(i) $1 + \dfrac{1}{2^2} + \dfrac{2^2}{3^3} + \dfrac{3^3}{4^4} + \dfrac{4^4}{5^5} + \cdots \infty$

(ii) $\sin\left(\dfrac{1}{1^{3/2}}\right) + \sin\left(\dfrac{1}{2^{3/2}}\right) + \sin\left(\dfrac{1}{3^{3/2}}\right) + \sin\left(\dfrac{1}{4^{3/2}}\right) + \cdots \infty$

(iii) $\left(\dfrac{2^2}{1^2} - \dfrac{2}{1}\right)^{-1} + \left(\dfrac{3^3}{2^3} - \dfrac{3}{2}\right)^{-2} + \left(\dfrac{4^4}{3^4} - \dfrac{4}{3}\right)^{-3} + \cdots \infty$.

(b) State D'Alembert's Ratio Test for infinite series of positive terms. Discuss the convergence of the series $\sum\limits_{n=1}^{\infty} n^4 e^{-n^2}$. 1+3

3. **(a)** If $y = \tan^{-1} x$, then prove that

$$(1+x^2)y_{n+1} + 2nxy_n + n(n-1)y_{n-1} = 0.$$

Also, find $y_n(0)$. 3+3

(b) Using Mean Value Theorem, prove that $\frac{\pi}{6} + \frac{\sqrt{3}}{15} < \sin^{-1}\left(\frac{3}{5}\right) < \frac{\pi}{6} + \frac{1}{8}$. 4

4. **(a)** Find the value of $\lim_{n\to\infty}\left\{\left(1+\frac{1}{n}\right)\left(1+\frac{2}{n}\right)\cdots\left(1+\frac{n}{n}\right)\right\}^{1/n}$. 4

(b) If $I_n = \int \frac{\cos n\theta}{\cos\theta} d\theta$, show that $(n-1)(I_n + I_{n-2}) = 2\sin(n-1)\theta$. Hence, evaluate $\int (4\cos^2\theta - 3)d\theta$. 4+2

5. **(a)** Find the whole length of the loop of the curve $9y^2 = (x-2)(x-5)^2$. 4

(b) Find the surface area generated by revolving the part of astroid $x^{2/3} + y^{2/3} = a^{2/3}$ in the first quadrant about the x-axis. 4

(c) If $f(x,y,z,w) = 0$, prove that $\frac{\partial x}{\partial y} \times \frac{\partial y}{\partial z} \times \frac{\partial z}{\partial w} \times \frac{\partial w}{\partial x} = 1$. 2

6. **(a)** Find the extrema of the function $x^3 + y^3 - 3x - 12y + 20$. 4

(b) If $f(v^2 - x^2,\ v^2 - y^2,\ v^2 - z^2) = 0$, where v is a function of x, y, z, show that

$$\frac{1}{x}\frac{\partial v}{\partial x} + \frac{1}{y}\frac{\partial v}{\partial y} + \frac{1}{z}\frac{\partial v}{\partial z} = \frac{1}{v}.$$ 3

(c) Evaluate $\iint_R \sqrt{4x^2 - y^2}\,dx\,dy$, where R is the triangular region bounded by the lines $y = 0,\ x = 1$ and $y = x$. 3

7. **(a)** Find the volume V of a solid bounded by $x = 0,\ y = 0,\ z = 0,\ x + y + z = 1$. 5

(b) Find the moment of inertia of the solid bounded in the first octant by the coordinate planes and $\frac{x}{a} + \frac{y}{b} + \frac{z}{c} = 1$, $(a > 0,\ b > 0,\ c > 0)$, (ρ is the constant density of the solid) about the x-axis. 5

8. **(a)** A variable plane passes through a fixed point (a,b,c) and meets the coordinate axes at A, B, C. Show that the locus of the point of intersection of the planes through A, B, C and parallel to the coordinate planes is $\frac{a}{x} + \frac{b}{y} + \frac{c}{z} = 1$. 5

(b) A straight line with direction ratios $2, 7, -5$ is drawn to intersect the lines $\frac{x-5}{3} = \frac{y-7}{-1} = \frac{z+2}{1}$ and $\frac{x+3}{-3} = \frac{y-3}{2} = \frac{z-6}{4}$. Find the coordinates of the points of intersection and length intercepted on it. 5

9. **(a)** Given two vectors $\vec{\alpha} = 3\hat{\imath} - \hat{\jmath} + 0\hat{k}$, and $\vec{\beta} = 2\hat{\imath} + \hat{\jmath} - 3\hat{k}$, express $\vec{\beta}$ in the form $\vec{\beta}_1 + \vec{\beta}_2$, where $\vec{\beta}_1$ is parallel to $\vec{\alpha}$ and $\vec{\beta}_2$ is perpendicular to $\vec{\alpha}$. 3

(b) Given three vectors $\vec{a}, \vec{b}, \vec{c}$, prove that

$$\vec{a} \times (\vec{b} \times \vec{c}) = (\vec{a}\cdot\vec{c})\vec{b} - (\vec{a}\cdot\vec{b})\vec{c}.$$ 4

(c) If $\vec{r} = x\hat{\imath} + y\hat{\jmath} + z\hat{k}$ and $r = |\vec{r}|$, show that grad $f(r) \times \vec{r} = \theta$, where θ is the null vector. 3

10. (a) Prove that curl (grad $(f)) = \theta$, where θ is the null vector. 2

(b) Verify Green's theorem in the plane for $\oint_\Gamma (x^2 dx + xy\, dy)$, where Γ is the square in the xy-plane given $x = 0,\ y = 0,\ x = a,\ y = a (a > 0)$ described in the positive sense. 5

(c) Evaluate by Divergence Theorem $\iint_S \{x^2 dydz + y^2 dzdx + 2z(xy - x - y)dxdy\}$, where S is the surface of the cube $0 \le x \le 1,\ 0 \le y \le 1,\ 0 \le z \le 1$. 3

11. (a) Show that

$$\iiint \frac{dx\,dy\,dz}{(x + y + z + 1)^3} = \frac{1}{2}\left[\log 2 - \frac{5}{8}\right]$$

integration being taken over the volume bounded by the coordinate planes and the plane $x + y + z = 1$. 5

(b) Find the Moment of Inertia of a thin uniform lamina in the form of an ellipse $\dfrac{x^2}{a^2} + \dfrac{y^2}{b^2} = 1$ about its major axis. 5

Answers

1. (ii) 1 (v) 1/2 (vii) $\pm 3/2$ (viii) $(2\hat{\imath} + \hat{\jmath} - \hat{k})/6$

2. (a) div. (ii) conv. (iii) conv.

3. (a) $y_n(0) = 0,$ when n is even,
$= (-1)^{(n-1)/2}(n-1)!,$ when n is odd

4. (a) $4/e$ (b) $-\theta + \sin 2\theta$ **5.** (a) $4\sqrt{3}$ units (b) $12\pi a^2/5$

6. (a) Minimum value 2 at $(1, 2)$ and maximum value 38 at $(-1, -2)$ (c) $(3\sqrt{3} + 2\pi)/18$

7. (a) 1/6 units (b) $abc(b^2 + c^2)\rho/60$

8. (b) The points of intersection with the first and second lines are respectively $A(2, 8-3)$ and $B(0, 1, 2)$. The distance $AB = \sqrt{78}$

9. (a) $\overline{\beta}_1 = \overline{\alpha} = 3\hat{\imath} - \hat{\jmath} + 0\hat{k}$ and $\overline{\beta}_2 = -\hat{\imath} + 2\hat{\jmath} - 3\hat{k}$ **10.** (c) 1/2

11. (b) $Mb^2/4$